# PyTorch
## 深度学习应用实战

陈昭明　洪锦魁　著

清华大学出版社
北　京

## 内 容 简 介

本书以统计学/数学为出发点，介绍深度学习必备的数理基础，讲解PyTorch的主体架构及最新的模块功能，包括常见算法与相关套件的使用方法，例如对象侦测、生成对抗网络、深度伪造、图像中的文字辨识、脸部辨识、BERT/Transformer、聊天机器人、强化学习、自动语音识别、知识图谱等。本书配有大量案例及图表说明，同时以程序设计取代定理证明，缩短学习过程，增加学习乐趣。

本书适合深度学习入门者、数据工程师、信息技术工作者阅读，也可作为高校计算机相关专业的教材。

本书封面贴有清华大学出版社防伪标签，无标签者不得销售。
版权所有，侵权必究。举报：010-62782989，beiqinquan@tup.tsinghua.edu.cn。

图书在版编目（CIP）数据

PyTorch深度学习应用实战 / 陈昭明，洪锦魁著. —北京：清华大学出版社，2023.9
ISBN 978-7-302-64510-8

Ⅰ. ①P… Ⅱ. ①陈… ②洪… Ⅲ. ①机器学习 Ⅳ. ①TP181

中国国家版本馆CIP数据核字(2023)第163004号

责任编辑：杜　杨
封面设计：郭二鹏
责任校对：胡伟民
责任印制：曹婉颖

出版发行：清华大学出版社
网　　址：https://www.tup.com.cn，https://www.wqxuetang.com
地　　址：北京清华大学学研大厦A座
邮　　编：100084
社 总 机：010-83470000
邮　　购：010-62786544
投稿与读者服务：010-62776969，c-service@tup.tsinghua.edu.cn
质 量 反 馈：010-62772015，zhiliang@tup.tsinghua.edu.cn

印 装 者：涿州汇美亿浓印刷有限公司
经　　销：全国新华书店
开　　本：170mm×240mm　　印　张：35　　字　数：855千字
版　　次：2023年11月第1版　　印　次：2023年11月第1次印刷
定　　价：139.00元

产品编号：099820-01

# 前　言

## 为何撰写本书

笔者从事机器学习教育训练已届五年，其间也在"IT 邦帮忙"撰写过上百篇的文章，从学员及读者的回馈获得许多宝贵意见，期望能将整个历程集结成册。同时，相关领域的进展也在飞速变化，过往的文章内容需要更新，因此借机再重整思绪，想一想如何能将算法的原理解释得更简易清晰，协助读者跨过 AI 的门槛。另外，也避免流于空谈，尽量增加应用范例，希望能达到即学即用，不要有过多理论的探讨。

AI 是一个将数据转化为知识的过程，算法就是过程中的生产设备，最后产出物是模型，再将模型植入各种硬件装置，例如计算机、手机、智能音箱、自动驾驶、医疗诊断仪器等，这些装置就拥有特殊专长的智能，再进一步整合各项技术就构建出智能制造、智能金融、智能交通、智慧医疗、智能城市、智能家居等应用系统。AI 的应用领域如此广阔，而个人精力有限，当然不可能具备十八般武艺，样样精通，唯有从基础扎根，再扩及有兴趣的领域。因此，笔者撰写本书的初衷非常单纯，就是希望在读者学习的过程中贡献一点微薄的力量。

## PyTorch 对比 TensorFlow

深度学习的初学者常会问"应该选择 PyTorch 还是 TensorFlow 框架"，依笔者个人看法，PyTorch、TensorFlow 好比倚天剑与屠龙刀，各有所长，两个框架的发展方向有所不同。例如，在侦测方面 PyTorch 比较容易，但用 TensorFlow/Keras 建模、训练、预测都只要一行程序。另外，对象侦测主流算法 YOLO 的第 4 版以 TensorFlow 开发，第 5 版以后则以 PyTorch 开发，若我们只懂 TensorFlow，那就无法使用最新版了。

PyTorch 与 TensorFlow 的基本设计概念是相通的，采用相同的学习路径，可以同时学习两个框架。本书主要讲解 PyTorch 开发，另一本《深度学习全书：公式＋推导＋代码＋TensorFlow 全程案例》则以 TensorFlow 为主，两本对照，可以发现要兼顾学习一点也不难，还可以比较彼此的优劣。

## 本书主要的特点

(1) 由于笔者身为统计人，希望能**"以统计学/数学为出发点"**，介绍深度学习必备的数理基础，但又不希望内文有太多数学公式的推导，让离开校园已久的读者看到一堆数学符号就心生恐惧，因此，尝试**"以程序设计取代定理证明"**，缩短学习过程，增进学习乐趣。

(2) PyTorch 版本迭代快速，几乎每一两个月就更新一个小版本，并且不断地推出新扩充模块，本书期望除对 PyTorch 主体架构做完整性的介绍外，也尽可能对最新的模块功能做深入探讨。

(3) 各种算法介绍以理解为主，辅以大量图表说明，摒弃长篇大论。

（4）完整的范例程序及各种算法的延伸应用，以实用为目的，希望能触发读者灵感，能在项目或产品内应用。

（5）介绍日益普及的算法与相关套件的使用，例如 YOLO(对象侦测)、GAN(生成对抗网络、DeepFake(深度伪造)或 OCR(辨识图像中的文字))、脸部辨识、BERT/Transformer、聊天机器人 (ChatBot)、强化学习 (Reinforcement Learning)、自动语音识别 (ASR)、知识图谱 (Knowledge Graph) 等。

## 目标对象

(1) 深度学习的入门者：必须熟悉 Python 程序语言及机器学习基本概念。
(2) 数据工程师：以应用系统开发为志向，希望能应用各种算法，进行实操。
(3) 信息技术工作者：希望能扩展深度学习知识领域。
(4) 从事其他领域的工作，希望能一窥深度学习奥秘者。

## 阅读重点

第 1 章介绍 AI 的发展趋势，鉴古知今，了解前两波 AI 失败的原因，比较第三波发展的差异性。

第 2 章介绍深度学习必备的统计学 / 数学基础，彻底理解神经网络求解的方法 ( 梯度下降法 ) 与原理。

第 3 章介绍 PyTorch 基础功能，包括张量 (Tensor) 运算、自动微分、神经层及神经网络模型。

第 4 章开始实操，依照机器学习 10 项流程，以 PyTorch 撰写完整的范例，包括各种损失函数、优化器、效能衡量指标。

第 5 章介绍 PyTorch 进阶功能，包括各种工具，如数据集 (Dataset)、数据加载器 (DataLoader)、前置处理、TensorBoard 及 TorchServe 部署工具，包括 Web、桌面程序。

第 6~10 章介绍图像 / 视频的算法及各种应用。
第 11~14 章介绍自然语言处理、语音及各种应用。
第 15 章介绍 AlphaGo 的基础——"强化学习"算法。
第 16 章介绍图神经网络 (Graph Neural Network, GNN)。
本书范例程序代码及素材、参考资料请扫描下方二维码下载。

程序代码及素材

参考资料

## 致谢

因作者水平有限，还是有许多议题成为遗珠之憾，仍待后续的努力。感谢清华大学出版社的大力支持，使本书得以顺利出版；谢谢家人的默默支持。

书中内容如有疏漏、谬误，欢迎来信指教 (duy@tup.tsinghua.edu.cn)。

作者

2023 年 10 月

# 目　录

## 第1篇　深度学习导论

### 第1章　深度学习介绍 ………… 2
- 1-1　人工智能历经的三波浪潮 ……… 2
- 1-2　AI 的学习地图 ……………… 3
- 1-3　TensorFlow 对比 PyTorch ……… 5
- 1-4　机器学习开发流程 …………… 6
- 1-5　开发环境安装 ………………… 7
- 1-6　免费云端环境开通 …………… 9

### 第2章　神经网络原理 ………… 12
- 2-1　必备的数学与统计知识 ……… 12
- 2-2　万般皆自"回归"起 ………… 13
- 2-3　神经网络 ……………………… 16
  - 2-3-1　神经网络概念 …………… 17
  - 2-3-2　梯度下降法 ……………… 19
  - 2-3-3　神经网络权重求解 ……… 22

## 第2篇　PyTorch 基础篇

### 第3章　PyTorch 学习路径与主要功能 ………… 25
- 3-1　PyTorch 学习路径 …………… 25
- 3-2　张量运算 ……………………… 26
  - 3-2-1　向量 ……………………… 26
  - 3-2-2　矩阵 ……………………… 30
  - 3-2-3　使用 PyTorch …………… 33
- 3-3　自动微分 ……………………… 36
- 3-4　神经网络层 …………………… 43
- 3-5　总结 …………………………… 50

### 第4章　神经网络实操 ………… 51
- 4-1　撰写第一个神经网络程序 …… 51
  - 4-1-1　最简短的程序 …………… 51
  - 4-1-2　程序强化 ………………… 53
  - 4-1-3　试验 ……………………… 62
- 4-2　模型种类 ……………………… 67
  - 4-2-1　Sequential model ……… 67
  - 4-2-2　Functional API ………… 68
- 4-3　神经层 ………………………… 71
  - 4-3-1　完全连接层 ……………… 72
  - 4-3-2　Dropout Layer ………… 72
- 4-4　激励函数 ……………………… 72
- 4-5　损失函数 ……………………… 74
- 4-6　优化器 ………………………… 75
- 4-7　效能衡量指标 ………………… 77

4-8 超参数调校 ……………………… 80

## 第 5 章 PyTorch 进阶功能 ……… 86

5-1 数据集及数据加载器 …………… 86
5-2 TensorBoard ……………………… 92
  5-2-1 TensorBoard 功能 ……… 92
  5-2-2 测试 ……………………… 94
5-3 模型部署与 TorchServe ………… 99
  5-3-1 自行开发网页程序 ……… 99
  5-3-2 TorchServe ……………… 100

## 第 6 章 卷积神经网络 …………… 103

6-1 卷积神经网络简介 ……………… 103
6-2 卷积 ……………………………… 104
6-3 各种卷积 ………………………… 107
6-4 池化层 …………………………… 110
6-5 CNN 模型实操 …………………… 111
6-6 影像数据增强 …………………… 119
6-7 可解释的 AI …………………… 126

## 第 7 章 预先训练的模型 ………134

7-1 预先训练模型的简介 …………… 134
7-2 采用完整的模型 ………………… 137
7-3 采用部分模型 …………………… 140
7-4 转移学习 ………………………… 144
7-5 Batch Normalization 说明……… 148

## 第 3 篇 进阶的影像应用

## 第 8 章 对象侦测 ……………… 153

8-1 图像辨识模型的发展 …………… 153
8-2 滑动窗口 ………………………… 154
8-3 方向梯度直方图 ………………… 156
8-4 R-CNN 对象侦测 ………………… 166
8-5 R-CNN 改良 ……………………… 170
8-6 YOLO 算法简介 ………………… 176
8-7 YOLO v5 测试 …………………… 179
8-8 YOLO v5 模型训练 ……………… 181
8-9 YOLO v7 测试 …………………… 181
8-10 YOLO 模型训练 ………………… 183
8-11 SSD 算法 ………………………… 186
8-12 对象侦测的效能衡量指标 …… 188
8-13 总结 …………………………… 188

## 第 9 章 进阶的影像应用 ……… 189

9-1 语义分割介绍 …………………… 189
9-2 自动编码器 ……………………… 190
9-3 语义分割实操 …………………… 202
9-4 实例分割 ………………………… 208
9-5 风格转换——人人都可以是
  毕加索 …………………………… 214
9-6 脸部辨识 ………………………… 220
  9-6-1 脸部侦测 ……………… 221
  9-6-2 MTCNN 算法 ………… 225
  9-6-3 脸部追踪 ……………… 230
  9-6-4 脸部特征点侦测 ……… 235
  9-6-5 脸部验证 ……………… 240
9-7 光学文字辨识 …………………… 242
9-8 车牌辨识 ………………………… 245
9-9 卷积神经网络的缺点 …………… 249

## 第 10 章　生成对抗网络 …… 251
10-1　生成对抗网络介绍 …… 251
10-2　生成对抗网络种类 …… 253
10-3　DCGAN …… 256
10-4　Progressive GAN …… 264
10-5　Conditional GAN …… 265
10-6　Pix2Pix …… 269
10-7　CycleGAN …… 277
10-8　GAN 挑战 …… 282
10-9　深度伪造 …… 284

# 第 4 篇　自然语言处理

## 第 11 章　自然语言处理的介绍 …… 287
11-1　词袋与 TF-IDF …… 287
11-2　词汇前置处理 …… 290
11-3　词向量 …… 295
11-4　GloVe 模型 …… 309
11-5　中文处理 …… 311
11-6　spaCy 套件 …… 313

## 第 12 章　自然语言处理的算法 …… 318
12-1　循环神经网络 …… 318
12-2　PyTorch 内建文本数据集 …… 330
12-3　长短期记忆网络 …… 335
12-4　自定义数据集 …… 338
12-5　时间序列预测 …… 340
12-6　Gate Recurrent Unit …… 344
12-7　股价预测 …… 346
12-8　注意力机制 …… 349
12-9　Transformer 架构 …… 361
　　12-9-1　Transformer 原理 …… 361
　　12-9-2　Transformer 效能 …… 363
12-10　BERT …… 364
　　12-10-1　Masked LM …… 365
　　12-10-2　Next Sentence Prediction …… 365
　　12-10-3　BERT 效能微调 …… 366
12-11　Transformers 框架 …… 368
　　12-11-1　Transformers 框架范例 …… 368
　　12-11-2　Transformers 框架效能微调 …… 377
　　12-11-3　Transformers 中文模型 …… 382
　　12-11-4　后续努力 …… 383
12-12　总结 …… 383

## 第 13 章　ChatBot …… 384
13-1　ChatBot 类别 …… 384
13-2　ChatBot 设计 …… 385
13-3　ChatBot 实操 …… 387
13-4　ChatBot 工具套件 …… 389
　　13-4-1　ChatterBot 实操 …… 389
　　13-4-2　chatbotAI 实操 …… 392
　　13-4-3　Rasa 实操 …… 395
13-5　Dialogflow 实操 …… 398
　　13-5-1　Dialogflow 基本功能 …… 400
　　13-5-2　履行 …… 405
　　13-5-3　整合 …… 408
13-6　总结 …… 410

## 第 14 章 语音识别 ………… 411

- 14-1 语音基本认识 ……………… 412
- 14-2 语音前置处理 ……………… 421
- 14-3 PyTorch 语音前置处理 ……… 430
- 14-4 PyTorch 内建语音数据集 …… 439
- 14-5 语音深度学习应用 …………… 443
- 14-6 自动语音识别 ………………… 454
- 14-7 自动语音识别实操 …………… 457
- 14-8 总结 ……………………………… 457

## 第 5 篇　强化学习

## 第 15 章　强化学习原理及应用… 460

- 15-1 强化学习的基础 ……………… 460
- 15-2 强化学习模型 ………………… 464
- 15-3 简单的强化学习架构 ………… 466
- 15-4 Gym 套件 ……………………… 476
- 15-5 Gym 扩充功能 ………………… 484
- 15-6 动态规划 ……………………… 486
- 15-7 值循环 ………………………… 495
- 15-8 蒙特卡罗 ……………………… 497
- 15-9 时序差分 ……………………… 505
  - 15-9-1 SARSA 算法 …………… 506
  - 15-9-2 Q-learning 算法 ………… 510
- 15-10 井字游戏 …………………… 513
- 15-11 连续型状态变量与 Deep Q-Learning 算法 ……… 519
- 15-12 Actor Critic 算法 …………… 524
- 15-13 实际应用案例 ……………… 525
- 15-14 其他算法 …………………… 528
- 15-15 总结 ………………………… 529

## 第 6 篇　图神经网络

## 第 16 章　图神经网络原理及应用 …………… 531

- 16-1 图形理论 ……………………… 531
- 16-2 PyTorch Geometric …………… 541
- 16-3 图神经网络 …………………… 545
- 16-4 总结 …………………………… 551

# 第1篇 深度学习导论

在正式迈进深度学习的殿堂之前,我们先来看看几个初学者经常会有的疑问:

1. 人工智能已历经三波浪潮,而这一波是否又即将进入寒冬?
2. 人工智能、数据科学、数据挖掘、机器学习、深度学习,上述概念彼此之间到底有何关联?
3. 机器学习的开发流程与一般应用系统开发有何差异?
4. 深度学习的学习路径是什么?从哪里开始?
5. 为什么要先学习数学与统计学,才能把深度学习学好?
6. 先学哪一套深度学习框架比较好?学 TensorFlow 或 PyTorch?
7. 如何准备开发环境?

本篇将解答以上问题。本书着重在程序的实操上,有别于强调理论基础的图书,能让读者快速掌握深度学习的应用领域,学以致用。

# 第 1 章
# 深度学习介绍

## ▌ 1-1 人工智能历经的三波浪潮

人工智能 (Artificial Intelligence, AI) 并非近几年才兴起，其实目前是它的第三波热潮，前两波都发展十余年后，由于一些因素而衰退，而最近这一波热潮延续至今也十年了，会不会高点已过又将走向谷底呢？要回答这问题，我们需要知道过去学者在研究上有了哪些突破与成果，以及导致前两次发展衰退的原因，所以先重点回顾一下近代 AI 发展史。

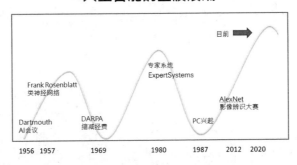

图 1.1 人工智能的三波浪潮

(1) 1956 年达特茅斯 (Dartmouth) 学院举办 AI 会议，确立第一波浪潮的开始。

(2) 1957 年 Frank Rosenblatt 创建感知器 (Perceptron)，即简易的神经网络，可惜当时还无法解决复杂多层的神经网络，直至 20 世纪 80 年代才想出解决办法。

(3) 1969 年美国国防部 (DARPA) 基于投资报酬率过低的理由，决定缩减 AI 研究经费，AI 迈入第一波寒冬。

(4) 1980 年专家系统 (Expert System) 兴起，企图将各行各业专家的内隐知识外显为一条条的规则，从而建立起专家系统，不过因不切实际，且需要使用大型且昂贵的计算机设备才能够建构相关系统，故而发展受挫。又不巧适逢个人计算机 (PC) 流行，相较之下，AI 的发展势头就被掩盖下去了，至此，AI 迈入第二波寒冬。

(5) 2012 年多伦多大学 Geoffrey Hinton 研发团队利用分布式计算环境及大量影像数据，结合过往的神经网络知识，开发了 AlexNet 神经网络，并参加 ImageNet 影像辨识大赛，结果一下子把错误率降低了十几个百分点，就此 AI 第三波浪潮兴起，至今方兴未艾。

回到前面的问题上，第三波热潮算下来也有十年的时间了，会不会又将迈入寒冬呢？其实仔细观察会发现，这波热潮相较于前两波，具备了以下几项优势。

(1) 基础功能架构较好，从影像、语音、文字辨识开始，再逐步往上构建各种的应用，例如自动驾驶 (Self-Driving)、对话机器人 (ChatBot)、机器人 (Robot)、智慧医疗、智慧城市等，这种由下往上的发展方式比较扎实。

(2) 硬件的快速发展。

①遵循摩尔定律快速发展：IC 上可容纳的晶体管数目，约每隔 18 个月至两年便会增加一倍，简单来说，就是 CPU 每隔两年便会增快一倍。过去 50 年均循此定律发展，此定律在未来十年也应该会继续适用，之后或许量子计算机 (Quantum Computer) 等新科技会继续接棒，它号称目前计算机要计算几百年的工作，量子计算机只需 30 分钟就搞定了，如果成真，到那时可能又是另一番光景了。

②云端数据中心的建立：各大 IT 公司在世界各地兴建大型的数据中心，收费模式采取 "用多少付多少" (Pay as you go)。由于模型训练时，通常需要大量运算，如果改用云端方案的话，一般企业就不须在前期购买昂贵设备，仅须支付必要的运算费用，也免去冗长的采购流程，只要几分钟就可以开通 (Provisioning) 所需设备，省钱省时。

③ GPU/ NPU 的开发：深度学习主要是以矩阵运算为主，而 GPU 在这方面的运算能力比 CPU 快非常多倍，所以专门生产 GPU 的美国公司英伟达 (NVIDIA) 备受瞩目，市值甚至超越 Intel [1]。当然其他硬件及系统厂商不会错失如此庞大的商机，纷纷积极抢食这块大饼，所以各种 NPU(Neural-network Processing Unit) 或 xPU 相继推出，使得指令周期越来越快，故而模型训练的时间能够大幅缩短，由于模型调试通常需要反复训练，所以能在短时间内得到答案的话，对数据科学家而言会是一大福音。另外，连接现场装置的计算机 (Raspberry pi、Jetson Nano、Auduino 等 ) 体积越来越小，运算能力也越来越强，对于 "边缘运算" 也有很大的帮助，例如路口监视器、无人机等。

(3) 算法推陈出新：计算能力提高后，许多无法在短时间完成训练的算法一一解封，尤其是神经网络，现在已经能够建构上百层的模型，包含高达上兆个参数，并且成功在短时间内调试出最佳模型，因此，模型设计就可以更复杂，算法逻辑也能更加完备。

(4) 大量数据的搜集及标注 (Labeling)：人工智能必须依赖大量数据，来让计算机学习，从中挖掘知识，近年来因因特网 (Internet) 及手机 (Mobile) 的盛行，企业除了通过社群媒体搜集大量数据之外，还可由物联网 (IoT) 的传感器产生源源不断的数据作为深度学习的养分 ( 训练数据 )，又加上这些大型网络公司的资金充足，只要雇用大量人力进行数据标注，来确保数据的质量，就能使得训练出来的模型越趋精准。

综合上述趋势发展所提供的迹象，与前两波相比，第三波热潮中 AI 研究成果已有一定程度的积累，硬件的发展也跟上了理论的脚步，而且大环境的支持也相对成熟，所以笔者推测目前这波热潮在短期内应该可以乐观对待。

# 1-2 AI 的学习地图

AI 发展史可划分成三个阶段，分别为人工智能 (Artificial Intelligence)、机器学习 (Machine Learning)、深度学习 (Deep Learning)，演进过程其实就是逐步缩小研究范围，聚焦在特定的算法。机器学习是人工智能的部分领域，而深度学习又是机器学习的部分算法。

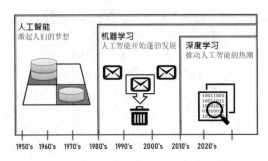

图 1.2　AI 三波热潮的重点

大部分的教育机构在规划 AI 的学习地图时，即依照这个轨迹逐步深入各项技术，通常分为四个阶段。

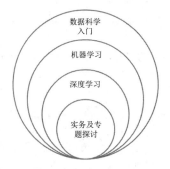

图 1.3　AI 学习地图

(1) 数据科学 (Data Science) 入门：内容包括 Python/R 程序语言、数据分析 (Data Analysis)、大数据平台 (Hadoop、Spark) 等。

(2) 机器学习 (Machine Learning)：包含一些典型的算法，如回归、罗吉斯回归、支持向量机 (SVM)、K-means 集群等，这些算法虽然简单，但却非常实用，容易在一般企业内普遍性应用。通常机器学习的分类如下图。

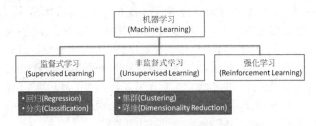

图 1.4　机器学习分类

最新的发展还有半监督学习 (Semi-supervised Learning)、自我学习 (Self Learning)、联合学习 (Federated Learning) 等，不一而足，千万不要被分类限制了你的想象。

另外，数据挖掘 (Data Mining) 与机器学习的算法大量重叠，其间的差异，有一种说法是数据挖掘着重在挖掘数据的隐藏状态 (Pattern)，而机器学习则着重于预测。

(3) 深度学习 (Deep Learning)：深度学习属于机器学习中的一环，所谓深度 (Deep) 是指多层式架构的模型，例如各种神经网络 (Neural Network) 模型、强化学习

(Reinforcement Learning, RL) 算法等，通过多层的神经层或 try-and-error 的方式来优化 (Optimization) 或反复求解。

(4) 实务及专题探讨 (Capstone Project)：将各种算法应用于各类领域 / 行业，强调专题探讨及产业应用实操。

## 1-3　TensorFlow 对比 PyTorch

深度学习 (Deep Learning) 的框架过去曾经百家争鸣，数量多达 20 多种，然而经过几年下来的厮杀，目前仅存在几个我们常用的主流框架了。

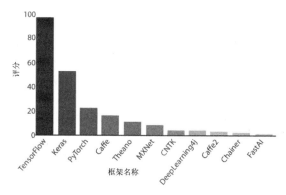

图 1.5　2018 年深度学习框架及评分

图片来源：Top 5 Deep Learning Frameworks to Watch in 2021 and Why TensorFlow [2]

目前比较常用的框架包括 Google TensorFlow、Facebook PyTorch、Apache MXNet、Berkeley Caffe，其中又以 TensorFlow、PyTorch 占有率较高，一般企业广泛使用的是 TensorFlow，而学术界则是偏好 PyTorch。两者的功能也是互相模仿与竞争，差异比较整理如表 1.1 所示。

表 1.1　TensorFlow、PyTorch 比较

| 比较项目 | TensorFlow | PyTorch |
| --- | --- | --- |
| 弹性 |  | 较好 |
| 效能 | 较好 |  |
| 简易训练指令 (fit) | Keras | PyTorch Lightning |
| 可视化接口工具 | TensorBoard | TensorBoardX |
| 部署工具 | TensorFlow Serving | TorchServe |
| TinyML | TensorFlow Lite | PyTorch Live |
| 预训模型 Hub | TensorFlow Hub | PyTorch Hub |

由上表可见两者的功能基本上大同小异，因此有人认为既然很相似，那就学习其中一种即可，但考虑到它们所专长的应用领域各有不同，并且网络上的扩充框架或范例程序常常只用其中一种语言开发，所以，同时熟悉 TensorFlow、PyTorch，会是一个比较周全的选择。

好在 TensorFlow、PyTorch 都是深度学习的框架，彼此间有共通的概念，只要遵循本书的学习路径就能一举两得，没有想象中困难。熟悉两个框架，还有助于设计概念的深入了解。因此本书介绍 PyTorch 的方式，会与另一本以 TensorFlow 为主题的《深度学习全书：公式＋推导＋代码＋TensorFlow 全程案例》[3] 相互对照。

根据官网说明及个人使用经验，PyTorch 有以下特点。

(1) Python First：PyTorch 与 Python 完美整合，在定义模型类别内可以任意加入侦错或转换的 Python 程序代码，TensorFlow/Keras 则需通过 Callback 才能在模型训练过程中传出信息，PyTorch 官方认为 Python 有丰富的套件，例如 NumPy、SciPy、Scikit-learn 等，无须另外发明轮子 (reinvent the wheel where appropriate)。

(2) 除错容易：TensorFlow/Keras 提供 fit 指令进行模型训练，虽然简单，但不易侦错，PyTorch 则须自行撰写优化程序，虽然烦琐，但可随时查看预测结果及损失函数变化，不必等到模型训练完后才能查看结果。

(3) GPU 内存管理较佳：笔者使用 GTX1050Ti，内存只有 4GB 时，执行 2 个以 TensorFlow 开发的 Notebook 文件时，常会发生内存不足的状况，但使用 PyTorch，即使 3、4 个 Notebook 文件也没有问题。

(4) 简洁快速：与 Intel MKL 和 NVIDIA (cuDNN, NCCL) 函数库整合，可提升执行的速度，程序可自由选择 CPU 或 GPU 运算，自由掌控内存的使用量。

(5) 无痛扩充：PyTorch 提供 C/C++ extension API，有效整合资源，不需桥接的包装程序 (wrapper)。

# 1-4　机器学习开发流程

一般来说，机器学习开发流程 (Machine Learning Workflow) 有多种建议的模型，例如数据挖掘 (Data Mining) 流程，包括 CRISP-DM (cross-industry standard process for data mining)、Google Cloud 建议的流程 [4] 等，个人偏好的流程如下。

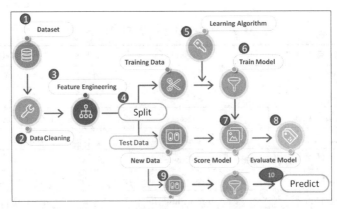

图 1.6　机器学习开发流程 (Machine Learning Workflow)

流程大概分为 10 个步骤 ( 不含较高层次的企业需求了解，只包括实际开发的步骤 )：

(1) 搜集数据，汇总为数据集 (Dataset)。

(2) 数据清理 (Data Cleaning)、数据探索与分析 (Exploratory Data Analysis, EDA)：EDA 通常通过描述统计量及统计图来观察数据的分布，了解数据的特性、极端值

(Outlier)、变量之间的关联性。

(3) 特征工程 (Feature Engineering)：原始搜集的数据未必是影响预测目标的关键因素，有时候需要进行数据转换，才得以找到关键的影响变量。

(4) 数据切割 (Data Split)：将数据切割为训练数据 (Training Data) 及测试数据 (Test Data)，一份数据供模型训练之用，另一份数据则用在衡量模型效能，例如准确度，切割的主要原因是确保测试数据不会参与训练，以维持其公正性，即 Out-of-Sample Test。

(5) 选择学习算法 (Learning Algorithms)：依据问题的类型选择适合的算法。

(6) 模型训练 (Model Training)：以算法及训练数据，进行训练产出模型。

(7) 模型计分 (Score Model)：计算准确度等效能指标，评估模型的准确性。

(8) 模型评估 (Evaluate Model)：比较多个参数组合、多个算法的准确度，找出最佳参数与算法。

(9) 部署 (Deploy)：复制最佳模型至正式环境 (Production Environment)，制作使用界面或提供 API，通常以网页服务 (Web Services) 方式开发。

(10) 预测 (Predict)：正式开始服务用户，用户传入新数据或文件后，输入至模型进行预测，并回传预测结果。

机器学习开发流程与一般应用系统开发流程有何差异？最大的差别如下：

(1) 一般应用系统利用输入数据与转换逻辑产生输出，例如撰写报表，根据转换规则将输入字段转换为输出字段，但机器学习先产生模型，再根据模型进行预测，故而重用性 (Reuse) 高。

(2) 机器学习不只使用输入数据，还会搜集大量的历史数据或从因特网中爬取出一堆数据，作为"饲料"。

(3) 新产生的数据可再返回模型，重新训练，自我学习，使模型更聪明。

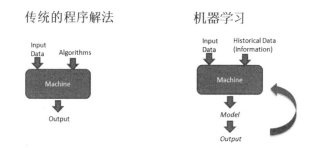

图 1.7　机器学习与一般应用系统开发流程的差异

## 1-5　开发环境安装

Python 是目前机器学习主流的程序语言，可以直接在本地安装开发环境，或使用云端环境。本书首先介绍本地安装的程序，建议依照以下顺序安装。

(1) 安装 Anaconda：建议安装此软件，它内含 Python 及上百个常用套件。先从 https://www.anaconda.com/products/individual 下载安装文件，在 Windows 操作系统安装时，建议执行到下列画面时，两者都勾选，就可将安装路径加入至环境变量 Path 内，这样在任何目录下均可执行 Python 程序。Mac/Linux 则须自行修改登录文件 (profile)，增加 Anaconda 安装路径。

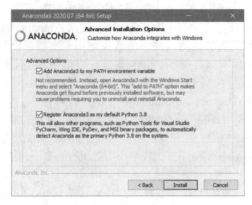

图 1.8　Anaconda 安装注意事项，将安装路径加入至环境变量 Path 内

(2) 安装 PyTorch 最新版：参阅 PyTorch 官网 (https://pytorch.org/get-started/locally/)，依次选择版本 (PyTorch Build)、操作系统 (Your OS)、指令 (Package)、语言 (Language)、CPU/GPU，选完后，会自动产生安装指令 ( 在最下面一列 )，指令的 pip3 应改为如下 pip。

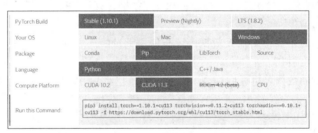

图 1.9　PyTorch 安装画面

(3) 在 Windows 操作系统安装时，开启 DOS 窗口，Mac/Linux 则须开启终端机，依次输入上一步骤产生的指令，例如：

pip install torch==1.10.1+cu113 torchvision==0.11.2+cu113

torchaudio===0.10.1+cu113 -f

https://download.pytorch.org/whl/cu113/torch_stable.html

** PyTorch 安装 GPU SDK 为 CUDA Toolkit 的子集合，与 TensoFlow 安装的 CUDA Toolkit 版本不同也不会互相冲突。

(4) 测试：安装完，在 DOS 窗口或终端机内输入 python，进入交互式环境，再输入以下指令测试：

>>>import torch

>>>torch.__version__

>>>exit()

会出现版本，例如：

1.10.0+cu113

注意，目前只支持 NVIDIA 独立显卡，若是较旧的显卡必须查阅驱动程序搭配的版本信息，请参考 NVIDIA 官网说明[5]，例如下图。

| CUDA Toolkit | Linux x86_64 Driver Version | Windows x86_64 Driver Version |
| --- | --- | --- |
| CUDA 11.1.1 Update 1 | >=455.32 | >=456.81 |
| CUDA 11.1 GA | >=455.23 | >=456.38 |
| CUDA 11.0.3 Update 1 | >= 450.51.06 | >= 451.82 |
| CUDA 11.0.2 GA | >= 450.51.05 | >= 451.48 |
| CUDA 11.0.1 RC | >= 450.36.06 | >= 451.22 |
| CUDA 10.2.89 | >= 440.33 | >= 441.22 |
| CUDA 10.1 (10.1.105 general release, and updates) | >= 418.39 | >= 418.96 |
| CUDA 10.0.130 | >= 410.48 | >= 411.31 |
| CUDA 9.2 (9.2.148 Update 1) | >= 396.37 | >= 398.26 |
| CUDA 9.2 (9.2.88) | >= 396.26 | >= 397.44 |
| CUDA 9.1 (9.1.85) | >= 390.46 | >= 391.29 |
| CUDA 9.0 (9.0.76) | >= 384.81 | >= 385.54 |
| CUDA 8.0 (8.0.61 GA2) | >= 375.26 | >= 376.51 |
| CUDA 8.0 (8.0.44) | >= 367.48 | >= 369.30 |
| CUDA 7.5 (7.5.16) | >= 352.31 | >= 353.66 |
| CUDA 7.0 (7.0.28) | >= 346.46 | >= 347.62 |

图 1.10　CUDA Toolkit 版本与驱动程序的搭配

奉劝各位读者，太旧的显卡若不能安装 PyTorch/TensorFlow 支持的版本，就不用安装 CUDA Toolkit、cuDNN SDK 了，因为显卡内存过小，执行 PyTorch/TensorFlow 时常会发生内存不足 (OOM) 的错误，徒增困扰。

## 1-6　免费云端环境开通

以上是本地的安装，接着再来谈谈云端环境的开通，几乎 Google GCP、AWS、Azure 都提供机器学习的开发环境，这里介绍免费的 Google 云端环境 Colaboratory，要有 Gmail 账号才能使用，它具备以下特点。

(1) 免安装，只须开通：常用的框架均已预安装，包括 TensorFlow、PyTorch。

(2) 免费的 GPU：使用 GPU 进行深度学习的模型训练会快上许多倍，Colaboratory 提供 NVIDIA Tesla K80 GPU 显卡，含 12GB 内存。

(3) 使用限制：Colaboratory 在使用时会实时开通 Docker container，只能连续使用 12 小时，逾时的话虚拟环境会被回收，虚拟机内的所有程序、数据一律会消失，要特别注意。

开通步骤如下。

(1) 使用 Google Chrome 浏览器，进入云端硬盘 (Google drive) 接口。

(2) 建立一个目录，例如 "0"，并切换至该目录。

(3) 在屏幕中间右击，单击"更多">"关联更多应用"。

(4) 在搜寻栏输入 Colaboratory，找到后单击该 App，单击"Connect"按钮即可开通。

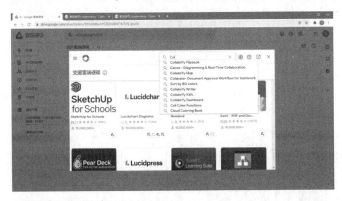

(5) 开通后，即可开始使用。可新增一个名为"Colaboratory"的文件。

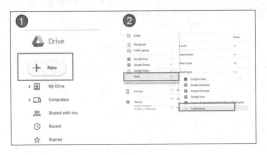

注意，Google Colaboratory 会自动开启虚拟环境，建立一个空白的 Jupyter Notebook 文件，文件名为 ipynb，几乎所有的云端环境及大数据平台 Databricks 都以 Notebook 为主要使用接口。

或者直接双击 Notebook 文件，也可自动开启虚拟环境，进行编辑与执行。本地的 Notebook 文件也可上传至云端硬盘，单击即可使用，完全不用转换，非常方便。

若要支持 GPU 可设定运行环境使用 GPU 或 TPU，TPU 为 Google 发明的 NPU。

Colaboratory 相关操作，可参考官网说明[6]。

本书所附的范例程序，一律为 Notebook 文件。Notebook 可以使用 Markdown 语法来撰写美观的说明，包括数学公式，并且程序也可以分格，实现单独执行，便于讲解，相关的用法可以参考 *Jupyter Notebook: An Introduction*[7]。

# 第 2 章
# 神经网络原理

## 2-1 必备的数学与统计知识

现在每天几乎都会看到几则有关 AI 的新闻，介绍 AI 的各种研发成果，一般人也许会基于好奇想一窥究竟，了解背后运用的技术与原理，但会发现涉及一堆数学符号及统计公式，可能就会产生疑问：要从事 AI 系统开发，非要搞定数学、统计不可吗？答案是肯定的，我们都知道机器学习是从数据中学习到知识 (Knowledge Discovery from Data, KDD)，而算法就是从数据中提取出知识，它必须以数学及统计为理论基础，才能证明其解法具有公信力与精准度。然而数学/统计理论都有局限，只有在假设成立的情况下，算法才是有效的，因此，如果不了解算法各个假设，随意套用公式，就好像无视交通规则，在马路上任意飙车一样危险。

因此，对于深度学习而言，我们至少需要熟悉以下学科：

(1) 线性代数 (Linear Algebra)；
(2) 微积分 (Calculus)；
(3) 统计与概率 (Statistics and Probability)；
(4) 线性规划 (Linear Programming)。

以神经网络权重求解的过程为例，四门学科就全部用上了，如下图所示。

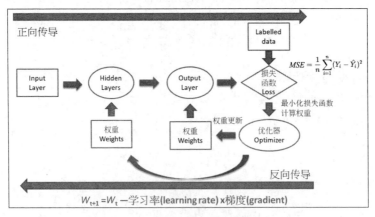

图 2.1　神经网络权重求解过程

(1) 正向传导：由"线性代数"计算误差及损失函数。

(2) 反向传导：通过"偏微分"计算梯度，同时，利用"线性规划"优化技巧寻找最优解。

(3)"统计"则串联整个环节，例如数据的探索与分析、损失函数定义、效能衡量指标，通通都基于统计的理论架构而成。

(4) 深度学习的推论以"概率"为基础，预测目标值。

四项学科相互为用，贯穿整个求解过程，因此，要通晓深度学习的运作原理，并且正确选用各种算法，甚至能够修改或创新算法，都必须对其背后的数学和统计有一定的基础认识，以免误用或滥用。

相关的数理基础可参考拙作《深度学习全书：公式 + 推导 + 代码 +TensorFlow 全程案例》第 2 章的介绍，本书直接切入主题，针对张量 (Tensor) 运算进行较详尽的说明。

## 2-2　万般皆自"回归"起

要探究神经网络优化的过程，要先了解简单线性回归求解，线性回归方程式如下：
$$y = wx + b$$
已知样本 $(x, y)$，要求解方程式中的参数权重 $(w)$、偏差 $(b)$。

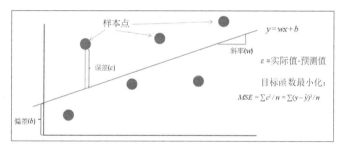

图 2.2　简单线性回归

一般求解方法有两种：

(1) 最小平方法 (Ordinary Least Square, OLS)；

(2) 最大似然估计法 (Maximum Likelihood Estimation, MLE)。

以最小平方法为例，首先定义目标函数 (Object Function) 或称损失函数 (Loss Function) 为均方误差 (MSE)，即预测值与实际值差距的平方和，MSE 当然越小越好，所以它是一个最小化的问题，我们可以利用偏微分推导出公式，过程如下。

(1) $MSE = \sum \varepsilon^2 / n = \sum (y - \hat{y})^2 / n$

其中 $\varepsilon$：误差，即实际值 $(y)$ 与预测值 $(\hat{y})$ 之差；

　　$n$：样本个数。

(2) $MSE = SSE / n$，$n$ 为常数，不影响求解，可忽略。
$$SSE = \sum \varepsilon^2 = \sum (y - \hat{y})^2 = \sum (y - wx - b)^2$$

(3) 分别对 $w$ 及 $b$ 偏微分，并且令一阶导数 =0，可以得到两个联立方程式，进而求得 $w$ 及 $b$。

(4) 先对 $b$ 偏微分，又因
$$f'(x) = g(x)g(x) = g'(x)g(x) + g(x)g'(x) = 2g(x)g'(x)$$

$$\frac{\partial SSE}{\partial b} = -2\sum_{i=1}^{n}(y - wx - b) = 0$$

➔ 两边同除以 2

$$\sum_{i=1}^{n}(y - wx - b) = 0$$

➔ 分解

$$\sum_{i=1}^{n}y - \sum_{i=1}^{n}wx - \sum_{i=1}^{n}b = 0$$

➔ 除以 $n$，$\bar{x}$、$\bar{y}$ 为 $x$，$y$ 的平均数

$$\bar{y} - w\bar{x} - b = 0$$

➔ 移项

$$b = \bar{y} - w\bar{x}$$

(5) 对 $w$ 偏微分：

$$\frac{\mathrm{d}SSE}{\mathrm{d}w} = -2\sum_{i=1}^{n}(y - wx - b)x = 0$$

➔ 两边同除以 -2

$$\sum_{i=1}^{n}(y - wx - b)x = 0$$

➔ 分解

$$\sum_{i=1}^{n}yx - \sum_{i=1}^{n}wx - \sum_{i=1}^{n}bx = 0$$

➔ 代入步骤 (4) 的计算结果 $b = \bar{y} - w\bar{x}$

$$\sum_{i=1}^{n}yx - \sum_{i=1}^{n}wx - \sum_{i=1}^{n}(\bar{y} - w\bar{x})x = 0$$

➔ 化简

$$\sum_{i=1}^{n}(y - \bar{y})x - w\sum_{i=1}^{n}(x^2 - \bar{x}x) = 0$$

$$w = \sum_{i=1}^{n}(y - \bar{y})x / \sum_{i=1}^{n}(x^2 - \bar{x}x)$$

$$w = \sum_{i=1}^{n}(y - \bar{y})x / \sum_{i=1}^{n}(x - \bar{x})^2$$

结论：

$$w = \sum_{i=1}^{n}(y - \bar{y})x / \sum_{i=1}^{n}(x - \bar{x})^2$$

$$b = \bar{y} - w\bar{x}$$

**范例 1.** 现有一个世界人口统计数据集，以年度 (year) 为 $x$，人口数为 $y$，按上述公式计算回归系数 $w$、$b$。

下列程序代码请参考【**02_01_ 线性回归 .ipynb**】。

(1) 使用 Pandas 相关函数计算，程序如下：

```
1  # 使用 OLS 公式计算 w、b
2  # 载入框架
3  import matplotlib.pyplot as plt
4  import numpy as np
5  import math
6  import pandas as pd
7
8  # 载入数据集
9  df = pd.read_csv('./data/population.csv')
10
11 w = ((df['pop'] - df['pop'].mean()) * df['year']).sum() \
12     / ((df['year'] - df['year'].mean())**2).sum()
13 b = df['pop'].mean() - w * df['year'].mean()
14
15 print(f'w={w}, b={b}')
```

执行结果：

$w$=0.061159358661557375，$b$=-116.35631056117687

(2) 改用 NumPy 的现成函数 polyfit 验算：

```
1  # 使用 NumPy 的现成函数 polyfit()
2  coef = np.polyfit(df['year'], df['pop'], deg=1)
3  print(f'w={coef[0]}, b={coef[1]}')
```

执行结果：答案相差不大。

$w$=0.061159358661554586，$b$=-116.35631056117121

(3) 上面公式，$x$ 只限一个，若以矩阵计算则更具通用性，多元回归亦可适用，即模型可以有多个特征 $(x)$，为简化模型，将 $b$ 视为 $w$ 的一环：

$$y=wx+b \Rightarrow y=wx+b\times 1 \Rightarrow y=[w\ b]\begin{bmatrix}X\\1\end{bmatrix} \Rightarrow y=w^{new}x^{new}$$

一样对 SSE 偏微分，一阶导数 =0 有最小值，公式推导如下：

$$SSE = \sum \varepsilon^2 = (y-\hat{y})^2 = (y-wx)^2 = yy' - 2wxy + w'x'xw$$

$$\frac{dSSE}{dw} = -2xy + 2wx'x = 0$$

→ 移项、整理

$$(xx')w = xy$$

→ 移项

$$w = (xx')^{-1}xy$$

(4) 使用 NumPy 相关函数计算，程序如下：

```
1  import numpy as np
2
3  X = df[['year']].values
4
5  # b = b * 1
6  one=np.ones((len(df), 1))
7
8  # 将 x 与 one 合并
9  X = np.concatenate((X, one), axis=1)
10
11 y = df[['pop']].values
12
13 # 求解
14 w = np.linalg.inv(X.T @ X) @ X.T @ y
15 print(f'w={w[0, 0]}, b={w[1, 0]}')
```

执行结果与上一段相同。

**范例2.** 再以Scikit-Learn的房价数据集为例，求解线性回归，该数据集有多个特征(*x*)。

(1) 以矩阵计算的方式，完全不变。

```
1  import numpy as np
2  from sklearn.datasets import load_boston
3
4  # 载入 Boston 房价数据集
5  X, y = load_boston(return_X_y=True)
6
7  # b = b * 1
8  one=np.ones((X.shape[0], 1))
9
10 # 将 x 与 one 合并
11 X = np.concatenate((X, one), axis=1)
12
13 # 求解
14 w = np.linalg.inv(X.T @ X) @ X.T @ y
15 w
```

执行结果如下：

```
array([-1.08011358e-01,  4.64204584e-02,  2.05586264e-02,  2.68673382e+00,
       -1.77666112e+01,  3.80986521e+00,  6.92224640e-04, -1.47556685e+00,
        3.06049479e-01, -1.23345939e-02, -9.52747232e-01,  9.31168327e-03,
       -5.24758378e-01,  3.64594884e+01])
```

(2) 以 Scikit-Learn 的线性回归类别验证答案。

```
1  from sklearn.linear_model import LinearRegression
2
3  X, y = load_boston(return_X_y=True)
4
5  lr = LinearRegression()
6  lr.fit(X, y)
7
8  lr.coef_, lr.intercept_
```

执行结果与采用矩阵计算的结果完全相同。

(3) PyTorch 自 v1.9 起提供线性代数函数库[1]，可直接调用，程序改写如下：

```
1  import numpy as np
2  from sklearn.datasets import load_boston
3  import torch
4
5  # 载入 Boston 房价数据集
6  X, y = load_boston(return_X_y=True)
7
8  X_tensor = torch.from_numpy(X)
9
10 # b = b * 1
11 one=torch.ones((X.shape[0], 1))
12
13 # 将 x 与 one 合并
14 X = torch.cat((X_tensor, one), axis=1)
15
16
17 # 求解
18 w = torch.linalg.inv(X.T @ X) @ X.T @ y
19 # w = (X.T @ X).inverse() @ X.T @ y # 也可以
20
21 w
```

执行结果与 NumPy 计算完全相同。

## 2-3 神经网络

有了以上的基础后，我们就可以进一步探讨神经网络 (Neural Network) 如何求解，这是进入深度学习领域非常重要的概念。

## 2-3-1 神经网络概念

神经网络是深度学习最重要的算法，它主要是模仿生物神经网络的传导系统，希望通过层层解析，归纳出预测的结果。

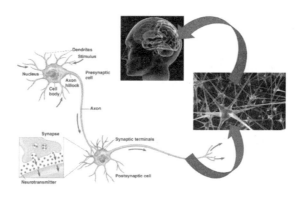

图 2.3　生物神经网络的传导系统

生物神经网络中表层的神经元接收到外界信号，归纳分析后，再通过神经末梢，将分析结果传给下一层的每个神经元，下一层神经元进行相同的动作，再往后传导，最后传至大脑，大脑做出最后的判断与反应。

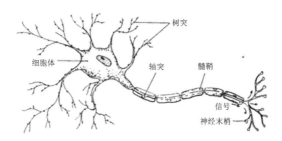

图 2.4　神经元结构

于是，AI 科学家将上述生物神经网络简化成下列的网络结构。

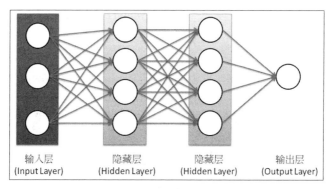

图 2.5　AI 神经网络

AI 神经网络最简单的连接方式称为完全连接 (Full Connected, FC)，亦即每个一神经元均连接至下一层的每个神经元，因此，我们可以把第二层以后的神经元均视为一条回归线的 $y$，它的特征变量 ($x$) 就是前一层的每一个神经元，例如下面的 $y_1$、$z_1$ 两条回归线。

$$y_1 = w_1x_1 + w_2x_2 + w_3x_3 + b$$
$$z_1 = w_1y_1 + w_2y_2 + w_3y_3 + b$$

所以，简单讲，一个神经网络可视为多条回归线组合而成的模型。

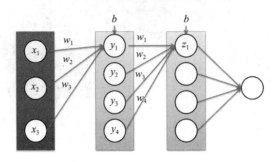

图 2.6　一个神经网络可视为多条回归线组合而成的模型

以上的回归线是线性的，为了支持更通用性的解决方案 (Generic Solution)，模型还会乘上一个非线性的函数，称为激励函数 (Activation Function)，期望也能解决非线性的问题，如下图所示。由于中译名称"激励函数"并不能明确表达其原意，故以下直接以英文 Activation Function 表示。

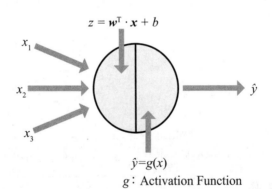

图 2.7　激励函数 (Activation Function)

如果不考虑 Activation Function，每一条线性回归线的权重 (Weight) 及偏差 (Bias) 可以通过最小平方法 (OLS) 求解，但乘上非线性的 Activation Function，就比较难用单纯的数学公式求解了，因此，学者就利用优化 (Optimization) 理论，针对权重、偏差各参数分别偏微分，沿着切线 ( 即梯度 ) 逐步逼近，找到最佳解，这种算法就称为"梯度下降法" (Gradient Descent)。

有一个很好的比喻来形容这个求解过程：当我们在山顶时，不知道下山的路，于是，就沿路往下走，遇到叉路时，就选择坡度最大的叉路走，直到抵达平地为止。所以梯度下降法利用偏微分 (Partial Differential) 求解斜率，沿斜率的方向，一步步往下走，逼近最佳解，直到损失函数没有显著改善为止，这时我们就认为已经找到最佳解了。

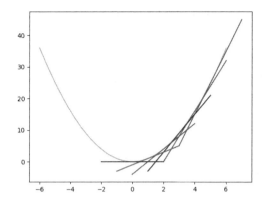

图 2.8　梯度下降法 (Gradient Descent) 示意图

## 2-3-2　梯度下降法

梯度其实就是斜率，单变量回归线的权重称为斜率，对于多变量回归线，须个别作偏微分求取权重值，就称为梯度。以下，先针对单变量求解，示范如何使用梯度下降法 (Gradient Descent) 求最小值。

**范例 1.** 假定损失函数 $f(x) = x^2$，而非 **MSE**，请使用梯度下降法求最小值。

注意，损失函数又称为目标函数或成本函数，在神经网络相关文献中大多称为损失函数，本书将统一以"损失函数"取代"目标函数"。

下列程序代码请参考【02_02_ 梯度下降法 .ipynb】。

(1) 定义函数 (func) 及其导数 (dfunc)。

```
1  # 载入套件
2  import numpy as np
3  import matplotlib.pyplot as plt
4
5  # 目标函数(损失函数):y=x^2
6  def func(x): return x ** 2 #np.square(x)
7
8  # 目标函数的一阶导数:dy/dx=2*x
9  def dfunc(x): return 2 * x
```

(2) 定义梯度下降法函数，反复更新 $x$，更新的公式如下，后面章节我们会推算公式的由来。

新的 $x$ = 目前的 $x$ - 学习率 (learning_rate) * 梯度 (gradient)

```
1  # 梯度下降
2  # x_start: x的起始点
3  # df: 目标函数的一阶导数
4  # epochs: 执行周期
5  # lr: 学习率
6  def GD(x_start, df, epochs, lr):
7      xs = np.zeros(epochs+1)
8      x = x_start
9      xs[0] = x
10     for i in range(epochs):
11         dx = df(x)
12         # x更新 x_new = x - learning_rate * gradient
13         x += - dx * lr
14         xs[i+1] = x
15     return xs
```

(3) 设定起始点、学习率 (lr)、执行周期数 (epochs) 等参数后，调用梯度下降法求解。

```
1   # 超参数(Hyperparameters)
2   x_start = 5      # 起始权重
3   epochs = 15      # 执行周期数
4   lr = 0.3         # 学习率
5
6   # 梯度下降法
7   w = GD(x_start, dfunc, epochs, lr=lr)
8
9   # 函数 y=x^2 绘图
10  plt.figure(figsize=(12,8))
11  t = np.arange(-6.0, 6.0, 0.01)
12  plt.plot(t, func(t), c='b')
13
14  # fix 中文乱码
15  from matplotlib.font_manager import FontProperties
16  plt.rcParams['font.sans-serif'] = ['Microsoft JhengHei'] # 正黑体
17  plt.rcParams['axes.unicode_minus'] = False # 矫正负号
18
19  plt.title('梯度下降法', fontsize=20)
20  plt.xlabel('w', fontsize=20)
21  plt.ylabel('Loss', fontsize=20)
22
23  color = list('rgbymr')  # 切线颜色
24  line_offset=2           # 切线长度
25  for i in range(5, -1, -1):
26      # 取相近两个点，画切线(tangent line)
27      z=np.array([i+0.001, i])
28      vec=np.vectorize(func)
29      cls = np.polyfit(z, vec(z), deg=1)
30      p = np.poly1d(cls)
31
32      # 画切线
33      x=np.array([i+line_offset, i-line_offset])
34      y=np.array([((i+line_offset)*p[1]+p[0], (i-line_offset)*p[1]+p[0]])
35      plt.plot(x, y, c=color[i-1])
36  plt.show()
```

执行结果：

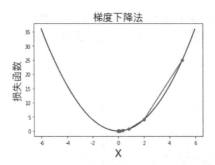

每一执行周期的损失函数如下，随着 $x$ 变化，损失函数逐渐收敛，即前后周期的损失函数差异逐渐缩小，最后当 $x=0$ 时，损失函数 $f(x)$ 等于 0，为函数的最小值，与最小平方法 (OLS) 的计算结果相同。

[5. 2, 0.8, 0.32, 0.13, 0.05, 0.02, 0.01, 0, 0, 0, 0, 0, 0, 0, 0]

如果改变起始点 (x_start) 为其他值，例如 −5，依然可以找到相同的最小值。

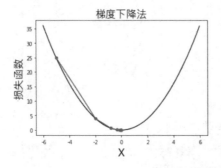

**范例2.** 假定损失函数 $f(x) = 2x^4 - 3x^2 + 2x - 20$，请使用梯度下降法求取最小值。

(1) 定义函数及其微分。

```
1  # 损失函数
2  def func(x): return 2*x**4-3*x**2+2*x-20
3
4  # 损失函数一阶导数
5  def dfunc(x): return 8*x**3-6*x+2
```

(2) 绘制损失函数。

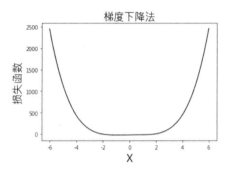

执行结果：

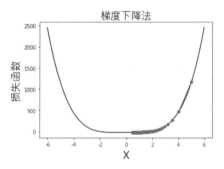

梯度下降法函数 (GD) 不变，执行程序，如果学习率不变 (lr = 0.3)，会出现错误信息：Result too large，原因是学习率过大，梯度下降过程错过最小值，往函数左方逼近，造成损失函数值越来越大，最后导致溢出。

```
------------------------------------------------------------
OverflowError                      Traceback (most recent call last)
C:\WINDOWS\TEMP/ipykernel_22480/2757641110.py in <module>
      6 # 梯度下降法
      7 # *** Function 可以直接当参数传递 ***
----> 8 w = GD(x_start, dfunc, epochs, lr=lr)
      9 print (np.around(w, 2))
     10

C:\WINDOWS\TEMP/ipykernel_22480/532521306.py in GD(x_start, df, epochs, lr)
      9     xs[0] = x
     10     for i in range(epochs):
---> 11         dx = df(x)
     12         # x更新 x_new = x - learning_rate * gradient
     13         x += - dx * lr

C:\WINDOWS\TEMP/ipykernel_22480/3069115439.py in dfunc(x)
      3
      4 # 损失函数一阶导数
----> 5 def dfunc(x): return 8*x**3-6*x+2

OverflowError: (34, 'Result too large')
```

修改学习率 (lr = 0.001)，同时增加执行周期数 (epochs = 15000)，避免还未逼近到最小值，就提早结束。

```
1  # 超参数(Hyperparameters)
2  x_start = 5       # 起始权重
3  epochs = 15000    # 执行周期数
4  lr = 0.001        # 学习率
```

执行结果：当 x=0.51 时，函数有最小值。

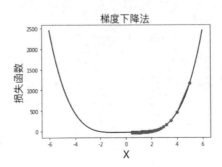

观察上述范例，不管函数为何，我们以相同的梯度下降法 (GD 函数 ) 都能够找到函数最小值，最重要的关键是"x 的更新公式"：

新的 $x$ = 目前的 $x$ - 学习率 (learning_rate) * 梯度 (gradient)

接着我们会说明此公式的由来，也就是神经网络求解的精华所在。

### 2-3-3 神经网络权重求解

神经网络权重求解是一个正向传导与反向传导反复执行的过程，如下图所示。

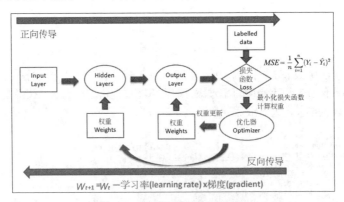

图 2.9　神经网络权重求解过程

(1) 由于神经网络是多条回归线的组合，建立模型的主要任务就是计算出每条回归线的权重 ($w$) 与偏差 ($b$)。

(2) 依上述范例的逻辑，一开始我们指定 $w$、$b$ 为任意值，建立回归方程式 $y=wx+b$，将特征值 ($x$) 代入方程式，可以求得预测值 ($\hat{y}$)，进而计算出损失函数，例如 MSE，这个过程称为正向传导 (Forward Propagation)。

(3) 通过最小化 MSE 的目标和偏微分，可以找到更好的 $w$、$b$，并依学习率来更新每一层神经网络的 $w$、$b$，此过程称为反向传导 (Back Propagation)。这部分可以由微分

的连锁率 (Chain Rule)，一次逆算出每一层神经元对应的 $w$、$b$，公式为：

$$W_{t+1} = W_t - 学习率 (learning\ rate) * 梯度 (gradient)$$

其中：

$$梯度 = -2 * x * (y - \hat{y})（稍后证明）$$

学习率是优化器事先设定的固定值或动能函数。

(4) 重复 (2)、(3) 步骤，一直到损失函数不再有明显改善为止。

梯度 (gradient) 公式证明如下：

(1) 损失函数 $MSE = \dfrac{\sum(y-\hat{y})^2}{n}$，因 $n$ 为常数，故仅考虑分子，即 $SSE$。

(2) $SSE = \sum(y-\hat{y})^2 = \sum(y-wx)^2 = \sum(y^2 - 2ywx + w^2x^2)$

(3) 以矩阵表示，$SSE = y^2 - 2ywx + w^2x^2$

(4) $\dfrac{\partial SSE}{\partial w} = -2yx + 2wx^2 = -2x(y-wx) = -2x(y-\hat{y})$

(5) 同理，$\dfrac{\partial SSE}{\partial b} = -2x^0(y-\hat{y}) = -2(y-\hat{y})$

(6) 为了简化公式，常把系数 2 拿掉。

(7) 最后公式为：

$$调整后权重 = 原权重 + (学习率 * 梯度)$$

(8) 有些文章将梯度负号拿掉，公式就修正为：

$$调整后权重 = 原权重 - (学习率 * 梯度)。$$

以上是以 MSE 为损失函数时的梯度计算公式，若使用其他损失函数，梯度计算结果也会有所不同，如果再加上 Activation Function，梯度公式计算就更加复杂了。还好，深度学习框架均提供自动微分 (Automatic Differentiation)、计算梯度的功能，我们就不用烦恼了。后续有些算法会自定义损失函数，因而产生意想不到的功能，例如风格转换 (Style Transfer) 可以合成两张图像，生成对抗网络 (Generative Adversarial Network, GAN)，产生几可乱真的图像。也因为如此关键，我们才花费了这么多的篇幅铺陈"梯度下降法"。

基础原理介绍到此，下一章，我们就以 PyTorch 实现自动微分、梯度下降、神经网络层，进而构建各种算法及相关的应用。

# 第 2 篇 | PyTorch 基础

PyTorch 是 Facebook AI 试验室 (Facebook's AI Research Lab, FAIR) 于 2016 年 9 月发布的深度学习框架，它是自 Torch 移植过来的 ( 以 Lua 程序语言开发 )，是深度学习最佳的入门框架之一，而且 PyTorch 官网很贴心地提供了中文版的说明文件。

本篇将介绍 PyTorch 的整体架构，包含下列内容：
(1) 从张量 (Tensor) 运算，到自动微分 (Automatic Differentiation)，再到神经层，最后构建完整的神经网络。
(2) 说明神经网络的各项函数，例如 Activation Function、损失函数 (Loss Function)、优化器 (Optimizer)、效能衡量指标 (Metrics)，并介绍运用梯度下降法找到最佳解的原理与过程。
(3) 示范 PyTorch 各项工具的使用，包含 TensorBoard 可视化工具、PyTorch Dataset/DataLoader、PyTorch Serve 部署等。
(4) 神经网络完整流程的实践。
(5) 卷积神经网络 (Convolutional Neural Network, CNN)。
(6) 预先训练的模型 (Pre-trained Model)。
(7) 转移学习 (Transfer Learning)。

# 第 3 章
# PyTorch 学习路径与主要功能

## 3-1 PyTorch 学习路径

梯度下降法是神经网络主要求解的方法,计算过程需要大量使用张量(Tensor)运算,另外,在反向传导的过程中,则要进行偏微分,计算梯度,求解多层结构的神经网络。因此,大多数的深度学习框架至少要具备下列功能:

(1) 张量运算:包括各种向量、矩阵运算。
(2) 自动微分 (Auto Differentiation):通过偏微分计算梯度。
(3) 各种神经层 (Layers) 及神经网络 (Neural Network) 模型构建。

所以学习的路径可以从简单的张量运算开始,再逐渐熟悉高阶的神经层函数,以奠定扎实的基础。

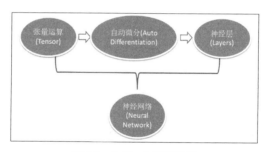

图 3.1 PyTorch 学习路径

掌握了 PyTorch 的核心之后,再外扩至支持工具 (TensorBoard)、移动装置 (Mobile)、部署工具 (TorchServe)、TorchScript、效能提升工具 (Profiler)、平行及分布式处理等。

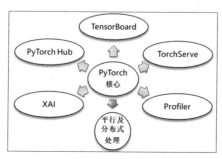

图 3.2 PyTorch 其他工具与扩充模块

## 3-2 张量运算

张量 (Tensor) 用于描述向量空间 (Vector Space) 中物体的特征，包括零维的标量 (Scalar)、一维的向量 (Vector)、二维的矩阵 (Matrix) 或更多维度的张量。线性代数则是说明张量如何进行各种运算，它被广泛应用于各种数值分析的领域。以下就以实例说明张量的概念与运算。PyTorch 线性代数函数库也都遵循 NumPy 套件的设计理念与语法，包括传播 (Broadcasting) 机制，甚至函数名称都相同。

深度学习模型的输入/输出格式以张量表示，所以，我们先熟悉向量、矩阵的相关运算及程序撰写语法。

**本节的程序代码请参阅【03_01_ 张量运算 .ipynb】。**

### 3-2-1 向量

向量 (Vector) 是一维的张量，它与线段的差别是除了长度 (Magnitude) 以外，还有方向 (Direction)，数学表示法为：

$$v = \begin{bmatrix} 2 \\ 1 \end{bmatrix}$$

以图形表示如下：

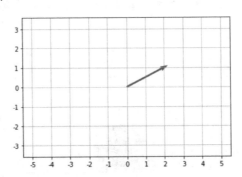

图 3.3　向量 (Vector) 长度与方向

(1) 长度 (Magnitude)：计算公式为欧几里得距离 (Euclidean Distance)。

$$\|v\| = \sqrt{v_1^2 + v_2^2} = \sqrt{5}$$

程序撰写如下：

```
1  # 向量(Vector)
2  v = np.array([2,1])
3
4  # 向量长度(magnitude) 计算
5  (v[0]**2 + v[1]**2) ** (1/2)
```

可以直接调用 np.linalg.norm() 计算向量长度：

```
1  # 使用 np.linalg.norm() 计算向量长度(magnitude)
2  import numpy as np
3
4  magnitude = np.linalg.norm(v)
5  print(magnitude)
```

也可以使用 PyTorch 的 torch.linalg.norm() 计算，但是要先将向量转换为 PyTorch 格式。

```
1  # 使用 PyTorch
2  torch.linalg.norm(torch.FloatTensor(v))
```

(2) 方向 (Direction)：使用 arctan() 函数计算。

$$\tan(\theta) = \frac{1}{2}$$

移项如下：

$$\theta = \arctan\left(\frac{1}{2}\right) \approx 26.57°$$

```
1   import math
2   import numpy as np
3   
4   # 向量(Vector)
5   v = np.array([2,1])
6   
7   vTan = v[1] / v[0]
8   print ('tan(θ) = 1/2')
9   
10  theta = math.atan(vTan)
11  print('弧度(radian) =', round(theta,4))
12  print('角度(degree) =', round(theta*180/math.pi, 2))
13  
14  # 也可以使用 math.degrees() 转换角度
15  print('角度(degree) =', round(math.degrees(theta), 2))
```

计算得到的单位为弧度，可转为大部分人比较熟悉的角度。

(3) 向量四则运算规则。

■ 加减乘除一个常数：常数直接对每个元素作加减乘除。

■ 加减乘除另一个向量：两个向量相同位置的元素作加减乘除，所以两个向量的元素个数须相等。

(4) 向量加减法：加减一个常数，长度、方向均改变。程序撰写如下：

```
1   # 载入套件
2   import numpy as np
3   import matplotlib.pyplot as plt
4   
5   # 向量(Vector) + 2
6   v = np.array([2,1])
7   v1 = np.array([2,1]) + 2
8   v2 = np.array([2,1]) - 2
9   
10  # 原点
11  origin = [0], [0]
12  
13  # 画有箭头的线
14  plt.quiver(*origin, *v1, scale=10, color='r')
15  plt.quiver(*origin, *v, scale=10, color='b')
16  plt.quiver(*origin, *v2, scale=10, color='g')
17  
18  plt.annotate('orginal vector',(0.025, 0.01), xycoords='data'
19               , fontsize=16)
20  
21  # 作图
22  plt.axis('equal')
23  plt.grid()
24  
25  plt.xticks(np.arange(-0.05, 0.06, 0.01), labels=np.arange(-5, 6, 1))
26  plt.yticks(np.arange(-3, 5, 1) / 100, labels=np.arange(-3, 5, 1))
27  plt.show()
```

执行结果：

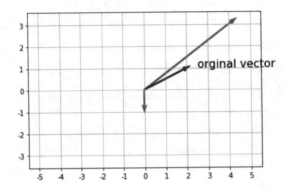

图 3.4 向量加减一个常数，长度、方向均改变

(5) 向量乘除法：乘除一个常数，长度改变，方向不改变。

```
1   # 载入套件
2   import numpy as np
3   import matplotlib.pyplot as plt
4
5   # 向量(Vector) * 2
6   v = np.array([2,1])
7   v1 = np.array([2,1]) * 2
8   v2 = np.array([2,1]) / 2
9
10  # 原点
11  origin = [0], [0]
12
13  # 画有箭头的线
14  plt.quiver(*origin, *v1, scale=10, color='r')
15  plt.quiver(*origin, *v, scale=10, color='b')
16  plt.quiver(*origin, *v2, scale=10, color='g')
17
18  plt.annotate('orginal vector',(0.025, 0.008), xycoords='data'
19               , color='b', fontsize=16)
20
21  # 作图
22  plt.axis('equal')
23  plt.grid()
24
25  plt.xticks(np.arange(-0.05, 0.06, 0.01), labels=np.arange(-5, 6, 1))
26  plt.yticks(np.arange(-3, 5, 1) / 100, labels=np.arange(-3, 5, 1))
27  plt.show()
```

执行结果：

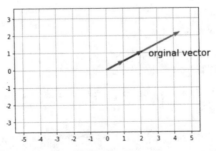

图 3.5 向量乘除一个常数，长度改变，方向不改变

(6) 向量加减乘除另一个向量：两个向量的相同位置的元素作加减乘除。

```
1  # 载入套件
2  import numpy as np
3  import matplotlib.pyplot as plt
4
5  # 向量(Vector) * 2
6  v = np.array([2,1])
7  s = np.array([-3,2])
8  v2 = v+s
9
10 # 原点
11 origin = [0], [0]
12
13 # 画有箭头的线
14 plt.quiver(*origin, *v, scale=10, color='b')
15 plt.quiver(*origin, *s, scale=10, color='b')
16 plt.quiver(*origin, *v2, scale=10, color='g')
17
18 plt.annotate('orginal vector',(0.025, 0.008), xycoords='data'
19              , color='b', fontsize=16)
20
21 # 作图
22 plt.axis('equal')
23 plt.grid()
24
25 plt.xticks(np.arange(-0.05, 0.06, 0.01), labels=np.arange(-5, 6, 1))
26 plt.yticks(np.arange(-3, 5, 1) / 100, labels=np.arange(-3, 5, 1))
27 plt.show()
```

执行结果：

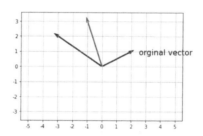

图 3.6　向量加另一个向量，长度、方向均会改变

(7)"内积"或称"点积"(Dot Product)计算公式如下：

$$\boldsymbol{v} \cdot \boldsymbol{s} = (v_1 \cdot s_1) + (v_2 \cdot s_2) + \cdots + (v_n \cdot s_n)$$

NumPy 是以 @ 作为内积的运算符号，而非 *。

```
1  # 载入套件
2  import numpy as np
3
4  # 向量(Vector)
5  v = np.array([2,1])
6  s = np.array([-3,2])
7
8  # 内积
9  d = v @ s  # 或 np.dot(v, s)、v.dot(s)
10
11 print (d)
```

(8) 计算两个向量的夹角，公式如下：

$$\boldsymbol{v} \cdot \boldsymbol{s} = \|\boldsymbol{v}\| \|\boldsymbol{s}\| \cos\theta$$

移项：

$$\cos(\theta) = \frac{\boldsymbol{v} \cdot \boldsymbol{s}}{\|\boldsymbol{v}\| \|\boldsymbol{s}\|}$$

再利用 arccos() 计算夹角 $\theta$。

```
1   # 载入套件
2   import math
3   import numpy as np
4
5   # 向量(Vector)
6   v = np.array([2,1])
7   s = np.array([-3,2])
8
9   # 计算长度(magnitudes)
10  vMag = np.linalg.norm(v)
11  sMag = np.linalg.norm(s)
12
13  # 计算 cosine(ϑ)
14  cos = (v @ s) / (vMag * sMag)
15
16  # 计算 ϑ
17  theta = math.degrees(math.acos(cos))
18
19  print(theta)
```

## 3-2-2 矩阵

矩阵是二维的张量，拥有行 (Row) 与列 (Column)，可用于表达一个平面的 $N$ 个点 ($N\times 2$)，或一个 3D 空间的 $N$ 个点 ($N\times 3$)，例如：

$$A = \begin{bmatrix} 1 & 2 & 3 \\ 4 & 5 & 6 \end{bmatrix}$$

(1) 矩阵加法／减法与向量相似，相同位置的元素作运算即可，但乘法运算通常是指内积，使用 @。试对两个矩阵相加：

$$\begin{bmatrix} 1 & 2 & 3 \\ 4 & 5 & 6 \end{bmatrix} + \begin{bmatrix} 6 & 5 & 4 \\ 3 & 2 & 1 \end{bmatrix} = \begin{bmatrix} 7 & 7 & 7 \\ 7 & 7 & 7 \end{bmatrix}$$

程序撰写如下：

```
1   # 载入套件
2   import numpy as np
3
4   # 矩阵
5   A = np.array([[1,2,3],
6                 [4,5,6]])
7   B = np.array([[6,5,4],
8                 [3,2,1]])
9
10  # 加法
11  print(A + B)
```

(2) 试对两个矩阵相乘：

$$\begin{bmatrix} 1 & 2 & 3 \\ 4 & 5 & 6 \end{bmatrix} \cdot \begin{bmatrix} 9 & 8 \\ 7 & 6 \\ 5 & 4 \end{bmatrix} = ?$$

左边矩阵的第 2 维须等于右边矩阵的第 1 维，即 (m, k) × (k, n) = (m, n)：

$$\begin{bmatrix} 1 & 2 & 3 \\ 4 & 5 & 6 \end{bmatrix} \cdot \begin{bmatrix} 9 & 8 \\ 7 & 6 \\ 5 & 4 \end{bmatrix} = \begin{bmatrix} 38 & 32 \\ 101 & 86 \end{bmatrix}$$

其中左上角的计算过程为 (1,2,3)×(9,7,5)=(1×9)+(2×7)+(3×5)=38，右上角的计算过程为 (1,2,3)×(8,6,4)=(1×8)+(2×6)+(3×4)=32，依此类推，如下图所示。

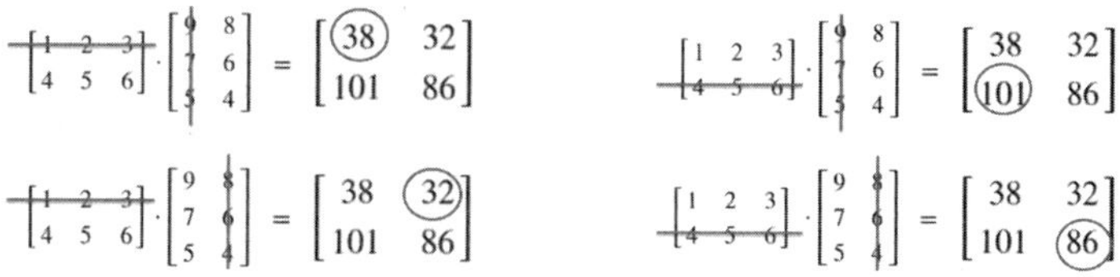

图 3.7　矩阵相乘

```
1  # 矩阵
2  A = np.array([[1,2,3],
3               [4,5,6]])
4  B = np.array([[9,8],
5               [7,6],
6               [5,4],
7               ])
8
9  # 乘法
10 print(A @ B)
```

(3) 矩阵 (***A***、***B***) 相乘，***A*** × ***B*** 是否等于 ***B*** × ***A***？

```
1  # 乘法：A x B != B x A
2
3  A = np.array([[1,2],
4               [4,5]])
5  B = np.array([[9,8],
6               [7,6],
7               ])
8
9  print(A @ B)
10 print()
11 print(B @ A)
12 print()
13 print('A x B != B x A')
```

执行结果：***A*** × ***B*** 不等于 ***B*** × ***A***。

```
[[23 20]
 [71 62]]

[[41 58]
 [31 44]]

A x B != B x A
```

(4) 矩阵在运算时，除了一般的加减乘除外，还有一些特殊的矩阵，包括转置矩阵 (Transpose)、逆矩阵 (Inverse)、对角矩阵 (Diagonal Matrix)、单位矩阵 (Identity Matrix) 等。

● 转置矩阵：列与行互换。

$$\begin{bmatrix} 1 & 2 & 3 \\ 4 & 5 & 6 \end{bmatrix}^T = \begin{bmatrix} 1 & 4 \\ 2 & 5 \\ 3 & 6 \end{bmatrix}$$

$(A^T)^T = A$：进行两次转置，会恢复成原来的矩阵。

(5) 对上述矩阵作转置，代码如下。

```
1  A = np.array([[1,2,3],
2                [4,5,6]])
3
4  # 转置矩阵
5  print(A.T) # 或 np.transpose(A)
```

也可以使用 np.transpose(A)。

(6) 逆矩阵 ($A^{-1}$)：$A$ 必须为方阵，即列数与行数须相等，且必须是非奇异矩阵 (Non-singular)，即每一列或行之间不可以相依于其他列或行。

```
1  import numpy as np
2
3  A = np.array([[1,2,3],
4                [4,5,6],
5                [7,8,9],
6                ])
7  print(np.linalg.inv(A))
```

执行结果：

```
[[ 3.15251974e+15 -6.30503948e+15  3.15251974e+15]
 [-6.30503948e+15  1.26100790e+16 -6.30503948e+15]
 [ 3.15251974e+15 -6.30503948e+15  3.15251974e+15]]
```

(7) 若 $A$ 为非奇异矩阵，则 $A @ A^{-1}$ = 单位矩阵 ($I$)。所谓的非奇异矩阵是任一行不能为其他行的倍数或多行的组合，包括各种四则运算。矩阵的列也须符合相同的规则。

(8) 试对下列矩阵验算 $A @ A^{-1}$ 是否等于单位矩阵 ($I$)。

$$A = \begin{bmatrix} 9 & 8 \\ 7 & 6 \end{bmatrix}$$

```
1  # A @ A逆矩阵 = 单位矩阵(I)
2  A = np.array([[9,8],
3                [7,6],
4                ])
5
6  print(np.around(A @ np.linalg.inv(A)))
```

执行结果：

```
[[1. 0.]
 [0. 1.]]
```

结果为单位矩阵，表示 $A$ 为非奇异矩阵。

(9) 试对下列矩阵验算 $A @ A^{-1}$ 是否等于单位矩阵 ($I$)。

$$A = \begin{bmatrix} 1 & 2 & 3 \\ 4 & 5 & 6 \\ 7 & 8 & 9 \end{bmatrix}$$

```
1  # A @ A逆矩阵 = 单位矩阵(I)
2  A = np.array([[1,2,4],
3                [4,7,6],
4                [7,8,9],
5                ])
6
7  print(np.around(A @ np.linalg.inv(A)))
```

执行结果：

```
[[ 0.  1. -0.]
 [ 0.  2. -1.]
 [ 0.  3.  2.]]
```

$A$ 为奇异 (Singular) 矩阵，原因如下：

第二列 = 第一列 + 1；

第三列 = 第一列 + 2；

故 $A @ A^{-1}$ 不等于单位矩阵。

## 3-2-3 使用 PyTorch

(1) 显示 PyTorch 版本。

```
1  # 载入框架
2  import torch
3
4  # 显示 PyTorch 版本
5  print(torch.__version__)
```

(2) 检查 GPU 及 CUDA Toolkit 是否存在。

```
1  # 检查 GPU 及 cuda toolkit 是否存在
2  torch.cuda.is_available()
```

执行结果为 True，表示侦测到 GPU，反之为 False。

(3) 如果要知道 GPU 的详细规格，可安装 PyCuda 套件。请注意在 Windows 环境下，无法以 pip install pycuda 安装，须在网站 Unofficial Windows Binaries for Python Extension Packages (https://www.lfd.uci.edu/~gohlke/pythonlibs/?cm_mc_uid=08085305845514542921829&cm_mc_sid_50200000=1456395916#pycuda) 下载对应 Python、Cuda Toolkit 版本的二进制文件。

举例来说，Python v3.8 安装 Cuda Toolkit v10.1 须下载 pycuda-2020.1+cuda101-cp38-cp38-win_amd64.whl，并执行 pip install pycuda-2020.1+cuda101-cp38-cp38-win_amd64.whl。

接着就可以执行本书所附的范例：python GpuQuery.py。

执行结果可显示 GPU 的详细规格，重要信息会排列在前面。

(4) 使用 torch.tensor 建立张量变量，PyTorch 会根据变量值决定数据类型，也可以声明特定类型，如整数 (torch.IntTensor)、长整数 (torch. LongTensor)、浮点数 (torch.FloatTensor)。

```
1   tensor = torch.tensor([[1, 2]])
2   print(tensor)
3
4   tensor2 = torch.IntTensor([[1, 2]])
5   print(tensor2)
6
7   tensor3 = torch.LongTensor([[1, 2]])
8   print(tensor3)
9
10  tensor4 = torch.FloatTensor([[1, 2]])
11  print(tensor4)
```

(5) 进行四则运算。

```
1  # 张量运算
2  A = torch.tensor([[1,2,3],
3                   [4,5,6]])
4  B = torch.tensor([[9,8,7],
5                   [7,6,5]
6                   ])
7
8  print(A + B)   # 加法
9  print(A - B)   # 减法
10 print(A * B)   # 乘法
11 print(A / B)   # 除法
12
13 # 内积
14 A = torch.tensor([[1,2,3],
15                  [4,5,6]])
16 B = torch.tensor([[9,8],
17                  [7,6],
18                  [5,4],
19                  ])
20 print(A @ B)
```

执行结果如下：

```
tensor([[10, 10, 10],
        [11, 11, 11]])
tensor([[-8, -6, -4],
        [-3, -1,  1]])
tensor([[ 9, 16, 21],
        [28, 30, 30]])
tensor([[0.1111, 0.2500, 0.4286],
        [0.5714, 0.8333, 1.2000]])
tensor([[ 38,  32],
        [101,  86]])
```

如果只要显示数值，可转为 NumPy 数组，例如：

x.numpy()

(6) NumPy 变量转为 PyTorch 张量变量。

```
1  import numpy as np
2
3  array = np.array([[1, 2]])
4  # Numpy -> PyTorch
5  tensor = torch.from_numpy(array)
6  tensor
```

(7) TensorFlow 常用的 reduce_sum 函数是沿着特定轴加总，输出会少一维，以 PyTorch 撰写可使用 sum() 替代：

```
1  # TensorFlow reduce_sum 的等式
2  A = torch.FloatTensor([[1,2,3],
3                        [4,5,6]])
4  A.sum(axis=1)
```

执行结果对每一行加总，输出会少一维：

tensor([ 6., 15.])

(8) 稀疏矩阵 (Sparse Matrix) 运算：稀疏矩阵是指矩阵内只有很少数的非零元素，如果按一般的矩阵存储会非常浪费内存，运算也是如此，因为大部分项目为零，不须浪费时间计算，所以，科学家针对稀疏矩阵设计了特殊的数据存储结构及算法，PyTorch 也支持此类数据类型。

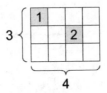

图 3.8　稀疏矩阵 (Sparse Matrix)

(9) TensorFlow 会自动决定变量在 CPU 或 GPU 运算，但 **PyTorch 稍麻烦一些**，必须手动将变量移动至 CPU 或 GPU 运算，不允许一个变量在 CPU，另一个变量在 GPU，进行运算会出现错误。程序撰写时要注意下列事项：

- 变量移动至 CPU/GPU，可使用 .to('cpu') 或 .to('cuda')，若有多张 GPU 卡，可指定移动至某一张，例如移动至第一张使用 .to('cuda:0')，也可以使用 .cuda()、.cpu()。
- 虽然手动移动比使用 TensorFlow 麻烦，不过，这种做法有助于内存管理，碰到 GPU 内存有限时，可以改在 CPU 运算，避免类似 TensorFlow 常见的内存不足 (OOM) 现象。
- 变量在 CPU/GPU 移动时会花费一点时间，若运算量不大且不复杂时，可直接在 CPU 运算。

```
1  # CPU -> GPU
2  tensor_gpu = tensor.cuda()
3  print(tensor_gpu)
4
5  # 若有多个 GPU，可指定 GPU 序号
6  tensor_gpu_2 = tensor.to('cuda:0')
7  print(tensor_gpu_2)
8
9  # GPU -> CPU
10 tensor_cpu = tensor_gpu.cpu()
11 print(tensor_cpu)
```

执行结果。

```
tensor([[1, 2]], device='cuda:0', dtype=torch.int32)
tensor([[1, 2]], device='cuda:0', dtype=torch.int32)
tensor([[1, 2]], dtype=torch.int32)
```

(10) CPU 与 GPU 变量不可混合运算。

```
1  tensor_gpu + tensor_cpu
```

执行结果如下。

```
---------------------------------------------------------------------------
RuntimeError                              Traceback (most recent call last)
<ipython-input-12-a07362b22321> in <module>
----> 1 tensor_gpu + tensor_cpu

RuntimeError: Expected all tensors to be on the same device, but found at least two devices, cuda:0 and cpu!
```

为了避免出错，要这样写才对：tensor_gpu + tensor_cpu.cuda()。

(11) 要在只有 GPU 或只有 CPU 的硬件均能执行，需要撰写如下：

```
1  device = 'cuda' if torch.cuda.is_available() else 'cpu'
2
3  tensor_gpu.to(device) + tensor_cpu.to(device)
```

(12) PyTorch 稀疏矩阵只须设定值的位置 (indices) 和数值 (values)，如下所示，i 为位置数组，v 为数值数组：

```
1  # 定义稀疏矩阵有值的(row, column)，例如第一个值在(0, 2)，第一/二列的第一栏
2  i = torch.LongTensor([[0, 1, 1],
3                        [2, 0, 2]])
4  # 稀疏矩阵的值
5  v = torch.FloatTensor([3, 4, 5])
6
7  # 定义稀疏矩阵的尺寸(2, 3)，并转为正常的矩阵
8  torch.sparse.FloatTensor(i, v, torch.Size([2,3])).to_dense()
```

执行结果如下,例如第一个值 (3) 在 (0, 2),第二个值 (4) 在 (1, 0),第三个值 (5) 在 (1, 2):

```
tensor([[0., 0., 3.],
        [4., 0., 5.]])
```

(13) 稀疏矩阵运算写法如下:

```
1  # 稀疏矩阵运算
2  a = torch.sparse.FloatTensor(i, v, torch.Size([2,3])) + \
3      torch.sparse.FloatTensor(i, v, torch.Size([2,3]))
4  a.to_dense()
```

执行结果:

```
tensor([[ 0., 0., 6.],
        [ 8., 0., 10.]])
```

(14) 要直接禁用 GPU 卡,可执行下列指令,注意,必须要在文件一开始就执行,否则无效。

```
1  # 载入框架
2  import torch
3  import os
4
5  os.environ["CUDA_VISIBLE_DEVICES"] = "-1"
6  # 检查 GPU 及 cuda toolkit 是否存在
7  print(torch.cuda.is_available())
```

(15) 若有多张 GPU 卡,可指定移动至某一张 GPU。

```
1  import os
2
3  os.environ["CUDA_VISIBLE_DEVICES"] = "0"
```

## 3-3 自动微分

反向传导时,会更新每一层的权重,这时就轮到偏微分运算派上用场,所以,深度学习框架的第二项主要功能就是自动微分 (Automatic Differentiation)。

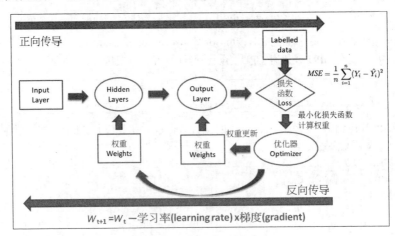

图 3.9　神经网络权重求解过程

**下列程序代码请参考【03_02_ 自动微分 .ipynb】。**

(1) 变量 y 会对 x 自动微分，代码如下。

```
1  import torch
2
3  # 设定 x 参与自动微分
4  x = torch.tensor(4.0, requires_grad=True)
5
6  y = x ** 2                # y = x^2
7
8  print(y)
9  print(y.grad_fn)          # y 梯度函数
10 y.backward()              # 反向传导
11 print(x.grad)             # 取得梯度
```

变量 x 要参与自动微分，须指定参数 requires_grad=True。

设定 y 是 x 的多项式，y.grad_fn 可取得 y 的梯度函数。

执行 y.backward()，会进行反向传导，即偏微分 $\dfrac{dy}{dx}$。

通过 x.grad 可取得梯度，若有多个变量也是如此。

执行结果中 $x^2$ 对 x 自动偏微分，得 $2x$，因 x=4，故 x 梯度 =8。

```
tensor(16., grad_fn=<PowBackward0>)
<PowBackward0 object at 0x0000024AA19EA7C0>
tensor(8.)
```

(2) 下列程序代码可取得自动微分相关的属性值。

```
1  # 设定变量值
2  x = torch.tensor(1.0, requires_grad = True)
3  y = torch.tensor(2.0)
4  z = x * y
5
6  # 显示自动微分相关属性
7  for i, name in zip([x, y, z], "xyz"):
8      print(f"{name}\ndata: {i.data}\nrequires_grad: {i.requires_grad}\n" +
9            "grad: {i.grad}\ngrad_fn: {i.grad_fn}\nis_leaf: {i.is_leaf}\n")
```

执行结果。

```
x
data: 1.0
requires_grad: True
grad: {i.grad}
grad_fn: {i.grad_fn}
is_leaf: {i.is_leaf}

y
data: 2.0
requires_grad: False
grad: {i.grad}
grad_fn: {i.grad_fn}
is_leaf: {i.is_leaf}

z
data: 2.0
requires_grad: True
grad: {i.grad}
grad_fn: {i.grad_fn}
is_leaf: {i.is_leaf}
```

(3) 来看一个较复杂的例子，以神经网络进行分类时，常使用交叉熵 (Cross Entropy) 作为损失函数，下图表达 Cross Entropy = $CE(y, wx + b)$。

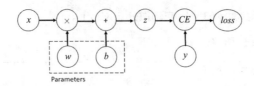

图 3.10　交叉熵 (Cross Entropy) 运算图

PyTorch 会依据程序，建构运算图 (Computational Graph)，描述梯度下降时，变量运算的顺序如下所示，先算 $x$，再算 $z$，最后计算 $loss$。

```
1  x = torch.ones(5)
2  y = torch.zeros(3)
3  w = torch.randn(5, 3, requires_grad=True)
4  b = torch.randn(3, requires_grad=True)
5  z = torch.matmul(x, w)+b
6  loss = torch.nn.functional.binary_cross_entropy_with_logits(z, y)
7
8  print('z 梯度函数：', z.grad_fn)
9  print('loss 梯度函数：', loss.grad_fn)
```

torch.nn.functional.binary_cross_entropy_with_logits 是 PyTorch 提供的交叉熵函数。

执行结果：

```
z 梯度函数： <AddBackward0 object at 0x0000020013478F10>
loss 梯度函数： <BinaryCrossEntropyWithLogitsBackward0 object at 0x0000020013478C40>
```

(4) 自动微分中，$z$ 是 $w$、$b$ 的函数，而 $loss$ 又是 $z$ 的函数，故只要对 $loss$ 进行反向传导即可。

```
1  loss.backward()
2  print(w.grad)              # w 梯度值
3  print(b.grad)              # b 梯度值
```

执行结果中，$w$、$b$ 梯度分别为：

```
tensor([[0.3148, 0.3112, 0.2117],
        [0.3148, 0.3112, 0.2117],
        [0.3148, 0.3112, 0.2117],
        [0.3148, 0.3112, 0.2117],
        [0.3148, 0.3112, 0.2117]])
tensor([0.3148, 0.3112, 0.2117])
```

(5) TensorFlow 使用常数 (Constant) 及变量 (Variable)，而 PyTorch 自 v0.4.0 起已弃用 Variable，直接使用 tensor 即可，但网络上依然常见 Variable，特此说明，详情请参阅本章参考文献 [1]。

```
1  # Variable 在 v0.4.0 已被弃用，直接使用 tensor 即可
2  from torch.autograd import Variable
3  x = Variable(torch.ones(1), requires_grad=True)
4  y = x + 1
5  y.backward()
6  print(x.grad)
```

上例以 torch.ones 替代 Variable。

```
1  # 替代 Variable
2  x2 = torch.ones(1, requires_grad=True)
3  y = x2 + 1
4  y.backward()
5  print(x2.grad)
```

(6) 模型训练时，会反复执行正向 / 反向传导，以找到最佳解，因此，梯度下降会

执行很多次，这时要注意两件事：
- y.backward 执行后，预设会将运算图销毁，y.backward 将无法再执行，故要保留运算图，须加参数 retain_graph=True。
- 梯度会不断累加，因此，执行 y.backward 后要重置 (Reset) 梯度，指令如下：x.grad.zero_()

(7) 不重置梯度代码如下：

```
1  x = torch.tensor(5.0, requires_grad=True)
2  y = x ** 3              # y = x^3
3
4  y.backward(retain_graph=True) # 梯度下降
5  print(f'一次梯度下降={x.grad}')
6
7  y.backward(retain_graph=True) # 梯度下降
8  print(f'二次梯度下降={x.grad}')
9
10 y.backward() # 梯度下降
11 print(f'三次梯度下降={x.grad}')
```

执行结果：
一次梯度下降 =75.0；
二次梯度下降 =150.0；
三次梯度下降 =225.0。
二次、三次梯度下降应该都是 75，结果都累加。

(8) 使用 x.grad.zero_() 梯度重置代码如下。

```
1  x = torch.tensor(5.0, requires_grad=True)
2  y = x ** 3              # y = x^3
3
4  y.backward(retain_graph=True) # 梯度下降
5  print(f'一次梯度下降={x.grad}')
6  x.grad.zero_()                # 梯度 reset
7
8  y.backward(retain_graph=True) # 梯度下降
9  print(f'二次梯度下降={x.grad}')
10 x.grad.zero_()                # 梯度 reset
11
12 y.backward() # 梯度下降
13 print(f'三次梯度下降={x.grad}')
```

(9) 多个变量梯度下降。

```
1  x = torch.tensor(5.0, requires_grad=True)
2  y = x ** 3              # y = x^3
3  z = y ** 2              # z = y^2
4
5  z.backward() # 梯度下降
6  print(f'x 梯度下降 ={x.grad}') # 6 * x^5
```

执行结果：z = 6 * (x^5) = 18750。

接着改写【02_02_ 梯度下降法 .ipynb】，将改用 PyTorch 函数微分。

(1) 只更动微分函数：原来是自己手动计算，现改用自动微分。

```
9  # 自动微分
10 def dfunc(x):
11     x = torch.tensor(float(x), requires_grad=True)
12     y = x ** 2 # 目标函数(损失函数)
13     y.backward()
14     return x.grad
```

执行结果：与【02_02_ 梯度下降法 .ipynb】相同。

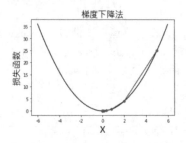

(2) 再将函数改为 $2x^4-3x^2+2x-20$。要缩小学习率 (0.001)，免得错过最小值，同时增大执行周期数 (15000)。

```
4  # 自动微分
5  def dfunc(x):
6      x = torch.tensor(float(x), requires_grad=True)
7      y = 2*x**4-3*x**2+2*x-20
8      y.backward()
9      return x.grad
```

执行结果：与【02_02_ 梯度下降法 .ipynb】相同。

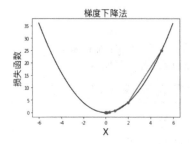

最后来操作一个完整范例，使用梯度下降法对线性回归求解，方程式如下，求 w、b 的最佳解。

$$y = wx + b$$

**下列程序代码请参考【03_03_ 简单线性回归 .ipynb】。**

(1) 载入套件。

```
1  # 载入套件
2  import numpy as np
3  import torch
```

(2) 定义训练函数：
- 刚开始 w、b 初始值均可设为任意值，这里使用正态分布之随机数。
- 定义损失函数 =MSE，公式见 11 行。
- 依照 2-3-3 节证明，权重更新公式如下：

  新权重 = 原权重 − 学习率 (learning_rate) × 梯度 (gradient)
- 权重更新必须"设定不参与梯度下降"才能运算，参见第 14~18 行。
- 每一训练周期的 w、b、损失函数都存储至数组，以利后续观察，参见第 22~24 行。
  要取得 w、b、损失函数的值，可以使用 .item() 转为常数，也可以使用 .detach().numpy() 转为 NumPy 数组，detach 作用是将变量脱离梯度下降的控制。
- 记得梯度重置，包括 w、b。

```python
1  def train(X, y, epochs=100, lr=0.0001):
2      loss_list, w_list, b_list=[], [], []
3  
4      # w、b 初始值均设为正态分布的随机数
5      w = torch.randn(1, requires_grad=True, dtype=torch.float)
6      b = torch.randn(1, requires_grad=True, dtype=torch.float)
7      for epoch in range(epochs):     # 执行训练周期
8          y_pred = w * X + b           # 预测值
9  
10         # 计算损失函数值
11         MSE = torch.square(y - y_pred).mean()
12         MSE.backward()
13  
14         # 设定不参与梯度下降，w、b才能运算
15         with torch.no_grad():
16             # 新权重 = 原权重 — 学习率(learning_rate) * 梯度(gradient)
17             w -= lr * w.grad
18             b -= lr * b.grad
19  
20         # 记录训练结果
21         if (epoch+1) % 1000 == 0 or epochs < 1000:
22             # detach：与运算图分离，numpy()：转成阵列
23             # w.detach().numpy()
24             w_list.append(w.item())  # w.item()：转成常数
25             b_list.append(b.item())
26             loss_list.append(MSE.item())
27  
28         # 梯度重置
29         w.grad.zero_()
30         b.grad.zero_()
31  
32     return w_list, b_list, loss_list
```

(3) 产生线性随机数据 100 笔，介于 0~50。

```python
1  # 产生线性随机数据100笔，介于 0-50
2  n = 100
3  X = np.linspace(0, 50, n)
4  y = np.linspace(0, 50, n)
5  
6  # 数据加一点噪声(noise)
7  X += np.random.uniform(-10, 10, n)
8  y += np.random.uniform(-10, 10, n)
```

(4) 执行训练。

```python
1  # 执行训练
2  w_list, b_list, loss_list = train(torch.tensor(X), torch.tensor(y))
3  
4  # 取得 w、b 的最佳解
5  print(f'w={w_list[-1]}, b={b_list[-1]}')
```

执行结果：$w=0.942326545715332$, $b=1.1824959516525269$。

(5) 执行训练 100000 次。

```python
1  # 执行训练
2  w_list, b_list, loss_list = train(torch.tensor(X), torch.tensor(y), epochs=100000)
3  
4  # 取得 w、b 的最佳解
5  print(f'w={w_list[-1]}, b={b_list[-1]}')
```

执行结果有差异：$w=0.8514814972877502$, $b=4.500218868255615$。

(6) 以 NumPy 验证。

```python
1  # 执行训练
2  coef = np.polyfit(X, y, deg=1)
3  
4  # 取得 w、b 的最佳解
5  print(f'w={coef[0]}, b={coef[1]}')
```

执行结果：$w=0.8510051491073364$, $b=4.517198474698629$。结果与梯度下降法训练 100000 次较相近，显示梯度下降法收敛较慢，需要较多执行周期的训练，这与默认的学习率 (lr) 有关，读者可以调整反复测试。所以说深度学习必须靠试验与经验，才能找到最佳参数值。

(7) 训练 100 次的模型绘图验证。

```
1  import matplotlib.pyplot as plt
2
3  plt.scatter(X, y, label='data')
4  plt.plot(X, w_list[-1] * X + b_list[-1], 'r-', label='predicted')
5  plt.legend()
```

执行结果：虽然训练次数不足，但回归线也确实在样本点的中线。

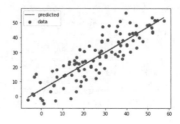

(8) NumPy 模型绘图验证。

```
1  # NumPy 求得的回归线
2  import matplotlib.pyplot as plt
3
4  plt.scatter(X, y, label='data')
5  plt.plot(X, coef[0] * X + coef[1], 'r-', label='predicted')
6  plt.legend()
```

执行结果：回归线在样本点的中线。

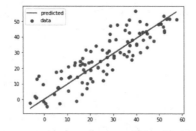

(9) 损失函数绘图验证。

```
1  plt.plot(loss_list)
```

执行结果：大约在第 10 个执行周期后就收敛了。

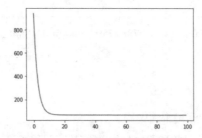

有了 PyTorch 自动微分的功能，反向传导变得非常简单，若要改用其他损失函数，只须修改一下公式，其他程序代码都照旧就可以了。这一节模型训练的程序架构非常重要，只要熟悉每个环节，后续复杂的模型也可以运用自如。

## 3-4 神经网络层

上一节运用自动微分实现一条简单线性回归线的求解,然而神经网络是多条回归线的组合,并且每一条回归线可能需再乘上非线性的 Activation Function,假如使用自动微分函数逐一定义每条公式,层层串连,程序可能需要很多个循环才能完成。所以为了简化程序开发的复杂度,PyTorch 直接建构了各种各样的神经层 (Layer) 函数,可以使用神经层组合神经网络的结构,我们只需要专注在算法的设计即可,轻松不少。

神经网络是多个神经层组合而成的,如下图,包括输入层 (Input Layer)、隐藏层 (Hidden Layer) 及输出层 (Output Layer),其中隐藏层可以有任意多层,广义来说,隐藏层大于或等于两层,即称为深度 (Deep) 学习。

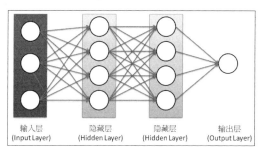

图 3.11　神经网络示意图

PyTorch 提供十多类神经层,每一类别又有很多种神经层,都定义在 torch.nn 命名空间下,可参阅官网说明[2]。

首先介绍完全神经层 (Linear Layers),参阅图 3.11,即上一层神经层的每一个神经元 ( 图中的圆圈 ) 都连接到下一层神经层的每一个神经元,所以也称为完全连接层,又分为 Linear、Bilinear、LazyLinear 等,可参阅维基百科[3],接下来开始实操完全连接层。

**范例1. 进行试验,熟悉完全连接层的基本用法。**

**下列程序代码请参考【03_04_ 完全连接层 .ipynb】。**

(1) 载入套件。

```
1  # 载入套件
2  import torch
```

(2) 产生随机随机数的输入数据,输出二维数据。

```
1  input = torch.randn(128, 20)
2  input.shape
```

(3) 建立神经层,Linear 参数依次为:
输入神经元个数。
输出神经元个数。
是否产生偏差项 (bias)。
设备:None、CPU('cpu') 或 GPU('cuda')。
数据类型。
Linear 神经层的转换为 $y = xA^T + b$,$y$ 为输出,$x$ 为输入。

```
1  # 建立神经层
2  # Linear参数依序为：输入神经元个数，输出神经元个数
3  layer1 = torch.nn.Linear(20, 30)
```

(4) 神经层计算：未训练 Linear 就是执行矩阵内积，维度 (128, 20) @ (20, 30) = (128, 30)。

```
1  output = layer1(input)
2  output.shape
```

执行结果：torch.Size([128, 30])。

再测试 Bilinear 神经层，Bilinear 有两个输入神经元个数，转换为 $y = x_1^T A^T x_2 + b$。

(1) 建立神经层。

```
1  layer2 = torch.nn.Bilinear(20, 30, 40)
2  input1 = torch.randn(128, 20)
3  input2 = torch.randn(128, 30)
```

(2) Bilinear 神经层计算：未训练 Linear 就是执行矩阵内积，维度 (128, 20) @ (20, 40) + (128, 30) @ (20, 40) = (128, 40)。

```
1  output = layer2(input1, input2)
2  output.shape
```

执行结果：torch.Size([128, 40])。

**范例2. 引进完全连接层，估算简单线性回归的参数 $w$、$b$。**

**下列程序代码请参考【03_05_简单线性回归_神经层 .ipynb】。**

(1) 载入套件。

```
1  # 载入套件
2  import numpy as np
3  import torch
```

(2) 产生随机数据，与上一节范例相同。

```
1  # 产生线性随机数据100笔，介于 0-50
2  n = 100
3  X = np.linspace(0, 50, n)
4  y = np.linspace(0, 50, n)
5
6  # 数据加一点噪声(noise)
7  X += np.random.uniform(-10, 10, n)
8  y += np.random.uniform(-10, 10, n)
```

(3) 定义模型函数：导入神经网络模型，简单的顺序型模型内含 Linear 神经层及扁平层 (Flatten)，扁平层参数设定哪些维度要转成一维，设定范围参数为 (0, -1) 可将所有维度转成一维。

```
1  # 定义模型
2  def create_model(input_feature, output_feature):
3      model = torch.nn.Sequential(
4          torch.nn.Linear(input_feature, output_feature),
5          torch.nn.Flatten(0, -1) # 所有维度转成一维
6      )
7      return model
```

测试扁平层如下，Flatten() 预设范围参数为 (1, -1)，故结果为二维，通常第一维为笔数，第二维为分类的预测答案。

```
1  # 测试扁平层(Flatten)
2  input = torch.randn(32, 1, 5, 5)
3  m = torch.nn.Sequential(
4      torch.nn.Conv2d(1, 32, 5, 1, 1),
5      torch.nn.Flatten()
6  )
7  output = m(input)
8  output.size()
```

执行结果：torch.Size([32, 288])。

(4) 定义训练函数。

定义模型：神经网络仅使用一层完全连接层，而且输入只有一个神经元，即 $x$，输出也只有一个神经元，即 $y$。偏差项 (bias) 默认值为 True，除了一个神经元输出外，还会有一个偏差项，设定其实就等于 $y=wx+b$。

定义损失函数：直接使用 MSELoss 函数取代 MSE 公式。

```
1  def train(X, y, epochs=2000, lr=1e-6):
2      model = create_model(1, 1)
3
4      # 定义损失函数
5      loss_fn = torch.nn.MSELoss(reduction='sum')
```

(5) 使用循环，反复进行正向 / 反向传导的训练。

计算 MSE：改为 loss_fn(y_pred, y)。

梯度重置：改由 model.zero_grad() 取代 $w$、$b$ 逐一设定。

权重更新：改用 model.parameters 取代 $w$、$b$ 逐一更新。

model[0].weight、model[0].bias 可取得权重、偏差项。

```
7       loss_list, w_list, b_list=[], [], []
8       for epoch in range(epochs):    # 执行训练周期
9           y_pred = model(X)          # 预测值
10
11          # 计算损失函数值
12          # print(y_pred.shape, y.shape)
13          MSE = loss_fn(y_pred, y)
14
15          # 梯度重置：改由model.zero_grad() 取代 w、b 逐一设定。
16          model.zero_grad()
17
18          # 反向传导
19          MSE.backward()
20
21          # 权重更新：改用 model.parameters 取代 w、b 逐一更新
22          with torch.no_grad():
23              for param in model.parameters():
24                  param -= lr * param.grad
25
26          # 记录训练结果
27          linear_layer = model[0]
28          if (epoch+1) % 1000 == 0 or epochs < 1000:
29              w_list.append(linear_layer.weight[:, 0].item())    # w.item()：转成常数
30              b_list.append(linear_layer.bias.item())
31              loss_list.append(MSE.item())
32
33      return w_list, b_list, loss_list
```

(6) 执行训练。

```
1  # 执行训练
2  X2, y2 = torch.FloatTensor(X.reshape(X.shape[0], 1)), torch.FloatTensor(y)
3  w_list, b_list, loss_list = train(X2, y2)
4
5  # 取得 w、b 的最佳解
6  print(f'w={w_list[-1]}, b={b_list[-1]}')
```

执行结果:$w$=0.847481906414032, $b$=3.2166433334350586。

(7) 以 NumPy 验证。

```
1  # 执行训练
2  coef = np.polyfit(X, y, deg=1)
3
4  # 取得 w、b 的最佳解
5  print(f'w={coef[0]}, b={coef[1]}')
```

执行结果与答案非常相近。

$w$=0.8510051491073364, $b$=4.517198474698629。

(8) 显示回归线。

```
1  import matplotlib.pyplot as plt
2
3  plt.scatter(X, y, label='data')
4  plt.plot(X, w_list[-1] * X + b_list[-1], 'r-', label='predicted')
5  plt.legend()
```

执行结果。

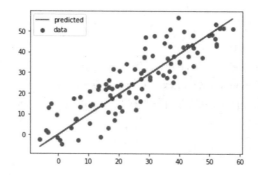

(9) NumPy 模型绘图验证。

```
1  # NumPy 求得的回归线
2  import matplotlib.pyplot as plt
3
4  plt.scatter(X, y, label='data')
5  plt.plot(X, coef[0] * X + coef[1], 'r-', label='predicted')
6  plt.legend()
```

执行结果非常相近。

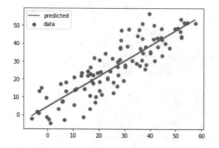

(10) 损失函数绘图。

```
2  plt.plot(loss_list)
```

执行结果：在第 25 个执行周期左右就已经收敛。

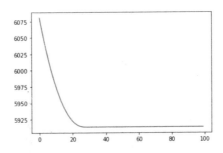

**范例3.** 接着我们再进一步，引进优化器(Optimizer)，操控学习率，参阅图3.9。

这里仅列出与范例 2 的差异，完整程序请参见程序【03_06_ 简单线性回归 _ 神经层 _ 优化器 .ipynb】。

(1) 定义训练函数。

定义优化器：之前为固定的学习率，改用 Adam 优化器，会采取动态衰减 (Decay) 的学习率，参见第 8 行。PyTorch 提供多种优化器，第 4 章会详细说明。

梯度重置：改由优化器 (Optimizer) 控制，optimizer.zero_grad() 取代 model.zero_grad()，参见第 19 行。

权重更新：改用 optimizer.step() 取代，$w$、$b$ 逐一更新，参见第 25 行。

```
1   def train(X, y, epochs=100, lr=1e-4):
2       model = create_model(1, 1)
3       
4       # 定义损失函数
5       loss_fn = torch.nn.MSELoss(reduction='sum')
6       
7       # 定义优化器
8       optimizer = torch.optim.Adam(model.parameters(), lr=lr)
9       loss_list, w_list, b_list=[], [], []
10      for epoch in range(epochs):    # 执行训练周期
11          y_pred = model(X)          # 预测值
12          
13          # 计算损失函数值
14          # print(y_pred.shape, y.shape)
15          MSE = loss_fn(y_pred, y)
16          
17          # 梯度重置：改由优化器(Optimizer)控制
18          optimizer.zero_grad()
19          
20          # 反向传导
21          MSE.backward()
22          
23          # 权重更新：改用 model.parameters 取代 w、b 逐一更新
24          optimizer.step()
25          
26          # 记录训练结果
27          if (epoch+1) % 1000 == 0 or epochs < 1000:
28              w_list.append(model[0].weight[:, 0].item())   # w.item():转成常数
29              b_list.append(model[0].bias.item())
30              loss_list.append(MSE.item())
31      
32      return w_list, b_list, loss_list
```

(2) 训练结果与范例 2 相同，**注意，若出现 $w$ 或 $b$=nan，损失等于无穷大 (inf)**，表示梯度下降可能错过最小值，继续往下寻找，碰到这种情况，可调低学习率。

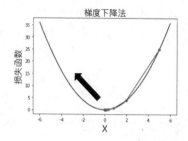

**范例4.** 除了回归之外，也可以处理分类(Classification)的问题，数据集采用Scikit-Learn套件内建的鸢尾花(Iris)。

下列程序代码请参考【03_07_IRIS 分类 .ipynb】。

(1) 载入套件。

```
1  # 载入套件
2  import numpy as np
3  import pandas as pd
4  from sklearn import datasets
5  import torch
```

(2) 载入 IRIS 数据集：有花萼长 / 宽、花瓣长 / 宽共 4 个特征。

```
1  dataset = datasets.load_iris()
2  df = pd.DataFrame(dataset.data, columns = dataset.feature_names)
3  df.head()
```

执行结果。

|   | sepal length (cm) | sepal width (cm) | petal length (cm) | petal width (cm) |
|---|---|---|---|---|
| 0 | 5.1 | 3.5 | 1.4 | 0.2 |
| 1 | 4.9 | 3.0 | 1.4 | 0.2 |
| 2 | 4.7 | 3.2 | 1.3 | 0.2 |
| 3 | 4.6 | 3.1 | 1.5 | 0.2 |
| 4 | 5.0 | 3.6 | 1.4 | 0.2 |

(3) 数据分割成训练及测试数据，测试数据占 20%。

```
1  from sklearn.model_selection import train_test_split
2  X_train, X_test, y_train, y_test = train_test_split(df.values,
3                                        dataset.target, test_size=0.2)
```

(4) 进行 one-hot encoding 转换：*y* 变成 3 个变量，代表 3 个品种的概率。

```
1  # one-hot encoding
2  y_train_encoding = pd.get_dummies(y_train)
3  y_test_encoding = pd.get_dummies(y_test)
```

也可以使用 PyTorch 函数。

```
1  # 使用 PyTorch 函数
2  torch.nn.functional.one_hot(torch.LongTensor(y_train))
```

(5) 转成 PyTorch Tensor。

```
1  # 转成 PyTorch Tensor
2  X_train = torch.FloatTensor(X_train)
3  y_train_encoding = torch.FloatTensor(y_train_encoding.values)
4  X_test = torch.FloatTensor(X_test)
5  y_test_encoding = torch.FloatTensor(y_test_encoding.values)
6  X_train.shape, y_train_encoding.shape
```

(6) 建立神经网络模型。

Linear(4, 3)：4 个输入特征；3 个品种预测值。

Softmax 激励函数会将预测值转为概率形式。

```
1  model = torch.nn.Sequential(
2      torch.nn.Linear(4, 3),
3      torch.nn.Softmax(dim=1)
4  )
```

(7) 定义损失函数、优化器：与之前相同。

```
1  loss_function = torch.nn.MSELoss(reduction='sum')
2  optimizer = torch.optim.Adam(model.parameters(), lr=0.01)
```

(8) 训练模型：第 9~10 行比较实际值与预测值相等的笔数，并转为百分比。

```
1  epochs=1000
2  accuracy = []
3  losses = []
4  for i in range(epochs):
5      y_pred = model(X_train)
6      loss = loss_function(y_pred, y_train_encoding)
7  
8      #print(np.argmax(y_pred.detach().numpy(), axis=1))
9      accuracy.append((np.argmax(y_pred.detach().numpy(), axis=1) == y_train)
10                     .sum()/y_train.shape[0]*100)
11     losses.append(loss.item())
12  
13     # 梯度重置
14     optimizer.zero_grad()
15  
16     # 反向传导
17     loss.backward()
18  
19     # 执行下一步
20     optimizer.step()
21  
22     if i%100 == 0:
23         print(loss.item())
```

(9) 绘制训练过程的损失及准确率趋势图。

```
1  import matplotlib.pyplot as plt
2  
3  # fix 中文乱码
4  from matplotlib.font_manager import FontProperties
5  plt.rcParams['font.sans-serif'] = ['Microsoft JhengHei']  # 微软正黑体
6  plt.rcParams['axes.unicode_minus'] = False
7  
8  plt.figure(figsize=(12,6))
9  plt.subplot(1,2,1)
10 plt.title('损失', fontsize=20)
11 plt.plot(range(0,epochs), losses)
12 
13 plt.subplot(1,2,2)
14 plt.title('准确率', fontsize=20)
15 plt.plot(range(0,epochs), accuracy)
16 plt.ylim(0,100)
17 plt.show()
```

执行结果：在第 400 个执行周期左右就已经收敛。

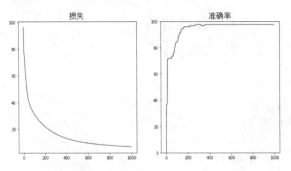

(10) 模型评估：评估测试数据的准确度。

```
1  predict_test = model(X_test)
2  _, y_pred = torch.max(predict_test, 1)
3
4  print(f'测试数据准确度: {((y_pred.numpy() == y_test).sum()/y_test.shape[0]):.2f}')
```

执行结果：准确率 100%，不过，数据分割是随机抽样，每次结果均不相同。

# 3-5 总结

本章介绍了张量基本运算、以自动微分实现梯度下降法，使用神经层、优化器构建神经网络，解决机器学习常见的回归与分类的问题，并逐步深入探讨神经网络的奥妙。

学到这里，大家可能会有许多疑问，如能否使用更多的神经层，甚至更复杂的神经网络结构？答案是肯定的，下一章我们将正式迈入深度学习的殿堂，学习如何应用 PyTorch 实现各种影像、自然语言的辨识，并且详细剖析各个函数的用法及参数说明。

# 第 4 章
# 神经网络实操

接下来,将开始以神经网络实现各种应用,内容着重于概念的澄清与程序的撰写。笔者会尽可能运用大量图解,帮助读者迅速掌握各种算法的原理,避免长篇大论。

首先通过"手写阿拉伯数字辨识"的案例,了解实操机器学习流程的 10 大步骤。其次,详细解说构建神经网络的函数用法及各项参数代表的意义。最后会撰写一个完整的窗口接口程序及网页程序,让终端用户 (End User) 亲自体验 AI 应用程序,期望激发用户对"企业导入 AI"有更多的想象。

## 4-1 撰写第一个神经网络程序

手写阿拉伯数字辨识,问题定义如下。

(1) 读取手写阿拉伯数字的影像,将影像中的每个像素当作一个特征。数据源为 MNIST 机构收集的 60000 笔训练数据,另含 10000 笔测试数据,每笔数据是一个阿拉伯数字,宽高为 (28, 28) 的位图形。

(2) 建立神经网络模型,利用梯度下降法,求解模型的参数值,一般称为权重 (Weight)。

(3) 依照模型推算每一个影像是 0~9 的概率,再以概率最大的影像为预测结果。

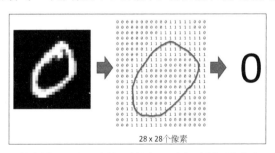

图 4.1 手写阿拉伯数字辨识,左图为输入的图形,中间为图形的像素,右图为预测结果

### 4-1-1 最简短的程序

TensorFlow v1.x 版使用会话 (Session) 及运算图 (Computational Graph) 的概念来编写,只是将两个张量相加就需写一大段程序,这被对手 Facebook PyTorch 嘲讽得体无完肤,于是 TensorFlow v2.x 为了回击对手,官网直接出大招,在文件首页展示一个超短程序,示范如何撰写"手写阿拉伯数字辨识"代码,这充分证明了改版后的

TensorFlow 确实超好用，现在我们就来看看这个程序。

**范例1. TensorFlow官网的手写阿拉伯数字辨识**[4]。

下列程序代码请参考【04_01_ 手写阿拉伯数字辨识.ipynb】。

```python
1  import tensorflow as tf
2  mnist = tf.keras.datasets.mnist
3
4  # 载入 MNIST 手写阿拉伯数字 训练数据
5  (x_train, y_train),(x_test, y_test) = mnist.load_data()
6
7  # 特征缩放至 (0, 1) 之间
8  x_train, x_test = x_train / 255.0, x_test / 255.0
9
10 # 建立模型
11 model = tf.keras.models.Sequential([
12   tf.keras.layers.Flatten(input_shape=(28, 28)),
13   tf.keras.layers.Dense(128, activation='relu'),
14   tf.keras.layers.Dropout(0.2),
15   tf.keras.layers.Dense(10, activation='softmax')
16 ])
17
18 # 设定优化器(optimizer)、损失函数(loss)、效能衡量指标(metrics)
19 model.compile(optimizer='adam',
20               loss='sparse_categorical_crossentropy',
21               metrics=['accuracy'])
22
23 # 模型训练，epochs: 执行周期，validation_split: 验证数据占 20%
24 model.fit(x_train, y_train, epochs=5, validation_split=0.2)
25
26 # 模型评估
27 model.evaluate(x_test, y_test)
```

执行结果如下。

```
Epoch 1/5
1500/1500 [==============================] - 5s 3ms/step - loss: 0.5336 - accuracy: 0.8432 - val_loss: 0.1558 - val_accuracy: 0.9555
Epoch 2/5
1500/1500 [==============================] - 5s 3ms/step - loss: 0.1676 - accuracy: 0.9505 - val_loss: 0.1134 - val_accuracy: 0.9657
Epoch 3/5
1500/1500 [==============================] - 5s 3ms/step - loss: 0.1203 - accuracy: 0.9646 - val_loss: 0.0975 - val_accuracy: 0.9704
Epoch 4/5
1500/1500 [==============================] - 5s 3ms/step - loss: 0.0981 - accuracy: 0.9703 - val_loss: 0.0968 - val_accuracy: 0.9717
Epoch 5/5
1500/1500 [==============================] - 5s 3ms/step - loss: 0.0786 - accuracy: 0.9758 - val_loss: 0.0958 - val_accuracy: 0.9713
313/313 [==============================] - 1s 3ms/step - loss: 0.0807 - accuracy: 0.9752
[0.08072374016046524, 0.9751999974250793]
```

上述程序扣除批注仅 10 多行，但辨识的准确率竟高达 97%~98%，TensorFlow 借此成功扳回一城。PyTorch 随后也开发出 Lightning 套件模仿 TensorFlow/Keras，训练也直接使用 model.fit，取代梯度重置、权重更新循环，双方攻防精彩万分。

**范例2. 以PyTorch Lightning撰写"手写阿拉伯数字辨识"的代码**，须先安装套件，指令如下：

pip install pytorch-lightning

官网首页的程序复制如【04_02_ 手写阿拉伯数字辨识_Lightning.ipynb】，程序不够简洁，笔者到官网教学范例中复制了一段更简洁的程序如下。

下列程序代码请参考【04_03_ 手写阿拉伯数字辨识_Lightning_short.ipynb】。

(1) 设定相关参数。

```python
1  # 设定参数
2  PATH_DATASETS = ""  # 预设路径
3  AVAIL_GPUS = min(1, torch.cuda.device_count())  # 使用GPU或CPU
4  BATCH_SIZE = 256 if AVAIL_GPUS else 64  # 批量
```

(2) 建立模型：指定另一种损失函数交叉熵(Cross Entropy)及 Adam 优化器，写法

与 TensorFlow/Keras 专家模式相似。

```python
1  # 建立模型
2  class MNISTModel(LightningModule):
3      def __init__(self):
4          super().__init__()
5          self.l1 = torch.nn.Linear(28 * 28, 10)  # 完全连接层
6
7      def forward(self, x):
8          # relu activation function + 完全连接层
9          return torch.relu(self.l1(x.view(x.size(0), -1)))
10
11     def training_step(self, batch, batch_nb):
12         x, y = batch
13         loss = F.cross_entropy(self(x), y)  # 交叉熵
14         return loss
15
16     def configure_optimizers(self):
17         return torch.optim.Adam(self.parameters(), lr=0.02)  # Adam 优化器
```

(3) 训练模型。

下载 MNIST 训练数据。

建立模型对象。

建立 DataLoader：DataLoader 是一种 Python Generator，一次只加载一批训练数据至内存，可节省内存的使用。

模型训练：与 TensorFlow/Keras 一样使用 fit。

```python
1  # 下载 MNIST 手写阿拉伯数字 训练数据
2  train_ds = MNIST(PATH_DATASETS, train=True, download=True,
3                   transform=transforms.ToTensor())
4
5  # 建立模型对象
6  mnist_model = MNISTModel()
7
8  # 建立 DataLoader
9  train_loader = DataLoader(train_ds, batch_size=BATCH_SIZE)
10
11 # 模型训练
12 trainer = Trainer(gpus=AVAIL_GPUS, max_epochs=3)
13 trainer.fit(mnist_model, train_loader)
```

**范例3.** "手写阿拉伯数字辨识" 完整范例请参阅【04_04_手写阿拉伯数字辨识_Lightning_accuracy.ipynb】，内含准确率的计算。由于我们的重点并不在这个范例上，所以这里就不再逐行说明了。

## 4-1-2　程序强化

上一节的范例 "手写阿拉伯数字辨识" 是官网为了炫技刻意缩短了程序，本节将会按照机器学习流程的 10 大步骤 ( 如图 4.2)，撰写此范例的完整程序，并对每个步骤仔细解析，大家务必理解每一行程序代表的意义。

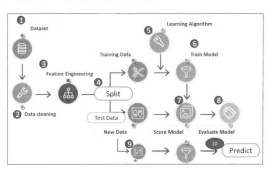

图 4.2　机器学习流程 10 大步骤

**范例4.** 依据图4.2所示的10大步骤撰写手写阿拉伯数字辨识程序。训练数据集采用 MNIST数据库,它的数据源是美国高中生及人口普查局员工手写的阿拉伯数字0~9,如图4.3所示。

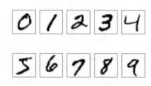

图 4.3　手写阿拉伯数字

下列程序代码请参考【**04_05_** 手写阿拉伯数字辨识 _ 完整版 **.ipynb**】。

(1) 载入框架。

```
1  import os
2  import torch
3  from torch import nn
4  from torch.nn import functional as F
5  from torch.utils.data import DataLoader, random_split
6  from torchmetrics import Accuracy
7  from torchvision import transforms
8  from torchvision.datasets import MNIST
```

(2) 设定参数,包括 GPU 的侦测,并决定是否使用 GPU。

PATH_DATASETS:数据集下载存放的路径,空字符串表示程序目前的文件夹。

```
1  PATH_DATASETS = ""  # 预设路径
2  BATCH_SIZE = 1024   # 批量
3  device = torch.device("cuda" if torch.cuda.is_available() else "cpu")
4  "cuda" if torch.cuda.is_available() else "cpu"
```

(3) 加载 MNIST 手写阿拉伯数字数据集。

```
1  # 下载 MNIST 手写阿拉伯数字 训练数据
2  train_ds = MNIST(PATH_DATASETS, train=True, download=True,
3                   transform=transforms.ToTensor())
4
5  # 下载测试数据
6  test_ds = MNIST(PATH_DATASETS, train=False, download=True,
7                  transform=transforms.ToTensor())
8
9  # 训练/测试数据的维度
10 print(train_ds.data.shape, test_ds.data.shape)
```

train=True 为训练数据;train=False 为测试数据。

执行结果:取得 60000 笔训练数据,10000 笔测试数据,每笔数据是一个阿拉伯数字,宽高各为 (28, 28) 的位图形,要注意数据的维度及大小,必须与模型的输入规格契合。

```
(60000, 28, 28) (60000,) (10000, 28, 28) (10000,)
```

(4) 对数据集进行探索与分析 (EDA),首先观察训练数据的目标值 (y),即影像的真实结果。

```
1  # 训练资料前10笔图片的数字
2  train_ds.targets[:10]
```

执行结果如下,每笔数据是一个阿拉伯数字。

```
array([5, 0, 4, 1, 9, 2, 1, 3, 1, 4], dtype=uint8)
```

(5) 显示第一笔训练数据的像素。

```
1  # 显示第1张图片内含值
2  train_ds.data[0]
```

执行结果如下，像素的值介于 (0, 255) 为灰度图，0 为白色，255 为最深的黑色，**注意，这与 RGB 色码刚好相反 (RGB 黑色为 0，白色为 255)**。

```
[[  0,   0,   0,   0,   0,   0,   0,   0,   0,   0,   0,   0,   0,
    0,   0,   0,   0,   0,   0,   0,   0,   0,   0,   0,   0,   0,
    0,   0],
 [  0,   0,   0,   0,   0,   0,   0,   0,   0,   0,   0,   0,   3,
   18,  18,  18, 126, 136, 175,  26, 166, 255, 247, 127,   0,   0,
    0,   0],
 [  0,   0,   0,   0,   0,   0,   0,   0,  30,  36,  94, 154, 170,
  253, 253, 253, 253, 253, 225, 172, 253, 242, 195,  64,   0,   0,
    0,   0],
 [  0,   0,   0,   0,   0,   0,   0,  49, 238, 253, 253, 253, 253,
  253, 253, 253, 253, 251,  93,  82,  82,  56,  39,   0,   0,   0,
    0,   0],
 [  0,   0,   0,   0,   0,   0,   0,  18, 219, 253, 253, 253, 253,
  253, 198, 182, 247, 241,   0,   0,   0,   0,   0,   0,   0,   0,
    0,   0],
 [  0,   0,   0,   0,   0,   0,   0,   0,  80, 156, 107, 253, 253,
  205,  11,   0,  43, 154,   0,   0,   0,   0,   0,   0,   0,   0,
    0,   0],
```

(6) 为了看清楚图片上手写的数字，将非 0 的数值转为 1，变为黑白两色的图片。

```
1  # 将非0的数字转为1，显示第1张图片
2  data = train_ds.data[0].clone()
3  data[data>0]=1
4  data = data.numpy()
5
6  # 将转换后二维内容显示出来，隐约可以看出数字为 5
7  text_image=[]
8  for i in range(data.shape[0]):
9      text_image.append(''.join(data[i].astype(str)))
10 text_image
```

执行结果如下，笔者以笔描绘 1 的范围，隐约可以看出是 5。

```
['000000000000000000000000000',
 '000000000000000000000000000',
 '000000000000000000000000000',
 '000000000000000000000000000',
 '000000000000000000000000000',
 '000000000000111111111110000',
 '000000000111111111111110000',
 '000000011111111111111100000',
 '000000011111111110000000000',
 '000000001111111010000000000',
 '000000000011110000000000000',
 '000000000001110000000000000',
 '000000000001110000000000000',
 '000000000001111000000000000',
 '000000000000111110000000000',
 '000000000000011111100000000',
 '000000000000001111100000000',
 '000000000000000111100000000',
 '000000000000001111100000000',
 '000000000001111111100000000',
 '000000001111111110000000000',
 '000000011111111110000000000',
 '000000111111111100000000000',
 '000011111111110000000000000',
 '000011111110000000000000000',
 '000000000000000000000000000',
 '000000000000000000000000000',
 '000000000000000000000000000']
```

(7) 显示第一笔训练数据图像，确认是 5。

```
1   # 显示第1张图片图像
2   import matplotlib.pyplot as plt
3
4   # 第一笔数据
5   X = train_ds.data[0]
6
7   # 绘制点阵图，cmap='gray'：灰阶
8   plt.imshow(X.reshape(28,28), cmap='gray')
9
10  # 隐藏刻度
11  plt.axis('off')
12
13  # 显示图形
14  plt.show()
```

执行结果如下。

(8) 使用 TensorFlow 进行特征工程，将特征缩放成 (0, 1) 之间，可提高模型准确度，并且可以加快收敛速度。但是，PyTorch 却会造成优化求解无法收敛，要特别注意，PyTorch 特征缩放是在训练时 DataLoader 加载数据才会进行，与 TensorFlow 不同，要设定特征缩放可在下载 MNIST 指令内设定，如下，请参阅【**04_07_ 手写阿拉伯数字辨识 _Normalize.ipynb**】。

```
1   transform=transforms.Compose([
2       transforms.ToTensor(),
3       transforms.Normalize((0.1307,), (0.3081,))
4       ])
5
6
7   # 下载 MNIST 手写阿拉伯数字 训练数据
8   train_ds = MNIST(PATH_DATASETS, train=True, download=True,
9                    transform=transform)
```

第 3 行指定标准化，平均数为 0.1307，标准偏差为 0.3081，这两个数字是官网范例建议的值，函数用法请参阅官网介绍[2]。

第 8 行套用特征缩放的转换 (transform=transform)。

(9) 数据分割为训练及测试数据，此步骤无须进行，因为加载 MNIST 数据时，已经切割好了。

(10) 建立模型结构，如图 4.4 所示。

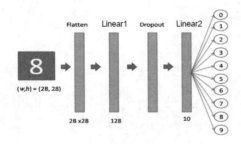

图 4.4　手写阿拉伯数字辨识的模型结构

PyTorch 与 TensorFlow/Keras 一样提供两类模型，即顺序型模型 (Sequential Model) 和 Functional API 模型，顺序型模型使用 torch.nn.Sequential 函数包覆各项神经层，适用于简单的结构，执行时神经层一层接一层地执行，另一种类型为 Functional API，使用 torch.nn.functional 函数，可以设计成较复杂的模型结构，包括多个输入层或多个输出层，也允许分叉，后续使用到时再详细说明。这里使用简单的顺序型模型，内含各种神经层。

```
1  # 建立模型
2  model = torch.nn.Sequential(
3      torch.nn.Flatten(),
4      torch.nn.Linear(28 * 28, 256),
5      torch.nn.Dropout(0.2),
6      torch.nn.Linear(256, 10),
7      # 使用nn.CrossEntropyLoss()时，不需要将输出经过softmax层，否则计算的损失会有误
8      # torch.nn.Softmax(dim=1)
9  ).to(device)
```

- 扁平层 (Flatten Layer)：将宽高各 28 个像素的图压扁成一维数组 (28 × 28 = 784 个特征 )。
- 完全连接层 (Linear Layer)：第一个参数为输入的神经元个数，通常是上一层的输出，TensorFlow/Keras 不用设定，但 PyTorch 需要设定，比较麻烦；第二个参数为输出的神经元个数，设定为 256 个神经元，即 256 条回归线，每一条回归线有 784 个特征。输出通常设定为 4 的倍数，并无建议值，可经由试验调校取得较佳的参数值。
- Dropout Layer：类似正则化 (Regularization)，希望避免过度拟合，在每一个训练周期随机丢弃一定比例 (0.2) 的神经元，一方面可以估计较少的参数，另一方面能够取得多个模型的均值，避免受极端值影响，借以校正过度拟合的现象。通常会在每一层完全连接层 (Linear) 后面加一个 Dropout，比例也无建议值，如图 4.5 所示。

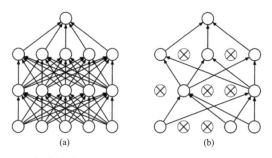

图 4.5　(a) 标准的神经网络，(b) Dropout Layer 形成的神经网络

- 第二个完全连接层 (Linear)：输出层，因为要辨识 0~9 十个数字，故输出要设成 10，一般通过 Softmax Activation Function，可以将输出转为概率形式，即预测 0~9 的个别概率，再从中选择最大概率者为预测值，不过，若使用交叉熵 (nn.CrossEntropyLoss)，因 PyTorch 已内含 Softmax 处理，不需额外再加 Softmax 层，否则计算的损失会有误，这与 TensorFlow/Keras 不同，要特别注意，详情请参阅 PyTorch 官网 CrossEntropyLoss 的说明 [3] 或《交叉熵损失，softmax 函数和 torch.nn.CrossEntropyLoss() 中文》[4]。
- 最后一行 to(device)：**要记得输入数据与模型必须一致，一律使用 GPU 或**

CPU，不可混用。

(11) 结合训练数据及模型，进行模型训练。

```python
epochs = 5
lr=0.1

# 建立 DataLoader
train_loader = DataLoader(train_ds, batch_size=600)

# 设定优化器(optimizer)
# optimizer = torch.optim.Adam(model.parameters(), lr=lr)
optimizer = torch.optim.Adadelta(model.parameters(), lr=lr)

criterion = nn.CrossEntropyLoss()

model.train()
loss_list = []
for epoch in range(1, epochs + 1):
    for batch_idx, (data, target) in enumerate(train_loader):
        data, target = data.to(device), target.to(device)
#         if batch_idx == 0 and epoch == 1: print(data[0])

        optimizer.zero_grad()
        output = model(data)
        loss = criterion(output, target)
        loss.backward()
        optimizer.step()

        if batch_idx % 10 == 0:
            loss_list.append(loss.item())
            batch = batch_idx * len(data)
            data_count = len(train_loader.dataset)
            percentage = (100. * batch_idx / len(train_loader))
            print(f'Epoch {epoch}: [{batch:5d} / {data_count}] ({percentage:.0f} %)' +
                  f'  Loss: {loss.item():.6f}')
```

第 1 行设定执行周期。

第 2 行设定学习率。

第 5 行设定 DataLoader，可一次取一批数据训练，节省内存。

第 8~9 行设定优化器 (Optimizer)：PyTorch 提供多种优化器，各有优点，后续会介绍，本例使用哪一种优化器均可。

第 11 行设定损失函数为交叉熵 (CrossEntropyLoss)，PyTorch 提供多种损失函数，后续会详细介绍。

第 13 行设定模型进入训练阶段，各神经层均会被执行，有别于评估阶段。

第 14~32 行进行模型训练，并显示训练过程与损失值，损失值多少不重要，需观察损失值是否随着训练逐渐收敛 ( 显著减少 )。

(12) 对训练过程的损失绘图。

```python
import matplotlib.pyplot as plt

plt.plot(loss_list, 'r')
```

执行结果：随着执行周期 (epoch) 次数的增加，损失越来越低。

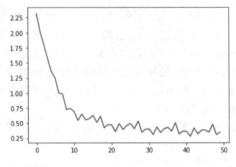

(13) 评分 (Score Model)：输入测试数据，计算出损失及准确率。

```
1   # 建立 Dataloader
2   test_loader = DataLoader(test_ds, shuffle=False, batch_size=BATCH_SIZE)
3
4   model.eval()
5   test_loss = 0
6   correct = 0
7   with torch.no_grad():
8       for data, target in test_loader:
9           data, target = data.to(device), target.to(device)
10          output = model(data)
11
12          # sum up batch loss
13          test_loss += criterion(output, target).item()
14
15          # 预测
16          pred = output.argmax(dim=1, keepdim=True)
17
18          # 正确笔数
19          correct += pred.eq(target.view_as(pred)).sum().item()
20
21  # 平均损失
22  test_loss /= len(test_loader.dataset)
23  # 显示测试结果
24  batch = batch_idx * len(data)
25  data_count = len(test_loader.dataset)
26  percentage = 100. * correct / data_count
27  print(f'平均损失: {test_loss:.4f}，准确率: {correct}/{data_count}' +
28        f' ({percentage:.0f}%)\n')
```

执行结果：平均损失为 0.3345； 准确率为 9057/10000 (91%)。

第 4 行设定模型进入评估阶段，Dropout 神经层不会被执行，因它抑制过度拟合，只用于训练阶段，这一行非常重要，否则预测会失准。

第 7~19 行预测所有测试数据，必须要包在 with torch.no_grad() 内，声明内嵌的程序代码不作梯度下降，否则程序会出现错误。

第 13 行是计算损失并累加。

第 16 行是预测并找最高概率的类别索引值，参数 keepdim=True 表示不修改 output 变量的维度。

第 19 行是计算准确笔数并累加。

(14) 实际比对测试数据的前 20 笔。

```
1   # 实际预测 20 笔数据
2   predictions = []
3   with torch.no_grad():
4       for i in range(20):
5           data, target = test_ds[i][0], test_ds[i][1]
6           data = data.reshape(1, *data.shape).to(device)
7           output = torch.argmax(model(data), axis=-1)
8           predictions.append(str(output.item()))
9
10  # 比对
11  print('actual    :', test_ds.targets[0:20].numpy())
12  print('prediction: ', ' '.join(predictions[0:20]))
```

执行结果如下，第 9 笔数据错误，其余全部正确。

```
actual     : [7 2 1 0 4 1 4 9 5 9 0 6 9 0 1 5 9 7 3 4]
prediction:  7 2 1 0 4 1 4 9 6 9 0 6 9 0 1 5 9 7 3 4
```

(15) 显示第 9 笔数据的概率。

```
1   # 显示第 9 笔的概率
2   import numpy as np
3
4   i=8
5   data = test_ds[i][0]
6   data = data.reshape(1, *data.shape).to(device)
7   #print(data.shape)
8   predictions = torch.softmax(model(data), dim=1)
9   print(f'0~9预测概率: {np.around(predictions.cpu().detach().numpy(), 2)}')
10  print(f'0~9预测概率: {np.argmax(predictions.cpu().detach().numpy(), axis=-1)}')
```

执行结果：发现 6 的概率高很多。

```
0~9预测概率: [[0.01 0.    0.06 0.    0.04 0.01 0.86 0.    0.01 0.  ]]
0~9预测概率: [6]
```

(16) 显示第 9 张图像。

```
1  # 显示第 9 张图像
2  X2 = test_ds[i][0]
3  plt.imshow(X2.reshape(28,28), cmap='gray')
4  plt.axis('off')
5  plt.show()
```

第 9 张图像如下，像 5 又像 6。

(17) 暂不进行效能评估，之后可调校相关超参数 (Hyperparameter) 及模型结构，寻找最佳模型和参数。超参数是指在模型训练前可以调整的参数，例如学习率、执行周期、权重初始值、训练批量等，但不含模型求算的参数如权重 (Weight) 或偏差项 (Bias)。

(18) 模型部署，将最佳模型存盘，再开发用户接口或提供 API，连同模型文件一并部署到上线环境 (Production Environment)。

```
1  # 模型存档
2  torch.save(model, 'model.pt')
3
4  # 模型载入
5  model = torch.load('model.pt')
```

扩展名通常以 .pt 或 .pth 存储，建议使用 .pt。

上述指令会将模型结构与权重一并存盘，如果只要存储权重，可执行以下指令，这部分概念与 TensorFlow 相同，但 TensorFlow 是存储至目录，而 PyTorch 是存储至文件，扩展名建议使用 .pth，与上述模型结构与权重一并存盘有所区别。

```
1  # 权重存盘
2  torch.save(model.state_dict(), 'model.pth')
3
4  # 权重载入
5  model.load_state_dict(torch.load('model.pth'))
```

model.parameters() 只含学习到的参数 (learnable parameters)，例如权重与偏差，model.state_dict() 会使用字典数据结构对照每一层神经层及其参数。

可显示每一层的 state_dict 维度。

```
1  # 显示每一层的 state_dict 维度
2  print("每一层的 state_dict:")
3  for param_tensor in model.state_dict():
4      print(param_tensor, "\t", model.state_dict()[param_tensor].size())
```

更详细的用法可参照 PyTorch 官网 Saving and Loading Models [8]。

(19) 系统上线，提供新数据预测，之前都是使用 MNIST 内建数据测试，但严格来说并不可靠，因为这些都是出自同一机构所收集的数据，所以建议读者利用绘图软件亲自撰写测试。我们准备了一些图片文件，放在 myDigits 目录内，读者可自行修改，再利用下列程序代码测试。注意，**从图片文件读入后要反转颜色**，颜色 0 为白色，与

RGB 色码不同，它的 0 为黑色。另外，使用 skimage 读取图片文件，像素会自动缩放至 [0,1] 区间，不需再进行转换。

```
1   # 使用小画家，绘制 0~9，实际测试看看
2   from skimage import io
3   from skimage.transform import resize
4   import numpy as np
5
6   # 读取影像并转为单色
7   for i in range(10):
8       uploaded_file = f'./myDigits/{i}.png'
9       image1 = io.imread(uploaded_file, as_gray=True)
10
11      # 缩为 (28, 28) 大小的影像
12      image_resized = resize(image1, (28, 28), anti_aliasing=True)
13      X1 = image_resized.reshape(1,28, 28)
14
15      # 反转颜色，颜色0为白色，与 RGB 色码不同，它的 0 为黑色
16      X1 = torch.FloatTensor(1.0-X1).to(device)
17
18      # 预测
19      predictions = torch.softmax(model(X1), dim=1)
20      # print(np.around(predictions.cpu().detach().numpy(), 2))
21      print(f'actual/prediction: {i} {np.argmax(predictions.detach().cpu().numpy())}')
```

(20) 显示模型的汇总信息。

```
1   print(model)
```

执行结果：包括每一神经层的名称及输出入参数的个数。

```
Sequential(
  (0): Flatten(start_dim=1, end_dim=-1)
  (1): Linear(in_features=784, out_features=256, bias=True)
  (2): Dropout(p=0.2, inplace=False)
  (3): Linear(in_features=256, out_features=10, bias=True)
)
```

也可以使用下列指令：

```
for name, module in model.named_children():
    print(f'{name}: {module}')
```

(21) 也可以安装 torchinfo 或 torch-summary 显示较美观的汇总信息。
以下列指令安装：

```
pip install torchinfo
```

以下列指令显示汇总信息，summary 第 2 个参数为输入数据的维度，含笔数。

```
1   from torchinfo import summary
2   summary(model, (60000, 28, 28))  # input dimension size
```

执行结果。

```
==========================================================================
Layer (type:depth-idx)              Output Shape          Param #
==========================================================================
Sequential                          --                    --
├─Flatten: 1-1                      [60000, 784]          --
├─Linear: 1-2                       [60000, 256]          200,960
├─Dropout: 1-3                      [60000, 256]          --
├─Linear: 1-4                       [60000, 10]           2,570
==========================================================================
Total params: 203,530
Trainable params: 203,530
Non-trainable params: 0
Total mult-adds (G): 12.21
==========================================================================
Input size (MB): 188.16
Forward/backward pass size (MB): 127.68
Params size (MB): 0.81
Estimated Total Size (MB): 316.65
==========================================================================
```

(22) PyTorch 无法像 TensorFlow 一样绘制如图 4.6 所示的模型图。

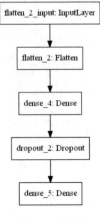

图 4.6　模型图

以上按机器学习流程的 10 大步骤撰写了一个完整的程序，之后的任何模型或应用都可遵循相同的流程完成。因此，熟悉流程的每个步骤非常重要。下一节我们会做些试验来说明建构模型的考虑。

## 4-1-3　试验

前一节我们完成了第一个神经网络程序，也见识到它的威力，扣除变量的检查，短短几行程序就能够辨识手写阿拉伯数字，且准确率达到 90% 左右，或许读者会心生一些疑问。

(1) 模型结构为什么要设计成两层完全连接层 (Linear)？更多层的准确率会提高吗？

(2) 第一层完全连接层 (Linear) 输出为什么要设为 256？设为其他值会有何影响？

(3) Activation Function 也可以抑制过度拟合，将 Dropout 改为 Activation Function ReLu 准确率有何不同？

(4) 优化器 (optimizer)、损失函数 (loss)、效能衡量指标 (metrics) 有哪些选择？设为其他值会有何影响？

(5) Dropout 比例为 0.2，设为其他值会更好吗？

(6) 影像为单色灰度，若是彩色可以辨识吗？如何修改？

(7) 执行周期 (epoch) 设为 5，设为其他值会更好吗？

(8) 准确率可以达到 100% 吗？

(9) 如果要辨识其他对象，程序要修改哪些地方？

(10) 如果要辨识多个数字，例如输入 4 位数，要如何辨识？

以上问题是这几年授课时学员常提出的疑惑，我们就来逐一试验，试着寻找答案。

问题 1. 模型结构为什么要设计成两层完全连接层 (Linear)？更多层的准确率会提高吗？
解答：
前面曾经说过，神经网络是多条回归线的组合，而且每一条回归线可能还会包在 Activation Function 内，变成非线性的函数，因此，要单纯以数学求解几乎不可能，只能以优化方法求得近似解,但是,只有凸集合的数据集,才能保证有全局最佳解(Global Minimization),以 MNIST 为例,总共有 28×28=784 个像素,每个像素视为一个特征,即 784 度空间,它是否为凸集合,是否存在最佳解?我们无从得知,因此严格地讲,到目前为止,神经网络依然是一个黑箱(Black Box)科学,我们只知道它威力强大,但如何达到较佳的准确率,仍旧需要经验与反复的试验。因此,模型结构并没有明确规定要设计成几层,要随着不同的问题及数据集进行试验,一次次地进行效能调校,找寻较佳的参数值。

理论上,越多层架构,回归线就越多条,预测越准确,像是后面介绍的 ResNet 模型就高达 150 层,然而,经过试验证实,一旦超过某一界限后,准确率就会下降,这跟训练数据量有关,如果只有少量的数据,要估算过多的参数(w、b),准确率自然不高。

我们就来小小试验一下,多一层完全连接层(Linear),准确率是否会提高?请参阅程序【04_05_ 手写阿拉伯数字辨识 _ 试验 2.ipynb】。

修改模型结构如下,加一对完全连接层(Linear)/Dropout,其余代码不变。

```
1  # 建立模型
2  model = torch.nn.Sequential(
3      torch.nn.Flatten(),
4      torch.nn.Linear(28 * 28, 256),
5      nn.Dropout(0.2),
6      torch.nn.Linear(256, 64),
7      nn.Dropout(0.2),
8      torch.nn.Linear(64, 10),
9  ).to(device)
```

执行结果:平均损失为 0.0003,准确率为 9069/10000 (91%),准确率稍微提升,不显著。

问题 2. 第一层完全连接层(Linear)输出为什么要设为 256?设为其他值会有何影响?
解答:
输出的神经元个数可以任意设定,一般来讲,会使用 4 的倍数,以下我们修改为 512,请参阅程序【04_05_ 手写阿拉伯数字辨识 _ 试验 3.ipynb】。

执行结果:平均损失为 0.0003,准确率为 9076/10000 (91%),准确率稍微提升,不显著。

同问题 1,照理来说,神经元个数越多,回归线就越多,特征也就越多,预测就会越准确,但经过验证,准确率并未显著提高。依据 *Deep Learning with TensorFlow 2 and Keras*[6] 一书测试,结果如图 4.7,也是有一个极限,超过就会不升反降。

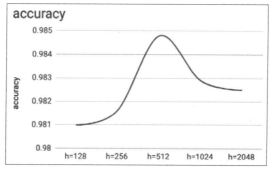

图 4.7　测试结果

神经元个数越多，训练时间就越长，如图 4.8 所示。

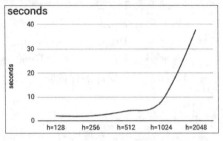

图 4.8　执行结果

问题 3. Activation Function 也可以抑制过度拟合，将 Dropout 改为 Activation Function ReLu，准确率有何不同？请参阅程序【**04_05_ 手写阿拉伯数字辨识 _ 试验 4.ipynb**】。

解答：

Activation Function 有很多种，后面会有详尽介绍，可先参阅维基百科[7]，部分表格截取如下，包括函数的名称、概率分配图形、公式及一阶导数，如图 4.9 所示。

| Name | Plot | Function, $f(x)$ | Derivative of $f$, $f'(x)$ |
|---|---|---|---|
| Identity | | $x$ | $1$ |
| Binary step | | $\begin{cases} 0 & \text{if } x < 0 \\ 1 & \text{if } x \geq 0 \end{cases}$ | $\begin{cases} 0 & \text{if } x \neq 0 \\ \text{undefined} & \text{if } x = 0 \end{cases}$ |
| Logistic, sigmoid, or soft step | | $\sigma(x) = \dfrac{1}{1+e^{-x}}$ [1] | $f(x)(1-f(x))$ |
| tanh | | $\tanh(x) = \dfrac{e^x - e^{-x}}{e^x + e^{-x}}$ | $1 - f(x)^2$ |
| Rectified linear unit (ReLU)[11] | | $\begin{cases} 0 & \text{if } x \leq 0 \\ x & \text{if } x > 0 \end{cases}$ $= \max\{0, x\} = x\mathbf{1}_{x>0}$ | $\begin{cases} 0 & \text{if } x < 0 \\ 1 & \text{if } x > 0 \\ \text{undefined} & \text{if } x = 0 \end{cases}$ |
| Gaussian error linear unit (GELU)[6] | | $\dfrac{1}{2}x\left(1 + \operatorname{erf}\left(\dfrac{x}{\sqrt{2}}\right)\right)$ $= x\Phi(x)$ | $\Phi(x) + x\phi(x)$ |
| Softplus[12] | | $\ln(1 + e^x)$ | $\dfrac{1}{1+e^{-x}}$ |

图 4.9　Activation Function 的种类

早期隐藏层大都使用 Sigmoid 函数，近几年发现 ReLU 准确率更高。

将 Dropout 改为 ReLU，如下所示。

```
1  # 建立模型
2  model = torch.nn.Sequential(
3      torch.nn.Flatten(),
4      torch.nn.Linear(28 * 28, 512),
5      torch.nn.ReLU(),
6      torch.nn.Linear(512, 10),
7  ).to(device)
```

执行结果：平均损失为 0.0003，准确率为 9109/10000 (91%)，准确率稍微提升，不显著。

新数据预测是否比较精准？答案是并不理想。

一般而言，神经网络使用 Dropout 比正则化 (Regularizer) 更理想。

问题 4. 优化器 (Optimizer)、损失函数 (Loss)、效能衡量指标 (Metrics) 有哪些选择？设为其他值会有何影响？

解答：

优化器有很多种，如最简单的固定值的学习率 (SGD)、复杂的动态改变的学习率以及能够自定义优化器，详情请参考 PyTorch 官网 [8]。优化器的选择，主要会影响收敛的速度，大多数状况下，Adam 优化器都有不错的表现，不过，在下一节介绍的 CNN 模型搭配 Adam 优化器却发生无法收敛的状况，而 TensorFlow 则不会，可见一个同名的优化器在不同套件上，开发的细节仍有所差异。

损失函数种类繁多，包括常见的 MSE、Cross Entropy，其他更多的信息请参考 PyTorch 官网 [9]。损失函数的选择影响着收敛的速度，另外，某些自定义损失函数有特殊功能，例如风格转换 (Style Transfer)，它能够制作影像合成的效果，生成对抗网络 (GAN) 更是发扬光大，后面章节会有详细的介绍。

效能衡量指标 (metrics)：除了准确率 (Accuracy)，还可以计算精确率 (Precision)、召回率 (Recall)、F1 等，尤其是面对不平衡的样本时。

问题 5. Dropout 比例为 0.2，设为其他值会更好吗？

解答：

将 Dropout 比例改为 0.5，测试看看，请参阅程序【04_05_ 手写阿拉伯数字辨识 _ 试验 5.ipynb】。

执行结果：平均损失为 0.0003，准确率为 9034/10000 (90%)，准确率略为降低。

抛弃比例过高时，准确率会陡降。

新数据预测是否比较准？结果并不理想，如图 4.10 所示。

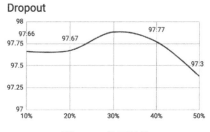

图 4.10　执行结果

问题 6. 目前 MNIST 影像为单色灰度，若是彩色可以辨识吗？如何修改？

解答：可以，若颜色有助于辨识，可以将 RGB 三通道分别输入辨识，后面我们讲到卷积神经网络 (Convolutional Neural Networks，CNN) 时会有范例说明。

问题 7. 执行周期 (epoch) 设为 5，设为其他值会更好吗？

解答：将执行周期 (epoch) 改为 10，请参阅程序【04_05_ 手写阿拉伯数字辨识 _ 试验 6.ipynb】。

```
1  epochs = 10
2  lr=0.1
```

执行结果：平均损失为 0.0003，准确率为 9165/10000 (92%)，准确率略为提高。

但损失率到后来已降不下去了，如图 4.11 所示。

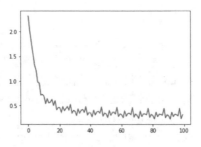

图 4.11　损失率结果

理论上，训练周期越多，准确率越高，然而，过多的训练周期会造成过度拟合 (Overfitting)，反而会使准确率降低，如图 4.12 所示。

图 4.12　准确率降低

问题 8. 准确率可以达到 100% 吗？

解答：很少的模型准确率能够达到 100%，除非是用数学推导出来的模型，优化只是求近似解而已，另外，由于神经网络是从训练数据中学习知识，而测试或预测数据并不参与训练，若与训练数据分布有所差异，甚至来自不同的概率分配，准确率很难能达到 100%。

问题 9. 如果要辨识其他对象，程序要修改哪些地方？

解答：只需修改很少的程序代码，就可以辨识其他对象，例如，Zalando 公司提供另一个类似的数据集 FashionMNIST，请参阅【04_06_FashionMNIST 辨识 _ 完整版 .ipynb】，除了加载数据的指令不同外，其他的程序代码几乎不变。这也说明了一点，神经网络并不是真的认识 0~9，它只是从像素数据中推估出来的模型，即所谓的从数据中学习到知识 (Knowledge Discovery from Data, KDD)，以 MNIST 而言，模型只是统计 0~9 这 10 个数字像素大部分分布在哪些位置而已。FashionMNIST 数据集下载指令如下。

```
2  train_ds = FashionMNIST(PATH_DATASETS, train=True, download=True,
3                          transform=transforms.ToTensor())
```

问题 10. 如果要辨识多个数字，例如输入 4 位数，要如何辨识？

解答：可以使用图像处理分割数字，再分别依次输入模型预测即可。或者更简单的方法，直接将视觉接口 (UI) 设计成 4 格，规定使用者只能在每个格子内各输入一个数字。

以上的试验大多只针对单一参数做比较，假如要同时比较多个变量，就必须跑遍所有参数组合，这样程序就会很复杂，别担心，有一些工具可以帮忙，包括 Keras Tuner、hyperopt、Ray Tune、Ax 等，在后续"超参数调校"一节有较详细的介绍。

由于 MNIST 的模型辨识率很高，要观察超参数调整对模型的影响，并不容易，建议找一些辨识率较低的模型进行相关试验，例如 FashionMNIST、CiFar 数据集，才能有比较显著的效果，FashionMNIST 的准确率只有 81% 左右，试验比较能观察出差异。

## 4-2 模型种类

PyTorch 与 TensorFlow/Keras 一样，提供两种模型结构：Sequential model 和 Functional API。

Sequential model：顺序型模型使用 torch.nn.Sequential 函数包覆各项神经层，适用于简单的结构，执行时神经层一层接一层地执行。

Functional API：使用 torch.nn.functional 函数，可以设计成较复杂的模型结构，包括多个输入层或多个输出层，也允许分叉及合并。PyTorch 并未正式命名为 Functional API，笔者为方便学习，自行采用 TensorFlow/Keras 的专门术语。

### 4-2-1 Sequential model

**下列程序代码请参考【04_09_Sequential_vs_Functional.ipynb】。**

(1) torch.nn.Sequential 包含各种的神经层，简洁的写法如下。

```
1  model = nn.Sequential(
2          nn.Linear(256,20),
3          nn.ReLU(),
4          nn.Linear(20,64),
5          nn.ReLU(),
6          nn.Softmax(dim=1),
7          )
```

(2) 可以为每一神经层命名，使用字典 (OrderedDict) 数据结构，设定名称及神经层种类，逐一配对。

```
1  # 使用 OrderedDict 可指定名称
2  from collections import OrderedDict
3  model = nn.Sequential(OrderedDict([
4          ('linear1', nn.Linear(256,20)),
5          ('relu1', nn.ReLU()),
6          ('linear2', nn.Linear(20,64)),
7          ('relu2', nn.ReLU()),
8          ('softmax', nn.Softmax(dim=1))
9          ]))
```

(3) 注意，上一层的输出神经元个数要等于下一层的输入神经元个数，PyTorch 必须同时设定输入与输出个数，而 TensorFlow/Keras 在第一层设定输入与输出个数，在第二层只要设定输出个数即可，因为输入可从上一层的输出得知，这点 TensorFlow/Keras 就略胜一筹，不要小看这一点，进阶神经层如 CNN、RNN 接到 Linear，要算 Linear 的输入个数，就要伤脑筋了。

(4) 可显示模型结构如下。

```
1  from torchinfo import summary
2  summary(model, (1, 256))
```

执行结果如下。

```
==========================================================
Layer (type:depth-idx)          Output Shape       Param #
==========================================================
Sequential                      --                 --
├─Linear: 1-1                   [1, 20]            5,140
├─ReLU: 1-2                     [1, 20]            --
├─Linear: 1-3                   [1, 64]            1,344
├─ReLU: 1-4                     [1, 64]            --
├─Softmax: 1-5                  [1, 64]            --
==========================================================
Total params: 6,484
Trainable params: 6,484
Non-trainable params: 0
Total mult-adds (M): 0.01
==========================================================
Input size (MB): 0.00
Forward/backward pass size (MB): 0.00
Params size (MB): 0.03
Estimated Total Size (MB): 0.03
==========================================================
```

## 4-2-2 Functional API

使用 torch.nn.functional 函数，可以设计成较复杂的模型结构，如包括多个输入层或输出层，也允许分叉及合并。相关函数与 torch.nn 几乎可以一一对照，只是 torch.nn.functional 函数名称均使用小写，可参阅 PyTorch 官网 torch.nn.functional 的说明 [10]。

下列程序代码请参考【04_09_Sequential_vs_Functional.ipynb】后半部。

**范例1. 先看一个简单的范例，看看torch.nn.functional.linear的用法。**

第 2 行：torch.nn.functional 通常会被取别名为 F。

第 6 行：torch.nn.functional 下的函数必须输入张量的值，而非张量的维度。

F.linear 的参数有两个：输入张量和权重，其中权重是初始值，训练过程会不断更新。

```
1  # linear 用法
2  from torch.nn import functional as F
3
4  inputs = torch.randn(100, 256)
5  weight = torch.randn(20, 256)
6  x = F.linear(inputs, weight)
```

**范例2. 将上述顺序型模型改写为Functional API。**

在第二层之后的完全连接层指定权重并不合理，因此，通常还是以 nn.Linear 取代 F.linear。

F.relu(x)：表示 relu 接在 x 的下一层。

```
1  inputs = torch.randn(100, 256)
2  x = nn.Linear(256,20)(inputs)
3  x = F.relu(x)
4  x = nn.Linear(20, 10)(x)
5  x = F.relu(x)
6  x = F.softmax(x, dim=1)
```

**范例3. 使用类别定义模型，这是一般Functional API定义模型的方式。**

类别至少要包含两个方法 (Method)：init 和 forward。

init 函数内声明要使用的神经层对象。

forward 函数内定义神经层的串连。

扁平层函数为 torch.flatten(x, 1)，命名空间不是 torch.nn.functional，第 2、第 3 个参数是打扁的起讫的维度，第 3 个参数预设为 -1，表示最后一个维度。

```python
class Net(nn.Module):
    def __init__(self):
        super(Net, self).__init__()
        self.fc1 = nn.Linear(784,256)
        self.fc2 = nn.Linear(256, 10)
        self.dropout1 = nn.Dropout(0.2)
        self.dropout2 = nn.Dropout(0.2)
        
    def forward(self, x):
        x = torch.flatten(x, 1)
        x = self.fc1(x)
        x = self.dropout1(x)
        x = self.fc2(x)
        x = self.dropout2(x)
        output = F.softmax(x, dim=1)
        return output
```

使用的指令如下。

model = Net()

可显示模型结构如下。

```python
from torchinfo import summary

model = Net()
summary(model, (1, 28, 28))
```

执行结果如下。

```
=================================================================
Layer (type:depth-idx)                   Output Shape         Param #
=================================================================
Net                                      --                   --
├─Linear: 1-1                            [1, 256]             200,960
├─Dropout: 1-2                           [1, 256]             --
├─Linear: 1-3                            [1, 10]              2,570
├─Dropout: 1-4                           [1, 10]              --
=================================================================
Total params: 203,530
Trainable params: 203,530
Non-trainable params: 0
Total mult-adds (M): 0.20
=================================================================
Input size (MB): 0.00
Forward/backward pass size (MB): 0.00
Params size (MB): 0.81
Estimated Total Size (MB): 0.82
=================================================================
```

**范例4. 使用类别定义模型，进行手写阿拉伯数字辨识(MNIST)。**

下列程序代码请参考【04_10_ 手写阿拉伯数字辨识 _ 专家模式 .ipynb 】。

以下只说明差异的程序代码。

定义模型。

```python
# 建立模型
class Net(nn.Module):
    def __init__(self):
        super().__init__()
        self.fc1 = torch.nn.Linear(28 * 28, 256) # 完全连接层
        self.dropout1 = nn.Dropout(0.2)
        self.fc2 = torch.nn.Linear(256, 10) # 完全连接层
        
    def forward(self, x):
        # 完全连接层 + dropout + 完全连接层 + dropout + log_softmax
        x = torch.flatten(x, 1)
        x = self.fc1(x)
        x = self.dropout1(x)
        x = self.fc2(x)
        return x

# 建立模型对象
model = Net().to(device)
```

训练模型：使用 CrossEntropyLoss()，模型不可加 SoftMax 层。

```
1   epochs = 5
2   lr=0.1
3
4   # 建立 DataLoader
5   train_loader = DataLoader(train_ds, batch_size=600)
6
7   # 设定优化器(optimizer)
8   optimizer = torch.optim.Adam(model.parameters(), lr=lr)
9
10  # 设定损失函数(Loss)
11  criterion = nn.CrossEntropyLoss()
12
13  model.train()
14  loss_list = []
15  for epoch in range(1, epochs + 1):
16      for batch_idx, (data, target) in enumerate(train_loader):
17          data, target = data.to(device), target.to(device)
18
19          optimizer.zero_grad()
20          output = model(data)
21          # 计算损失(Loss)
22          loss = criterion(output, target)
23          loss.backward()
24          optimizer.step()
25
26          if batch_idx % 10 == 0:
27              loss_list.append(loss.item())
28              batch = batch_idx * len(data)
29              data_count = len(train_loader.dataset)
30              percentage = (100. * batch_idx / len(train_loader))
31              print(f'Epoch {epoch}: [{batch:5d} / {data_count}] ({percentage:.0f} %)' +
32                    f'  Loss: {loss.item():.6f}')
```

结果与顺序模型差异不大。

**范例5.** 使用另一种损失函数——**Negative Log Likelihood Loss**，进行手写阿拉伯数字辨识(MNIST)，函数介绍可参阅官网说明[11]。

下列程序代码请参考【04_11_ 手写阿拉伯数字辨识 _ 专家模式 _NLL_LOSS.ipynb】。

以下只说明差异的程序代码。

定义模型：使用 F.log_softmax 取代 F.softmax( 第 17 行 )，可避免损失为负值。

```
1   # 建立模型
2   class Net(nn.Module):
3       def __init__(self):
4           super().__init__()
5           self.fc1 = torch.nn.Linear(28 * 28, 256) # 完全连接层
6           self.dropout1 = nn.Dropout(0.2)
7           self.fc2 = torch.nn.Linear(256, 10) # 完全连接层
8           self.dropout2 = nn.Dropout(0.2)
9
10      def forward(self, x):
11          # 完全连接层 + dropout + 完全连接层 + dropout + log_softmax
12          x = torch.flatten(x, 1)
13          x = self.fc1(x)
14          x = self.dropout1(x)
15          x = self.fc2(x)
16          x = self.dropout2(x)
17          output = F.log_softmax(x, dim=1)
18          return output
19
20  # 建立模型物件
21  model = Net().to(device)
```

训练模型：使用 F.nll_loss 取代 CrossEntropyLoss()，才可加 SoftMax 层 ( 第 19 行 )。

```
1   epochs = 5
2   lr=0.1
3
4   # 建立 DataLoader
5   train_loader = DataLoader(train_ds, batch_size=600)
6
7   # 设定优化器(optimizer)
8   optimizer = torch.optim.Adam(model.parameters(), lr=lr)
9
```

```
10  model.train()
11  loss_list = []
12  for epoch in range(1, epochs + 1):
13      for batch_idx, (data, target) in enumerate(train_loader):
14          data, target = data.to(device), target.to(device)
15  
16          optimizer.zero_grad()
17          output = model(data)
18          # 计算损失(Loss)
19          loss = F.nll_loss(output, target)
20          loss.backward()
21          optimizer.step()
22  
23          if batch_idx % 10 == 0:
24              loss_list.append(loss.item())
25              batch = batch_idx * len(data)
26              data_count = len(train_loader.dataset)
27              percentage = (100. * batch_idx / len(train_loader))
28              print(f'Epoch {epoch}: [{batch:5d} / {data_count}] ({percentage:.0f} %)' +
29                    f'  Loss: {loss.item():.6f}')
```

评分：使用 F.nll_loss(output, target, reduction='sum').item()，计算多笔测试数据的损失和 ( 第 13 行 )。

```
1   # 建立 DataLoader
2   test_loader = DataLoader(test_ds, batch_size=600)
3   
4   model.eval()
5   test_loss = 0
6   correct = 0
7   with torch.no_grad():
8       for data, target in test_loader:
9           data, target = data.to(device), target.to(device)
10          output = model(data)
11  
12          # sum up batch loss
13          test_loss += F.nll_loss(output, target, reduction='sum').item()
14  
15          # 预测
16          pred = output.argmax(dim=1, keepdim=True)
17  
18          # 正确笔数
19          correct += pred.eq(target.view_as(pred)).sum().item()
20  
21  # 平均损失
22  test_loss /= len(test_loader.dataset)
23  # 显示测试结果
24  batch = batch_idx * len(data)
25  data_count = len(test_loader.dataset)
26  percentage = 100. * correct / data_count
27  print(f'平均损失: {test_loss:.4f}, 准确率: {correct}/{data_count}' +
28        f' ({percentage:.0f}%)\n')
```

## 4-3 神经层

神经层是神经网络的主要成员，PyTorch 有各种各样的神经层，详情可参阅官网[12]，如下列举了一些比较常见的类别，随着算法的发明，还会不断地增加，也可以自定义神经层 (Custom layer)。

- 完全连接层 (Linear Layers)；
- 卷积神经层 (Convolution Layers)；
- 池化神经层 (Pooling Layers)；
- 常态化神经层 (Normalization Layers)；
- 循环神经层 (Recurrent Layers)；
- Transformer Layers；
- Dropout Layers。

由于中文翻译大部分都不贴切原意，建议读者尽可能使用英文术语。

### 4-3-1 完全连接层

完全连接层 (Linear) 是最常见的神经层,上一层每个输出的神经元 (y) 都会完全连接到下一层的每个输入的神经元 (x),即 $y = wx + b$。语法如下:

torch.nn.Linear(in_features, out_features, bias=True, device=None, dtype=None)

in_features:输入神经元个数。
out_features:输出神经元个数。
bias:训练出的模型是否含偏差项 (bias)。
device:以 CPU 或 GPU 设备训练。
dtype:输出、输入张量的数据类型。
通常只设置前面两个参数。

请参阅程序【03_04_ 完全连接层 .ipynb】。

### 4-3-2 Dropout Layer

Dropout Layer 在每一次 epoch/step 训练时,会随机丢弃设定比例的输入神经元,避免过度拟合,只在训练时运作,预测时会忽视 Dropout,不会有任何作用。语法如下:

torch.nn.Dropout(p=0.5, inplace=False)

参数说明如下:
p:丢弃的比例,介于 (0, 1) 之间。
inplace=True:对输入直接修改,不另产生变量,节省内存,通常不设定。

根据大部分学者的经验,在神经网络中使用 Dropout 会比正则化 (Regularizer) [13] 效果来得好。

Dropout 不会影响输出神经元个数。

```
1  # 载入框架
2  import torch
3
4  m = torch.nn.Dropout(p=0.2)
5  input = torch.randn(20, 16)
6  output = m(input)
7  output.shape
```

执行结果:输入维度为 (20, 16),输出维度仍是 (20, 16)。
其他的神经层在后续算法使用到时再说明。

## 4-4 激励函数

激励函数 (Activation Function) 是将方程式乘上非线性函数,变成非线性模型,目的是希望能提供更通用的解决方案,而非单纯的线性回归。

$$Output = activation\ function(x_1w_1 + x_2w_2 + \cdots + x_nw_n + bias)$$

PyTorch 提供非常多种的 Activation Function 函数,可参阅官网中的 Activation Functions 介绍 [14]。

**范例.** 列举常用的 **Activation Function** 函数并进行测试。

**下列程序代码请参考【04_12_Activation_Functions.ipynb】。**

(1) ReLU：早期隐藏层使用 Sigmoid 函数，近年发现 ReLU 效果更好。

公式：ReLU(x) = max(0,x)，将小于 0 的数值转换为 0，即过滤掉负值的输入。

```
1  # 载入框架
2  import torch
3
4  m = torch.nn.ReLU()
5  input = torch.tensor([5, 2, 0, -10])
6  output = m(input)
7  output
```

执行结果：[5, 2, 0, 0]。

(2) LeakyReLU：ReLU 将小于 0 的数值转换为 0，会造成某些特征 (x) 失效，为保留所有的特征，LeakyReLU 将小于 0 的数值转换为非常小的负值，而非 0。

公式：LeakyReLU(x)=max(0, x) + negative_slope × min(0, x)。

```
1  m = torch.nn.LeakyReLU()
2  input = torch.tensor([5, 2, 0, -10, -100], dtype=float)
3  output = m(input)
4  output
```

执行结果：[5.0000, 2.0000, 0.0000, -0.1000, -1.0000]。

(3) Sigmoid：罗吉斯回归 (Logistic regression)，将输入值转换为 [0, 1]，适合二分类。

公式：$Sigmoid(x) = \dfrac{1}{1+e^{-x}}$

```
1  m = torch.nn.Sigmoid()
2  input = torch.tensor([5, 2, 0, -10, -100], dtype=float)
3  output = m(input)
4  output
```

执行结果：[9.9331e-01, 8.8080e-01, 5.0000e-01, 4.5398e-05, 3.7201e-44]，均介于 [0, 1] 之间。

(4) Tanh：将输入值转换为 [-1, 1]，适合二分类。

公式：$Tanh(x) = \dfrac{e^x - e^{-x}}{e^x + e^{-x}}$

```
1  m = torch.nn.Tanh()
2  input = torch.tensor([5, 2, 0, -10, -100], dtype=float)
3  output = m(input)
4  output
```

执行结果：[0.9999, 0.9640, 0.0000, -1.0000, -1.0000]，均介于 [-1, 1] 之间。

(5) Softmax：将输入转换为概率，总和为 1，通常使用在最后一层，方便比较每一类的预测概率大小。

公式：$Softmax(x) = \dfrac{e^{x_i}}{\sum_j e^{x_j}}$

```
1  m = torch.nn.Softmax(dim=1)
2  input = torch.tensor([[1.0, 2.0, 3.0, 4.0]], dtype=float)
3  output = m(input)
4  output
```

执行结果：[0.0321, 0.0871, 0.2369, 0.6439]，总和为 1。

## 4-5 损失函数

损失函数 (Loss Functions) 又称为目标函数 (Objective Function)、成本函数 (Cost Function)，算法以损失最小化为目标，估算模型所有的参数，即权重与偏差，例如回归的损失函数为均方误差 (MSE)，我们也可以定义不同的损失函数，产生各种各样的应用，例如后续会讨论的风格转换 (Style Transfer)、生成对抗网络 (GAN) 等。

PyTorch 支持许多损失函数，可参阅官网关于损失函数的介绍[9]。

**范例. 列举常用的损失函数并进行测试。**

下列程序代码请参考【04_13_ 损失函数 .ipynb】。

(1) 均方误差 (MSE)：通常用于预测连续型的变量 (y)。语法如下：torch.nn.MSELoss(reduction='mean')，若 reduction='sum'，得到误差平方和 (SSE)。

```
1  # 载入框架
2  import torch
3
4  loss = torch.nn.MSELoss()        # 产生MSE物件
5  input = torch.randn(3, 5, requires_grad=True)
6  target = torch.randn(3, 5)       # 目标值
7  output = loss(input, target)     # 计算预测值与目标值之均方误差
8  output
```

执行结果：tensor(1.8842, grad_fn=<MseLossBackward0>)。

损失的参数应为预测值与目标值，上述的程序只是测试，将 input 直接代入。

(2) 绝对误差 (MAE)：使用 torch.nn.L1Loss。

(3) CrossEntropyLoss：交叉熵 (Cross Entropy)，通常用于预测离散型的变量 (y)。语法如下：

torch.nn.CrossEntropyLoss(weight=None, reduction='mean', label_smoothing=0.0)

weight：每一类别占的权重，若为 None，每一类别的权重均等。

reduction：与 MSELoss 相同，多一种选项 None，表示不作加总或平均。

label_smoothing：计算后的损失函数是否平滑化，可设范围为 [0, 1]，0 表不平滑化，1 表完全平滑化。

```
1  loss = torch.nn.CrossEntropyLoss()   # 产生框架
2  input = torch.randn(3, 5, requires_grad=True)
3  target = torch.empty(3, dtype=torch.long).random_(5)  # 目标值
4  output = loss(input, target)     # 计算预测值与目标值之均方误差
5  output.backward()
6  output
```

执行结果：tensor(2.4065, grad_fn=<NllLossBackward0>)。

**注意**，TensorFlow 使用交叉熵时，目标值 (y) 要先进行 One-hot encoding 转换，将单一变量变成 n 个变量，n 为 y 的类别数，例如辨识 0~9，n=10。或者采用 SparseCategoricalCrossentropy，y 即可不转换，但 PyTorch 很贴心，可接受原本的 y 或经 One-hot encoding 的 y，两者均可。

PyTorch 还有很多损失函数，可用于语音、自然语言处理上，后续如有使用，再详细介绍。

## 4-6 优化器

优化器是神经网络中反向传导的求解方法，着重在两方面：
(1) 设定学习率的变化，加速求解的收敛速度。
(2) 避开马鞍点 (Saddle Point) 等局部最小值，并且找到全局的最小值 (Global Minimum)。

优化的过程如图 4.13，随着训练的过程，沿着等高线逐步逼近圆心，权重不断更新，最终得到近似最佳解。

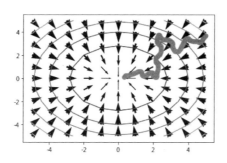

图 4.13　随机梯度下降法 (Stochastic Gradient Descent, SGD) 求解图示

PyTorch 支持很多种不同的优化器，可参阅官网中关于优化器介绍[8]，大部分都是动态调整的学习率，一开始离最佳解很远时，学习率可加大，越接近最佳解，学习率就越小，以免错过最佳解。常见的优化器如下：

- SGD；
- Adam；
- RMSprop；
- Adadelta；
- Adagrad；
- Adamax；
- Nadam；
- AMSGrad。

各种优化器的公式可参考 *Gradient Descent Optimizers*[15] 或 *10 Stochastic Gradient Descent Optimisation Algorithms + Cheat Sheet*[16]，优缺点比较可参考 *Various Optimization Algorithms For Training Neural Network*[17]。

**范例.** 列举常用的优化器并进行测试。

**下列程序代码请参考【04_14_ 优化器 .ipynb】。**

(1) 随机梯度下降法 (Stochastic Gradient Descent, SGD)：是最常见、最单纯的优化器，语法为：
torch.optim.SGD(model.parameters(), lr, , momentum=0, dampening=0, weight_decay=0, nesterov=False)
可以设定为：
model.parameters()：模型的参数 ( 权重 )。

lr：学习率，为必填参数，在未设定其他参数时，学习率为固定值。

momentum：学习率变化速率的动能。

weight_decay：L2 惩罚项的权重衰减率。

nesterov：是否使用 Nesterov momentum，默认值是 False。要了解技术细节可参阅 "Understanding Nesterov Momentum (NAG)" [18]。

```
1   # 载入框架
2   import torch
3
4   # 建立模型
5   model = torch.nn.Sequential(
6       torch.nn.Flatten(),
7       torch.nn.Linear(28 * 28, 256),
8       torch.nn.Dropout(0.2),
9       torch.nn.Linear(256, 10),
10  )
11
12  criterion = torch.nn.CrossEntropyLoss()
13
14  # 随机梯度下降法(SGD)
15  optimizer = torch.optim.SGD(model.parameters(), lr=0.1, momentum=0.9)
16
17  optimizer.zero_grad()
18  input = torch.randn(3, 28 * 28, requires_grad=True)
19  target = torch.empty(3, dtype=torch.long).random_(5) # 目标值
20  loss = criterion(model(input), target)
21  loss.backward()
22  optimizer.step()
```

第 14 行：建立随机梯度下降法 (SGD) 优化器。

第 22 行：优化器执行一个步骤，反向传导，更新权重。

(2) Adam(Adaptive Moment Estimation)：是最常用的优化器，这里引用 Kingma 等学者于 2014 年发表的 *Adam: A Method for Stochastic Optimization*[19] 一文所作的评论 "Adam 计算效率高、内存耗费少，适合大数据集及参数个数很多的模型"。

Adam 语法：torch.optim.Adam(model.parameters(), lr, betas, eps, weight_decay, amsgrad)

model.parameters()：模型的参数 ( 权重 )。

lr：学习率，为必填参数，若未设定其他参数，学习率为固定值。

betas：计算平均梯度及其平方项的系数。

eps：公式分母的加项，以改善优化的稳定性。

weight_decay：L2 惩罚项的权重衰减率。

amsgrad：是否使用 AMSGrad，技术细节可参阅《一文告诉你 Adam、AdamW、Amsgrad 区别和联系》[20]。

(3) 另外还有几种常用的优化器：

Adagrad(Adaptive Gradient-based optimization)：设定每个参数的学习率更新频率不同，较常变动的特征使用较小的学习率，较少调整，反之，使用较大的学习率，比较频繁地调整，主要是针对稀疏的数据集。

RMSprop：每次学习率更新是除以均方梯度 (average of squared gradients)，以指数的速度衰减。

Adadelta：是 Adagrad 改良版，学习率更新会配合过去的平均梯度调整。

各种优化器会在一些比较特殊的状况下，突破马鞍点，顺利找到全局的最小值，一般情况下采用 Adam 及预设参数值即可，大致都可以达到梯度下降的效果。网络上也有许多优化器的比较和动画，有兴趣的读者可参阅 *Alec Radford's animations for optimization algorithms*[21]。

不管是神经层、Activation Function、损失函数或优化器，Functional API 都有对应的函数，都在 torch.nn.functional 命名空间内，与 torch.nn 无差别，根据我们要采取哪一类的模型而定。相关 Functional API 函数可参阅官网 torch.nn.functional 介绍 [10]。

## 4-7 效能衡量指标

效能衡量指标 (Performance Metrics) 是定义模型优劣的衡量标准，要了解各种效能衡量指标，先要理解混淆矩阵 (Confusion Matrix)，以二分类而言，如图 4.14。

|  | | 真实 | |
|---|---|---|---|
|  | | 真(True) | 假(False) |
| 预测 | 阳性(Positive) | TP | FP |
|  | 阴性(Negative) | TN | FN |

图 4.14　混淆矩阵 (Confusion Matrix)

(1) 横轴为预测结果，分为阳性 (Positive, 简称 P)、阴性 (Negative, 简称 N)。
(2) 纵轴为真实状况，分为真 (True, 简称 T)、假 (False, 简称 F)。
(3) 依预测结果及真实状况的组合，共分为四种状况：
TP( 真阳性 )：预测为阳性，且预测正确。
TN( 真阴性 )：预测为阴性，且预测正确。
FP( 伪阳性 )：预测为阳性，但预测错误，又称型一误差 (Type I Error)，或 α 误差。
FN( 伪阴性 )：预测为阴性，但预测错误，又称型二误差 (Type II Error)，或 β 误差。
(4) 有了 TP/TN/FP/FN 之后，我们就可以定义各种效能衡量指标，常见的有四种：
准确率 (Accuracy)= (TP+TN)/(TP+FP+FN+TN)，即预测正确数 / 总数。
精确率 (Precision)= TP/(TP+FP)，即正确预测阳性数 / 总阳性数。
召回率 (Recall)= TP/(TP+FN)，即正确预测阳性数 / 实际为真的总数。
F1 = 精确率与召回率的调和平均数，即 1 / [ (1 / Precision) + (1 / Recall)]。
(5) FP( 伪阳性 ) 与 FN( 伪阴性 ) 是相冲突的，以 Covid-19 检验为例，如果降低阳性认定值，可以尽最大可能找到所有的确诊者，减少伪阴性，避免传染病扩散，但有些没染疫的人因而被误判，伪阳性相对增加，导致资源的浪费，更严重的可能造成医疗体系崩溃，得不偿失。
(6) 除了准确率之外，为什么还需要参考其他指标？
以医疗检验设备来举例，假设某疾病实际染病的比率为 1%，这时我们拿一个故障的检验设备，不管有无染病，都判定为阴性，这时计算设备准确率，结果竟然是 99%。会有这样离谱的统计，是因为在此案例中，验了 100 个样本，确实只错一个。所以，碰到真假比例悬殊的不平衡 (Imbalanced) 样本，必须使用其他指标来衡量效能。
精确率：再以医疗检验设备为例，我们只关心被验出来的阳性病患，有多少比例是真的染病，而不去关心验出为阴性者，因为验出为阴性，通常不会再被复检，或者不放心又跑到其他医院复检，医院其实很难追踪他们是否真的患病。
召回率：以 Covid-19 为例，我们关心的是所有的染病者有多少比例被验出阳性，因为一旦有漏网之鱼 ( 伪阴性 )，可能就会造成小区传染。

(7) 针对二分类，还有一种较客观的指标称为 ROC/AUC 曲线，它是在各种检验门槛值下，以假阳率为 X 轴，真阳率为 Y 轴，绘制出来的曲线，称为 ROC。覆盖的面积(AUC)越大，表示模型在各种门槛值下的平均效能越好，这个指标有别于一般预测固定以 0.5 当作判断真假的基准。

TensorFlow/Keras 的效能衡量指标可参阅 Keras 官网[22]，但 PyTorch 不提供相关效能衡量指标的函数，可以使用 NumPy 或 Scikit-Learn 的函数，可参阅 "Scikit-Learn 文件"[23]，如果要一律采用 PyTorch 相关的套件，有兴趣的读者可以参阅 "TorchMetrics 文件"[24]，本文采用 Scikit-Learn。

下列三个范例的程序代码请参考【04_15_效能衡量指标.ipynb】。

**范例1. 假设有8笔数据如下，请计算混淆矩阵(Confusion Matrix)。**

实际值 = [0, 0, 0, 1, 1, 1, 1, 1]
预测值 = [0, 1, 0, 1, 0, 1, 0, 1]

(1) 加载相关套件。

```
1  import numpy as np
2  import matplotlib.pyplot as plt
3  from sklearn.metrics import accuracy_score, classification_report
4  from sklearn.metrics import precision_score, recall_score, confusion_matrix
```

(2) Scikit-Learn 提供混淆矩阵 (Confusion Matrix) 函数，程序代码如下。

```
1  from sklearn.metrics import confusion_matrix
2
3  y_true = [0, 0, 0, 1, 1, 1, 1, 1] # 实际值
4  y_pred = [0, 1, 0, 1, 0, 1, 0, 1] # 预测值
5
6  # 混淆矩阵(Confusion Matrix)
7  tn, fp, fn, tp = confusion_matrix(y_true, y_pred).ravel()
8  print(f'TP={tp}, FP={fp}, TN={tn}, FN={fn}')
```

注意，Scikit-Learn 提供的混淆矩阵，回传值与图 4.14 位置不同。
实际值与预测值上下比较，TP 为 (1, 1)、FP 为 (0, 1)、TN 为 (0, 0)、FN 为 (1, 0)。
执行结果：TP=3, FP=1, TN=2, FN=2。

(3) 绘图。

```
1  # 修正中文问题
2  plt.rcParams['font.sans-serif'] = ['Microsoft JhengHei']
3  plt.rcParams['axes.unicode_minus'] = False
4
5  # 显示矩阵
6  fig, ax = plt.subplots(figsize=(2.5, 2.5))
7
8  # 1:蓝色, 0:白色
9  ax.matshow([[1, 0], [0, 1]], cmap=plt.cm.Blues, alpha=0.3)
10
11 # 标示文字
12 ax.text(x=0, y=0, s=tp, va='center', ha='center')
13 ax.text(x=1, y=0, s=fp, va='center', ha='center')
14 ax.text(x=0, y=1, s=tn, va='center', ha='center')
15 ax.text(x=1, y=1, s=fn, va='center', ha='center')
16
17 plt.xlabel('实际', fontsize=20)
18 plt.ylabel('预测', fontsize=20)
19
20 # x/y 标签
21 plt.xticks([0,1], ['T', 'F'])
22 plt.yticks([0,1], ['P', 'N'])
23 plt.show()
```

执行结果。

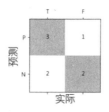

**范例2. 按上述数据计算效能衡量指标。**

(1) 准确率。

```
1  print(f'准确率:{accuracy_score(y_true, y_pred)}')
2  print(f'验算={(tp+tn) / (tp+tn+fp+fn)}')
```

执行结果：0.625。

(2) 计算精确率。

```
1  print(f'精确率:{precision_score(y_true, y_pred)}')
2  print(f'验算={(tp) / (tp+fp)}')
```

执行结果：0.75。

(3) 计算召回率。

```
1  print(f'召回率:{recall_score(y_true, y_pred)}')
2  print(f'验算={(tp) / (tp+fn)}')
```

执行结果：0.6。

**范例3. 按数据文件data/auc_data.csv计算AUC。**

(1) 读取数据文件。

```
1  # 读取数据文件
2  import pandas as pd
3  df=pd.read_csv('./data/auc_data.csv')
4  df
```

执行结果。

|    | predict | actual |
|----|---------|--------|
| 0  | 0.11    | 0      |
| 1  | 0.35    | 0      |
| 2  | 0.72    | 1      |
| 3  | 0.10    | 1      |
| 4  | 0.99    | 1      |
| 5  | 0.44    | 1      |
| 6  | 0.32    | 0      |
| 7  | 0.80    | 1      |
| 8  | 0.22    | 1      |
| 9  | 0.08    | 0      |
| 10 | 0.56    | 1      |

(2) 以 Scikit-Learn 函数计算 AUC。

```
1  from sklearn.metrics import roc_curve, roc_auc_score, auc
2
3  # fpr：假阳率，tpr：真阳率，threshold：各种决策门槛
4  fpr, tpr, threshold = roc_curve(df['actual'], df['predict'])
5  print(f'假阳率={fpr}\n\n真阳率={tpr}\n\n决策门槛={threshold}')
```

执行结果。

```
假阳率=[0.         0.         0.         0.14285714 0.14285714 0.28571429
 0.28571429 0.57142857 0.57142857 0.71428571 0.71428571 1.                    ]
真阳率=[0.         0.09090909 0.27272727 0.27272727 0.63636364 0.63636364
 0.81818182 0.81818182 0.90909091 0.90909091 1.         1.                    ]
决策门槛=[1.99 0.99 0.8  0.73 0.56 0.48 0.42 0.32 0.22 0.11 0.1  0.03]
```

(3) 绘制 AUC。

```
1  # 绘图
2  auc1 = auc(fpr, tpr)
3  ## Plot the result
4  plt.title('ROC/AUC')
5  plt.plot(fpr, tpr, color = 'orange', label = 'AUC = %0.2f' % auc1)
6  plt.legend(loc = 'lower right')
7  plt.plot([0, 1], [0, 1],'r--')
8  plt.xlim([0, 1])
9  plt.ylim([0, 1])
10 plt.ylabel('True Positive Rate')
11 plt.xlabel('False Positive Rate')
12 plt.show()
```

执行结果。

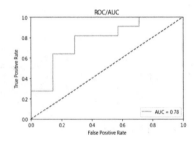

# 4-8 超参数调校

这一节来研究超参数 (Hyperparameters) 对效能的影响，在 4-1-3 节只对单一变量进行调校，假如要同时调校多个超参数，有一些套件可以帮忙，包括 Ray Tune、Keras Tuner、hyperopt、Ax 等。

PyTorch 官网推荐 Ray Tune[25]，Ray 是一个非常强大的平行处理的套件，其中的一个模块 Ray Tune 是用于效能调校，以下就介绍其基本的用法。首先安装套件，指令如下：pip install ray。

**范例. 使用Ray Tune对神经网络进行超参数调校。**

下列程序代码请参考【04_16_ 超参数调校 .ipynb】。

(1) 载入套件。

```
1  import numpy as np
2  import torch
3  import torch.optim as optim
4  import torch.nn as nn
5  from torchvision import datasets, transforms
6  from torch.utils.data import DataLoader
7  import torch.nn.functional as F
8  from ray import tune
9  from ray.tune.schedulers import ASHAScheduler
```

(2) 判断是否有 GPU，若有则使用 GPU。

```
1  device = torch.device("cuda" if torch.cuda.is_available() else "cpu")
2  "cuda" if torch.cuda.is_available() else "cpu"
```

(3) 建立模型。

```
1  class ConvNet(nn.Module):
2      def __init__(self):
3          super(ConvNet, self).__init__()
4          # In this example, we don't change the model architecture
5          # due to simplicity.
6          self.conv1 = nn.Conv2d(1, 3, kernel_size=3)
7          self.fc = nn.Linear(192, 10)
8  
9      def forward(self, x):
10         x = F.relu(F.max_pool2d(self.conv1(x), 3))
11         x = x.view(-1, 192)
12         x = self.fc(x)
13         return F.log_softmax(x, dim=1)
```

(4) 定义模型训练及测试函数，测试函数要回传效能衡量指标给 Ray 做判断，以决定最佳超参数组合。

```
1  # 训练周期
2  EPOCH_SIZE = 5
3  
4  # 定义模型训练函数
5  def train(model, optimizer, train_loader):
6      model.train()
7      for batch_idx, (data, target) in enumerate(train_loader):
8          data, target = data.to(device), target.to(device)
9          optimizer.zero_grad()
10         output = model(data)
11         loss = F.nll_loss(output, target)
12         loss.backward()
13         optimizer.step()
14  
15 # 定义模型测试函数
16 def test(model, data_loader):
17     model.eval()
18     correct = 0
19     total = 0
20     with torch.no_grad():
21         for batch_idx, (data, target) in enumerate(data_loader):
22             data, target = data.to(device), target.to(device)
23             outputs = model(data)
24             # 准确数计算
25             _, predicted = torch.max(outputs.data, 1)
26             total += target.size(0)
27             correct += (predicted == target).sum().item()
28  
29     return correct / total
```

(5) 定义特征缩放函数：采用标准化，平均数为 0.1307，标准偏差为 0.3081。

```
1  mnist_transforms = transforms.Compose(
2      [transforms.ToTensor(),
3       transforms.Normalize((0.1307, ), (0.3081, ))
4      ])
```

(6) 定义数据加载及模型训练函数，还包括：

优化器：使用组态参数，提供多种组合的测试，config 为组态参数内容。

训练结果交回给 Ray Tune：tune.report(mean_accuracy=acc)，指定 acc 为平均准确率，作为效能比较的基准。

每 5 周期存盘一次。

```
 1  def train_mnist(config):
 2      # 载入 MNIST 手写阿拉伯数字资料
 3      train_loader = DataLoader(
 4          datasets.MNIST("", train=True, transform=mnist_transforms),
 5          batch_size=64,
 6          shuffle=True)
 7      test_loader = DataLoader(
 8          datasets.MNIST("", train=False, transform=mnist_transforms),
 9          batch_size=64,
10          shuffle=True)
11
12      # 建立模型
13      model = ConvNet().to(device)
14
15      # 优化器，使用组态参数
16      optimizer = optim.SGD(model.parameters(),
17                      lr=config["lr"], momentum=config["momentum"])
18      # 训练 10 周期
19      for i in range(10):
20          train(model, optimizer, train_loader)
21          # 测试
22          acc = test(model, test_loader)
23
24          # 训练结果交回给 Ray Tune
25          tune.report(mean_accuracy=acc)
26
27          # 每 5 周期存盘一次
28          if i % 5 == 0:
29              torch.save(model.state_dict(), "./model.pth")
```

(7) 定义参数调校的组态。

学习率 (learning rate) 测试选项含 0.01、0.1、0.5，使用 grid_search 表示每一选项都要测试，若使用 choice，则是多选一，sample_from 则是随机抽样。

学习率动能 (momentum)：采用均匀分配抽样。

除了上述优化器参数外，调校任何超参数及模型的神经元个数均可。

Ray Tune 提供非常多的随机分配，详情请参考 "Search Space API"[26]。

第 12 行：实际执行参数调校，若无 GPU，请移除 resources_per_trial={'gpu': 1}，另外，参数调校预设执行 10 个回合，可以加入 num_samples 参数，指定回合数。

```
 1  # 参数组合
 2  search_space = {
 3      #"lr": tune.sample_from(lambda spec: 10**(-10 * np.random.rand())),
 4      "lr": tune.grid_search([0.01, 0.1, 0.5]),   # 每一选项都要测试
 5      "momentum": tune.uniform(0.1, 0.9)          # 均匀分配抽样
 6  }
 7
 8  # 加下一行，采用分散式处理
 9  # ray.init(address="auto")
10
11  # 执行参数调校
12  analysis = tune.run(train_mnist, config=search_space, resources_per_trial={'gpu': 1})
```

执行结果如下，按平均准确率 (acc) 降序排列，最佳参数组合为 lr: 0.01, momentum: 0.620798，平均准确率为 0.9644。

| Trial name | status | loc | lr | momentum | acc | iter | total time (s) |
|---|---|---|---|---|---|---|---|
| train_mnist_632aa_00000 | TERMINATED | 127.0.0.1:21828 | 0.01 | 0.620798 | 0.9644 | 10 | 142.228 |
| train_mnist_632aa_00001 | TERMINATED | 127.0.0.1:22708 | 0.1 | 0.409923 | 0.9625 | 10 | 137.528 |
| train_mnist_632aa_00002 | TERMINATED | 127.0.0.1:21072 | 0.5 | 0.651037 | 0.1135 | 10 | 138.692 |

(8) 对训练过程的准确率绘图。

## 第 4 章 | 神经网络实操

```
1  import matplotlib.pyplot as plt
2
3  # 取得实验的参数
4  config_list = []
5  for i in analysis.get_all_configs().keys():
6      config_list.append(analysis.get_all_configs()[i])
7
8  # 绘图
9  plt.figure(figsize=(12,6))
10 dfs = analysis.trial_dataframes
11 for i, d in enumerate(dfs.values()):
12     plt.subplot(1,3,i+1)
13     plt.title(config_list[i])
14     d.mean_accuracy.plot()
15 plt.tight_layout()
16 plt.show()
```

执行结果如图 4.15，第一个组合准确率最高，且接近收敛。

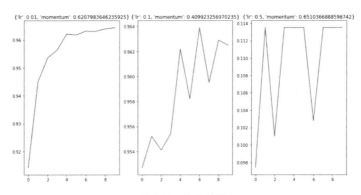

图 4.15　执行结果

(9) 显示详细调校内容：例如每一回合 (Trial) 训练过程的准确率及运行时间。

```
1  for i in dfs.keys():
2      parameters = i.split("\\")[-1]
3      print(f'{parameters}\n', dfs[i][['mean_accuracy', 'time_total_s']])
```

截取两个参数组合的执行结果。

```
train_mnist_632aa_00000_0_lr=0.01,momentum=0.6208_2022-01-02_21-10-41
   mean_accuracy   time_total_s
0      0.9142         23.599039
1      0.9450         36.718488
2      0.9536         49.732055
3      0.9564         62.833749
4      0.9621         75.938153
5      0.9618         89.218834
6      0.9632        102.366150
7      0.9631        115.633676
8      0.9640        128.915956
9      0.9644        142.228140
train_mnist_632aa_00001_1_lr=0.1,momentum=0.40992_2022-01-02_21-10-42
   mean_accuracy   time_total_s
0      0.9527         21.514920
1      0.9552         34.425164
2      0.9541         47.275895
3      0.9554         60.217369
4      0.9622         73.019098
5      0.9582         85.883695
6      0.9639         98.806184
7      0.9595        111.709021
8      0.9629        124.651322
9      0.9625        137.528468
```

其他字段可参考《Ray Tune 使用手册》[27]。

- `config`: The hyperparameter configuration
- `date`: String-formatted date and time when the result was processed
- `done`: True if the trial has been finished, False otherwise
- `episodes_total`: Total number of episodes (for RLLib trainables)
- `experiment_id`: Unique experiment ID
- `experiment_tag`: Unique experiment tag (includes parameter values)
- `hostname`: Hostname of the worker
- `iterations_since_restore`: The number of times `tune.report()`/`trainable.train()` has been called after restoring the worker from a checkpoint
- `node_ip`: Host IP of the worker
- `pid`: Process ID (PID) of the worker process
- `time_since_restore`: Time in seconds since restoring from a checkpoint
- `time_this_iter_s`: Runtime of the current training iteration in seconds (i.e. one call to the trainable function or to `_train()` in the class API.
- `time_total_s`: Total runtime in seconds.
- `timestamp`: Timestamp when the result was processed
- `timesteps_since_restore`: Number of timesteps since restoring from a checkpoint
- `timesteps_total`: Total number of timesteps
- `training_iteration`: The number of times `tune.report()` has been called
- `trial_id`: Unique trial ID

(10) 显示各组合的执行结果。

```
1  analysis.results_df
```

执行结果如下。

| trial_id | mean_accuracy | time_this_iter_s | done | timesteps_total | episodes_total | training_iteration | experiment_id | date | timestamp |
|---|---|---|---|---|---|---|---|---|---|
| 632aa_00000 | 0.9644 | 13.312184 | True | None | None | 10 | e635b4cc50164158a2385aeed6bf481e | 2022-01-02_21-13-06 | 1641129186 |
| 632aa_00001 | 0.9625 | 12.877146 | True | None | None | 10 | 6fe5ee93539148baa431cb0473c73e0d | 2022-01-02_21-15-26 | 1641129326 |
| 632aa_00002 | 0.1135 | 13.038788 | True | None | None | 10 | 8705d0f4763f47c1be798a6faf64cc65 | 2022-01-02_21-17-52 | 1641129472 |

(11) 取得最佳模型参数：取最大 (max) 的平均准确率 (mean_accuracy)，若有多笔，取最后一笔。

```
1  best_trial = analysis.get_best_trial("mean_accuracy", "max", "last")
2  best_trial.config
```

执行结果：{'lr': 0.01,'momentum': 0.6207983646235925}

(12) 加载最佳模型。

```
1  logdir = analysis.get_best_logdir("mean_accuracy", mode="max")
2  state_dict = torch.load(os.path.join(logdir, "model.pth"))
3
4  model = ConvNet().to(device)
5  model.load_state_dict(state_dict)
```

(13) 使用最佳模型测试。

```
1  test_ds = datasets.MNIST('', train=False, download=True, transform=mnist_transforms)
2
3  # 建立 DataLoader
4  test_loader = DataLoader(test_ds, shuffle=False, batch_size=1000)
5
6  model.eval()
```

```
 7   correct = 0
 8   with torch.no_grad():
 9       for data, target in test_loader:
10           data, target = data.to(device), target.to(device)
11           output = model(data)
12   
13           # 正确笔数
14           _, predicted = torch.max(output, 1)
15           correct += (predicted == target).sum().item()
16   
17   # 显示测试结果
18   data_count = len(test_loader.dataset)
19   percentage = 100.0 * correct / data_count
20   print(f'准确率: {correct}/{data_count} ({percentage:.0f}%)\n')
```

执行结果：准确率非常高，达到 9722/10000 (97%)。

上述程序只是简单的范例，Ray Tune 还有更多进阶的函数及参数可供使用，详情可参阅 Pytorch 官网范例[25] 及 Ray Tune 官网[28]。

参数调校是深度学习中非常重要的步骤，因为深度学习是一个黑箱科学，加上我们对高维数据的联合概率分配也是一无所知，唯有通过大量的试验，才能获得较佳的模型。但困难的是，模型训练的执行非常耗时，如何通过各种方法或套件的协助，平行处理或分散至多台机器执行，缩短调校时间，是建构 AI 模型时须认真思考的课题。

# 第 5 章
# PyTorch 进阶功能

除了建构模型外，PyTorch 还提供各种的工具和指令，在程序开发流程中使用，包括数据集 (Dataset)、数据加载器 (DataLoader)、前置处理、TensorBoard、除错等功能，认识这些功能可以使程序执行更有效率，也比较容易找出错误。

## 5-1 数据集及数据加载器

torch.utils.data.Dataset 是 PyTorch 内建数据结构，可同时存储特征 (x) 及目标 (y)，包含一些内建的数据集：

(1) 影像数据集：例如 MNIST、FashionMNIST 等，可参考 Pytorch 官网 torchvision.datasets[1]。

(2) 语音数据集：可参考 Pytorch 官网 torchaudio.datasets[2]。

(3) 文字数据集：可参考 Pytorch 官网 torchtext.datasets[3]。

(4) 除此之外，还可以自定义数据集。

**范例. 加载数据集，并读取相关数据。**

下列程序代码请参考【05_01_Datasets.ipynb】。

(1) 载入套件。

```
1  import os
2  import torch
3  from torchvision.datasets import MNIST, FashionMNIST
4  from torch.utils.data import DataLoader, random_split
5  from torchvision import transforms
```

(2) 检查是否有 GPU。

```
1  device = torch.device("cuda" if torch.cuda.is_available() else "cpu")
2  "cuda" if torch.cuda.is_available() else "cpu"
```

(3) 加载 MNIST 手写阿拉伯数字数据。MNIST 等数据集都有 5 个参数。
根路径 (root)：数据集下载后存储的目录，空字符串表示目前文件夹。
train：True 表示下载训练数据集；False 表示下载测试数据集。
download：True 表示数据集不存在则自网络下载；False 表示不会自动下载。
transform：数据集读入后特征 (x) 要做何种转换，至少要转成 PyTorch Tensor。各种转换可参考 Pytorch 官网 torchvision.transforms[4]。
target_transform：数据集读入后，目标 (y) 要做何种转换。

```
1  # 下载 MNIST 手写阿拉伯数字 训练数据
2  train_ds = MNIST("", train=True, download=True,
3                  transform=transforms.ToTensor())
4
5  # 下载测试数据
6  test_ds = MNIST("", train=False, download=True,
7                 transform=transforms.ToTensor())
8
9  # 训练/测试数据的维度
10 print(train_ds.data.shape, test_ds.data.shape)
```

(4) 读取数据：直接指定索引值，例如 train_ds.data[0]，即可读取第一笔数据。**注意，使用 train_ds.data[0] 读取数据并不会应用到 Transform 函数，必须使用 DataLoader 读取数据，Transform 函数才会发生效果。**

```
1  # 显示第1张图片图像
2  import matplotlib.pyplot as plt
3
4  # 第一笔数据
5  X = train_ds.data[0]
6
7  # 绘制点阵图，cmap='gray'：灰阶
8  plt.imshow(X.reshape(28,28), cmap='gray')
9
10 # 隐藏刻度
11 plt.axis('off')
12
13 # 显示图形
14 plt.show()
```

执行结果。

(5) 再看另一个数据集 FashionMNIST，同时说明数据转换 (Transform) 及自定义数据集 (Custom Dataset) 的用法。

```
1  training_data = FashionMNIST(
2      root="data",
3      train=True,
4      download=True,
5      transform=transforms.ToTensor()
6  )
7
8  test_data = FashionMNIST(
9      root="data",
10     train=False,
11     download=True,
12     transform=transforms.ToTensor()
13 )
```

(6) 任意抽样 9 笔数据显示：labels_map 是目标值与名称的对照。

```
1  labels_map = {
2      0: "T-shirt",
3      1: "Trouser",
4      2: "Pullover",
5      3: "Dress",
6      4: "Coat",
7      5: "Sandal",
8      6: "Shirt",
9      7: "Sneaker",
10     8: "Bag",
11     9: "Ankle Boot",
```

```
12  }
13  figure = plt.figure(figsize=(8, 8))
14  cols, rows = 3, 3
15  for i in range(1, cols * rows + 1):
16      sample_idx = torch.randint(len(training_data), size=(1,)).item()
17      img, label = training_data[sample_idx]
18      figure.add_subplot(rows, cols, i)
19      plt.title(labels_map[label])
20      plt.axis("off")
21      plt.imshow(img.squeeze(), cmap="gray")
22  plt.show()
```

执行结果。

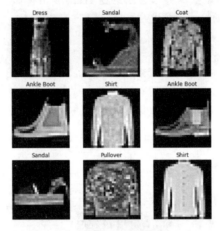

(7) 数据转换 (Transform)：PyTorch 提供非常多的转换函数，包括转换成 PyTorch Tensor、放大/缩小、剪裁、彩色转灰度、各种数据增强 (Data Augmentation) 的效果等，可减少数据前置处理的负担，TensorFlow 目前缺乏类似的功能。我们先来看单张图片的转换，程序代码修改自 Pytorch 官网 Illustration of transforms[5]。

(8) 读取范例图片文件：使用 skimage 套件内建的女航天员图像。

```
1  import skimage
2
3  orig_img = skimage.data.astronaut()
4  skimage.io.imsave('images_test/astronaut.jpg', orig_img)
5  plt.imshow(orig_img)
```

执行结果。

(9) 转换输入须为 Pillow 格式，再以 Pillow 函数读取图片文件。

```
1  # 转换输入须为 Pillow 格式
2  from PIL import Image
3
4  orig_img = Image.open('images_test/astronaut.jpg')
```

(10) 定义绘图函数。

```python
from PIL import Image
from pathlib import Path
import matplotlib.pyplot as plt
import numpy as np
import torchvision.transforms as T

def plot(imgs, with_orig=True, row_title=None, **imshow_kwargs):
    if not isinstance(imgs[0], list):
        # Make a 2d grid even if there's just 1 row
        imgs = [imgs]

    num_rows = len(imgs)
    num_cols = len(imgs[0]) + with_orig
    fig, axs = plt.subplots(nrows=num_rows, ncols=num_cols, squeeze=False)
    for row_idx, row in enumerate(imgs):
        row = [orig_img] + row if with_orig else row
        for col_idx, img in enumerate(row):
            ax = axs[row_idx, col_idx]
            ax.imshow(np.asarray(img), **imshow_kwargs)
            ax.set(xticklabels=[], yticklabels=[], xticks=[], yticks=[])

    if with_orig:
        axs[0, 0].set(title='Original image')
        axs[0, 0].title.set_size(8)
    if row_title is not None:
        for row_idx in range(num_rows):
            axs[row_idx, 0].set(ylabel=row_title[row_idx])

    plt.tight_layout()
```

(11) 图片放大 / 缩小。

```python
# resize
resized_imgs = [T.Resize(size=size)(orig_img) for size in (30, 50, 100, orig_img.size)]
plot(resized_imgs)
```

执行结果：第 1 张为原图，之后为缩小成 30%、50%、100%、原比例的图，可以看到缩小后再经 ax.imshow 放大，显示就变模糊了。

(12) 自中心裁剪。

```python
center_crops = [T.CenterCrop(size=size)(orig_img) for size in (30, 50, 100, orig_img.size)]
plot(center_crops)
```

执行结果：第 1 张为原图，之后为裁剪成 30%、50%、100%、原比例的图，可以看到以中心点为参考点，向外裁剪。

(13) FiveCrop：以左上、右上、左下、右下及中心点为参考点，一次裁剪 5 张图。

```python
(top_left, top_right, bottom_left, bottom_right, center) = T.FiveCrop(size=(100, 100))(orig_img)
plot([top_left, top_right, bottom_left, bottom_right, center])
```

执行结果。

(14) 转灰度。

```
1  gray_img = T.Grayscale()(orig_img)
2  plot([gray_img], cmap='gray')
```

执行结果。

(15) 旁边补零：指定补零宽度为 3、10、30、50。

```
1  padded_imgs = [T.Pad(padding=padding)(orig_img) for padding in (3, 10, 30, 50)]
2  plot(padded_imgs)
```

执行结果：观察边框的宽窄。

总共超过 20 种转换，中文说明可参考《PyTorch 学习笔记（三）：transforms 的二十二个方法》[6]，这些效果都可以任意组合至 transforms.Compose 函数内。

另外，处理图像时常会做特征缩放，在 TensorFlow 范例中会采取正规化 (Normalization)，公式为 $(x-min)/(max-min)$，使 $x$ 的范围介于 [0,1] 之间，而 PyTorch 并未提供此转换，通常采用标准化，但却命名为 Normalize，与 Normalization 有点混淆，公式为 $(x-\mu)/\delta$，请特别注意。一般而言，标准化是假设 $x$ 是常态分配，但像素颜色 0~255，应该属均匀分配，采用正规化似乎比较合理，但 PyTorch 官网采用常态分配，我们在此不计较这些。

程序代码如下，含两组参数，第一组为 RGB 三色的平均数 ($\mu$)，第二组为 RGB 三色的标准偏差 ($\delta$)，这是从 ImageNet 大量数据集统计的结果：

transforms.Normalize((0.485, 0.456, 0.406), (0.229, 0.224, 0.225))

若图像为单色，程序代码如下：

transforms.Normalize((0.1307,), (0.3081,))

可参考程序【**04_07_手写阿拉伯数字辨识_Normalize.ipynb**】。

若要采取正规化，完整范例可参考程序【**05_02_手写阿拉伯数字辨识_**

**MinMaxScaler.ipynb】**，辨识率不佳，可见 PyTorch 与 TensorFlow 在图像的细部处理上是有所差异的。

（16）接着，我们实操一个范例，并同时示范自定义数据集的做法，将一目录下的所有文件制作成数据集，并转换为正确的输入格式。

（17）先制作一个目标名称与代码的对照表，之后将文件名转换为目标代码。

```
1  # 目标名称→目标代码
2  labels_code = {v.lower():k for k, v in labels_map.items()}
```

（18）自定义数据集：自定义数据集类别必须包含三个方法：\_\_init\_\_、\_\_len\_\_、\_\_getitem\_\_，作用分别为初始化、总笔数、取得下一笔数据，这种方式不必一次加载所有图像，可以节省内存的耗用。

```
1   import os
2   import pandas as pd
3   from torchvision.io import read_image
4   from torch.utils.data import Dataset
5   import re
6
7   class CustomImageDataset(Dataset):
8       def __init__(self, img_dir, transform=None, target_transform=None):
9           self.img_labels = [file_name for file_name in os.listdir(img_dir)]
10          self.img_dir = img_dir
11          self.transform = transform
12          self.target_transform = target_transform
13
14      def __len__(self):
15          return len(self.img_labels)
16
17      def __getitem__(self, idx):
18          # 组合文件完整路径
19          img_path = os.path.join(self.img_dir, self.img_labels[idx])
20          # 读取图文件
21          image = read_image(img_path)
22          # 去除副文件名
23          label = self.img_labels[idx].split('.')[0]
24          # 将文件名数字去除
25          label = re.sub('[0-9]','', label)
26
27          # 转换
28          if self.transform:
29              image = self.transform(image)
30          if self.target_transform:
31              label = self.target_transform(label)
32
33          # 将三维转为二维
34          image = image.reshape(*image.shape[1:])
35          # 反转颜色，颜色0为白色，与 RGB 色码不同，它的 0 为黑色
36          image = 1.0-image
37          label = labels_code[label.lower()]
38
39          return image, label
```

（19）加载【04_06_FashionMNIST 辨识 _ 完整版 .ipynb】存储的模型。

```
1  # 模型载入
2  model = torch.load('./FashionMNIST.pt')
```

（20）建立转换：依次转灰度、缩放、居中、转 PyTorch Tensor。

```
1  # 建立 transforms
2  transform = transforms.Compose([
3      transforms.Grayscale(),
4      transforms.Resize((28, 28)),
5      transforms.CenterCrop(28),
6      # transforms.PILToTensor(),
7      transforms.ConvertImageDtype(torch.float),
8  ])
```

(21) 建立 DataLoader：加载自定义数据集，进行测试。

```
10  # 建立 DataLoader
11  test_loader = DataLoader(CustomImageDataset('./fashion_test_data', transform)
12                          , shuffle=False, batch_size=10)
13
14  model.eval()
15  criterion = nn.CrossEntropyLoss()
16  test_loss = 0
17  correct = 0
18  with torch.no_grad():
19      for data, target in test_loader:
20          data, target = data.to(device), target.to(device)
21          output = model(data)
22          # sum up batch loss
23          test_loss += criterion(output, target).item()
24
25          # 预测
26          pred = output.argmax(dim=1, keepdim=True)
27
28          # 正确笔数
29          correct += pred.eq(target.view_as(pred)).sum().item()
30
31  # 平均损失
32  test_loss /= len(test_loader.dataset)
33  # 显示测试结果
34  data_count = len(test_loader.dataset)
35  percentage = 100. * correct / data_count
36  print(f'平均损失: {test_loss:.4f}, 准确率: {correct}/{data_count}' +
37        f' ({percentage:.0f}%)\n')
```

执行结果：与【04_06_FashionMNIST 辨识 _ 完整版 .ipynb】测试结果相同。

Dataset 一次获取一笔数据，使用 DataLoader 则可以一次获取一"批"数据，方便我们做批量测试，加快训练及测试速度，参数如下：

第一个参数：Dataset。

batch_size：批量。

shuffle：读取数据前是否先洗牌。

不通过循环，一次获取一"批"数据。

```
1  # 一次获取一"批"数据
2  data, target = next(iter(test_loader))
3  print(data.shape, target)
```

执行结果：torch.Size([7, 28, 28]) tensor([8, 5, 5, 6, 0, 1, 1])，取出 7 笔数据 ( 不足 10 笔 )。

有关语音及文字数据集在后续章节再作介绍。

# 5-2 TensorBoard

TensorBoard 是一种可视化的诊断工具，功能非常强大，可以显示模型结构、训练过程，显示包括图片、文字和音频数据。在训练的过程中启动 TensorBoard，能够实时观看训练过程。TensorBoard 虽是 Tensorflow 团队开发的，但 PyTorch 也极力推荐使用，因此，若未安装 Tensorflow，可独立安装 TensorBoard，指令如下：

pip install tensorboard

## 5-2-1 TensorBoard 功能

TensorBoard 包含下列功能。

(1) 追踪损失和准确率等效能衡量指标 (Metrics)，并以可视化呈现。

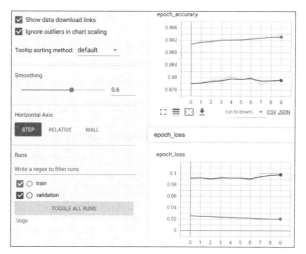

(2) 显示运算图 (Computational Graph)：包括张量运算 (tensor operation) 和神经层 (layers)。

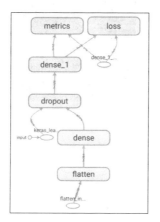

(3) 直方图 (Histogram)：显示训练过程中的权重 (weights)、偏差 (bias) 的概率分配。

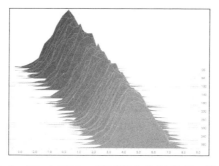

(4) 词嵌入 (Word Embedding) 展示：把词嵌入向量降维，投影到三维空间来显示。画面右边可输入任意单词，例如 King，就会出现下图，将与其相近的单词显示出来，原理是通过词向量 (Word2Vec) 将每个单词转为向量，再利用 Cosine_Similarity 计算相似性，详情会在后续章节介绍。

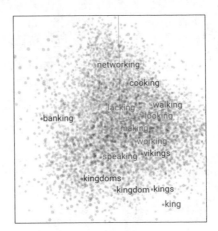

(5) 显示图片、文字和音频数据。

## 5-2-2 测试

首先在程序中必须将信息写入工作记录文件(Log)，之后，可以启动 TensorBoard 观看工作记录文件(Log)，我们直接以范例展示。

**范例1. 先介绍写入工作记录文件的API，包括影像、语音均可写入，甚至嵌入 (Embedding)向量(后续章节说明)。**

下列程序代码请参考【05_04_TensorBoard.ipynb】。

(1) 载入套件。

```
1  import matplotlib.pyplot as plt
2  import numpy as np
3
4  import torch
5  import torchvision
6  import torchvision.transforms as transforms
7
8  import torch.nn as nn
9  import torch.nn.functional as F
10 import torch.optim as optim
```

(2) 使用内建数据集 FashionMNIST，建立 transform、trainset、trainloader，设定批量 =4。

```python
1   # transforms
2   transform = transforms.Compose(
3       [transforms.ToTensor(),
4       transforms.Normalize((0.5,), (0.5,))])
5   
6   # datasets
7   trainset = torchvision.datasets.FashionMNIST('.',
8       download=True,
9       train=True,
10      transform=transform)
11  testset = torchvision.datasets.FashionMNIST('.',
12      download=True,
13      train=False,
14      transform=transform)
15  
16  # dataloaders
17  trainloader = torch.utils.data.DataLoader(trainset, batch_size=4,
18                                          shuffle=True, num_workers=2)
19  
20  
21  testloader = torch.utils.data.DataLoader(testset, batch_size=4,
22                                          shuffle=False, num_workers=2)
```

(3) 设定 log 目录，开启 log 文件。

```python
1   from torch.utils.tensorboard import SummaryWriter
2   
3   # 设定工作记录文件目录
4   writer = SummaryWriter('runs/fashion_mnist_experiment_1')
```

(4) 写入图片。

```python
1   # 读取数据
2   dataiter = iter(trainloader)
3   images, labels = dataiter.next()
4   
5   # 建立图像方格
6   img_grid = torchvision.utils.make_grid(images)
7   
8   # 写入 tensorboard
9   writer.add_image('four_fashion_mnist_images', img_grid)
```

(5) 这时可先启动 TensorBoard，观看执行结果。单击 "IMAGES" 标签，如果没有出现标签，可至下拉式菜单单击，出现 4 张图像。

启动 TensorBoard：需指定 Log 目录，如下。

tensorboard --logdir=runs

也可以在 Jupyter Notebook 启动：先加载 TensorBoard notebook 扩充程序 (Extension)，即可在 Jupyter notebook 启动 TensorBoard。

%load_ext tensorboard

%tensorboard --logdir=runs

启动后即可使用网页浏览器观看：http://localhost:6006/。

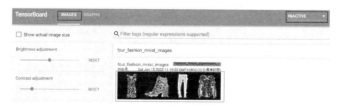

(6) 语音也可以写入 Log，要注意下列事项：

需另外安装套件，以支持语音处理，Windows 操作系统要安装 PySoundFile，Linux 操作系统则要安装 sox。

使用 DataLoader，要读取一批多笔语音数据时，要注意每一笔数据要等长，否则 next 指令会出错，故常设定 batch_size=1。

语音写入 Log，要加参数采样率 (sample_rate)，否则播放会变异音：writer.add_audio('audio', waveform, sample_rate=sample_rate)。

(7) 下载语音数据集：multiprocessing.Process 可平行下载 yes/no 内建数据集。

```
1  import torchaudio
2  import os
3  import multiprocessing
4
5  # 建立目录
6  _SAMPLE_DIR = "_sample_data"
7  YESNO_DATASET_PATH = os.path.join(_SAMPLE_DIR, "yes_no")
8  os.makedirs(YESNO_DATASET_PATH, exist_ok=True)
9
10 # 读取数据
11 def _download_yesno():
12     if os.path.exists(os.path.join(YESNO_DATASET_PATH, "waves_yesno.tar.gz")):
13         return
14     torchaudio.datasets.YESNO(root=YESNO_DATASET_PATH, download=True)
15
16 YESNO_DOWNLOAD_PROCESS = multiprocessing.Process(target=_download_yesno)
17 YESNO_DOWNLOAD_PROCESS.start()
18 YESNO_DOWNLOAD_PROCESS.join()
```

(8) 语音写入 Log：含播放语音。

```
1  from IPython.display import Audio, display
2
3  # 播放语音函数
4  def play_audio(waveform, sample_rate):
5      waveform = waveform.numpy()
6
7      num_channels, num_frames = waveform.shape
8      if num_channels == 1: # 单声道
9          display(Audio(waveform[0], rate=sample_rate))
10     elif num_channels == 2: # 立体声道
11         display(Audio((waveform[0], waveform[1]), rate=sample_rate))
12
13 # 读取语音数据集
14 dataset = torchaudio.datasets.YESNO(YESNO_DATASET_PATH, download=True)
15
16 # 读取 3 笔数据
17 for i in [1, 3, 5]:
18     waveform, sample_rate, label = dataset[i]
19     # 写入 tensorboard
20     writer.add_audio('audio_'+str(i), waveform, sample_rate=sample_rate)
21     # 播放语音
22     play_audio(waveform, sample_rate)
```

(9) 使用 TensorBoard，单击"GRAPHS"标签，观看执行结果，可单击"Play"(三角形符号) 播放语音，每个文件有 8 个音频。

(10) 若使用 DataLoader 将语音写入 Log，程序代码如下。

```
1  # datasets
2  trainset = torchaudio.datasets.YESNO(YESNO_DATASET_PATH,
3      download=True)
4
5  # dataloaders, batch_size必须为1, 否则 next 会出错, 因为每笔语音长度不一致
6  trainloader = torch.utils.data.DataLoader(trainset, batch_size=1,
7                                            shuffle=True)
```

(11) 将语音写入 Log：注意需加 [0]，因 next() 回传的是数组。

```
1  # 读取数据
2  dataiter = iter(trainloader)
3  # 下一行会出错, 因为每笔语音长度不一致, 可能要使用 transform
4  waveform, sample_rate, label = dataiter.next()
5
6  # 写入 TensorBoard
7  writer.add_audio('audio', waveform[0], sample_rate=sample_rate.numpy()[0])
```

(12) 模型也可以写入 Log：先建立模型。

```
1  class Net(nn.Module):
2      def __init__(self):
3          super(Net, self).__init__()
4          self.conv1 = nn.Conv2d(1, 6, 5)
5          self.pool = nn.MaxPool2d(2, 2)
6          self.conv2 = nn.Conv2d(6, 16, 5)
7          self.fc1 = nn.Linear(16 * 4 * 4, 120)
8          self.fc2 = nn.Linear(120, 84)
9          self.fc3 = nn.Linear(84, 10)
10
11     def forward(self, x):
12         x = self.pool(F.relu(self.conv1(x)))
13         x = self.pool(F.relu(self.conv2(x)))
14         x = x.view(-1, 16 * 4 * 4)
15         x = F.relu(self.fc1(x))
16         x = F.relu(self.fc2(x))
17         x = self.fc3(x)
18         return x
19
20 net = Net()
```

(13) 写入模型：第一个参数为模型对象，第二个参数为模型输入。

```
1  writer.add_graph(net, images)
```

(14) 使用 TensorBoard，单击"GRAPHS"标签，观看执行结果。
双击 (Double click) Net 方块，可看到详细模型结构。
双击 (Double click) input/output 方块，可看到输入 / 输出规格。

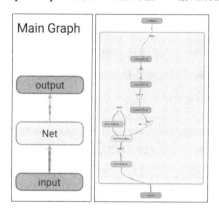

(15) 显示嵌入向量投影机 (Projector)：将影像转换为向量，连同类别名称一并写入 Log。不过，指令 writer.add_embedding 会发生错误，需先用下列指令修正。

```
1  # 修正 writer.add_embedding 错误
2  import tensorflow as tf
3  import tensorboard as tb
4  tf.io.gfile = tb.compat.tensorflow_stub.io.gfile
```

(16) 随机抽样 100 笔数据，转为二维向量，写入 Log。

```
1  # 随机抽样函数
2  def select_n_random(data, labels, n=100):
3      perm = torch.randperm(len(data))
4      return data[perm][:n], labels[perm][:n]
5
6  # 随机抽样
7  images, labels = select_n_random(trainset.data, trainset.targets)
8
9  # 类别名称
```

```
10  classes = ('T-shirt/top', 'Trouser', 'Pullover', 'Dress', 'Coat',
11             'Sandal', 'Shirt', 'Sneaker', 'Bag', 'Ankle Boot')
12
13  # 转换类别名称
14  class_labels = [classes[lab] for lab in labels]
15
16  # 转为二维向量,以利显示
17  features = images.view(-1, 28 * 28)
18
19  # 将 embeddings 写入 Log
20  writer.add_embedding(features, metadata=class_labels,
21                       label_img=images.unsqueeze(1))
```

单击"PROJECTOR"标签,观看执行结果:可看到相同类别的影像会聚集在一起。

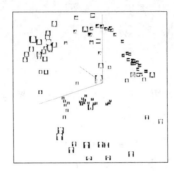

(17) 最后记得将缓冲区清空,并关闭 Log。

```
1  writer.flush()
2  writer.close()
```

**范例2.** 记录训练过程的损失:以 MNIST 辨识作测试,前面加载数据与建立模型程序代码的流程不变,在训练时将步骤序号及损失写入 Log。以下仅列出关键的程序代码。

下列程序代码请参考【05_06_ 手写阿拉伯数字辨识 _TensorBoard.ipynb】。

(1) 训练时将步骤序号 (n) 及损失 (loss) 写入 Log。writer.add_scalar 可写入单一变数,第一个参数为变量名称,第二个参数为变数值。

```
25   # 将损失写入Log
26   n+=1
27   writer.add_scalar("Loss/train", loss, n)
```

(2) 启动 TensorBoard:tensorboard --logdir=runs_2。
(3) 启动后即可单击"SCALARS"标签,使用网页浏览器观看:http://localhost:6006/。

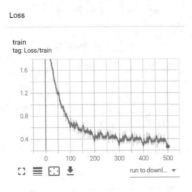

TensorBoard 随着时间增加的功能越来越多,以上我们只做了很简单的试验,如果需要更详细的信息,可以参阅 TensorBoard 官网的指南 [7]。

TensorFlow 与 TensorBoard 完全整合,不过 PyTorch 并没有提供所有的功能。

## 5-3 模型部署与 TorchServe

一般深度学习的模型安装的选项如下:

(1) 本地服务器 (Local Server)。

(2) 云端服务器 (Cloud Server)。

(3) 边缘运算 (IoT Hub):譬如要侦测全省的温度,我们会在各县市安装上千个传感器,每个 IoT Hub 会负责多个传感器的信号接收、初步过滤和分析,分析完成后再将数据送到数据中心。

呈现的方式可能是网页、手机 App 或桌面程序,以下先就网页开发作一说明。

### 5-3-1 自行开发网页程序

若是自行开发网页程序,并且安装在本地服务器的话,可以运用 Python 套件,例如 Django、Flask 或 Streamlit,快速建立网页。其中 Streamlit 最为简单,不需要懂 HTML/CSS/Javascript,只靠 Python 一招半式就可以搞定一个初阶的网站,以下我们实际建立一个手写阿拉伯数字的辨识网站。

(1) 安装 Streamlit 套件:pip install streamlit。

(2) 执行此 Python 程序,必须以 streamlit run 开头,而非 python 执行,例如,streamlit run 05_07_web.py。

(3) 网页显示后,拖曳 myDigits 目录内的任一文件至画面中的上传图片文件区域,就会显示辨识结果,也可以使用绘图软件书写数字。

程序代码说明如下。

**完整程序请参阅【05_07_web.py】。**

(1) 加载相关套件。

```
3  import streamlit as st
4  from skimage import io
5  from skimage.transform import resize
6  import numpy as np
7  import torch
```

(2) 模型加载：其中 @st.cache 可以将模型存储至快取 (Cache)，避免每一次请求都至硬盘读取，拖慢预测速度。

```
9   # 模型载入
10  device = torch.device("cuda" if torch.cuda.is_available() else "cpu")
11  @st.cache(allow_output_mutation=True)
12  def load_model():
13      return torch.load('./model.pt').to(device)
14
15  model = load_model()
```

(3) 上传图文件。

```
20  # 上传图文件
21  uploaded_file = st.file_uploader("上传图片(.png)", type="png")
```

(4) 文件上传后，执行下列工作。

第 24~28 行：把图像缩小成宽高各为 (28, 28)。

第 31 行：RGB 的白色为 255，但训练数据 MNIST 的白色为 0，故需反转颜色。

第 34 行：辨识上传文件。

```
22  if uploaded_file is not None:
23      # 读取上传图片文件
24      image1 = io.imread(uploaded_file, as_gray=True)
25
26      # 缩为 (28, 28) 大小的影像
27      image_resized = resize(image1, (28, 28), anti_aliasing=True)
28      X1 = image_resized.reshape(1,28, 28) #/ 255.0
29
30      # 反转颜色，颜色0为白色，与 RGB 色码不同，它的 0 为黑色
31      X1 = torch.FloatTensor(1-X1).to(device)
32
33      # 预测
34      predictions = torch.softmax(model(X1), dim=1)
35
36      # 显示预测结果
37      st.write(f'### 预测结果:{np.argmax(predictions.detach().cpu().numpy())}')
38
39      # 显示上传图片文件
40      st.image(image1)
```

## 5-3-2 TorchServe

TorchServe 与 TensorFlow Serving 类似，直接提供一个具有弹性且强大的网页服务，支持平行处理、分散处理及批处理，架构如图 5.1 所示。

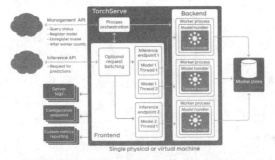

图 5.1 TorchServe 系统架构，图片来源：TorchServe GitHub [8]

TorchServe 安装非常复杂，笔者花了一天才测试成功，详细说明可参阅 TorchServe 官网[9]。依照 TorchServe GitHub 说明，安装及使用程序如下。

(1) 安装 Java Run Time(JRE)，必须为 v.11 版以上 (https://download.java.net/java/GA/jdk11/9/GPL/openjdk-11.0.2_windows-x64_bin.zip)，并在环境变量 Path 中加入 C:\Program Files\Java\jdk-11.0.2\bin，在环境变量 CLASSPATH 中加入 C:\Program Files\Java\jdk-11.0.2\lib。

(2) 安装 TorchServe，指令如下。**注意，官网漏列 captum 套件安装，未安装时会出现错误信息"Load model failed"**：

pip install torchserve torch-model-archiver torch-workflow-archiver captum

(3) 下载 TorchServe GitHub 原始码：https://github.com/pytorch/serve.git。

(4) 复制模型：建立 model_store 子目录，自 https://download.pytorch.org/models/densenet161-8d451a50.pth 下载训练好的 DenseNet 模型至 model_store 子目录，注意，必须是 torch.save(model.state_dict(), "model.pt")，而非 torch.save(model, 'model.pt')。

(5) 在 serve 目录下开启 cmd 或终端机，产生模型存盘 (Archive)：会产生 densenet161.mar 文件，笔者已将下列指令存成 archive.bat。

torch-model-archiver --force --model-name densenet161 --version 1.0 --model-file examples\image_classifier\densenet_161\model.py --serialized-file model_store\densenet161-8d451a50.pth --extra-files examples\image_classifier\index_to_name.json --handler image_classifier --export-path=model_store

(6) 启动 TorchServe 服务器端：注意有无错误信息，默认会启动三个 worker，笔者已将下列指令存成 run.bat，参数 --ncs 会忽略上次执行的组态文件，预设为背景 (Background) 执行：

torchserve --start --ncs --model-store model_store --models densenet161.mar

(7) 再安装 Google RPC (GRPC) 相关套件，以利联机 TorchServe 服务器：

pip install -U grpcio protobuf grpcio-tools

(8) 产生 GRPC 客户端组态文件，笔者已将下列指令存成 generate_inference_client.bat：

python -m grpc_tools.protoc --proto_path=frontend/server/src/main/resources/proto/ --python_out=ts_scripts --grpc_python_out=ts_scripts frontend/server/src/main/resources/proto/inference.proto frontend/server/src/main/resources/proto/management.proto

(9) 预测：指定一张图片 (kitten.jpg)，送出请求，笔者已将下列指令存成 inference.bat：

python ts_scripts/torchserve_grpc_client.py infer densenet161 examples/image_classifier/kitten.jpg

执行结果：如下，预测为虎斑猫 (tabby)，概率为 46%。

{
 "tabby": 0.4666188061237335,
 "tiger_cat": 0.46449077129364014,
 "Egyptian_cat": 0.0661403015255928,
 "lynx": 0.0012924385955557227,
 "plastic_bag": 0.00022909721883479506
}

(10) 停止 TorchServe 服务器执行。

torchserve --stop

**范例. 使用模型辨识MNIST，说明如何运作自定义的模型服务。**

程序参阅 serve\examples\image_classifier\mnist\README.md，整理如下，相关文件均位于 serve\examples\image_classifier\mnist 目录：

(1) 建立模型结构：内容如 mnist.py。

(2) 准备训练好的模型文件：mnist_cnn.pt。

(3) 输入数据处理：内容如 mnist_handler.py。

(4) 产生模型存盘 (Archive)：会产生 mnist.mar 文件。

torch-model-archiver --model-name mnist --version 1.0 --model-file examples/image_classifier/mnist/mnist.py --serialized-file examples/image_classifier/mnist/mnist_cnn.pt --handler examples/image_classifier/mnist/mnist_handler.py --export-path=model_store

(5) 启动 TorchServe 服务器端：注意有无错误信息。

torchserve --start --ncs --model-store model_store --models mnist.mar

(6) 预测：指定一张图片 (2.png)，送出请求。

python ts_scripts/torchserve_grpc_client.py infer mnist examples\image_classifier\mnist\test_data\2.png

执行结果：2，辨识无误，如果要使用自己的图像，注意要黑白反转。

由上述范例可以看出 Server 端完全不必编写程序，非常方便。但是，前置安装要细心处理，忽略一个步骤，可能就会发生错误，serve\examples 目录下还有很多的范例模型，读者可以自己试试看。

# 第 6 章
# 卷积神经网络

第三波人工智能浪潮在自然用户接口 (Natural User Interface, NUI) 上有突破性的进展，包括影像 (Image、Video)、语音 (Voice) 与文字 (Text) 的辨识 / 生成 / 分析，机器学会人类日常生活中所使用的沟通方式，与使用者的互动不仅更具亲和力，也能对周围的环境做出更合理、更有智慧的判断与反应。将这种能力附加到产品上，可使产品的应用发展爆发无限可能，包括自动驾驶 (Self-Driving)、无人机 (Drone)、智能家居 (Smart Home)、制造 / 服务机器人 (Robot)、聊天机器人 (ChatBot) 等，不胜枚举。

从这一章开始，我们逐一来探讨影像 (Image、Video)、语音 (Voice)、文字 (Text) 的相关算法。

## 6-1 卷积神经网络简介

之前程序辨识手写阿拉伯数字，是使用像素 (Pixel) 作为特征，与人类辨识图形的方式有所差异，我们通常不会逐点辨识图形内的数字，以像素辨识图形有以下缺点。

(1) 手写阿拉伯数字，通常都会将字写在中央，所以中央的像素重要性应远大于周边的像素。

(2) 像素之间有所关联，而非互相独立，比如 1，为一垂直线。

(3) 人类辨识数字应该是观察线条或轮廓，而非逐个像素检视。

因此，卷积神经网络 (Convolutional Neural Network, CNN) 引进了卷积层 (Convolution Layer)，先进行特征提取 (Feature Extraction)，将像素转换为各种线条特征，再交给完全连接层 (Linear) 辨识，也就是图 1.7 机器学习流程的第 3 步骤——特征工程 (Feature Engineering)。

卷积 (Convolution) 简单说就是将图形抽样化 (Abstraction)，把不必要的信息删除，例如色彩、背景等，图 6.1 经过三层卷积后，有些图依稀可辨识出人脸的轮廓了，因此，模型即可依据这些线条辨识出是人、车或其他动物。

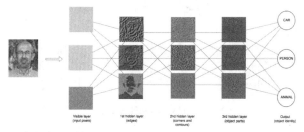

图 6.1　卷积神经网络 (Convolutional Neural Network, CNN) 的特征提取

卷积神经网络 (Convolutional Neural Network)，以下简称 CNN，它的模型结构如图 6.2 所示。

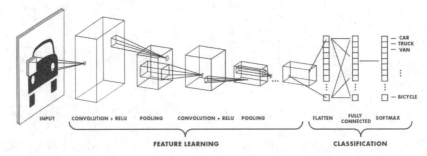

图 6.2　卷积神经网络 (Convolutional Neural Network, CNN) 的模型结构

(1) 先输入一张图像，可以是彩色的，每个色彩通道 (Channel) 分别卷积再合并。

(2) 图像经过卷积层 (Convolution Layer) 运算，变成特征图 (Feature Map)，卷积可以指定很多个，卷积矩阵内容不是预定的，而是在训练过程中由反向传导推估出来的，这与传统的图像处理不同，另外，卷积层后面通常会附加 ReLU Activation Function。

(3) 卷积层后面还会接一个池化层 (Pooling)，作下采样 (Down Sampling)，以降低模型的参数个数，避免模型过于庞大。

(4) 最后把特征图 (Feature Map) 压扁 (Flatten) 成一维，交给完全连接层辨识。

## 6-2　卷积

卷积定义一个滤波器 (Filter) 或称卷积核 (Kernel)，对图像进行乘积和运算，如图 6.3 所示，计算步骤如下：

(1) 将输入图像依照滤波器裁切相同尺寸的部分图像。
(2) 裁切的图像与滤波器相同的位置进行相乘。
(3) 加总所有格的数值，即为输出的第一格数值。
(4) 逐步向右滑动窗口 ( 如图 6.4)，回到步骤 (1)，计算下一格的值。
(5) 滑到最右边后，再往下滑动窗口，继续进行。

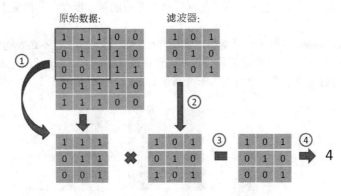

图 6.3　卷积计算 (1)

第 6 章 | 卷积神经网络

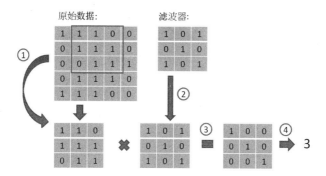

图 6.4　卷积计算 (2)

网络上有许多动画或影片可以参考，例如 *Convolutional Neural Networks—Simplified* [1] 一文中卷积计算的 GIF 动画 [2]。

**范例 1. 使用程序计算卷积。**

**下列程序代码请参考【06_01_convolutions.ipynb】。**

(1) 准备数据及滤波器 (Filter)。

```python
import numpy as np

# 测试数据
source_map = np.array(list('1110001110001110011001100')).astype(np.int)
source_map = source_map.reshape(5,5)
print('原始数据:')
print(source_map)

# 滤波器(Filter)
filter1 = np.array(list('101010101')).astype(np.int).reshape(3,3)
print('\n滤波器:')
print(filter1)
```

执行结果。

```
原始数据：
[[1 1 1 0 0]
 [0 1 1 1 0]
 [0 0 1 1 1]
 [0 0 1 1 0]
 [0 1 1 0 0]]

滤波器：
[[1 0 1]
 [0 1 0]
 [1 0 1]]
```

(2) 计算卷积。

```python
# 计算卷积
# 初始化计算结果的矩阵
width = height = source_map.shape[0] - filter1.shape[0] + 1
result = np.zeros((width, height))

# 计算每一格
for i in range(width):
    for j in range(height):
        value1 =source_map[i:i+filter1.shape[0], j:j+filter1.shape[1]] * filter1
        result[i, j] = np.sum(value1)
print(result)
```

执行结果。

```
[4. 3. 4.]
[2. 4. 3.]
[2. 3. 4.]
```

(3) 使用 SciPy 套件提供的卷积函数验算，执行结果一致。

```
1  # 使用 SciPy 计算卷积
2  from scipy.signal import convolve2d
3
4  # convolve2d：二维卷积
5  convolve2d(source_map, filter1, mode='valid')
```

卷积计算时，其实还有两个参数：

(1) 补零 (Padding)：上面的卷积计算会使得图像尺寸变小，因为，滑动窗口时，裁切的窗口会不足 2 个 ( 滤波器宽度 3-1=2)，PyTorch 预设为不补零，即图像尺寸会变小，若要补零直接指定个数即可。反观 TensorFlow 的 Padding 只有两个选项：

Padding='same'：在图像周围补上不足的列与行，使计算结果的矩阵尺寸不变(same)，与原始图像尺寸相同，如图 6.5 所示。

Padding='valid'：不补零，即 Padding=0。

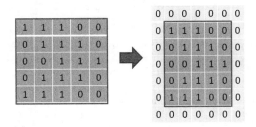

图 6.5　Padding='same'，在图像周围补上不足的列与行

(2) 滑动窗口的步数 (Stride)：图 6.4 是 Stride=1，图 6.6 是 Stride=2，可减少要估算的参数个数。

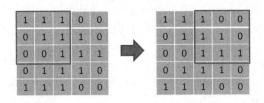

图 6.6　Stride=2，一次滑动 2 格窗口

以上是二维的卷积 (Conv2d) 的运作，通常应用在图像上。PyTorch/TensorFlow 还提供 Conv1d、Conv3d，其中 Conv1d 因只考虑上下文 (Context Sensitive)，所以可应用于语音或文字方面，Conv3d 则可应用于立体的对象。还有 nn.ConvTransposeNd 提供反卷积 (Deconvolution) 或称上采样 (Up Sampling) 的功能，由特征图重建图像。卷积和反卷积两者相结合，可以组合成 AutoEncoder 模型，它是许多生成模型的基础算法，可以去除噪声，生成干净的图像。

## 6-3 各种卷积

虽然 CNN 会自动配置卷积的种类，不过我们还是来看看各种卷积的图像处理效果，进而加深大家对 CNN 的理解。

**下列程序代码请参考【06_01_convolutions.ipynb】后半部。**

(1) 首先定义一个卷积的影像转换函数，代码如下。

```python
1   # 卷积的影像转换函数，padding='same'
2   from skimage.exposure import rescale_intensity
3
4   def convolve(image, kernel):
5       # 取得图像与滤波器的宽高
6       (iH, iW) = image.shape[:2]
7       (kH, kW) = kernel.shape[:2]
8
9       # 计算 padding='same' 单边所需的补零行数
10      pad = int((kW - 1) / 2)
11      image = cv2.copyMakeBorder(image, pad, pad, pad, pad, cv2.BORDER_REPLICATE)
12      output = np.zeros((iH, iW), dtype="float32")
13
14      # 卷积
15      for y in np.arange(pad, iH + pad):
16          for x in np.arange(pad, iW + pad):
17              roi = image[y - pad:y + pad + 1, x - pad:x + pad + 1]   # 裁切图像
18              k = (roi * kernel).sum()                                 # 卷积计算
19              output[y - pad, x - pad] = k                             # 更新计算结果的矩阵
20
21      # 调整影像色彩深浅范围至 (0, 255)
22      output = rescale_intensity(output, in_range=(0, 255))
23      output = (output * 255).astype("uint8")
24
25      return output      # 回传结果影像
```

(2) 需要安装 Python OpenCV 套件，OpenCV 是一个图像处理的套件。
pip install opencv-python

(3) 将影像灰度化：skimage 全名为 scikit-image，也是一个图像处理的套件，功能较 OpenCV 简单。

```python
1   # pip install opencv-python
2   import skimage
3   import cv2
4
5   # 自 skimage 取得内建的图像
6   image = skimage.data.chelsea()
7   cv2.imshow("original", image)
8
9   # 灰阶化
10  gray = cv2.cvtColor(image, cv2.COLOR_BGR2GRAY)
11  cv2.imshow("gray", gray)
12
13  # 按 Enter 键关闭窗口
14  cv2.waitKey(0)
15  cv2.destroyAllWindows()
```

执行结果。
原图：

灰度化：

(4) 模糊化 (Blur)：滤波器设定为周围点的平均，就可以让图像模糊化，一般用于消除红眼现象或噪声。

```
1   # 小模糊 filter
2   smallBlur = np.ones((7, 7), dtype="float") * (1.0 / (7 * 7))
3
4   # 卷积
5   convoleOutput = convolve(gray, smallBlur)
6   opencvOutput = cv2.filter2D(gray, -1, smallBlur)
7   cv2.imshow("little Blur", convoleOutput)
8
9   # 大模糊
10  largeBlur = np.ones((21, 21), dtype="float") * (1.0 / (21 * 21))
11
12  # 卷积
13  convoleOutput = convolve(gray, largeBlur)
14  opencvOutput = cv2.filter2D(gray, -1, largeBlur)
15  cv2.imshow("large Blur", convoleOutput)
16
17  # 按 Enter 键关闭窗口
18  cv2.waitKey(0)
19  cv2.destroyAllWindows()
```

小模糊：7×7 矩阵。

大模糊：21×21 矩阵，滤波器尺寸越大，影像越模糊。

(5) 锐化 (sharpen)：可使图像的对比更加明显。

```
1   # sharpening filter
2   sharpen = np.array((
3       [0, -1, 0],
4       [-1, 5, -1],
5       [0, -1, 0]), dtype="int")
6
```

```
 7  # 卷积
 8  convoleOutput = convolve(gray, sharpen)
 9  opencvOutput = cv2.filter2D(gray, -1, sharpen)
10  cv2.imshow("sharpen", convoleOutput)
11
12  # 按 Enter 键关闭窗口
13  cv2.waitKey(0)
14  cv2.destroyAllWindows()
```

执行结果：卷积凸显中间点，使图像特征越明显。

(6) Laplacian 边缘侦测：可侦测图像的轮廓。

```
 1  # Laplacian filter
 2  laplacian = np.array((
 3      [0, 1, 0],
 4      [1, -4, 1],
 5      [0, 1, 0]), dtype="int")
 6
 7  # 卷积
 8  convoleOutput = convolve(gray, laplacian)
 9  opencvOutput = cv2.filter2D(gray, -1, laplacian)
10  cv2.imshow("laplacian edge detection", convoleOutput)
11
12  # 按 Enter 键关闭窗口
13  cv2.waitKey(0)
14  cv2.destroyAllWindows()
```

执行结果：卷积凸显边缘，显现图像外围线条。

(7) Sobel X 轴边缘侦测：沿着 X 轴侦测边缘，故可侦测垂直线特征。

```
 1  # Sobel x-axis filter
 2  sobelX = np.array((
 3      [-1, 0, 1],
 4      [-2, 0, 2],
 5      [-1, 0, 1]), dtype="int")
 6
 7  # 卷积
 8  convoleOutput = convolve(gray, sobelX)
 9  opencvOutput = cv2.filter2D(gray, -1, sobelX)
10  cv2.imshow("x-axis edge detection", convoleOutput)
11
12  # 按 Enter 键关闭窗口
13  cv2.waitKey(0)
14  cv2.destroyAllWindows()
```

执行结果：卷积行由小至大，显现图像垂直线条。

(8) Sobel Y 轴边缘侦测：沿着 Y 轴侦测边缘，故可侦测水平线特征。

```
1  # Sobel y-axis filter
2  sobelY = np.array((
3      [-1, -2, -1],
4      [0, 0, 0],
5      [1, 2, 1]), dtype="int")
6  
7  # 卷积
8  convoleOutput = convolve(gray, sobelY)
9  opencvOutput = cv2.filter2D(gray, -1, sobelY)
10 cv2.imshow("y-axis edge detection", convoleOutput)
11 
12 # 按 Enter 键关闭窗口
13 cv2.waitKey(0)
14 cv2.destroyAllWindows()
```

执行结果：卷积列由小至大，显现图像水平线条。

# 6-4 池化层

通常卷积层的滤波器个数会设定为 4 的倍数，总输出 = 笔数 × W_out × H_out × 滤波器个数，会使输出尺寸变得很大，因此会通过池化层 (Pooling Layer) 进行下采样 (Down Sampling)，只取滑动窗口内的最大值或平均值，换句话说，就是将整个滑动窗口转化为一个点，这样就能有效降低每一层输入的尺寸，同时也能保有每个窗口的特征。我们来举个例子说明会比较清楚。

以最大池化层 (Max Pooling) 为例：

(1) 图 6.7 左边为原始图像。
(2) 假设滤波器尺寸为 (2, 2)、Stride = 2。
(3) 滑动窗口取 (2, 2)，如图 6.7 左上角的框，取最大值 =6。
(4) 接着再滑动 2 步，如图 6.8，取最大值 =8。

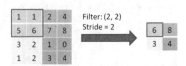

图 6.7　最大池化层 (Max Pooling)

图 6.8 最大池化层——滑动 2 步

## 6-5 CNN 模型实操

一般卷积会采用 3×3 或 5×5 的滤波器，尺寸越大，可以提取越大的特征，但相对地，较小的特征就容易被忽略。而池化层通常会采用 2×2，stride=2 的滤波器，使用越大的尺寸，会使得参数个数减少很多，但提取到的特征也相对减少。

以下就先以 CNN 模型实操 MNIST 辨识。

**范例1. 将手写阿拉伯数字辨识的模型改用CNN。**

**完整程序请参阅【06_02_MNIST_CNN.ipynb】，以下仅介绍差异的程序代码。**

(1) 改用 CNN 模 型：使用两组 Conv2d/MaxPool2d，比较特别的是使用 2 个 nn.Sequential 各包一组 Conv2d/MaxPool2d。Conv2d 参数依次如下：

in-channel：输入通道数，第一个Conv2d要指定颜色数，单色为1，彩色为3(R/G/B)。
out-channel：输出通道数，为要产生的滤波器个数。
kernel size：滤波器尺寸。
Stride：滑动的步数。
Padding：补零的行数。

```python
1   # 建立模型
2   class ConvNet(nn.Module):
3       def __init__(self, num_classes=10):
4           super(ConvNet, self).__init__()
5           self.layer1 = nn.Sequential(
6               # Conv2d 参数：in-channel, out-channel, kernel size, Stride, Padding
7               nn.Conv2d(1, 16, kernel_size=5, stride=1, padding=2),
8               nn.BatchNorm2d(16),
9               nn.ReLU(),
10              nn.MaxPool2d(kernel_size=2, stride=2))
11          self.layer2 = nn.Sequential(
12              nn.Conv2d(16, 32, kernel_size=5, stride=1, padding=2),
13              nn.BatchNorm2d(32),
14              nn.ReLU(),
15              nn.MaxPool2d(kernel_size=2, stride=2))
16          self.fc = nn.Linear(7*7*32, num_classes)
17  
18      def forward(self, x):
19          out = self.layer1(x)
20          out = self.layer2(out)
21          out = out.reshape(out.size(0), -1)
22          out = self.fc(out)
23          out = F.log_softmax(out, dim=1)
24          return out
25  
26  model = ConvNet().to(device)
```

(2) 之前讲过 PyTorch 的完全连接层需要指定输入及输出参数，这时就麻烦了，从卷积/池化层输出的神经元个数是多少呢？亦即第16行的"7*7*32"是怎么计算出来的？因 Padding/Stride 设定，输出会有所差异，因此，我们必须仔细计算，才能作为下一个完全连接层的输入参数。

卷积层输出图像宽度公式如下：
$$W\_out = (W-F+2P)/S+1$$

其中：

$W$：输入图像宽度。

$F$：滤波器 (Filter) 宽度。

$P$：补零的行数 (Padding)。

$S$：滑动的步数 (Stride)。

池化层输出图像宽度的公式与卷积层相同。

卷积/池化层输出神经元的公式：

$$输出图像宽度 \times 输出图像高度 \times 滤波器个数$$

TensorFlow 完全连接层只须填输出参数，完全没有这方面的困扰。

笔者将相关计算公式写成多个函数如下。

```python
# 卷积/池化层公式计算
import math

# W, F, P, S: image Width, Filter width, Padding, Stride
def Conv_Width(W, F, P, S):
    return math.floor(((W - F + 2 * P) / S) + 1)

def Conv_Output_Volume(W, F, P, S, out):
    return Conv_Width(W, F, P, S) ** 2 * out

# C: no of channels
def Conv_Parameter_Count(F, C, out):
    return F ** 2 * C * out

def Pool_Width(W, F, P, S):
    return Conv_Width(W, F, P, S)

# filter_count: no of filter in last conv
# stride count default value = Filter width
def Pool_Output_Volume(W, F, P, S, filter_count):
    return Conv_Output_Volume(W, F, P, S, filter_count)

def Pool_Parameter_Count(W, F, S):
    return 0
```

(3) 测试。

```python
# Conv2d/MaxPool2d/Conv2d/MaxPool2d
c1_Width = Conv_Width(28, 5, 2, 1)
p1_Width = Pool_Width(c1_Width, 2, 0, 2)
c2_Width = Conv_Width(p1_Width, 5, 2, 1)
p2_out = Pool_Output_Volume(c2_Width, 2, 0, 2, 32)
p2_out, 7*7*32
```

执行结果：1568，即 $7 \times 7 \times 32$。

如果模型定义错误，会产生错误信息，同时也会出现正确的输出个数，届时再依据错误信息更正也可以，这是投机的小技巧，例如：

RuntimeError: mat1 and mat2 shapes cannot be multiplied (1000x12544 and 9216x128)

表示要把 9216 改成 12544。

(4) 验证：显示模型的汇总信息，内含各层输入及输出参数。

```python
# 显示模型的汇总信息
for name, module in model.named_children():
    print(f'{name}: {module}')
```

执行结果：如下程序代码最后一行，1568，即 7×7×32。

```
layer1: Sequential(
  (0): Conv2d(1, 16, kernel_size=(5, 5), stride=(1, 1), padding=(2, 2))
  (1): BatchNorm2d(16, eps=1e-05, momentum=0.1, affine=True, track_running_stats=True)
  (2): ReLU()
  (3): MaxPool2d(kernel_size=2, stride=2, padding=0, dilation=1, ceil_mode=False)
)
layer2: Sequential(
  (0): Conv2d(16, 32, kernel_size=(5, 5), stride=(1, 1), padding=(2, 2))
  (1): BatchNorm2d(32, eps=1e-05, momentum=0.1, affine=True, track_running_stats=True)
  (2): ReLU()
  (3): MaxPool2d(kernel_size=2, stride=2, padding=0, dilation=1, ceil_mode=False)
)
fc: Linear(in_features=1568, out_features=10, bias=True)
```

(5) 测试数据评分。

```
1  # 建立 DataLoader
2  test_loader = DataLoader(test_ds, shuffle=False, batch_size=BATCH_SIZE)
3
4  model.eval()
5  test_loss = 0
6  correct = 0
7  with torch.no_grad():
8      for data, target in test_loader:
9          data, target = data.to(device), target.to(device)
10         output = model(data)
11
12         # sum up batch loss
13         test_loss += F.nll_loss(output, target).item()
14
15         # 预测
16         output = model(data)
17
18         # 计算正确数
19         _, predicted = torch.max(output.data, 1)
20         correct += (predicted == target).sum().item()
21
22 # 平均损失
23 test_loss /= len(test_loader.dataset)
24 # 显示测试结果
25 batch = batch_idx * len(data)
26 data_count = len(test_loader.dataset)
27 percentage = 100. * correct / data_count
28 print(f'平均损失: {test_loss:.4f}, 准确率: {correct}/{data_count}' +
29       f' ({percentage:.2f}%)\n')
```

执行结果：准确率约 97.11%，实际预测前 20 笔数据，完全正确。

(6) 以笔者手写的文件测试。

```
1  # 读取影像并转为单色
2  for i in range(10):
3      uploaded_file = f'./myDigits/{i}.png'
4      image1 = io.imread(uploaded_file, as_gray=True)
5
6      # 缩为 (28, 28) 大小的影像
7      image_resized = resize(image1, tuple(data_shape)[2:], anti_aliasing=True)
8      X1 = image_resized.reshape(*data_shape)
9
10     # 反转颜色，颜色0为白色，与 RGB 色码不同，它的 0 为黑色
11     X1 = 1.0-X1
12
13     X1 = torch.FloatTensor(X1).to(device)
14
15     # 预测
16     predictions = torch.softmax(model(X1), dim=1)
17     # print(np.around(predictions.cpu().detach().numpy(), 2))
18     print(f'actual/prediction: {i} {np.argmax(predictions.detach().cpu().numpy())}')
```

执行结果：比之前的效果好多了，只有数字 9 辨识错误，不过，笔者把训练周期加大为 10，因为只训练 5 周期，损失似乎还未趋于收敛。

```
actual/prediction: 0 0
actual/prediction: 1 1
actual/prediction: 2 2
actual/prediction: 3 3
actual/prediction: 4 4
actual/prediction: 5 5
actual/prediction: 6 6
actual/prediction: 7 7
actual/prediction: 8 8
actual/prediction: 9 7
```

从卷积层运算观察，CNN 模型有两个特点：

(1) 部分连接 (Locally Connected or Sparse Connectivity)：完全连接层 (Linear) 每个输入的神经元完全连接 (Full Connected) 至每个输出的神经元，但卷积层的输出神经元则只连接滑动窗口神经元，如图 6.9 所示。想象一下，假设在手臂上拍打一下，手臂以外的神经元应该不会收到信号，既然没收到信号，理所当然就不必往下一层传送信号了，所以，下一层的神经元只会收到上一层少数神经元的信号，接收到的范围称为感知域 (Reception Field)。

由于部分连接的关系，神经层中每条回归线的输入特征大幅减少，要估算的权重个数也就少了很多。

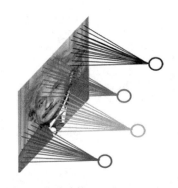

图 6.9　部分连接 (Locally Connected)

(2) 权重共享 (Weight Sharing)：单一滤波器应用到滑动窗口时，卷积矩阵值都是一样的，如图 6.10 所示，基于这个假设，要估计的权重个数就减少许多，模型复杂度因而进一步简化了。

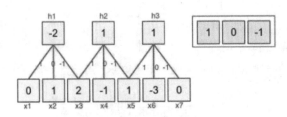

图 6.10　权重共享 (Weight Sharing)

基于以上两个假设，CNN 模型参数会比较少。

另外为什么 CNN 模型输入数据要加入色彩信道 (Channel)？这是因为有些情况加入色彩，会比较容易辨识，比如狮子毛发大部分是金黄色的，又或者侦测是否有戴口罩，只要图像上有一块白色的矩形，我们应该就能假定有戴口罩，当然目前口罩颜色已经五花八门，需要更多的训练数据，才能正确辨识。

**范例2.** 加入标准化 (Normalize) 前置处理，通常训练可以较快速收敛，且准确度可以提高。

下列程序代码请参考【06_03_MNIST_CNN_Normalize.ipynb】。

(1) 标准化：其中的平均数 / 标准偏差是依据 ImageNet 数据集统计出来的最佳值 ( 第 3 行 )，并在 DataSet 中设定转换函数 ( 第 8 行 )。

```
1  transform=transforms.Compose([
2      transforms.ToTensor(),
3      transforms.Normalize(mean=(0.1307,), std=(0.3081,))
4      ])
5
6  # 下载 MNIST 手写阿拉伯数字 训练数据
7  train_ds = MNIST(PATH_DATASETS, train=True, download=True,
8                   transform=transform)
9
10 # 下载测试数据
11 test_ds = MNIST(PATH_DATASETS, train=False, download=True,
12                  transform=transform)
13
14 # 训练/测试数据的维度
15 print(train_ds.data.shape, test_ds.data.shape)
```

(2) 以测试数据评分的准确率为 98.39%，无明显帮助，笔者写的数字还是没有全对。这部分与 TensorFlow 测试结果有所差异，TensorFlow 采取 MinMaxScaler 准确率会明显提高。

**范例3.** 前面的数据集都是单色的图像，我们也使用彩色的图像测试看看。

直接复制【06_03_MNIST_CNN_Normalize.ipynb】，稍作修改即可。

以下仅说明关键程序代码，完整程序参阅【06_04_Cifar_RGB_CNN.ipynb】。

(1) 载入套件：PyTorch 对影像、语音及文字提供个别的命名空间 (Namespace)，分别为 torchvision、torchaudio、torchtext，相关的函数和数据集都涵盖在内。

```
1  import torch
2  import torchvision
3  import torchvision.transforms as transforms
```

(2) 加载 Cifar10 数据，与单色的图像辨识相同，但不需转换为单色，标准化 (Normalize) 的平均数 / 标准偏差也可以依据 ImageNet 数据集统计出来的最佳值设定，这里单纯设定为 0.5，读者可试试有无差异。

```
1  # 数据转换
2  transform = transforms.Compose(
3      [transforms.ToTensor(),
4      # 读入图像范围介于[0, 1]之间，将其转换为 [-1, 1]
5      transforms.Normalize((0.5, 0.5, 0.5), (0.5, 0.5, 0.5))])
6      # ImageNet
7      # transforms.Normalize((0.485, 0.456, 0.406), (0.229, 0.224, 0.225))
8      ])
9
10 # 批量
11 batch_size = 1000
12
13 # 载入数据集，如果出现 BrokenPipeError 错误，将 num_workers 改为 0
14 train_ds = torchvision.datasets.CIFAR10(root='./CIFAR10', train=True,
15                  download=True, transform=transform)
16
```

```
17  train_loader = torch.utils.data.DataLoader(train_ds, batch_size=batch_size,
18                                              shuffle=True, num_workers=2)
19
20  test_ds = torchvision.datasets.CIFAR10(root='./CIFAR10', train=False,
21                                         download=True, transform=transform)
22
23  test_loader = torch.utils.data.DataLoader(test_ds, batch_size=batch_size,
24                                             shuffle=False, num_workers=2)
25
26  # 训练/测试数据的维度
27  print(train_ds.data.shape, test_ds.data.shape)
```

执行结果如下：训练数据共 50000 笔，测试数据共 10000 笔，图像尺寸为 (32, 32)，有三个颜色 (R/G/B)。

(50000, 32, 32, 3) (10000, 32, 32, 3)

(3) CIFAR 10 数据集共 10 种类别。

```
1  classes = ('plane', 'car', 'bird', 'cat',
2             'deer', 'dog', 'frog', 'horse', 'ship', 'truck')
```

(4) 观察数据内容：显示 8 张图片。

注意，自 DataLoader 读出的维度为 [8, 3, 32, 32]，颜色放在第 2 维，这是配合 Conv2d 的输入要求，要显示图像，必须将颜色放在最后一维 ( 第 9 行 )。

以 DataLoader 读取的图像范围介于 [0, 1] 之间，经过转换后数据介于 [-1, 1]，要显示图像将数据还原 ( 第 6 行 )，公式为 $(x \times \delta) + \mu$。

```
1   import matplotlib.pyplot as plt
2   import numpy as np
3
4   # 图像显示函数
5   def imshow(img):
6       img = img * 0.5 + 0.5  # 还原图像
7       npimg = img.numpy()
8       # 颜色换至最后一维
9       plt.imshow(np.transpose(npimg, (1, 2, 0)))
10      plt.axis('off')
11      plt.show()
12
13
14  # 取一笔数据
15  batch_size_tmp = 8
16  train_loader_tmp = torch.utils.data.DataLoader(train_ds, batch_size=batch_size_tmp)
17  dataiter = iter(train_loader_tmp)
18  images, labels = dataiter.next()
19  print(images.shape)
20
21  # 显示图像
22  plt.figure(figsize=(10,6))
23  imshow(torchvision.utils.make_grid(images))
24  # 显示类别
25  print(' '.join(f'{classes[labels[j]]:5s}' for j in range(batch_size_tmp)))
```

执行结果：分辨率不是很高，图像有点模糊，下面有对应的标注 (Label)。

(5) 建立 CNN 模型：第 8 行 Conv2d 第 1 个参数为 3，即颜色通道数，表 R/G/B 三颜色。

```python
class Net(nn.Module):
    def __init__(self):
        super().__init__()
        # 颜色要放在第1维，3:RGB三颜色
        self.conv1 = nn.Conv2d(3, 6, 5)
        self.pool = nn.MaxPool2d(2, 2)
        self.conv2 = nn.Conv2d(6, 16, 5)
        self.fc1 = nn.Linear(16 * 5 * 5, 120)
        self.fc2 = nn.Linear(120, 84)
        self.fc3 = nn.Linear(84, 10)

    def forward(self, x):
        x = self.pool(F.relu(self.conv1(x)))
        x = self.pool(F.relu(self.conv2(x)))
        x = torch.flatten(x, 1)
        x = F.relu(self.fc1(x))
        x = F.relu(self.fc2(x))
        x = self.fc3(x)
#       output = F.log_softmax(x, dim=1)
        return x
```

(6) 定义训练函数：同前，只是逐步将相关程序代码模块化，以提高生产力。

```python
def train(model, device, train_loader, criterion, optimizer, epoch):
    model.train()
    loss_list = []
    for batch_idx, (data, target) in enumerate(train_loader):
        data, target = data.to(device), target.to(device)

        optimizer.zero_grad()
        output = model(data)
        loss = criterion(output, target)
        loss.backward()
        optimizer.step()

        if (batch_idx+1) % 10 == 0:
            loss_list.append(loss.item())
            batch = (batch_idx+1) * len(data)
            data_count = len(train_loader.dataset)
            percentage = (100. * (batch_idx+1) / len(train_loader))
            print(f'Epoch {epoch}: [{batch:5d} / {data_count}] ' +
                  f'({percentage:.0f} %)  Loss: {loss.item():.6f}')
    return loss_list
```

(7) 定义测试函数：同前。

```python
def test(model, device, test_loader):
    model.eval()
    test_loss = 0
    correct = 0
    with torch.no_grad():
        for data, target in test_loader:
            data, target = data.to(device), target.to(device)
            output = model(data)
            _, predicted = torch.max(output.data, 1)
            correct += (predicted == target).sum().item()

    # 平均损失
    test_loss /= len(test_loader.dataset)
    # 显示测试结果
    data_count = len(test_loader.dataset)
    percentage = 100. * correct / data_count
    print(f'准确率: {correct}/{data_count} ({percentage:.2f}%)')
```

(8) 执行训练：笔者使用之前的损失函数 (F.nll_loss) 及优化器 (Adadelta)，但训练效果不佳，后来依照原文使用 nn.CrossEntropyLoss、SGD，才得到比较合理的模型。另外，调用 train 函数时，若使用 nn.CrossEntropyLoss，后面要加 ()，因为它是类别，要先建立对象才能计算损失，其他损失函数以 F. 开头的，可直接使用。

```
1  epochs = 10
2  lr=0.1
3
4  # 建立模型
5  model = Net().to(device)
6
7  # 定义损失函数
8  # 注意，nn.CrossEntropyLoss是类别，要先建立物件，要加 ()，其他损失函数不需要
9  criterion = nn.CrossEntropyLoss() # F.nll_loss
10
11 # 设定优化器(optimizer)
12 #optimizer = torch.optim.Adadelta(model.parameters(), lr=lr)
13 optimizer = torch.optim.SGD(model.parameters(), lr=lr, momentum=0.9)
14
15 loss_list = []
16 for epoch in range(1, epochs + 1):
17     loss_list += train(model, device, train_loader, criterion, optimizer, epoch)
18     #test(model, device, test_loader)
19     #optimizer.step()
```

(9) 评分。

```
1  test(model, device, test_loader)
```

执行结果：准确率为 57.31%，比单色图像辨识准确率低很多，因为背景图案不是单纯的白色。

(10) 测试一批数据。

```
1  batch_size=8
2  test_loader = torch.utils.data.DataLoader(test_ds, batch_size=batch_size)
3  dataiter = iter(test_loader)
4  images, labels = dataiter.next()
5
6  # 显示图像
7  plt.figure(figsize=(10,6))
8  imshow(torchvision.utils.make_grid(images))
9
10 print('真实类别: ', ' '.join(f'{classes[labels[j]]:5s}'
11                          for j in range(batch_size)))
12
13 # 预测
14 outputs = model(images.to(device))
15
16 _, predicted = torch.max(outputs, 1)
17
18 print('预测类别: ', ' '.join(f'{classes[predicted[j]]:5s}'
19                          for j in range(batch_size)))
```

执行结果：有一半辨识错误。

真实类别: cat  ship ship plane frog frog car  frog
预测类别: cat  car  car  ship  deer frog car  frog

(11) 计算各类别的准确率：观察是否有特别难以辨识的类别，可针对该类别进一步处理。

```
1  # 初始化各类别的正确数
2  correct_pred = {classname: 0 for classname in classes}
3  total_pred = {classname: 0 for classname in classes}
4
5  # 预测
6  batch_size=1000
7  test_loader = torch.utils.data.DataLoader(test_ds, batch_size=batch_size)
8  model.eval()
9  with torch.no_grad():
```

```
10      for data, target in test_loader:
11          data, target = data.to(device), target.to(device)
12          outputs = model(data)
13          _, predictions = torch.max(outputs, 1)
14          # 计算各类别的正确数
15          for label, prediction in zip(target, predictions):
16              if label == prediction:
17                  correct_pred[classes[label]] += 1
18              total_pred[classes[label]] += 1
19
20
21  # 计算各类别的准确率
22  for classname, correct_count in correct_pred.items():
23      accuracy = 100 * float(correct_count) / total_pred[classname]
24      print(f'{classname:5s}: {accuracy:.1f} %')
```

执行结果：cat、dog 准确率偏低，后续我们会试着改善模型准确率。

```
plane: 41.1 %
car  : 66.8 %
bird : 50.0 %
cat  : 27.5 %
deer : 51.5 %
dog  : 30.5 %
frog : 72.8 %
horse: 74.0 %
ship : 76.6 %
truck: 82.3 %
```

另外，数据前置处理补充说明如下：

影像 / 视频数据可使用 Pillow、scikit-image、OpenCV 套件存取。

语音数据可使用 SciPy、librosa 套件存取。

文字数据可使用 NLTK、SpaCy 套件存取。

## 6-6　影像数据增强

之前的辨识手写阿拉伯数字程序，还有以下缺点：

(1) 使用 MNIST 的测试数据，辨识率达 98%，但如果以绘图软件里使用鼠标书写的文件测试，辨识率就差很多。这是因为 MNIST 的训练数据与鼠标书写的样式有所差异，MNIST 数据是请受测者先写在纸上，再扫描存盘，所以图像会有深浅不一的灰度和锯齿状，所以，如果要实际应用，还是须自行收集训练数据，准确率才会提升。

(2) 若要自行收集数据，找上万个测试者书写，可能不太容易，又加上有些人书写字体可能会有歪斜、偏一边或大小不同，都会影响预测准确度，这时可以借由数据增强 (Data Augmentation) 技术，自动产生各种变形的训练数据，让模型更强健 (Robust)。

数据增强可将一张正常图像转换成各种的图像，例如旋转、偏移、拉近 / 拉远、亮度等效果，将这些数据当作训练数据训练出来的模型，就能较好地辨识有缺陷的图像。

PyTorch 提供的数据增强函数很多元，可参阅 *TRANSFORMING AND AUGMENTING IMAGES* [3]，我们已在 5-1 节测试过，详情请参阅程序【05_01_Datasets.ipynb】。

**范例1.** 将数据增强函数整合至【06_03_MNIST_CNN_Normalize.ipynb】中。

完整程序请参阅【06_05_Data_Augmentation_MNIST.ipynb】。

(1) 程序几乎不需改变，只要在数据转换方式加上随机转换 (Random*)，注意，数据增强函数很多，但阿拉伯数字有书写方向，有些随机转换不可采用，例如水平转换

(RandomHorizontalFlip)，若使用的话，3 就变成 ε 了。相关的效果可参阅 PyTorch 官网 *Illustration of transforms*[4]。

```
1   image_width = 28
2   train_transforms = transforms.Compose([
3       #transforms.ColorJitter(), # 亮度、饱和度、对比数据增强
4       # 裁切部分图像,再调整图像尺寸
5       transforms.RandomResizedCrop(image_width, scale=(0.8, 1.0)),
6       transforms.RandomRotation(degrees=(-10, 10)), # 旋转10°
7       #transforms.RandomHorizontalFlip(), # 水平翻转
8       #transforms.RandomAffine(10), # 仿射
9       transforms.ToTensor(),
10      transforms.Normalize(mean=(0.1307,), std=(0.3081,))
11  ])
12
13  test_transforms = transforms.Compose([
14      transforms.Resize((image_width, image_width)), # 调整图像尺寸
15      transforms.ToTensor(),
16      transforms.Normalize(mean=(0.1307,), std=(0.3081,))
17  ])
```

训练数据要数据增强，测试数据不需要。

(2) 以测试数据评分，准确率为 98.54%，并未显著提高。

(3) 测试自行书写的数字，原来的模型无法正确辨识笔者写的 9，经过数据增强后，已经可以正确辨识了。注意使用不同套件，读出的像素值区间会有所不同。

使用 torchvision.io.read_image 读取文件，像素介于 [0, 1]，回传的数据类型为 torch.Tensor。

使用 PIL 读取文件，像素介于 [0, 255]，回传的数据类型为 image，须以 np.array() 转换成 NumPy ndarray，才能进行运算，之后，再利用 Image.fromarray() 转换回 Image。

使用 scikit-image 读取文件，像素介于 [0, 1]，回传的数据类型为 NumPy ndarray。

(4) 先使用 PIL 读取文件，测试自行书写的数字。

```
1   # 使用PIL读取文件,像素介于[0, 255]
2   import PIL.Image as Image
3
4   data_shape = data.shape
5
6   for i in range(10):
7       uploaded_file = f'./myDigits/{i}.png'
8       image1 = Image.open(uploaded_file).convert('L')
9
10      # 缩为 (28, 28) 大小的影像
11      image_resized = image1.resize(tuple(data_shape)[2:])
12      X1 = np.array(image_resized).reshape([1]+list(data_shape)[1:])
13      # 反转颜色,颜色0为白色,与 RGB 色码不同,它的 0 为黑色
14      X1 = 1.0-(X1/255)
15
16      # 图像转换
17      X1 = (X1 - 0.1307) / 0.3081
18
19      # 显示转换后的图像
20      # imshow(X1)
21
22      X1 = torch.FloatTensor(X1).to(device)
23
24      # 预测
25      output = model(X1)
26      # print(output, '\n')
27      _, predicted = torch.max(output.data, 1)
28      print(f'actual/prediction: {i} {predicted.item()}')
```

执行结果：准确率为 100%。

(5) 使用 scikit-image 读取文件，测试自行书写的数字。

```
1   # 使用 skimage 读取文件,像素介于[0, 1]
2   from skimage import io
3   from skimage.transform import resize
4
5   # 读取影像并转为单色
6   for i in range(10):
7       uploaded_file = f'./myDigits/{i}.png'
8       image1 = io.imread(uploaded_file, as_gray=True)
9
10      # 缩为 (28, 28) 大小的影像
11      image_resized = resize(image1, tuple(data_shape)[2:], anti_aliasing=True)
12      X1 = image_resized.reshape([1]+list(data_shape)[1:])
13      # 反转颜色,颜色0为白色,与 RGB 色码不同,它的 0 为黑色
14      X1 = 1.0-X1
15
16      # 图像转换
17      X1 = (X1 - 0.1307) / 0.3081
18
19      # 显示转换后的图像
20      # imshow(X1)
21
22      X1 = torch.FloatTensor(X1).to(device)
23
24      # 预测
25      output = model(X1)
26      _, predicted = torch.max(output.data, 1)
27      print(f'actual/prediction: {i} {predicted.item()}')
```

执行结果:预测结果相同。

(6) 自定义数据集:可与训练数据采取一致的转换,不易出错。若是以次目录名称为标注,可直接使用 torchvision.datasets.ImageFolder,不必自定义数据集。

```
1   class CustomImageDataset(torch.utils.data.Dataset):
2       def __init__(self, img_dir, transform=None, target_transform=None
3                    , to_gray=False, size=28):
4           self.img_labels = [file_name for file_name in os.listdir(img_dir)]
5           self.img_dir = img_dir
6           self.transform = transform
7           self.target_transform = target_transform
8           self.to_gray = to_gray
9           self.size = size
10
11      def __len__(self):
12          return len(self.img_labels)
13
14      def __getitem__(self, idx):
15          # 组合文件完整路径
16          img_path = os.path.join(self.img_dir, self.img_labels[idx])
17          # 读取图片文件
18          mode = 'L' if self.to_gray else 'RGB'
19          image = Image.open(img_path, mode='r').convert(mode)
20          image = Image.fromarray(1.0-(np.array(image)/255))
21
22          # print(image.shape)
23          # 去除副文件名
24          label = int(self.img_labels[idx].split('.')[0])
25
26          # 转换
27          if self.transform:
28              image = self.transform(image)
29          if self.target_transform:
30              label = self.target_transform(label)
31
32          return image, label
```

(7) 使用自定义数据集预测。

```
1   ds = CustomImageDataset('./myDigits', to_gray=True, transform=test_transforms)
2   data_loader = torch.utils.data.DataLoader(ds, batch_size=10, shuffle=False)
3
4   test(model, device, data_loader)
```

执行结果:准确率为 100%。

(8) 验证：修改 test 函数，显示每一笔数据的实际值与预测值。

```
1  model.eval()
2  test_loss = 0
3  correct = 0
4  with torch.no_grad():
5      for data, target in data_loader:
6          print(target)
7          data, target = data.to(device), target.to(device)
8
9          # 预测
10         output = model(data)
11         _, predicted = torch.max(output.data, 1)
12         correct += (predicted == target).sum().item()
13         print(predicted)
```

执行结果：

tensor([0, 1, 2, 3, 4, 5, 6, 7, 8, 9])

tensor([0, 1, 2, 3, 4, 5, 6, 7, 8, 9], device='cuda:0')

**范例2.** 单色图像结果非常完美，我们进一步编写书写接口，试试看模型是否可以派上用场。

程序：**cnn_desktop\main.py**。

(1) 先复制【06_05_Data_Augmentation_MNIST.ipynb】程序产生的 cnn_augmentation_model.pt 模型至本程序所在目录。

(2) 载入套件。

```
1  from tkinter import *
2  from tkinter import filedialog
3  from PIL import ImageDraw, Image, ImageGrab
4  import numpy as np
5  from skimage import color
6  from skimage import io
7  import os
8  import io
9  import torch
10 import torch
11 from torch import nn
12 from torch.nn import functional as F
```

(3) 加载模型。

```
149  def loadModel():
150      model = torch.load('cnn_augmentation_model.pt').to(device)
151      return model
```

(4) 预测。

```
111      # 图像转为 PyTorch 张量
112      img = np.reshape(img, (1, 1, 28, 28))
113      data = torch.FloatTensor(img).to(device)
114
115      # 预测
116      output = model(data)
117      # Get index with highest probability
118      _, predicted = torch.max(output.data, 1)
119      #print(pred)
120      self.prediction_text.delete("1.0", END)
121      self.prediction_text.insert(END, predicted.item())
```

(5) 复制模型结构，会发现使用 Functional API 的 Class 也需放入程序中，才能顺利使用模型。

```
123  class Net(nn.Module):
124      def __init__(self):
125          super(Net, self).__init__()
126          self.conv1 = nn.Conv2d(1, 32, 3, 1)
127          self.conv2 = nn.Conv2d(32, 64, 3, 1)
128          self.dropout1 = nn.Dropout(0.25)
129          self.dropout2 = nn.Dropout(0.5)
130          self.fc1 = nn.Linear(9216, 128)
131          self.fc2 = nn.Linear(128, 10)
132
133      def forward(self, x):
134          x = self.conv1(x)
135          x = F.relu(x)
136          x = self.conv2(x)
137          x = F.relu(x)
138          x = F.max_pool2d(x, 2)
139          x = self.dropout1(x)
140          x = torch.flatten(x, 1)
141          x = self.fc1(x)
142          x = F.relu(x)
143          x = self.dropout2(x)
144          x = self.fc2(x)
145          output = F.log_softmax(x, dim=1)
146          return output
```

(6) 窗口接口使用 Tkinter，细节请参考程序文件。

(7) 执行 python main.py：以鼠标书写后，单击"辨识"按钮，辨识结果就会出现在右下文字框中，测试结果非常好。

**范例3. 接着再试试CIFAR彩色图像的数据增强。**

下列程序代码请参考【06_06_Data_Augmentation_CIFAR.ipynb】。

(1) 数据转换加上水平翻转 (RandomHorizontalFlip)、随机裁切 (RandomCrop)，其他转换也可以添加，不过 CIFAR 图像分辨率过低，而且侦测的对象均占满整个图片，因此添加其他转换似乎并无太大帮助。

```
1  image_width = 32
2  train_transforms = transforms.Compose([
3      # 裁切部分图像，再调整图像尺寸
4      #transforms.RandomResizedCrop(image_width, scale=(0.8, 1.0)),
5      #transforms.RandomRotation(degrees=(-10, 10)), # 旋转10°
6      #transforms.RandomHorizontalFlip(), # 水平翻转
7      transforms.RandomHorizontalFlip(p=0.5),
8      transforms.RandomCrop(image_width, padding=4),
9      #transforms.ColorJitter(), # 亮度、饱和度、对比数据增强
10     #transforms.RandomAffine(10), # 仿射
11     transforms.ToTensor(),
12     transforms.Normalize(mean=(0.1307,), std=(0.3081,))
13  ])
14
15  test_transforms = transforms.Compose([
16      transforms.Resize((image_width, image_width)), # 调整图像尺寸
17      transforms.ToTensor(),
18      transforms.Normalize(mean=(0.1307,), std=(0.3081,))
19  ])
```

(2) 改用 CIFAR 数据集。

```
1  # 载入数据集，如果出现 BrokenPipeError 错误，将 num_workers 改为 0
2  train_ds = torchvision.datasets.CIFAR10(root='./CIFAR10', train=True,
3                      download=True, transform=train_transforms)
4
5  train_loader = torch.utils.data.DataLoader(train_ds, batch_size=BATCH_SIZE,
6                      shuffle=True, num_workers=2)
7
8  test_ds = torchvision.datasets.CIFAR10(root='./CIFAR10', train=False,
9                      download=True, transform=test_transforms)
10
11 test_loader = torch.utils.data.DataLoader(test_ds, batch_size=BATCH_SIZE,
12                     shuffle=False, num_workers=2)
13
14 # 训练/测试数据的维度
15 print(train_ds.data.shape, test_ds.data.shape)
```

(3) 训练模型：发现添加转换后，一个训练周期的数据量仍然是 50000 笔，并未增加，也就是产生了更多样化的数据，但并未随同原来的训练数据一起被取出训练，只是以转换后的增强数据取代原数据，因此，笔者增加训练周期(10➔20)，以增加数据被广泛抽中的概率，同时也调小学习率，希望以更小的步幅寻求最佳解。

```
1  epochs = 20
2  lr=0.01
3
4  # 建立模型
5  model = Net().to(device)
6
7  # 定义损失函数
8  # 注意，nn.CrossEntropyLoss是类别，要先建立对象，要加 ()，其他损失函数不需要
9  criterion = nn.CrossEntropyLoss() # F.nll_loss
10
11 # 设定优化器(optimizer)
12 #optimizer = torch.optim.Adadelta(model.parameters(), lr=lr)
13 optimizer = torch.optim.SGD(model.parameters(), lr=lr, momentum=0.9)
14
15 loss_list = []
16 for epoch in range(1, epochs + 1):
17     loss_list += train(model, device, train_loader, criterion, optimizer, epoch)
18     #test(model, device, test_loader)
19     optimizer.step()
```

执行结果：观察下图，损失尚未收敛，准确率为 49.34%，反而下降了，原因应该是背景过于复杂、训练数据不足，就算再多的转换也无济于事。

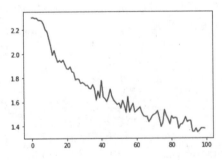

笔者在网络上搜寻其他先进的做法，发现两篇文章可供参考：

(1) *PyTorch Implementation of CIFAR-10 Image Classification Pipeline Using VGG Like Network*[5]，较复杂的 VGG 模型(后续会介绍)，使用数据增强，分别训练 40/80/120/160/300 周期，发现训练 160 周期后，准确率可达 90% 以上，如再训练更多周期，则会产生过度拟合的现象，即验证数据的准确率会背离训练的准确率，如下图。

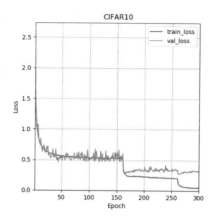

(2) *How Data Augmentation Improves your CNN performance?*[6] 一文使用 ResNet 模型 ( 后续会介绍 )，训练 15 周期，未使用数据增强，准确率可达 75%；使用数据增强，准确率可达 83%。

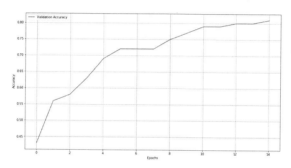

从以上的试验可知，数据增强并不重要，较复杂的模型及更多训练周期，才是提高 CIFAR 辨识准确率的关键因素。

TensorFlow 提供的数据增强功能效果比较明显，读者可以比较看看，另外还有其他的函数库，提供更多的数据增强效果，比如 Albumentations[7]，包含的类型多达 70 种，很多都是 TensorFlow/PyTorch 所没有的效果，例如下图的颜色数据增强。

## 6-7 可解释的 AI

虽然前文提到过深度学习是黑箱科学，但是科学家依然试图解释模型是如何辨识的，这方面的研究领域统称为"可解释的 AI"(eXplainable AI, XAI)，研究目的如下：

(1) 确认模型辨识的结果是合理的：深度学习永远不会跟你说错，垃圾进、垃圾出 (Garbage In, Garbage Out)，就算是很离谱的输入，模型还是会给你一个答案，因此确认模型推估的合理性相当重要。

(2) 改良算法：唯有知其所以然，才能有较大的进步，光靠参数的调校，只能有微幅的改善。目前机器学习还只能从数据中学习到知识 (Knowledge Discovery from Data, KDD)，要想进阶到机器具有智慧 (Wisdom) 及感知 (Feeling) 能力，实现真正的人工智能，势必要有突破性的发展。

目前 XAI 用可视化的方式呈现特征对模型的影响力，例如：
(1) 使用卷积层提取图像的线条特征，可以观察到转换后的结果吗？
(2) 更进一步，我们可以知道哪些线条对辨识最有帮助吗？

接下来我们以两个实例展示相关的做法。

**范例1.** 重建卷积层处理后的影像：观察每一次的卷积层/池化层处理后图像有何种变化。

此范例部分程序代码参考 How to Visualize Filters and Feature Maps in Convolutional Neural Networks[8]、Extracting Features from an Intermediate Layer of a Pretrained ResNet Model in PyTorch[9]。

**下列程序代码请参阅【06_07_XAI.ipynb】。**

(1) 载入套件：torchsummary 套件可显示模型结构信息，安装指令如下：pip install torchsummary。

```
1  import torch
2  from torchvision import models
3  from torch import nn
4  import numpy as np
5  from torchsummary import summary
```

(2) 加载 ResNet 18 模型：为求程序代码简洁，采用预先训练好的模型 ResNet，它含有多组卷积层/池化层，我们可以观察多次卷积后的效果，也可以采用其他预先训练的模型或自建模型。

```
1  rn18 = models.resnet18(pretrained=True)
```

(3) 显示神经层名称：以下指令只能显示神经层区块 (Layer Block)，每一区块又内含很多神经层。

```
1  children_counter = 0
2  for n,c in rn18.named_children():
3      print("Children Counter: ",children_counter," Layer Name: ",n)
4      children_counter+=1
```

执行结果：layer1/ layer2/ layer3/ layer4 层内含卷积层/池化层。

```
Children Counter:  0    Layer Name:  conv1
Children Counter:  1    Layer Name:  bn1
Children Counter:  2    Layer Name:  relu
Children Counter:  3    Layer Name:  maxpool
Children Counter:  4    Layer Name:  layer1
Children Counter:  5    Layer Name:  layer2
Children Counter:  6    Layer Name:  layer3
Children Counter:  7    Layer Name:  layer4
Children Counter:  8    Layer Name:  avgpool
Children Counter:  9    Layer Name:  fc
```

(4) 显示神经层明细。

```
1  rn18._modules
```

执行结果：layer1 层内含卷积层 / BatchNorm2d 层，等等。

```
OrderedDict([('conv1',
              Conv2d(3, 64, kernel_size=(7, 7), stride=(2, 2), padding=(3, 3), bias=False)),
             ('bn1',
              BatchNorm2d(64, eps=1e-05, momentum=0.1, affine=True, track_running_stats=True)),
             ('relu', ReLU(inplace=True)),
             ('maxpool',
              MaxPool2d(kernel_size=3, stride=2, padding=1, dilation=1, ceil_mode=False)),
             ('layer1',
              Sequential(
                (0): BasicBlock(
                  (conv1): Conv2d(64, 64, kernel_size=(3, 3), stride=(1, 1), padding=(1, 1), bias=False)
                  (bn1): BatchNorm2d(64, eps=1e-05, momentum=0.1, affine=True, track_running_stats=True)
                  (relu): ReLU(inplace=True)
                  (conv2): Conv2d(64, 64, kernel_size=(3, 3), stride=(1, 1), padding=(1, 1), bias=False)
                  (bn2): BatchNorm2d(64, eps=1e-05, momentum=0.1, affine=True, track_running_stats=True)
                )
                (1): BasicBlock(
                  (conv1): Conv2d(64, 64, kernel_size=(3, 3), stride=(1, 1), padding=(1, 1), bias=False)
                  (bn1): BatchNorm2d(64, eps=1e-05, momentum=0.1, affine=True, track_running_stats=True)
                  (relu): ReLU(inplace=True)
                  (conv2): Conv2d(64, 64, kernel_size=(3, 3), stride=(1, 1), padding=(1, 1), bias=False)
                  (bn2): BatchNorm2d(64, eps=1e-05, momentum=0.1, affine=True, track_running_stats=True)
                )
              )),
             ('layer2',
```

(5) 另一角度观察模型：使用 torchsummary。

```
1  from torchsummary import summary
2
3  summary(rn18.to(device), input_size=(3, 224, 224))
```

执行结果：共 68 层，完整信息请参考程序执行结果。

```
        Layer (type)               Output Shape         Param #
================================================================
            Conv2d-1         [-1, 64, 112, 112]           9,408
       BatchNorm2d-2         [-1, 64, 112, 112]             128
              ReLU-3         [-1, 64, 112, 112]               0
         MaxPool2d-4           [-1, 64, 56, 56]               0
            Conv2d-5           [-1, 64, 56, 56]          36,864
       BatchNorm2d-6           [-1, 64, 56, 56]             128
              ReLU-7           [-1, 64, 56, 56]               0
            Conv2d-8           [-1, 64, 56, 56]          36,864
       BatchNorm2d-9           [-1, 64, 56, 56]             128
             ReLU-10           [-1, 64, 56, 56]               0
       BasicBlock-11           [-1, 64, 56, 56]               0
           Conv2d-12           [-1, 64, 56, 56]          36,864
      BatchNorm2d-13           [-1, 64, 56, 56]             128
             ReLU-14           [-1, 64, 56, 56]               0
           Conv2d-15           [-1, 64, 56, 56]          36,864
      BatchNorm2d-16           [-1, 64, 56, 56]             128
             ReLU-17           [-1, 64, 56, 56]               0
       BasicBlock-18           [-1, 64, 56, 56]               0
           Conv2d-19          [-1, 128, 28, 28]          73,728
      BatchNorm2d-20          [-1, 128, 28, 28]             256
             ReLU-21          [-1, 128, 28, 28]               0
           Conv2d-22          [-1, 128, 28, 28]         147,456
      BatchNorm2d-23          [-1, 128, 28, 28]             256
           Conv2d-24          [-1, 128, 28, 28]           8,192
      BatchNorm2d-25          [-1, 128, 28, 28]             256
             ReLU-26          [-1, 128, 28, 28]               0
```

(6) 移除 layer1 后面的神经层,可视化卷积结果,观察重建后的图像。
new_model 类别:可指定神经层名称,以移除后面的部分神经层。
第 23 行:建立该类别的对象。

```python
1  class new_model(nn.Module):
2      def __init__(self, output_layer):
3          super().__init__()
4          self.output_layer = output_layer
5          self.pretrained = models.resnet18(pretrained=True)
6          self.children_list = []
7          # 依次取得每一层
8          for n,c in self.pretrained.named_children():
9              self.children_list.append(c)
10             # 找到特定层即终止
11             if n == self.output_layer:
12                 print('found !!')
13                 break
14
15         # 构建新模型
16         self.net = nn.Sequential(*self.children_list)
17         self.pretrained = None
18
19     def forward(self,x):
20         x = self.net(x)
21         return x
22
23 model = new_model(output_layer = 'layer1')
24 model = model.to(device)
```

(7) 使用 torchsummary,再观察模型:只剩 18 层。

```python
1  from torchsummary import summary
2
3  summary(rn18.to(device), input_size=(3, 224, 224))
```

执行结果。

```
        Layer (type)               Output Shape         Param #
================================================================
            Conv2d-1         [-1, 64, 112, 112]           9,408
       BatchNorm2d-2         [-1, 64, 112, 112]             128
              ReLU-3         [-1, 64, 112, 112]               0
         MaxPool2d-4           [-1, 64, 56, 56]               0
            Conv2d-5           [-1, 64, 56, 56]          36,864
       BatchNorm2d-6           [-1, 64, 56, 56]             128
              ReLU-7           [-1, 64, 56, 56]               0
            Conv2d-8           [-1, 64, 56, 56]          36,864
       BatchNorm2d-9           [-1, 64, 56, 56]             128
             ReLU-10           [-1, 64, 56, 56]               0
       BasicBlock-11           [-1, 64, 56, 56]               0
           Conv2d-12           [-1, 64, 56, 56]          36,864
      BatchNorm2d-13           [-1, 64, 56, 56]             128
             ReLU-14           [-1, 64, 56, 56]               0
           Conv2d-15           [-1, 64, 56, 56]          36,864
      BatchNorm2d-16           [-1, 64, 56, 56]             128
             ReLU-17           [-1, 64, 56, 56]               0
       BasicBlock-18           [-1, 64, 56, 56]               0
================================================================
```

(8) 预测模型:以猫的图片为例。

```python
1  from PIL import Image
2  import matplotlib.pyplot as plt
3  import torchvision.transforms as transforms
4
5  img = Image.open("./images_test/cat.jpg")
6  plt.imshow(img)
7  plt.axis('off')
8  plt.show()
9
10 resize = transforms.Resize([224, 224])
11 img = resize(img)
12
13 to_tensor = transforms.ToTensor()
14 img = to_tensor(img).to(device)
15 img = img.reshape(1, *img.shape)
16 out = model(img)
17 out.shape
```

执行结果：ResNet 18 模型的输入宽/高为 (224, 224)、彩色，输出格式为 512 个 (7, 7) 的矩阵。

torch.Size([1, 512, 7, 7])

(9) 重建 8×8 格图像：只显示部分卷积结果。

```
1   # 重建 8×8 图像
2   def show_grid(out):
3       square = 8
4       plt.figure(figsize=(12, 10))
5       for fmap in out.cpu().detach().numpy():
6           # plot all 64 maps in an 8x8 squares
7           ix = 1
8           for _ in range(square):
9               for _ in range(square):
10                  # specify subplot and turn of axis
11                  ax = plt.subplot(square, square, ix)
12                  ax.set_xticks([])
13                  ax.set_yticks([])
14                  # plot filter channel in grayscale
15                  plt.imshow(fmap[ix-1, :, :], cmap='gray')
16                  ix += 1
17          # show the figure
18          plt.show()
19  
20  show_grid(out)
```

执行结果：可以看见第一层图像处理结果，有的滤波器可以抓到轮廓，有的则是漆黑一片。

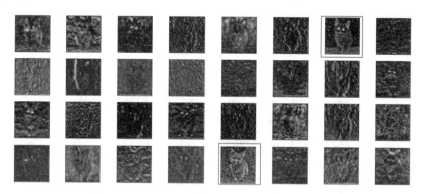

(10) 重建第 2 层卷积层的输出图像。

```
1   model = new_model(output_layer = 'layer2').to(device)
2   out = model(img)
3   show_grid(out)
```

执行结果：逐渐抽象化，已经认不出来是猫了。

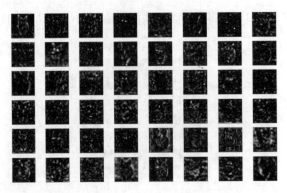

(11) 重建第 3 层卷积层的输出图像。

```
1  model = new_model(output_layer = 'layer3').to(device)
2  out = model(img)
3  show_grid(out)
```

执行结果。

(12) 重建第 4 层卷积层的输出图像。

```
1  model = new_model(output_layer = 'layer4').to(device)
2  out = model(img)
3  show_grid(out)
```

执行结果：完全认不出来是猫了。

从以上的试验，可以很清楚地看到 CNN 的处理过程，虽然我们不明白计算机辨识的逻辑，但是至少能够观察到整个模型处理的过程。有人曾举例说明多个卷积层的作用：第一层侦测线条；第二层将侦测线条组合、几何形状；第三层侦测出耳朵、鼻子，例如猫的耳朵是尖的，所以，就能辨识出猫或狗了，这就是逐渐抽象化的过程。

**范例2. 使用SHAP套件，观察图像的哪些位置对辨识最有帮助。**

SHAP (SHapley Additive exPlanations) 套件是由 Scott Lundberg 及 Su-In Lee 所开发的，提供 Shapley value 的计算，并具有可视化的接口，目的是希望能解释各种机器学习模型。套件使用说明可参考介绍说明 [10]，以下仅说明神经网络的应用。

Shapley Value 是由多人赛局理论 (Game Theory) 发展而来的，原本是用来分配利益给团队中的每个人时所使用的分配函数，如今沿用到了机器学习领域，则被应用在特征对预测结果的个别影响力评估。详细的介绍可参考维基百科 [11]。

SHAP 套件安装：pip install shap。

此程序修改自 *PyTorch Deep Explainer MNIST example* [12]。

**下列程序代码请参考【06_08_Shap_MNIST.ipynb】。**

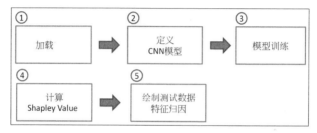

(1) 载入套件。

```
1  import torch, torchvision
2  from torchvision import datasets, transforms
3  from torch import nn, optim
4  from torch.nn import functional as F
5
6  import numpy as np
7  import shap
```

(2) 载入 MNIST 数据集。

```
1  from torchvision.datasets import MNIST
2
3  batch_size = 128
4  num_epochs = 2
5
6  # 下载 MNIST 手写阿拉伯数字 训练数据
7  train_ds = MNIST('.', train=True, download=True,
8                  transform=transforms.ToTensor())
9
10 # 下载测试数据
11 test_ds = MNIST('.', train=False, download=True,
12                 transform=transforms.ToTensor())
13
14 # 训练/测试数据的维度
15 print(train_ds.data.shape, test_ds.data.shape)
16
17 train_loader = torch.utils.data.DataLoader(
18     datasets.MNIST('mnist_data', train=True, download=True,
19                    transform=transforms.Compose([
20                        transforms.ToTensor()
21                    ])),
```

```
22          batch_size=batch_size, shuffle=True)
23
24 test_loader = torch.utils.data.DataLoader(
25     datasets.MNIST('mnist_data', train=False, transform=transforms.Compose([
26                     transforms.ToTensor()
27                    ])),
28          batch_size=batch_size, shuffle=True)
```

(3) 定义 CNN 模型：也可使用其他模型做测试。

```
1  class Net(nn.Module):
2      def __init__(self):
3          super(Net, self).__init__()
4
5          self.conv_layers = nn.Sequential(
6              nn.Conv2d(1, 10, kernel_size=5),
7              nn.MaxPool2d(2),
8              nn.ReLU(),
9              nn.Conv2d(10, 20, kernel_size=5),
10             nn.Dropout(),
11             nn.MaxPool2d(2),
12             nn.ReLU(),
13         )
14         self.fc_layers = nn.Sequential(
15             nn.Linear(320, 50),
16             nn.ReLU(),
17             nn.Dropout(),
18             nn.Linear(50, 10),
19             nn.Softmax(dim=1)
20         )
21
22     def forward(self, x):
23         x = self.conv_layers(x)
24         x = x.view(-1, 320)
25         x = self.fc_layers(x)
26         return x
27
28 model = Net().to(device)
```

(4) 定义训练/测试函数。

```
1  # 训练函数
2  def train(model, device, train_loader, optimizer, epoch):
3      model.train()
4      for batch_idx, (data, target) in enumerate(train_loader):
5          data, target = data.to(device), target.to(device)
6          optimizer.zero_grad()
7          output = model(data)
8          loss = F.nll_loss(output.log(), target)
9          loss.backward()
10         optimizer.step()
11         if batch_idx % 100 == 0:
12             print('Train Epoch: {} [{}/{} ({:.0f}%)]\tLoss: {:.6f}'.format(
13                 epoch, batch_idx * len(data), len(train_loader.dataset),
14                 100. * batch_idx / len(train_loader), loss.item()))
15
16 # 测试函数
17 def test(model, device, test_loader):
18     model.eval()
19     test_loss = 0
20     correct = 0
21     with torch.no_grad():
22         for data, target in test_loader:
23             data, target = data.to(device), target.to(device)
24             output = model(data)
25             test_loss += F.nll_loss(output.log(), target).item()
26             pred = output.max(1, keepdim=True)[1]
27             correct += pred.eq(target.view_as(pred)).sum().item()
28
29     test_loss /= len(test_loader.dataset)
30     print('\nTest set: Average loss: {:.4f}, Accuracy: {}/{} ({:.0f}%)\n'.format(
31         test_loss, correct, len(test_loader.dataset),
32         100. * correct / len(test_loader.dataset)))
```

(5) 模型训练。

```
1  optimizer = optim.SGD(model.parameters(), lr=0.01, momentum=0.5)
2
3  for epoch in range(1, num_epochs + 1):
4      train(model, device, train_loader, optimizer, epoch)
5      test(model, device, test_loader)
```

(6)Shapley Value 计算：以前面 100 笔为背景值，计算 Shapley Value，测试 5 笔数据。

```
1  # 以前面 100 笔为背景值，计算 Shapley Value
2  batch = next(iter(test_loader))
3  images, _ = batch
4  images = images.to(device)
5
6  background = images[:100]
7  test_images = images[100:105]
8
9  e = shap.DeepExplainer(model, background)
10 shap_values = e.shap_values(test_images)
```

(7)绘制 5 笔测试数据的特征归因：红色区块 ( 请参看程序 ) 代表贡献率较大的区域。

```
1  shap_numpy = [np.swapaxes(np.swapaxes(s, 1, -1), 1, 2) for s in shap_values]
2  test_numpy = np.swapaxes(np.swapaxes(test_images.cpu().numpy(), 1, -1), 1, 2)
3
4  # plot the feature attributions
5  shap.image_plot(shap_numpy, -test_numpy)
```

执行结果：每一列第一个数字为真实的标记，后面为预测每个数字贡献率较大的区域。

从 SHAP 套件的功能，可以很容易判断出中央位置是辨识的重点区域，这与我们认知是一致的。另一个名为 LIME[13] 的套件，与 SHAP 套件齐名，读者如果对该领域感兴趣，可以由此深入研究。

还有一篇论文 *Learning Deep Features for Discriminative Localization*[14] 提出了 Class Activation Mapping 概念，可以描绘辨识的热区，如图 6.11 所示。Kaggle 也有一个很不错的实操案例[15]，值得大家好好欣赏一番。

图 6.11 左上角的图像为原图，左下角的图像显示了辨识热区，即猴子的头部和颈部都是辨识的主要区域

通过以上可视化的辅助，不仅可以帮助我们更了解 CNN 模型的运作，也能够让我们在收集数据时，有较明确的方向知道重点应该放在哪里，当然，如果未来有更创新的想法，来改良算法，那就可以开香槟庆祝了。

# 第 7 章
# 预先训练的模型

通过 CNN 模型和数据增强的强化，我们已经能够建立准确度还不错的模型，然而，与近几年影像辨识竞赛中的冠、亚军模型相比较，只能算是小巫见大巫了，冠、亚军模型的神经层数量有些多达 100 多层，若要自行训练这些模型需要花上几天甚至几个星期的时间。难道缩短训练时间的办法，只有购置企业级服务器这个选项吗？

幸好 PyTorch、TensorFlow/Keras 等深度学习框架早已为我们设想好了对策，直接提供事先训练好的模型，我们可以直接套用，也可以只采用部分模型，再接上自定义的神经层，进行其他对象的辨识，这些预先训练好的模型称为 Pre-trained Model。

## 7-1 预先训练模型的简介

近几年在 ImageNet 举办的竞赛 (ILSVRC) 中所产生的冠、亚军大都是 CNN 模型的变形，整个演进过程非常精彩，简述如下。

2012 年冠军 AlexNet 一举将错误率减少 10% 以上，且首度导入 Dropout 层。

2014 年亚军 VGGNet 承袭 AlexNet 思路，建立更多层的模型，VGG 16/19 分别包括 16 层及 19 层卷积层及池化层。

2014 年图像分类冠军 GoogNet & Inception 同时导入多种不同尺寸的 Kernel，让系统决定最佳 Kernel 尺寸。Inception 引入 Batch Normalization 等概念，参见 *Batch Normalization: Accelerating Deep Network Training by Reducing Internal Covariate Shift* [1]。

2015 年冠军 ResNet 发现到 20 层以上的模型在其前面几层会发生退化 (degradation)，因而提出以残差 (Residual) 的方法来解决问题，参见 *Deep Residual Learning for Image Recognition* [2]。

ImageNet 竞赛 (ILSVRC) 历年冠亚军如图 7.1 所示。

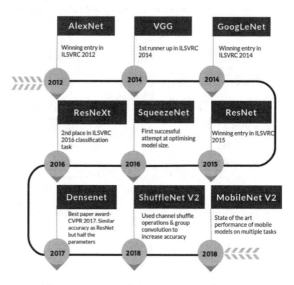

图 7.1 ImageNet 竞赛 (ILSVRC) 历年冠亚军

图片来源：*PyTorch for Beginners: Image Classification using Pre-trained models* [3]

PyTorch 收录许多预先训练的模型 [4]，随着版本的更新，提供的模型越来越多，目前 (2022 年) 包括：

AlexNet；

VGG；

ResNet；

SqueezeNet；

DenseNet；

Inception v3；

GoogLeNet；

ShuffleNet v2；

MobileNetV2；

MobileNetV3；

ResNeXt；

Wide ResNet；

MNASNet；

EfficientNet；

RegNet。

每种模型又细分为多个不同层数的模型，例如 VGG-11、VGG-13、VGG-16 及 VGG-19，PyTorch 官网还列出它们的源代码，准确率、参数个数及层数也整理在表格中，Keras 官网整理得比较详尽，因此，在此引用它的表格 ( 如表 7.1 所示 )，虽然准确率与 PyTorch 有些微差异。

表 7.1　Keras 提供的预先训练模型

| Model | Size | Top-1 Accuracy | Top-5 Accuracy | Parameters | Depth |
|---|---|---|---|---|---|
| Xception | 88 MB | 0.790 | 0.945 | 22,910,480 | 126 |
| VGG16 | 528 MB | 0.713 | 0.901 | 138,357,544 | 23 |
| VGG19 | 549 MB | 0.713 | 0.900 | 143,667,240 | 26 |
| ResNet50 | 98 MB | 0.749 | 0.921 | 25,636,712 | - |
| ResNet101 | 171 MB | 0.764 | 0.928 | 44,707,176 | - |
| ResNet152 | 232 MB | 0.766 | 0.931 | 60,419,944 | - |
| ResNet50V2 | 98 MB | 0.760 | 0.930 | 25,613,800 | - |
| ResNet101V2 | 171 MB | 0.772 | 0.938 | 44,675,560 | - |
| ResNet152V2 | 232 MB | 0.780 | 0.942 | 60,380,648 | - |
| InceptionV3 | 92 MB | 0.779 | 0.937 | 23,851,784 | 159 |
| InceptionResNetV2 | 215 MB | 0.803 | 0.953 | 55,873,736 | 572 |
| MobileNet | 16 MB | 0.704 | 0.895 | 4,253,864 | 88 |
| MobileNetV2 | 14 MB | 0.713 | 0.901 | 3,538,984 | 88 |
| DenseNet121 | 33 MB | 0.750 | 0.923 | 8,062,504 | 121 |
| DenseNet169 | 57 MB | 0.762 | 0.932 | 14,307,880 | 169 |
| DenseNet201 | 80 MB | 0.773 | 0.936 | 20,242,984 | 201 |
| NASNetMobile | 23 MB | 0.744 | 0.919 | 5,326,716 | - |
| NASNetLarge | 343 MB | 0.825 | 0.960 | 88,949,818 | - |
| EfficientNetB0 | 29 MB | - | - | 5,330,571 | |
| EfficientNetB1 | 31 MB | - | - | 7,856,239 | |
| EfficientNetB2 | 36 MB | - | - | 9,177,569 | |
| EfficientNetB3 | 48 MB | - | - | 12,320,535 | |
| EfficientNetB4 | 75 MB | - | - | 19,466,823 | |
| EfficientNetB5 | 118 MB | - | - | 30,562,527 | |
| EfficientNetB6 | 166 MB | - | - | 43,265,143 | |
| EfficientNetB7 | 256 MB | - | - | 66,658,687 | |

来源：Keras Applications [5]

上述表格的字段说明如下：

Size：模型文件大小。

Top-1 Accuracy：预测一次就正确的概率。

Top-5 Accuracy：预测五次中有一次正确的概率。

Parameters：模型参数 (权重、偏差) 的数目。

Depth：模型层数。

PyTorch、Keras 研发团队将这些模型先进行训练与参数调校，并且存储，使用者不用自行训练，直接套用即可，故称为预先训练的模型 (Pre-trained Model)。

这些模型主要应用在图像辨识，各模型结构的复杂度和准确率有所差异，如图 7.2 所示是各模型的比较，这里提供一个简单的选用原则，如果是注重准确率，可选择准确率较高的模型，例如 ResNet 152，反之，如果要部署在手机上，就可考虑使用文件较小的模型，例如 MobileNet。

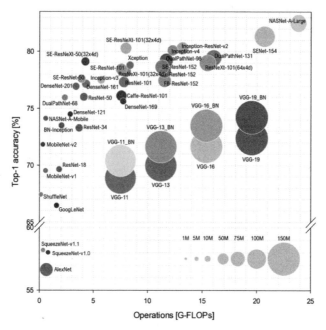

图 7.2　预先训练模型的准确率与计算速度的比较

图片来源：*How to Choose the Best Keras Pre-Trained Model for Image Classification* [6]

这些模型使用 ImageNet 100 万张图片作为训练数据集，内含 1000 种类别，类别内容请参考 *Yagnesh Revar GitHub* [7]，几乎涵盖了日常生活中能看到的对象类别，例如动物、植物、交通工具等，所以如果要辨识的对象属于这 1000 种，就可以直接套用模型，反之，如果要辨识这 1000 种以外的对象，就需要接上自定义的输入层及完全连接层 (Linear)，只利用预先训练模型的中间层提取特征。

因此应用这些预先训练的模型，有以下三种方式：
(1) 采用完整的模型，可辨识 ImageNet 所提供的 1000 种对象。
(2) 采用部分的模型，只提取特征，不作辨识。
(3) 采用部分的模型，并接上自定义的输入层和完全连接层 (Linear)，即可辨识这 1000 种以外的对象。

以下我们依照这三种方式各实操一个范例。

# 7-2　采用完整的模型

预先训练的模型的第一种用法，是采用完整的模型来辨识 1000 种对象，直截了当。

**范例1. 使用VGG16模型进行对象的辨识，也借此熟悉预先训练模型的结构与用法。**

使用 VGG16 模型进行对象辨识的相关步骤如图 7.3 所示。

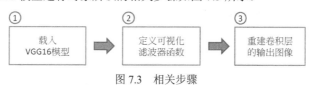

图 7.3　相关步骤

程序：【07_01_pretrained_model.ipynb】。

(1) 载入套件。

```
1  import torch
2  from torchvision import models
3  from torch import nn
4  import numpy as np
5  from torchsummary import summary
```

(2) 加载 VGG16 模型。

```
1  model = models.vgg16(pretrained=True)
```

pretrained=True：会把训练好的权重一并载入，反之，只加载模型结构。

(3) 显示神经层名称。

```
1  model = models.vgg16(pretrained=True)
```

执行结果：VGG 16 使用多组的卷积 / 池化层，共有 3 个区块。

```
Children Counter:  0  Layer Name:  features
Children Counter:  1  Layer Name:  avgpool
Children Counter:  2  Layer Name:  classifier
```

(4) 显示神经层明细。

```
1  model.modules
```

执行结果：可以看到 3 个区块内的神经层。

```
<bound method Module.modules of VGG(
  (features): Sequential(
    (0): Conv2d(3, 64, kernel_size=(3, 3), stride=(1, 1), padding=(1, 1))
    (1): ReLU(inplace=True)
    (2): Conv2d(64, 64, kernel_size=(3, 3), stride=(1, 1), padding=(1, 1))
    (3): ReLU(inplace=True)
    (4): MaxPool2d(kernel_size=2, stride=2, padding=0, dilation=1, ceil_mode=False)
    (5): Conv2d(64, 128, kernel_size=(3, 3), stride=(1, 1), padding=(1, 1))
    (6): ReLU(inplace=True)
    (7): Conv2d(128, 128, kernel_size=(3, 3), stride=(1, 1), padding=(1, 1))
    (8): ReLU(inplace=True)
    (9): MaxPool2d(kernel_size=2, stride=2, padding=0, dilation=1, ceil_mode=False)
    (10): Conv2d(128, 256, kernel_size=(3, 3), stride=(1, 1), padding=(1, 1))
    (11): ReLU(inplace=True)
    (12): Conv2d(256, 256, kernel_size=(3, 3), stride=(1, 1), padding=(1, 1))
    (13): ReLU(inplace=True)
    (14): Conv2d(256, 256, kernel_size=(3, 3), stride=(1, 1), padding=(1, 1))
    (15): ReLU(inplace=True)
    (16): MaxPool2d(kernel_size=2, stride=2, padding=0, dilation=1, ceil_mode=False)
    (17): Conv2d(256, 512, kernel_size=(3, 3), stride=(1, 1), padding=(1, 1))
```

torch.nn.Sequential(*list(model.children())[:])：也可以达到相同的效果。
使用 torchsummary 的 summary(model, input_size=(3, 224, 224)) 只显示明细。
model._modules.keys()：只显示 3 个区块名称。
model.features：显示 features 区块的明细。
model.features[0]：显示 features 区块内的第一个神经层。
model.classifier[-1].out_features：显示 classifier 区块内的最后一个神经层的输出。

(5) 任选一张图片预测所属类别。

```
1  from PIL import Image
2  from torchvision import transforms
3
4  filename = './images_test/cat.jpg'
5  input_image = Image.open(filename)
6
7  transform = transforms.Compose([
8      transforms.Resize((224, 224)),
9      transforms.ToTensor(),
10     transforms.Normalize(mean=[0.485, 0.456, 0.406],
11                          std=[0.229, 0.224, 0.225])
12 ])
13 input_tensor = transform(input_image)
14 input_batch = input_tensor.unsqueeze(0).to(device) # 增加一维(笔数)
15
16 # 预测
17 model.eval()
18 with torch.no_grad():
19     output = model(input_batch)
20
21 # 转成概率
22 probabilities = torch.nn.functional.softmax(output[0], dim=0)
23 print(probabilities)
```

执行结果：显示所有类别的概率。

```
tensor([9.6767e-08, 4.2125e-07, 7.1025e-09, 4.2960e-08, 3.6617e-09, 5.4819e-07,
        7.3084e-10, 6.0334e-07, 2.4721e-04, 1.4279e-06, 5.4144e-06, 8.3407e-06,
        6.0837e-06, 1.8260e-06, 2.4115e-06, 1.5520e-06, 5.5406e-06, 1.5232e-05,
        4.1942e-07, 1.0240e-06, 5.1658e-07, 7.9180e-06, 1.8201e-08, 4.9936e-07,
        1.7920e-07, 2.8996e-07, 3.1131e-07, 4.0312e-08, 1.3995e-07, 3.7367e-07,
        8.9198e-07, 8.6777e-07, 3.2260e-06, 2.2061e-08, 1.4229e-08, 2.1457e-07,
        5.5169e-07, 1.8037e-07, 7.1420e-06, 2.2228e-08, 1.2892e-06, 7.9648e-07,
        9.2213e-06, 9.4030e-06, 2.4247e-06, 3.0765e-06, 1.3920e-06, 7.1425e-06,
        4.1772e-07, 5.0599e-06, 7.0444e-08, 3.5808e-06, 9.1419e-07, 4.3269e-07,
        6.3600e-06, 1.0685e-06, 1.6913e-07, 5.3218e-08, 6.5058e-07, 2.2943e-06,
        7.0011e-06, 4.6382e-07, 2.0770e-06, 1.6658e-05, 3.8818e-07, 2.0196e-07,
        4.4105e-06, 3.9283e-07, 2.6178e-06, 5.3054e-08, 1.9659e-07, 2.6170e-07,
```

(6) 显示最大概率的类别代码。

```
1  # 显示最大机率的类别代码
2  print(f'{torch.argmax(probabilities).item()}: {torch.max(probabilities).item()}')
```

执行结果：索引值为 285，概率为 0.71。

285: 0.7133557200431824

(7) 显示最大概率的类别名称，torchvision 源代码[8] 有类别列表，也可以从 https://raw.githubusercontent.com/pytorch/hub/master/imagenet_classes.txt 下载。

```
1  # 显示最大概率的类别名称
2  with open("imagenet.categories", "r") as f:
3      # 取第一栏
4      categories = [s.strip().split(',')[0] for s in f.readlines()]
5  categories[torch.argmax(probabilities).item()]
```

执行结果：埃及猫 (Egyptian cat)。

(8) 改用其他模型亦可。

```
1  # 载入 resnet50 模型
2  model = models.resnet50(pretrained=True).to(device)
3
4  # 预测
5  filename = './images_test/cat.jpg'
6  input_image = Image.open(filename)
7
8  transform = transforms.Compose([
9      transforms.Resize((224, 224)),
```

```
10      transforms.ToTensor(),
11      transforms.Normalize(mean=[0.485, 0.456, 0.406],
12                           std=[0.229, 0.224, 0.225])
13  ])
14  input_tensor = transform(input_image)
15  input_batch = input_tensor.unsqueeze(0).to(device) # 增加一维(笔数)
16
17  model.eval()
18  with torch.no_grad():
19      output = model(input_batch)
20
21  # 转成机率
22  probabilities = torch.nn.functional.softmax(output[0], dim=0)
23  max_item = torch.argmax(probabilities).item()
24  print(f'{max_item} {categories[max_item]}: {torch.max(probabilities).item()}')
```

执行结果：结果相同。

若使用官网程序转换，先 Resize(256)，再 CenterCrop(224)，执行结果为虎斑猫 (tabby)，竟然与前面不同，辨识结果应该不对，需进一步确认。

281 tabby: 0.2819097936153412

(9) 显示前 5 名。

```
2  top5_prob, top5_catid = torch.topk(probabilities, 5)
3  for i in range(top5_prob.size(0)):
4      print(f'{categories[top5_catid[i]]:12s}:{top5_prob[i].item()}')
```

执行结果：第三名才是埃及猫 (Egyptian cat)。

```
tabby        :0.2819097936153412
tiger cat    :0.19214917719364166
Egyptian cat :0.18028706312179565
lynx         :0.17349961400032043
hamper       :0.01312144286930561
```

各个预先训练模型预测结果竟然有所差异，与图 7.2 比较，resnet50 比 vgg16 准确率高，但实测结果并不相符，笔者猜测应该是跟 MNIST 类似，以本身的测试数据较准，但自行收集的数据预测就没么好了 (TensorFlow 也有这方面的问题，请参阅【07_02_Keras_applications.ipynb】)，因此，还是要多多收集数据，仔细测试与使用。

# 7-3 采用部分模型

预先训练的模型的第二种用法是采用部分模型，只提取特征，不做辨识。例如，一个 3D 模型的网站，提供模型搜寻功能 ( 如图 7.4 所示 )，首先用户上传要搜寻的图文件，网站实时比对出相似的图片文件，显示在网页上让用户勾选下载。操作请参考 Sketchfab 网站[9]，类似的功能亦适用于许多场域，例如比对嫌疑犯、商品推荐，等等。

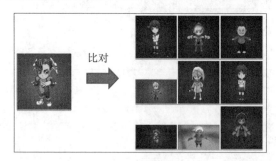

图 7.4  3D 模型搜寻

图片来源：*Using Keras' Pretrained Neural Networks for Visual Similarity Recommendations* [10]

## 范例2. 使用VGG16模型进行对象的辨识。

使用 VGG16 模型进行对象辨识的相关步骤如图 7.5 所示。

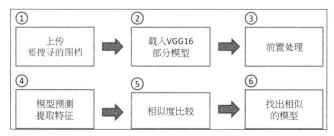

图 7.5　相关步骤

**程序：【07_03_cosine_similarity.ipynb】。**

(1) 载入套件。

```
1  import torch
2  from torchvision import models
3  from torch import nn
4  from torchsummary import summary
5  import numpy as np
```

(2) 加载 VGG 16 模型，并显示模型结构。

```
1  model = models.vgg16(pretrained=True)
2  model._modules
```

执行结果：最后一个区块为辨识层。

```
OrderedDict([('features',
              Sequential(
                (0): Conv2d(3, 64, kernel_size=(3, 3), stride=(1, 1), padding=(1, 1))
                (1): ReLU(inplace=True)
                (2): Conv2d(64, 64, kernel_size=(3, 3), stride=(1, 1), padding=(1, 1))
                (3): ReLU(inplace=True)
                (4): MaxPool2d(kernel_size=2, stride=2, padding=0, dilation=1, ceil_mode=False)
                (5): Conv2d(64, 128, kernel_size=(3, 3), stride=(1, 1), padding=(1, 1))
                (6): ReLU(inplace=True)
                (7): Conv2d(128, 128, kernel_size=(3, 3), stride=(1, 1), padding=(1, 1))
                (8): ReLU(inplace=True)
                (9): MaxPool2d(kernel_size=2, stride=2, padding=0, dilation=1, ceil_mode=False)
                (10): Conv2d(128, 256, kernel_size=(3, 3), stride=(1, 1), padding=(1, 1))
                (11): ReLU(inplace=True)
                (12): Conv2d(256, 256, kernel_size=(3, 3), stride=(1, 1), padding=(1, 1))
                (13): ReLU(inplace=True)
                (14): Conv2d(256, 256, kernel_size=(3, 3), stride=(1, 1), padding=(1, 1))
                (15): ReLU(inplace=True)
                (16): MaxPool2d(kernel_size=2, stride=2, padding=0, dilation=1, ceil_mode=False)
                (17): Conv2d(256, 512, kernel_size=(3, 3), stride=(1, 1), padding=(1, 1))
                (18): ReLU(inplace=True)
                (19): Conv2d(512, 512, kernel_size=(3, 3), stride=(1, 1), padding=(1, 1))
                (20): ReLU(inplace=True)
                (21): Conv2d(512, 512, kernel_size=(3, 3), stride=(1, 1), padding=(1, 1))
                (22): ReLU(inplace=True)
                (23): MaxPool2d(kernel_size=2, stride=2, padding=0, dilation=1, ceil_mode=False)
                (24): Conv2d(512, 512, kernel_size=(3, 3), stride=(1, 1), padding=(1, 1))
                (25): ReLU(inplace=True)
                (26): Conv2d(512, 512, kernel_size=(3, 3), stride=(1, 1), padding=(1, 1))
                (27): ReLU(inplace=True)
                (28): Conv2d(512, 512, kernel_size=(3, 3), stride=(1, 1), padding=(1, 1))
                (29): ReLU(inplace=True)
                (30): MaxPool2d(kernel_size=2, stride=2, padding=0, dilation=1, ceil_mode=False)
              )),
             ('avgpool', AdaptiveAvgPool2d(output_size=(7, 7))),
             ('classifier',
              Sequential(
                (0): Linear(in_features=25088, out_features=4096, bias=True)
                (1): ReLU(inplace=True)
                (2): Dropout(p=0.5, inplace=False)
                (3): Linear(in_features=4096, out_features=4096, bias=True)
                (4): ReLU(inplace=True)
                (5): Dropout(p=0.5, inplace=False)
                (6): Linear(in_features=4096, out_features=1000, bias=True)
              ))])
```

(3) 移除最后一个区块，因为这个范例不进行辨识。

```python
1   class new_model(nn.Module):
2       def __init__(self, pretrained, output_layer):
3           super().__init__()
4           self.output_layer = output_layer
5           self.pretrained = pretrained
6           self.children_list = []
7           # 依次取得每一层
8           for n,c in self.pretrained.named_children():
9               self.children_list.append(c)
10              # 找到特定层即终止
11              if n == self.output_layer:
12                  print('found !!')
13                  break
14
15          # 建构新模型
16          self.net = nn.Sequential(*self.children_list)
17          self.pretrained = None
18
19      def forward(self,x):
20          x = self.net(x)
21          return x
22
23  model = new_model(model, 'avgpool')
24  model = model.to(device)
25  model._modules
```

执行结果：最后一个区块已被移除。

(4) 提取特征：任选一张图片，例如老虎侧面照，取得图片文件的特征向量。

```python
1   # 任选一张图片，例如老虎侧面照，取得图片文件的特征向量
2   from PIL import Image
3   from torchvision import transforms
4
5   filename = './images_test/tiger2.jpg'
6   input_image = Image.open(filename)
7
8   transform = transforms.Compose([
9       transforms.Resize((224, 224)),
10      transforms.ToTensor(),
11      transforms.Normalize(mean=[0.485, 0.456, 0.406],
12                           std=[0.229, 0.224, 0.225])
13  ])
14  input_tensor = transform(input_image)
15  input_batch = input_tensor.unsqueeze(0).to(device) # 增加一维（笔数）
16
17  # 预测
18  model.eval()
19  with torch.no_grad():
20      output = model(input_batch)
21  output
```

执行结果：得到图片文件的特征向量。

```
tensor([[[[0.0000, 0.0000, 0.0000,  ..., 0.0000, 0.0000, 0.0000],
          [0.0000, 0.0000, 0.0000,  ..., 0.2543, 0.0000, 0.0000],
          [0.0000, 0.0000, 0.0000,  ..., 0.0000, 0.0000, 0.0000],
          ...,
          [2.1993, 0.0000, 0.0000,  ..., 0.0000, 0.0000, 0.2719],
          [1.5349, 0.0000, 0.0000,  ..., 0.0000, 1.7577, 5.2424],
          [0.0000, 0.0000, 0.0000,  ..., 0.4238, 2.0388, 5.9582]],

         [[0.0000, 0.0000, 0.0000,  ..., 0.0000, 0.6102, 1.0531],
          [0.7704, 0.0000, 0.0000,  ..., 0.0000, 2.1455, 2.3483],
          [2.3654, 0.4831, 0.0000,  ..., 0.0000, 0.0000, 1.4597],
          ...,
          [0.0000, 2.7323, 5.3333,  ..., 1.9977, 2.2498, 1.4196],
          [0.0000, 3.2158, 3.4539,  ..., 0.5091, 1.0910, 0.5725],
          [0.0000, 0.0000, 0.0000,  ..., 0.0000, 0.0000, 0.0000]],
```

(5) 先取得 images_test 目录下所有后缀为 .jpg 文件名。

```
1  from os import listdir
2  from os.path import isfile, join
3
4  # 取得 images_test 目录下所有后缀为.jpg文件名
5  img_path = './images_test/'
6  image_files = np.array([f for f in listdir(img_path)
7          if isfile(join(img_path, f)) and f[-3:] == 'jpg'])
8  image_files
```

执行结果。

```
array(['astronaut.jpg', 'bird.jpg', 'bird2.jpg', 'cat.jpg', 'daisy1.jpg',
       'daisy2.jpg', 'deer.jpg', 'elephant.jpg', 'elephant2.jpg',
       'lion1.jpg', 'lion2.jpg', 'panda1.jpg', 'panda2.jpg', 'panda3.jpg',
       'rose2.jpg', 'tiger1.jpg', 'tiger2.jpg', 'tiger3.jpg'],
      dtype='<U13')
```

(6) 取得 images_test 目录下所有后缀为 .jpg 文件的像素，并转换及合并。

```
1  import os
2
3  # 合并所有图片文件
4  model.eval()
5  X = torch.tensor([])
6  for filename in image_files:
7      input_image = Image.open(os.path.join(img_path, filename))
8      input_tensor = transform(input_image)
9      input_batch = input_tensor.unsqueeze(0).to(device) # 增加一维(笔数)
10     if len(X.shape) == 1:
11         # print(input_batch.shape)
12         X = input_batch
13     else:
14         # print(input_batch.shape)
15         X = torch.cat((X, input_batch), dim=0)
```

(7) 预测：取得所有图片文件的特征向量。

```
1  # 预测所有图片文件
2  with torch.no_grad():
3      features = model(X)
4  features.shape
```

执行结果：18 张图像的维度为 [18, 512, 7, 7]。

(8) 相似度比较：使用 cosine_similarity 比较特征向量。Cosine Similarity 计算两个向量的夹角，如图 7.6 所示，判断两个向量的方向是否近似，Cosine 介于 (-1，1) 之间，越接近 1，表示方向越相近。

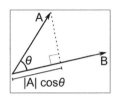

图 7.6　夹角与 Cosine 函数

```
1  from sklearn.metrics.pairwise import cosine_similarity
2
3  # 比较 Tiger2.jpg 与其他图片文件特征向量
4  no=-2
5  print(image_files[no])
6
7  # 转为二维向量，类似扁平层(Flatten)
8  features2 = features.cpu().reshape((features.shape[0], -1))
9
```

## 第 2 篇 | PyTorch 基础

```
10  # 排除 Tiger2.jpg 的其他图片文件特征向量
11  other_features = np.concatenate((features2[:no], features2[no+1:]))
12
13  # 使用 cosine_similarity 计算 Cosine 函数
14  similar_list = cosine_similarity(features2[no:no+1], other_features,
15                                   dense_output=False)
16
17  # 显示相似度，由大排到小
18  print(np.sort(similar_list[0])[::-1])
19
20  # 依相似度，由大排到小，显示文件名
21  image_files2 = np.delete(image_files, no)
22  image_files2[np.argsort(similar_list[0])[::-1]]
```

执行结果：与 tiger2.jpg 比较的相似度，由大排到小。

[0.28911456 0.2833875  0.23362085 0.18441561 0.17196876 0.16713579
 0.14983664 0.12871663 0.11995038 0.11563288 0.10740422 0.09983709
 0.09405126 0.08491081 0.08096127 0.06599604 0.04436902]

对应的文件名。

['tiger1.jpg', 'tiger3.jpg', 'lion1.jpg', 'lion2.jpg',
 'elephant2.jpg', 'cat.jpg', 'elephant.jpg', 'panda1.jpg',
 'bird2.jpg', 'panda3.jpg', 'bird.jpg', 'panda2.jpg', 'deer.jpg',
 'daisy2.jpg', 'rose2.jpg', 'astronaut.jpg', 'daisy1.jpg'],

结果如预期一样是正确的。再比对 bird.jpg，结果不如预期，可能是因为图像尺寸未等比例缩放。无论如何，利用这种方式，不只能够比较 ImageNet 1000 类中的对象，也可以比较其他对象，不限于既有的对象范围，因为我们只借用预先训练模型提取特征的能力。

## 7-4 转移学习

预先训练模型的第三种用法，是采用部分模型，再加上自定义的输入层和辨识层，如此就能够不受限于模型原先辨识的对象，这就是所谓的转移学习 (Transfer Learning) 或者翻译为迁移学习。其实不使用预先训练的模型，直接建构 CNN 模型，也是可以辨识出任何对象的，然而为什么要使用预先训练的模型呢？原因归纳如下。

(1) 预先训练模型使用大量高质量的数据 (ImageNet 为普林斯顿大学与斯坦福大学所主导的项目，有名校做保证！)，又加上设计较复杂的模型结构，例如 ResNet 多达 150 层神经层，准确率因此大大提高。

(2) 使用较少的训练数据：因为模型前半段已经训练好了。

(3) 训练速度比较快：只需要重新训练自定义的辨识层即可。

一般的转移学习分为两阶段：

(1) 建立预先训练的模型 (Pre-trained Model)：包括目前介绍的视觉应用的模型，也包含后面章节会谈到的自然语言模型——Transformer、BERT，它们利用大量的训练数据和复杂的模型结构，取得通用性的图像与自然语言特征向量。

(2) 微调 (Fine Tuning)：依照特定应用领域的需求，微调模型并训练，例如本节所述，利用预先训练模型的前半段，加入自定义的神经层，进行特殊类别的辨识。

## 第 7 章 | 预先训练的模型

**范例3.** 使用**ResNet18**模型，辨识自定义数据集，程序源自*Transfer learning for computer vision tutorial* [11]，笔者进行了一些修改和批注。

使用 ResNet18 模型，辨识自定义数据集的相关步骤如图 7.7 所示。

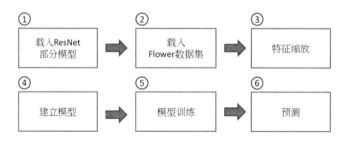

图 7.7　相关步骤

程序：【07_04_transfer_learning.ipynb】。

(1) 载入套件。

```
1  import torch
2  import torch.nn as nn
3  import torch.optim as optim
4  from torch.optim import lr_scheduler
5  import numpy as np
6  import torchvision
7  from torchvision import datasets, models, transforms
8  import matplotlib.pyplot as plt
9  import time
10 import os
11 import copy
```

(2) 载入 hymenoptera 数据集：只含蜜蜂 (Bee) 及蚂蚁 (Ant) 两个类别。

```
1  # 训练资料进行资料增补，验证资料不需要
2  data_transforms = {
3      'train': transforms.Compose([
4          transforms.RandomResizedCrop(224), # 资料增补
5          transforms.RandomHorizontalFlip(),
6          transforms.ToTensor(),
7          transforms.Normalize([0.485, 0.456, 0.406], [0.229, 0.224, 0.225])
8      ]),
9      'val': transforms.Compose([
10         transforms.Resize(256),
11         transforms.CenterCrop(224),
12         transforms.ToTensor(),
13         transforms.Normalize([0.485, 0.456, 0.406], [0.229, 0.224, 0.225])
14     ]),
15 }
16
17 # 使用 ImageFolder 可方便转换为 dataset
18 data_dir = './hymenoptera_data'
19 image_datasets = {x: datasets.ImageFolder(os.path.join(data_dir, x),
20                                           data_transforms[x])
21                   for x in ['train', 'val']}
22 dataloaders = {x: torch.utils.data.DataLoader(image_datasets[x], batch_size=4,
23                                                shuffle=True, num_workers=4)
24                for x in ['train', 'val']}
25
26 # 取得资料笔数
27 dataset_sizes = {x: len(image_datasets[x]) for x in ['train', 'val']}
28
29 # 取得类别
30 class_names = image_datasets['train'].classes
```

执行结果：共 244 笔训练数据、153 笔验证数据。

(3) 取得一批数据显示图像。

```python
def imshow(inp, title=None):
    inp = inp.numpy().transpose((1, 2, 0))
    mean = np.array([0.485, 0.456, 0.406])
    std = np.array([0.229, 0.224, 0.225])
    inp = std * inp + mean
    inp = np.clip(inp, 0, 1)
    plt.axis('off')
    plt.imshow(inp)
    if title is not None:
        plt.title(title)
    plt.pause(0.001)  # pause a bit so that plots are updated

# 取得一批数据
inputs, classes = next(iter(dataloaders['train']))

# 显示一批数据
out = torchvision.utils.make_grid(inputs)
imshow(out, title=[class_names[x] for x in classes])
```

执行结果。

(4) 定义模型训练函数。

```python
# 同时含训练/评估
def train_model(model, criterion, optimizer, scheduler, num_epochs=25):
    since = time.time()
    
    best_model_wts = copy.deepcopy(model.state_dict())
    best_acc = 0.0
    
    for epoch in range(num_epochs):
        print('Epoch {}/{}'.format(epoch, num_epochs - 1))
        print('-' * 10)
        
        # Each epoch has a training and validation phase
        for phase in ['train', 'val']:
            if phase == 'train':
                model.train()  # Set model to training mode
            else:
                model.eval()   # Set model to evaluate mode
            
            running_loss = 0.0
            running_corrects = 0
            
            # 逐批训练或验证
            for inputs, labels in dataloaders[phase]:
                inputs = inputs.to(device)
                labels = labels.to(device)
                
                # zero the parameter gradients
                optimizer.zero_grad()
                # 训练时需要梯度下降
                with torch.set_grad_enabled(phase == 'train'):
                    outputs = model(inputs)
                    _, preds = torch.max(outputs, 1)
                    loss = criterion(outputs, labels)
                    
                    # 训练时需要 backward + optimize
                    if phase == 'train':
                        loss.backward()
                        optimizer.step()
                
                # 统计损失
                running_loss += loss.item() * inputs.size(0)
                running_corrects += torch.sum(preds == labels.data)
```

```python
44        if phase == 'train':
45            scheduler.step()
46
47        epoch_loss = running_loss / dataset_sizes[phase]
48        epoch_acc = running_corrects.double() / dataset_sizes[phase]
49
50        print('{} Loss: {:.4f} Acc: {:.4f}'.format(
51            phase, epoch_loss, epoch_acc))
52
53        # 如果是评估阶段，且准确率创新高即存入 best_model_wts
54        if phase == 'val' and epoch_acc > best_acc:
55            best_acc = epoch_acc
56            best_model_wts = copy.deepcopy(model.state_dict())
57    print()
58
59 time_elapsed = time.time() - since
60 print('Training complete in {:(time_elapsed // 60):.0f}m {(time_elapsed % 60):
61 print(f'Best val Acc: {best_acc:4f}')
62
63 # 载入最佳模型
64 model.load_state_dict(best_model_wts)
65 return model
```

(5) 定义显示预测结果的函数。

```python
1 def imshow2(inp, title=None):
2     inp = inp.numpy().transpose((1, 2, 0))
3     mean = np.array([0.485, 0.456, 0.406])
4     std = np.array([0.229, 0.224, 0.225])
5     inp = std * inp + mean
6     inp = np.clip(inp, 0, 1)
7     plt.imshow(inp)
```

```python
1  def visualize_model(model, num_images=6):
2      was_training = model.training
3      model.eval()
4      images_so_far = 0
5      fig = plt.figure()
6
7      with torch.no_grad():
8          for i, (inputs, labels) in enumerate(dataloaders['val']):
9              inputs = inputs.to(device)
10             labels = labels.to(device)
11
12             outputs = model(inputs)
13             _, preds = torch.max(outputs, 1)
14
15             for j in range(inputs.size()[0]):
16                 images_so_far += 1
17                 plt.subplot(num_images//4+1, 4, images_so_far)
18                 plt.axis('off')
19                 plt.title(class_names[preds[j]])
20                 imshow2(inputs.cpu().data[j])
21
22                 if images_so_far == num_images:
23                     model.train(mode=was_training)
24                     return
25         model.train(mode=was_training)
26     plt.tight_layout()
27     plt.show()
```

(6) 建立模型结构：使用 ResNet18 加上自定义的辨识层，直接将最后一层改为只辨识两类，或在最后一层加上一个辨识层，优化器采用排程 (scheduler)，随着执行周期，学习率逐渐降低，以追求更精准的最佳解，同时兼顾训练时间的缩短。

```python
1  model_ft = models.resnet18(pretrained=True)
2  num_ftrs = model_ft.fc.in_features
3  # 改为自订辨识层
4  model_ft.fc = nn.Linear(num_ftrs, 2)
5
6  model_ft = model_ft.to(device)
7
8  # 定义损失函数
9  criterion = nn.CrossEntropyLoss()
10
11 # 定义优化器
12 optimizer_ft = optim.SGD(model_ft.parameters(), lr=0.001, momentum=0.9)
13
14 # 每7个执行周期，学习率降 0.1
15 exp_lr_scheduler = lr_scheduler.StepLR(optimizer_ft, step_size=7, gamma=0.1)
```

(7) 模型训练：CPU 训练时间需 15~20 分钟，GPU 训练时间少于 5 分钟。

```
1  model_ft = train_model(model_ft, criterion, optimizer_ft, exp_lr_scheduler,
2                         num_epochs=25)
```

执行结果：训练时间 4 分钟，最佳准确率 (Best val Acc) 为 0.934641。

(8) 显示预测结果。

```
1  visualize_model(model_ft)
```

执行结果：很准确。

(9) 改进：可设定预先训练的模型不用重新训练，CPU 训练时间可以减半。

```
1  model_conv = torchvision.models.resnet18(pretrained=True)
2  for param in model_conv.parameters():
3      # 不用重新训练
4      param.requires_grad = False
```

进阶的技巧可参考 Pytorch 官网 *Quantized Transfer Learning for Computer Vision Tutorial* [12]，其中说明了如何调换神经层。

## 7-5 Batch Normalization 说明

上一节我们使用复杂的 ResNet18 模型，其中内含许多 Batch Normalization 神经层，它在神经网络的反向传导时可消除梯度消失 (Gradient Vanishing) 或梯度爆炸 (Gradient Exploding) 现象，所以，我们花点时间研究其原理与应用时机。

当神经网络包含很多神经层时，经常会在其中放置一些 Batch Normalization 层，顾名思义，它的用途应该是特征缩放，然而，内部究竟是如何运作的？有哪些好处？运用的时机？摆放的位置如何？

Sergey Ioffe 与 Christian Szegedy 在 2015 年首次提出 Batch Normalization，论文标题为 *Batch Normalization: Accelerating Deep Network Training by Reducing Internal Covariate Shift*[13]。简单来说，Batch Normalization 即为特征缩放，将前一层的输出标准化后，再转至下一层。

标准化的好处就是让收敛速度快一点，假如没有标准化，模型通常会针对梯度较大的变数先优化，进而造成收敛路线曲折前进，如图 7.8 所示，左图是特征未标准化的优化路径，右图则是标准化后的优化路径。

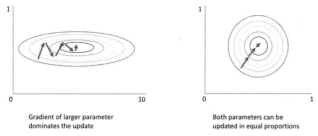

图 7.8　未标准化对比标准化优化过程的示意图

图片来源：Why Batch Normalization Matters? [14]

Batch Normalization 另外再引进两个变量 $\gamma$、$\beta$，分别控制规模缩放 (Scale) 和偏移 (Shift)，如图 7.9 所示。

**Input:** Values of $x$ over a mini-batch: $\mathcal{B} = \{x_{1...m}\}$;
Parameters to be learned: $\gamma, \beta$
**Output:** $\{y_i = \mathrm{BN}_{\gamma,\beta}(x_i)\}$

$$\mu_{\mathcal{B}} \leftarrow \frac{1}{m}\sum_{i=1}^{m} x_i \qquad \text{// mini-batch mean}$$

$$\sigma_{\mathcal{B}}^2 \leftarrow \frac{1}{m}\sum_{i=1}^{m} (x_i - \mu_{\mathcal{B}})^2 \qquad \text{// mini-batch variance}$$

$$\widehat{x}_i \leftarrow \frac{x_i - \mu_{\mathcal{B}}}{\sqrt{\sigma_{\mathcal{B}}^2 + \epsilon}} \qquad \text{// normalize}$$

$$y_i \leftarrow \gamma \widehat{x}_i + \beta \equiv \mathrm{BN}_{\gamma,\beta}(x_i) \qquad \text{// scale and shift}$$

图 7.9　Batch Normalization 公式

图片来源：Why Batch Normalization Matters? [14]

补充说明：

(1) 标准化是在训练时逐批处理的，而非同时将所有数据一起标准化，通常加在 Activation Function 之前。

(2) $\varepsilon$ 是为了避免分母为 0 而加上的一个微小正数。

(3) $\gamma$、$\beta$ 值是由训练过程中计算出来的，并不是事先设定好的。

假设我们要建立小狗的辨识模型，收集黄狗的图片进行训练，模型完成后，却拿有斑纹的狗的图片来辨识，效果想当然会变差，要改善的话则必须重新收集数据再训练一次，这种现象就称为 Covariate Shift，正式的定义是“假设我们要使用 $X$ 预测 $Y$ 时，当 $X$ 的分配随着时间有所变化时，模型就会逐渐失效”。股价预测也是类似的情形，当股价长期趋势上涨时，原来的模型就慢慢失准了，除非纳入最新的数据重新训练模型。

由于神经网络的权重会随着反向传导不断更新，每一层的输出都会受到上一层的输出影响，这是一种回归的关系，随着神经层越多，整个神经网络的输出有可能会逐渐偏移，此种现象称为 Internal Covariate Shift。

而 Batch Normalization 就可以校正 Internal Covariate Shift 现象，它在输出至下一层的神经层时，每批数据都会先被标准化，这使得输入数据的分布全属于 $N(0, 1)$ 的标准常态分配，因此，不管有多少层神经层，都不用担心发生输出逐渐偏移的问题。

至于什么是梯度消失和梯度爆炸？这是由于 CNN 模型共享权值 (Shared Weights) 的关系，使得梯度逐渐消失或爆炸，原因如下，相同的 $W$ 值若是经过很多层：

(1) 如果 $W<1$ ➜ 模型前几层的 $n$ 越大，$W^n$ 会趋近于 0，则影响力逐渐消失，即梯度消失 (Gradient Vanishing)。

(2) 如果 $W>1$ ➜ 模型前几层的 $n$ 越大，$W^n$ 会趋近于 ∞，则造成模型优化无法收敛，即梯度爆炸 (Gradient Explosion)。

只要经过 Batch Normalization，将每一批标准化后，梯度都会重新计算，这样就不会有梯度消失或梯度爆炸的状况发生了。除此之外，根据原作者的说法，Batch Normalization 还有以下优点：

(1) 优化收敛速度快 (Train faster)。

(2) 可使用较大的学习率 (Use higher learning rates)，加快训练过程。

(3) 权重初始化较容易 (Parameter initialization is easier)。

(4) 不使用 Batch Normalization 时，Activation function 容易在训练过程中消失或提早停止学习，但如果经过 Batch Normalization 则又会再复活 (Makes activation functions viable by regulating the inputs to them)。

(5) 准确率全面性提升 (Better results overall)。

(6) 类似 Dropout 的效果，可防止过度拟合 (It adds noise which reduces overfitting with a regularization effect)，所以，当使用 Batch Normalization 时，就不需要加 Dropout 层了，为避免效果加乘过强，反而造成低度拟合 (Underfitting)。

有一篇文章 *On The Perils of Batch Norm* [15] 做了一个很有趣的试验，使用两个数据集模拟 Internal Covariate Shift 现象，一个是 MNIST 数据集，背景是单纯白色，另一个则是 SVHN 数据集，有复杂的背景，试验过程如下。

首先合并两个数据集来训练第一种模型，如图 7.10 所示。

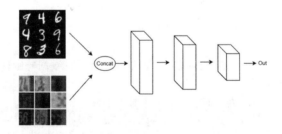

图 7.10　合并两个数据集来训练一个模型

再使用两个数据集各自分别训练模型，但共享权值，为第二种模型，如图 7.11 所示。

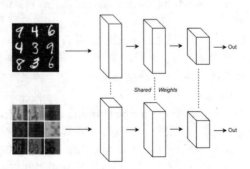

图 7.11　使用两个数据集个别训练模型，但共享权值

两种模型都有插入 Batch Normalization，比较结果，前者即单一模型准确度较高，因为 Batch Normalization 可以校正 Internal Covariate Shift 现象。后者则由于数据集内容的不同，两个模型共享权值本来就不合理，如图 7.12 所示。

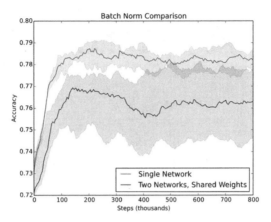

图 7.12　两种模型准确率比较

第三种模型：使用两个数据集训练两个模型，个别作 Batch Normalization，但不共享权值。比较结果，第三种模型效果最好，如图 7.13 所示。

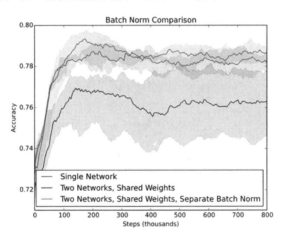

图 7.13　三种模型准确率的比较

# 第 3 篇 进阶的影像应用

恭喜各位读者通过卷积神经网络 (CNN) 的关卡，越过一座高山，本篇就来好好秀一下努力的成果，展现 CNN 在各领域应用上有哪些厉害的功能吧！

本篇包括下列主题：
- 对象侦测 (Object Detection)。
- 语义分割 (Semantic Segmentation)。
- 人脸辨识 (Facial Recognition)。
- 风格转换 (Style Transfer)。
- 光学文字辨识 (Optical Character Recognition, OCR)。

# 第 8 章
# 对象侦测

前面介绍的图像辨识模型，一张图片中仅含有一个对象，接下来要讲的对象侦测可以同时侦测多个对象，并且标示出对象的位置。标示位置有什么用处呢？图 8.1 是现今最热门的对象侦测算法 YOLO 的发明人 Joseph Redmon 给出的一张有趣的照片。

图 8.1　机器人制作煎饼

图片来源：*Real-Time Grasp Detection Using Convolutional Neural Networks* [1]

机器人要能完成制作煎饼的任务，它必须知道煎饼的所在位置，才能够将饼翻面，如果有两张以上的饼，还需要知道要翻哪一张。不只机器人工作时需要计算机视觉，其他领域也会用到对象侦测。

(1) 自动驾驶车 (Self-Driving)：需要实时掌握前方路况及闪避障碍物。

(2) 智能交通：车辆侦测，利用一辆车在两个时间点的位置，计算车速，进而可以推算道路拥堵的状况，也可以用来侦测违规车辆。

(3) 玩具、无人机、导弹等都可以做类似的应用。

(4) 异常侦测 (Anomaly Detection)：可以用在生产在线架设摄影机，实时侦测异常的瑕疵，如印制电路板、产品外观等。

(5) 无人商店的购物篮扫描，自动结账。

## 8-1　图像辨识模型的发展

综观历年 ImageNet ILSVRC 挑战赛 (Large Scale Visual Recognition Challenge) 的竞赛题目，从 2011 年的影像分类 (Classification) 与定位 (Classification with Localization) [2]，到 2017 年，题目扩展至对象定位 (Object Localization)、对象侦测 (Object Detection)、影片对象侦测 (Object Detection from Video) [3]。我们可从中观察到图像辨识模型的发展

史，了解到整个技术的演进。目前图像辨识大概分为四大类型，如图 8.2 所示。

图 8.2　对象侦测类型

图片来源：Detection and Segmentation [4]

语义分割 (Semantic Segmentation)：按照对象类别来划分像素区域，但不区分实例 (Instance)。拿图 8.2 的第 4 张照片为例，照片中有 2 只狗，都使用同一种颜色表达，即是语义分割，2 只狗使用不同颜色来表示，区分实例，则称为实例分割。

定位 (Classification + Localization)：标记单一对象 (Single Object) 的类别与所在的位置。

对象侦测 (Object Detection)：标注多个对象 (Multiple Object) 的类别与所在的位置。

实例分割 (Instance Segmentation)：标记实例 (Instance)，同一类的对象可以区分，并标示个别的位置，尤其是对象之间有重叠时。

接下来逐一介绍上述四类算法，并说明如何利用 PyTorch 实操。

## 8-2　滑动窗口

对象侦测要能够同时辨识对象的类别与位置，如果拆开来看就是两项任务 (Task)。

分类 (Classification)：辨识对象的类别。

回归 (Regression)：找到对象的位置，包括对象左上角的坐标和宽度 / 高度。

在使用神经网络前，我们先介绍对象侦测的传统方法，结合滑动窗口 (Sliding Window)、影像金字塔 (Image Pyramid) 及方向梯度直方图 (Histogram of Oriented Gradient，HOG)，步骤如图 8.3 所示。

图 8.3　相关步骤

(1) 设定某一尺寸的窗口，比如宽高各为 128 像素，由原图左上角起裁剪成窗口大小。

(2) 辨识窗口内是否有对象存在。

(3) 滑动窗口，再次裁剪，并回到步骤 (2)，直到全图扫描完为止。

(4) 缩小原图尺寸后，再重新回到步骤 (1)，辨识窗口保持不变，这样就可以寻找较大尺寸的对象。

这种将原图缩小成各种尺寸的方式称为影像金字塔 (Image Pyramid)，详情请参阅 *Image Pyramids with Python and OpenCV* [5] 一文，如图 8.4 所示。

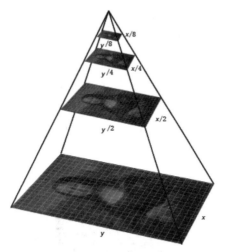

图 8.4　影像金字塔 (Image Pyramid)，最下层为原图，往上逐步缩小原图尺寸

图片来源：IIPImage [6]

**范例1. 先实操滑动窗口及影像金字塔。** 本范例程序修改自 *Sliding Windows for Object Detection with Python and OpenCV* [7]。

**程序：**【08_01_Sliding_Window_And_Image_Pyramid.ipynb】。

(1) 加载套件，需先安装 OpenCV、imutils，imutils 为一简单版的图像处理套件。

```
import cv2
import time
import imutils
```

(2) 定义影像金字塔操作函数：逐步缩小原图尺寸，以便找到较大尺寸的对象。

```
# 影像金字塔操作
# image：原图，scale：每次缩小倍数，minSize：最小尺寸
def pyramid(image, scale=1.5, minSize=(30, 30)):
    # 第一次回传原图
    yield image
    
    while True:
        # 计算缩小后的尺寸
        w = int(image.shape[1] / scale)
        # 缩小
        image = imutils.resize(image, width=w)
        # 直到最小尺寸为止
        if image.shape[0] < minSize[1] or image.shape[1] < minSize[0]:
            break
        # 回传缩小后的图像
        yield image
```

(3) 定义滑动窗口函数。

```
# 滑动窗口
def sliding_window(image, stepSize, windowSize):
    for y in range(0, image.shape[0], stepSize):      # 向下滑动 stepSize 格
        for x in range(0, image.shape[1], stepSize):  # 向右滑动 stepSize 格
            # 回传裁剪后的窗口
            yield (x, y, image[y:y + windowSize[1], x:x + windowSize[0]])
```

(4) 测试。

```
1   # 读取一个图文件
2   image = cv2.imread('./images_Object_Detection/lena.jpg')
3
4   # 视窗尺寸
5   (winW, winH) = (128, 128)
6
7   # 取得影像金字塔各种尺寸
8   for resized in pyramid(image, scale=1.5):
9       # 滑动窗口
10      for (x, y, window) in sliding_window(resized, stepSize=32,
11                                            windowSize=(winW, winH)):
12          # 窗口尺寸不合即放弃，滑动至边缘时，尺寸过小
13          if window.shape[0] != winH or window.shape[1] != winW:
14              continue
15          # 标示滑动的窗口
16          clone = resized.copy()
17          cv2.rectangle(clone, (x, y), (x + winW, y + winH), (0, 255, 0), 2)
18          cv2.imshow("Window", clone)
19          cv2.waitKey(1)
20          # 暂停
21          time.sleep(0.025)
22
23  # 结束时关闭窗口
24  cv2.destroyAllWindows()
```

执行结果。

由于 Jupyter Notebook 不适合播放动画，请执行以下指令 (py 文件内容与【08_01_Sliding_Window_And_Image_Pyramid.ipynb】完全相同)：python 08_01_Sliding_Window_And_Image_Pyramid.py

## 8-3 方向梯度直方图

方向梯度直方图 (Histogram of Oriented Gradient, HOG) 是抓取图像轮廓线条的算法，先将图片切成很多个区域 (Cell)，从每个区域中找出方向梯度，并把它描绘出来，就形成了对象的轮廓，与其他边缘提取的算法比起来，它对环境的变化有较强 (Robust) 的适应力，例如不受光线影响，如图 8.5 所示。有关 HOG 的详细内容可参阅《方向梯度直方图 (HOG)》[8] 一文。

图 8.5　HOG 处理：左图为原图；右图为 HOG 处理过后的图，可抓到对象的轮廓

根据 *Histogram of Oriented Gradients and Object Detection* [9] 一文的介绍，结合了 HOG 的对象侦测，流程如图 8.6 所示。

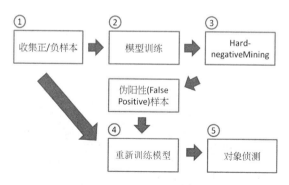

图 8.6　结合 HOG 的对象侦测的流程图

(1) 收集正样本 (Positive Set)：集结目标对象的各种图像样本，包括不同视角、尺寸、背景的图像。

(2) 收集负样本 (Negative Set)：集结无目标对象的各种图像样本，若找到相近的对象则更好，可增加辨识准确度。

(3) 使用以上正/负样本与分类算法训练二分类模型，判断是否包含目标对象，一般使用支持向量机 (SVM) 算法。

(4) Hard-Negative Mining：扫描负样本，使用滑动窗口的技巧，将每个窗口输入模型来预测，如果侦测到目标对象，即是伪阳性 (False Positive)，接着将这些图像加到训练数据集中重新进行训练，这个步骤可以重复很多次，能够有效地提高模型准确率，类似 Boosting 整体学习算法。

(5) 使用最后的模型进行对象侦测：将目标对象的图像用滑动窗口与影像金字塔技巧，输入模型进行辨识，找出合格的窗口。

(6) 筛选合格的窗口：使用 Non-Maximum Suppression (NMS) 算法，剔除多余重叠的窗口。

**范例2. 使用HOG、滑动窗口及SVM进行对象侦测。**

程序：【08_02_HOG-Face-Detection.ipynb】，修改自 scikit-image 的范例。

(1) 载入套件：本例使用 scikit-image 套件，OpenCV 也支持类似的函数。

```
2  import numpy as np
3  import matplotlib.pyplot as plt
4  from skimage.feature import hog
5  from skimage import data, exposure
```

(2) HOG 测试：使用 scikit-image 内建的女航天员图像来测试 HOG 的效果。

```
1   # 测试图片
2   image = data.astronaut()
3
4   # 取得图片的HOG
5   fd, hog_image = hog(image, orientations=8, pixels_per_cell=(16, 16),
6                       cells_per_block=(1, 1), visualize=True, multichannel=True)
7
8   # 原图与HOG图比较
9   fig, (ax1, ax2) = plt.subplots(1, 2, figsize=(12, 6), sharex=True, sharey=True)
10
11  ax1.axis('off')
```

```
12  ax1.imshow(image, cmap=plt.cm.gray)
13  ax1.set_title('Input image')
14
15  # 调整对比度,让显示比较清楚
16  hog_image_rescaled = exposure.rescale_intensity(hog_image, in_range=(0, 10))
17
18  ax2.axis('off')
19  ax2.imshow(hog_image_rescaled, cmap=plt.cm.gray)
20  ax2.set_title('Histogram of Oriented Gradients')
21  plt.show()
```

执行结果:原图与 HOG 处理过后的图比较。

(3) 收集正样本 (Positive Set):使用 scikit-learn 内建的人脸数据集作为正样本,共有 13233 笔。

```
1  # 收集正样本 (positive set)
2  # 使用 scikit-learn 的人脸数据集
3  from sklearn.datasets import fetch_lfw_people
4  faces = fetch_lfw_people()
5  positive_patches = faces.images
6  positive_patches.shape
```

(4) 观察正样本中部分的图片。

```
2  fig, ax = plt.subplots(4,6)
3  for i, axi in enumerate(ax.flat):
4      axi.imshow(positive_patches[500 * i], cmap='gray')
5      axi.axis('off')
```

执行结果:每张图片宽高为 (62, 47)。

(5) 收集负样本 (Negative Set):使用 scikit-image 内建的数据集,共有 9 笔。

```
1  # 收集负样本 (negative set)
2  # 使用 scikit-image 的非人脸数据
3  from skimage import data, transform, color
4
5  imgs_to_use = ['hubble_deep_field', 'text', 'coins', 'moon',
6                 'page', 'clock','coffee','chelsea','horse']
7  images = [color.rgb2gray(getattr(data, name)())
8            for name in imgs_to_use]
9  len(images)
```

(6) 增加负样本笔数：将负样本转换为不同的尺寸，也可以使用数据增强技术。

```python
1  # 将负样本转换为不同的尺寸
2  from sklearn.feature_extraction.image import PatchExtractor
3
4  # 转换为不同的尺寸
5  def extract_patches(img, N, scale=1.0, patch_size=positive_patches[0].shape):
6      extracted_patch_size = tuple((scale * np.array(patch_size)).astype(int))
7      # PatchExtractor：产生不同尺寸的图像
8      extractor = PatchExtractor(patch_size=extracted_patch_size,
9                                 max_patches=N, random_state=0)
10     patches = extractor.transform(img[np.newaxis])
11     if scale != 1:
12         patches = np.array([transform.resize(patch, patch_size)
13                             for patch in patches])
14     return patches
15
16 # 产生 27000 张图像
17 negative_patches = np.vstack([extract_patches(im, 1000, scale)
18                               for im in images for scale in [0.5, 1.0, 2.0]])
19 negative_patches.shape
```

执行结果：产生 27000 张图像。

(7) 观察负样本中部分的图片。

```python
2  fig, ax = plt.subplots(4,6)
3  for i, axi in enumerate(ax.flat):
4      axi.imshow(negative_patches[600 * i], cmap='gray')
5      axi.axis('off')
```

执行结果。

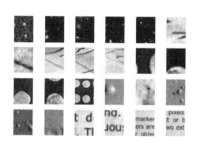

(8) 合并正样本与负样本。

```python
2  from skimage import feature    # To use skimage.feature.hog()
3  from itertools import chain
4
5  X_train = np.array([feature.hog(im)
6                      for im in chain(positive_patches,
7                                      negative_patches)])
8  y_train = np.zeros(X_train.shape[0])
9  y_train[:positive_patches.shape[0]] = 1
```

(9) 使用 SVM 进行二分类的训练：使用 GridSearchCV 寻求最佳参数值。

```python
2  from sklearn.svm import LinearSVC
3  from sklearn.model_selection import GridSearchCV
4
5  # C为矫正过度拟合强度的倒数，使用 GridSearchCV 寻求最佳参数值
6  grid = GridSearchCV(LinearSVC(dual=False), {'C': [1.0, 2.0, 4.0, 8.0]},cv=3)
7  grid.fit(X_train, y_train)
8  grid.best_score_
```

执行结果：最佳模型准确率为 98.77%。

(10) 取得最佳参数值。

```python
2  grid.best_params_
```

(11) 按最佳参数值再训练一次，取得最终模型。

```
2  model = grid.best_estimator_
3  model.fit(X_train, y_train)
```

(12) 新图像测试：需先转为灰度图。

```
1  # 获取新图像测试
2  test_img = data.astronaut()
3  test_img = color.rgb2gray(test_img)
4  test_img = transform.rescale(test_img, 0.5)
5  test_img = test_img[:120, 60:160]
6
7
8  plt.imshow(test_img, cmap='gray')
9  plt.axis('off');
```

执行结果。

(13) 定义滑动窗口函数。

```
2  def sliding_window(img, patch_size=positive_patches[0].shape,
3                    istep=2, jstep=2, scale=1.0):
4      Ni, Nj = (int(scale * s) for s in patch_size)
5      for i in range(0, img.shape[0] - Ni, istep):
6          for j in range(0, img.shape[1] - Ni, jstep):
7              patch = img[i:i + Ni, j:j + Nj]
8              if scale != 1:
9                  patch = transform.resize(patch, patch_size)
10             yield (i, j), patch
```

(14) 计算 HOG：使用滑动窗口来计算每一滑动窗口的 HOG，输入模型辨识。

```
1  # 使用滑动窗口计算每一窗口的 HOG
2  indices, patches = zip(*sliding_window(test_img))
3  patches_hog = np.array([feature.hog(patch) for patch in patches])
4
5  # 辨识每一窗口
6  labels = model.predict(patches_hog)
7  labels.sum()  # 侦测到的总数
```

执行结果：共有 55 个合格窗口。

(15) 显示这 55 个合格窗口。

```
1  # 将每一个侦测到的窗口显示出来
2  fig, ax = plt.subplots()
3  ax.imshow(test_img, cmap='gray')
4  ax.axis('off')
5
6  # 取得左上角坐标
7  Ni, Nj = positive_patches[0].shape
8  indices = np.array(indices)
9
10 # 显示
11 for i, j in indices[labels == 1]:
12     ax.add_patch(plt.Rectangle((j, i), Nj, Ni, edgecolor='red',
13                                alpha=0.3, lw=2, facecolor='none'))
```

执行结果。

(16) 筛选合格窗口：使用Non-Maximum Suppression(NMS)算法，剔除多余的窗口。以下采用 Non-Maximum Suppression for Object Detection in Python [10] 一文的程序代码。

定义NMS算法函数：这是由Pedro Felipe Felzenszwalb等学者发明的算法，执行速度较慢，Tomasz Malisiewicz [11] 因而提出改善的算法。函数的重叠比例门槛(OverlapThresh)参数一般设为0.3~0.5。

```python
1   # Non-Maximum Suppression演算法 by Felzenszwalb et al.
2   # boxes：所有候选的窗口，overlapThresh：窗口重叠的比例门槛
3   def non_max_suppression_slow(boxes, overlapThresh=0.5):
4       if len(boxes) == 0:
5           return []
6   
7       pick = []                    # 存储筛选的结果
8       x1 = boxes[:,0]              # 取得候选的窗口的左/上/右/下 坐标
9       y1 = boxes[:,1]
10      x2 = boxes[:,2]
11      y2 = boxes[:,3]
12  
13      # 计算候选窗口的面积
14      area = (x2 - x1 + 1) * (y2 - y1 + 1)
15      idxs = np.argsort(y2)        # 依窗口的底Y坐标排序
16  
17      # 比对重叠比例
18      while len(idxs) > 0:
19          # 最后一笔
20          last = len(idxs) - 1
21          i = idxs[last]
22          pick.append(i)
23          suppress = [last]
24  
25          # 比对最后一笔与其他窗口重叠的比例
26          for pos in range(0, last):
27              j = idxs[pos]
28  
29              # 取得所有窗口的涵盖范围
30              xx1 = max(x1[i], x1[j])
31              yy1 = max(y1[i], y1[j])
32              xx2 = min(x2[i], x2[j])
33              yy2 = min(y2[i], y2[j])
34              w = max(0, xx2 - xx1 + 1)
35              h = max(0, yy2 - yy1 + 1)
36  
37              # 计算重叠比例
38              overlap = float(w * h) / area[j]
39  
40              # 如果大于门槛值，则存储起来
41              if overlap > overlapThresh:
42                  suppress.append(pos)
43  
44          # 删除合格的窗口，继续比对
45          idxs = np.delete(idxs, suppress)
46  
47      # 回传合格的窗口
48      return boxes[pick]
```

(17) 调用 non_max_suppression_slow 函数，剔除多余的窗口。

```
1  # 使用 non-maximum suppression 演算法，剔除多余的窗口。
2  candidate_boxes = []
3  for i, j in indices[labels == 1]:
4      candidate_boxes.append([j, i, Nj, Ni])
5  final_boxes = non_max_suppression_slow(np.array(candidate_boxes).reshape(-1, 4))
6
7  # 将每一个合格的窗口显示出来
8  fig, ax = plt.subplots()
9  ax.imshow(test_img, cmap='gray')
10 ax.axis('off')
11
12 # 显示
13 for i, j, Ni, Nj in final_boxes:
14     ax.add_patch(plt.Rectangle((i, j), Ni, Nj, edgecolor='red',
15                                alpha=0.3, lw=2, facecolor='none'))
```

执行结果：得到两个合格窗口。

以上范例的过程中省略了一些细节，例如 Hard-Negative Mining、影像金字塔，这个例子无法侦测多个不同实体 (Instance) 与不同尺寸的对象。所以我们再来看另一个范例，可使用任何 CNN 模型结合影像金字塔，进行多对象、多实体的侦测。

**范例3. 预先训练模型可以辨识1000种对象，我们以ResNet50辨识裁剪的图片是否含对象，取代前例的HOG。**

以 ResNet50 辨识裁剪的图片是否含对象的相关步骤如图 8.7 所示。

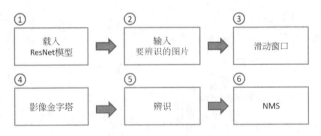

图 8.7　相关步骤

程序：【08_03_Object_Detection.ipynb】。

(1) 载入套件。

```
1  import torch
2  import torch.nn as nn
3  import torch.optim as optim
4  from torch.optim import lr_scheduler
5  import torchvision
6  from torchvision import datasets, models, transforms
7  import numpy as np
8  import time
9  import cv2
```

(2) 参数设定：此范例是辨识自行车的图像 (bike.jpg)。

```
2  WIDTH = 600              # 图像缩放为 (600, 600)
3  PYR_SCALE = 1.5          # 影像金字塔缩放比例
4  WIN_STEP = 16            # 窗口滑动步数
5  ROI_SIZE = (250, 250)    # 窗口大小
6  INPUT_SIZE = (224, 224)  # CNN的输入尺寸
```

(3) 加载 ResNet50 模型：注意，本例一次预测数百个窗口会造成 GPU 内存不足，故改用 CPU，直接令 device = "cpu"。

```
1  model = models.resnet50(pretrained=True).to(device)
```

(4) 读取要辨识的图片。

```
1  from PIL import Image
2
3  filename = './images_Object_Detection/bike.jpg'
4  orig = Image.open(filename)
5  # 等比例缩放图片
6  orig = orig.resize((WIDTH, int(orig.size[1] / orig.size[0] * WIDTH)))
7  Width_Height_ratio = orig.size[1] / orig.size[0]
8  orig.size
```

(5) 定义滑动窗口和影像金字塔函数，这部分与范例 1 的流程相同。

```
1   # 滑动窗口函数
2   def sliding_window(image, step, ws):
3       for y in range(0, image.size[1] - ws[1], step):    # 向下滑动 stepSize 格
4           for x in range(0, image.size[0] - ws[0], step): # 向右滑动 stepSize 格
5               # 回传裁剪后的窗口
6               yield (x, y, image.crop((x, y, x + ws[0], y + ws[1])))
7
8   # 影像金字塔函数
9   # image：原图，scale：每次缩小倍数，minSize：最小尺寸
10  def image_pyramid(image, scale=1.5, minSize=(224, 224)):
11      # 第一次回传原图
12      yield image
13
14      # keep looping over the image pyramid
15      while True:
16          # 计算缩小后的尺寸
17          w = int(image.size[0] / scale)
18          image = image.resize((w, int(Width_Height_ratio * w)))
19
20          # 直到最小尺寸为止
21          if image.size[0] < minSize[1] or image.size[1] < minSize[0]:
22              break
23
24          # 回传缩小后的图像
25          yield image
```

(6) 定义转换函数：PyTorch 预设支持 PIL 格式，但其像素操作较不方便，故定义格式转换函数。

```
1   # 转换函数
2   transform = transforms.Compose([
3       transforms.Resize(INPUT_SIZE),
4       transforms.ToTensor(),
5       transforms.Normalize(mean=[0.485, 0.456, 0.406],
6                            std=[0.229, 0.224, 0.225])
7   ])
8
9   # PIL 格式转换为 OpenCV 格式
10  def PIL2CV2(orig):
11      pil_image = orig.copy()
12      open_cv_image = np.array(pil_image)
13      return open_cv_image[:, :, ::-1].copy()
```

(7) 经由影像金字塔与滑动窗口操作，取得每一个要侦测的窗口。

```
1   # 产生影像金字塔
2   pyramid = image_pyramid(orig, scale=PYR_SCALE, minSize=ROI_SIZE)
3   rois = torch.tensor([])        # 候选框
4   locs = []                      # 位置
5   for image in pyramid:
6       # 框与原图的比例
7       scale = WIDTH / float(image.size[0])
8       print(image.size, 1/scale)
9
10      # 滑动窗口
11      for (x, y, roiOrig) in sliding_window(image, WIN_STEP, ROI_SIZE):
12          # 取得候选框
13          x = int(x * scale)
14          y = int(y * scale)
15          w = int(ROI_SIZE[0] * scale)
16          h = int(ROI_SIZE[1] * scale)
17
18          # 缩放图形以符合模型输入规格
19          roi = transform(roiOrig)
20          roi = roi.unsqueeze(0) # 增加一维(笔数)
21
22          # 加入输出变数中
23          if len(rois.shape) == 1:
24              rois = roi
25          else:
26              rois = torch.cat((rois, roi), dim=0)
27          locs.append((x, y, x + w, y + h))
28
29  rois = rois.to(device)
```

(8) 预测。

```
1   # 读取类别列表
2   with open("imagenet_classes.txt", "r") as f:
3       categories = [s.strip() for s in f.readlines()]
4
5   # 预测
6   model.eval()
7   with torch.no_grad():
8       output = model(rois)
9
10  # 转成概率
11  probabilities = torch.nn.functional.softmax(output, dim=1)
12
13  # 取得第一名
14  top_prob, top_catid = torch.topk(probabilities, 1)
15  probabilities
```

(9) 检查预测结果：只选出辨识概率须大于设定值且辨识结果为自行车，代码为 671。

```
1   MIN_CONFIDENCE = 0.4   # 辨识概率门槛值
2
3   labels = {}
4   for (i, p) in enumerate(zip(top_prob.numpy().reshape(-1),
5                               top_catid.numpy().reshape(-1))):
6       (prob, imagenetID) = p
7       label = categories[imagenetID]
8
9       # 概率大于设定值，则放入候选名单
10      if prob >= MIN_CONFIDENCE:
11          # 只侦测自行车(671)
12          if imagenetID != 671: continue  # bike
13          # 放入候选名单
14          box = locs[i]
15          L = labels.get(label, [])
16          L.append((box, prob))
17          labels[label] = L
18
19  labels.keys()
```

(10) 定义 NMS 函数：使用程序【08_02_HOG-Face-Detection.ipynb】的 non_max_suppression_slow 函数，也可以使用 nms_pytorch 函数，程序来自 *Non Maximum*

Suppression: Theory and Implementation in PyTorch [12],由于程序代码过长,故不列出。

(11) 进行 NMS,并对侦测到的对象画框。

```
1    # 扫描每一个类别
2    for label in labels.keys():
3        #if label != categories[671]: continue # bike
4        
5        # 复制原图
6        open_cv_image = PIL2CV2(orig)
7        
8        # 画框
9        for (box, prob) in labels[label]:
10           (startX, startY, endX, endY) = box
11           cv2.rectangle(open_cv_image, (startX, startY), (endX, endY),
12               (0, 255, 0), 2)
13       
14       # 显示 NMS(non-maxima suppression) 前的框
15       cv2.imshow("Before NMS", open_cv_image)
16       
17       # NMS
18       open_cv_image2 = PIL2CV2(orig)
19       boxes = np.array([p[0] for p in labels[label]])
20       proba = np.array([p[1] for p in labels[label]])
21       # print(boxes.shape, proba.shape)
22       # boxes = nms_pytorch(torch.cat((torch.tensor(boxes),
23       #     torch.tensor(proba).reshape(proba.shape[0], -1)), dim=1) ,
24       #     MIN_CONFIDENCE) # non max suppression
25       boxes = non_max_suppression_slow(boxes, MIN_CONFIDENCE) # non max suppression
26       
27       color_list=[(0, 255, 0), (255, 0, 0), (255, 255, 0), (0, 0, 0), (0, 255, 255)]
28       for i, x in enumerate(boxes):
29           # startX, startY, endX, endY, label = x.numpy()
30           startX, startY, endX, endY = x #.numpy()
31           # 画框及类别
32           cv2.rectangle(open_cv_image2, (int(startX), int(startY)), (int(endX), int(endY))
33               , color_list[i%len(color_list)], 2)
34           startY = startY - 15 if startY - 15 > 0 else startY + 15
35           cv2.putText(open_cv_image2, str(label), (int(startX), int(startY)),
36               cv2.FONT_HERSHEY_SIMPLEX, 0.45, (0, 0, 255), 2)
37       
38       # 显示
39       cv2.imshow("After NMS", open_cv_image2)
40       cv2.waitKey(0)
41       
42   cv2.destroyAllWindows()     # 关闭所有视窗
```

执行结果:左图显示所有候选的窗口,右图显示 NMS 过滤后的窗口。

这个程序有以下缺点:

(1) 滑动窗口及影像金字塔要侦测的窗口很多,侦测耗时。

(2) 使用 ResNet 或其他辨识单一对象的算法,不管哪一窗口都会侦测一种类别,就算里面没有对象,也会有一类概率最大,这并不是我们所希望的。

(3) 如果有重叠的自行车图像,无法被侦测到。

后续对象侦测专用的算法可以改善这些缺点。由于对象侦测的应用范围非常广泛,有许多学者前仆后继地提出各种改良的算法,试图提高准确率并加快辨识速度,接下来我们就沿着学者们的研究轨迹,逐步深入探讨。

## 8-4 R-CNN 对象侦测

结合滑动窗口、影像金字塔及 HOG 的算法虽然好用，但它还是有以下的缺点：
(1) 滑动窗口加上影像金字塔，需要检查的窗口个数太多了，耗时过久。
(2) 一个 SVM 分类器只能侦测一个对象。
(3) 通用性的 CNN 模型辨识并不准确，尤其是重叠的对象。
从 2014 年开始，每年都有改良的算法出现，如图 8.8 所示。

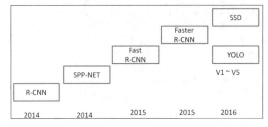

图 8.8　对象侦测算法的发展过程

2014 年 Ross B. Girshick 等学者在 *Rich feature hierarchies for accurate object detection and semantic segmentation* [13] 一文中提出 Regions with CNN 算法，以下简称 R-CNN。

R-CNN 架构及步骤如下，如图 8.9 所示：
(1) 读取要辨识的图片。
(2) 使用区域推荐 (Region Proposal) 算法，找到 2000 个候选区域 (Candidate Region)。
(3) 使用 CNN 提取每一个候选区域特征。
(4) 将上一步骤的输出交由 SVM 辨识。

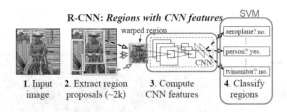

图 8.9　R-CNN 架构

图片来源：Rich feature hierarchies for accurate object detection and semantic segmentation [13]

更详细的架构如图 8.10 所示。

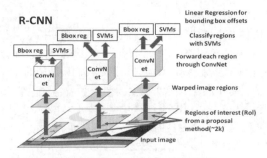

图 8.10　另一视角的 R-CNN 架构

程序处理流程如图 8.11 所示。

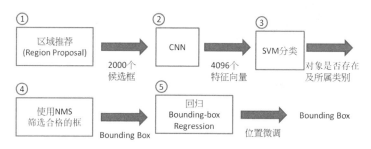

图 8.11　R-CNN 处理流程

区域推荐 (Region Proposal)：用途为改善滑动窗口的过程检查过多窗口的问题，使用区域推荐算法，只找出 2000 个候选区域 (Candidate Region) 输入到模型。区域推荐 (Region Proposal) 不止一种算法，R-CNN 所采用的是 Selective Search，它会依据颜色 (Color)、纹理 (Texture)、规模 (Scale)、空间关系 (Enclosure) 来进行合并，接着再选取 2000 个最有可能包含对象的区域，称为候选区域，如图 8.12 所示。

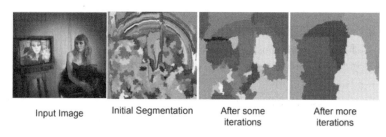

图 8.12　区域推荐 (Region Proposal)：最左边的图为原图，
将颜色、纹理、规模、空间关系相近的区域合并，最后变成最右边图的区域

特征提取 (Feature Extractor)：将 2000 个候选区域使用影像变形转换 (Image Warping)，转成固定尺寸 227 × 227 的图像，输入 CNN 进行特征提取，每个候选区域转换成 4096 个特征向量。

SVM 分类器：比对特征向量，侦测对象是否存在于所属的类别，注意，一种类使用一个二分类 SVM。

使用 Non-Maximum Suppression (NMS) 筛选合格的框：选取可信度较高的候选区域为基准，计算与基准框的 IoU(Intersection-over Union )，如图 8.13 所示，高 IoU 值表示高度重叠，可以把它们过滤掉，类似上一节的做法。

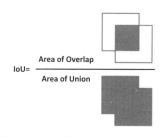

图 8.13　IoU：分母为与目标框联集的面积，分子为与目标框交集的面积

位置微调：利用回归 (Bounding-box Regression) 微调预测区域的位置。

利用回归计算候选区域的四个变量：中心点 ($P_x$, $P_y$) 与宽高 ($P_w$, $P_h$)，其微调公式如下 (G 为预估值。推论过程有点复杂，详情可参考原文附录 C)。

$$\hat{G}_x = P_w d_x(P) + P_x$$
$$\hat{G}_y = P_h d_y(P) + P_y$$
$$\hat{G}_w = P_w \exp(d_w(P))$$
$$\hat{G}_h = P_h \exp(d_h(P))$$

损失函数如下，采用 Ridge Regression，以最小平方法估算出权重：

$$W_\star = \underset{\hat{w}_\star}{\operatorname{argmin}} \sum_i^N (t_\star^i - \hat{w}_\star^{\mathrm{T}} \phi_5(P^i))^2 + \lambda \|\hat{w}_\star\|^2$$

微调后的目标值 $t_\star$：

$$t_x = (G_x - P_x)/P_w$$
$$t_y = (G_y - P_y)/P_h$$
$$t_w = \log(G_w/P_w)$$
$$t_h = \log(G_h/P_h)$$

整个 R-CNN 处理流程涉及相当多的算法，包括：
(1) 区域推荐 (Region Proposal)：Selective Search。
(2) 特征提取 (Feature Extractor)：AlexNet，也可采取 VGG 或者其他 CNN 模型。
(3) SVM 分类器。
(4) Non-Maximum Suppression (NMS)。
(5) Bounding-box Regression。

**范例4. 区域推荐**：OpenCV扩展版支持Selective Search算法。此范例程序修改自 *Selective Search for Object Recognition* [14]。

**程序**：【08_04_selective_search_test.py】。

**执行**：【python 08_04_selective_search_test.py】< 图片文件 > < 算法类别 >

(1) 安装 OpenCV 扩展版：先卸载 OpenCV 一般版，再安装 OpenCV 扩展版，一般版与扩展版只能择其一。

pip uninstall opencv-contrib-python opencv-python

pip install opencv-contrib-python

(2) 算法有三种类别，差异不大，有兴趣的读者可详阅 OpenCV 官网说明：

SingleStrategy。

SelectiveSearchQuality。

SelectiveSearchFast。

(3) 操作：一开始会呈现 10 个框，可利用下列按键增减和结束程序。

+：增加 10 个框。

−：减少 10 个框。

q：程序结束。

(4) Selective Search 程序代码如下：rects 会包含所有的候选区域。

```
12    cv2.setUseOptimized(True)
13    cv2.setNumThreads(8)
14    gs = cv2.ximgproc.segmentation.createSelectiveSearchSegmentation()
15    gs.setBaseImage(img)
16    gs.switchToSelectiveSearchFast()
17    rects = gs.process()
```

**范例5.** 修改【08_03_Object_Detection.ipynb】，以 **Selective Search** 取代滑动窗口及影像金字塔，以下仅说明差异的程序代码。

**完整程序请参阅**【08_05_Object_Detection_with_selective_search.ipynb】。

(1) 以 Selective Search 取代滑动窗口及影像金字塔，取得每一个要侦测的候选区域。

```
1   # 产生 Selective Search 影像
2   import matplotlib.pyplot as plt
3
4   plt.figure(figsize=(16, 16))
5   def Selective_Search(img_path):
6       img = cv2.imread(img_path)
7       img = cv2.resize(img, (WIDTH, int(orig.size[1] / orig.size[0] * WIDTH))
8                       , interpolation=cv2.INTER_AREA)
9       img=cv2.cvtColor(img,cv2.COLOR_BGR2RGB)
10
11      # 执行 Selective Search
12      cv2.setUseOptimized(True)
13      cv2.setNumThreads(8)
14      gs = cv2.ximgproc.segmentation.createSelectiveSearchSegmentation()
15      gs.setBaseImage(img)
16      gs.switchToSelectiveSearchFast()
17      rects = gs.process()
18      # print(rects)
19
20      rois = torch.tensor([])       # 候选框
21      locs = []                     # 位置
22      j=1
23      for i in range(len(rects)):
24          x, y, w, h = rects[i]
25          if w < 100 or w > 400 or h < 100: continue
26
27          # 框与原图的比例
28          scale = WIDTH / float(w)
29
30          # 缩放图形以符合模型输入规格
31          crop_img = img[y:y+h, x:x+w]
32          crop_img = Image.fromarray(crop_img)
33          if j <= 100:
34              plt.subplot(10, 10, j)
35              plt.imshow(crop_img)
36          j+=1
37
38          roi = transform(crop_img)
39          roi = roi.unsqueeze(0)  # 增加一维(笔数)
40
41          # 加入输出变数中
42          if len(rois.shape) == 1:
43              rois = roi
44          else:
45              rois = torch.cat((rois, roi), dim=0)
46          locs.append((x, y, x + w, y + h))
47
48      return rois.to(device), locs
49
50  rois, locs = Selective_Search(filename)
51  plt.tight_layout()
```

执行结果：候选区域大小不一，为加快执行速度，删除过小区域，同时删除过大区域，避免最后 NMS 会删除其他区域，请参见程序代码第 25 行。

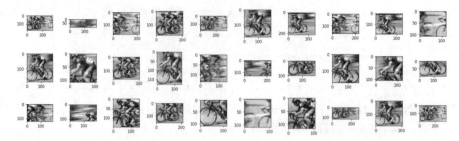

(2) 其他程序代码几乎不用修改。

执行结果：左图是初步筛选的候选区域，右图是 NMS 筛选后的候选区域，抓到两个对象。

以上的范例并未使用 Bounding-box Regression，原作者有提供 R-CNN 程序代码[15]，但须安装 MATLAB、Caffe 才能执行，超出本书范围，后面针对改良的 Faster R-CNN 进行测试。

R-CNN 依然不尽如人意的原因如下：

(1) 每张图经由区域推荐处理后，各会产生出 2000 个候选区域，每个区域都需经过辨识，运行时间还是过长，而且区域推荐也不具备自我学习能力。

(2) 每个区域经由 CNN 模型提取 4096 个特征向量，合计有 2000 × 4096 = 8 192 000 个特征向量，内存消耗也很大。

(3) 每笔数据都要经过 CNN、SVM、回归三个模型的训练与预测，过于复杂。

总体而论，对象侦测不只追求高准确率，更要求能够实时侦测，像骑自动驾驶，总不能等撞到障碍物后才侦测到障碍物，那就悲剧了。原作者虽然以 Caffe(C++) 开发 R-CNN，希望缩短侦测时间，但仍需要 40 多秒才能侦测一张图像，因此引发一波算法的改良浪潮。接下来，我们就来介绍各个改良算法。

## 8-5 R-CNN 改良

首先 Kaiming He 等学者提出 SPP-Net(Spatial Pyramid Pooling in Deep Convolutional Networks for Visual Recognition) 算法，针对 R-CNN 把每个候选区域都需要变形转换才能输入 CNN 的缺点进行改良，做法如下：

(1) R-CNN 每一个尺寸候选区域都需要转换为固定尺寸才能输入 CNN 模型，各个区域的长宽不一，非等比例的转换会造成准确度降低，SPP-Net 作者所提出 Spatial pyramid pooling(SPP) 神经层，各种尺寸的输入图像都能产生一个固定长度的输出，作者在最后一个卷积层后增加了一个 SPP 层，负责转换成固定长度的特征提取。

(2) 其他的处理与 R-CNN 类似。

R-CNN 与 SPP-Net 的模型结构比较，如图 8.14 所示。

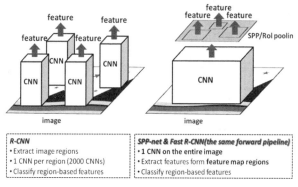

图 8.14　R-CNN 与 SPP-Net 模型结构比较

SPP 还是有以下缺点：

(1) 虽然解决了 CNN 计算过多的状况，但没有处理分类 (SVM) 与回归过慢的问题。

(2) 特征向量太占内存空间。

详细处理流程可参阅 *Spatial Pyramid Pooling in Deep Convolutional Networks for Visual Recognition*[16] 一文，中文说明可参阅《SPP-Net 论文详解》[17] 的内容。

参酌 PyTorch[18] 或 Keras[19] 程序代码，可帮助理解细节。以 PyTorch 为例，模型架构如下，摘录自程序【cnn_with_spp.py】，作者在第 46 行增加一层 SPP 神经层。

```
34        x = self.conv1(x)
35        x = self.LReLU1(x)
36
37        x = self.conv2(x)
38        x = F.leaky_relu(self.BN1(x))
39
40        x = self.conv3(x)
41        x = F.leaky_relu(self.BN2(x))
42
43        x = self.conv4(x)
44        # x = F.leaky_relu(self.BN3(x))
45        # x = self.conv5(x)
46        spp = spatial_pyramid_pool(x,1,[int(x.size(2)),int(x.size(3))]
47                                  ,self.output_num)
48        # print(spp.size())
49        fc1 = self.fc1(spp)
50        fc2 = self.fc2(fc1)
51        s = nn.Sigmoid()
52        output = s(fc2)
53        return output
```

SPP 神经层转换逻辑请参考程序【spp_layer.py】，将各种输入尺寸转为统一的尺寸。

之后 Ross B. Girshick 陆续提出 Fast R-CNN、Faster R-CNN 等算法。

Fast R-CNN 做法：

(1) 将整个图像直接经由 CNN 转成特征向量，不再使用 2000 个候选区域输入 CNN。

(2) 自定义一个 RoI(Region of Interest) 池化层，通过候选区域在整个图像的所在位置，换算出每个候选区域的特征向量。

(3) 其他流程与 R-CNN 类似。
Fast R-CNN 模型结构如图 8.15 所示。

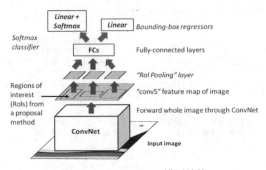

图 8.15　Fast R-CNN 模型结构

优点：
(1) CNN 模型只需训练原图就好，不用训练 2000 个候选区域。
(2) 通过 RoI 池化层得到固定尺寸的特征后，只要连接辨识层进行分类即可。

由于使用区域推荐算法找 2000 张候选区域，还是太耗时，Ross B. Girshick 决定放弃使用 Selective Search，引进 RPN (Region Proposal Network) 神经层，开发 Faster R-CNN 模型，CNN 输出的特征图 (Feature Map) 同时提供 RPN 及分类器使用，可以同步处理，大大提高执行速度，可参照图 8.16 说明。

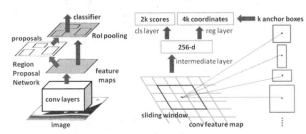

图 8.16　Faster R-CNN 模型结构

其中 RPN 会依据 CNN 输出的特征图产生固定几种尺寸的 Anchor Box，作为候选窗口，不再使用 Selective Search 费力地寻找 2000 个候选区域，如图 8.17 所示。

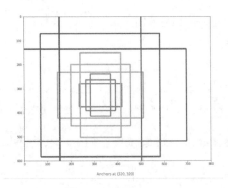

图 8.17　Anchor Box

网络上有许多关于 Faster R-CNN 的解说，其中《捋一捋 PyTorch 官方 FasterRCNN 代码》[20] 一文中有非常详尽的说明，有兴趣的读者可参阅。也可以参考 Train your own object detector with Faster-RCNN & PyTorch [21] 的 Github 程序代码【faster_RCNN.py】，使用 PyTorch 函数建构 Faster R-CNN 的模型架构。

虽然 Ross B. Girshick 在 GitHub 中放上 Faster R-CNN 程序代码[22]，但安装不仅复杂，执行环境的要求也很高 (Caffe/C++)，还好，网络上还有许多 TensorFlow/PyTorch 改写的程序代码，以下先参考 Faster R-CNN Object Detection with PyTorch [23] 一文测试。

**范例6.** 使用**Faster R-CNN算法进行对象侦测**。本范例程序修改自*Faster R-CNN Object Detection with PyTorch* [23]。

程序：【08_06_Faster_RCNN.ipynb】。

(1) 载入套件。

```
1  from PIL import Image
2  import matplotlib.pyplot as plt
3  import torch
4  import torchvision.transforms as T
5  import torchvision
6  import torch
7  import numpy as np
8  import cv2
9  import os
```

(2) 加载模型：要测试的候选区域很多，为防止 GPU 内存不足，请将 device 设为 cpu。

```
1  model = torchvision.models.detection.fasterrcnn_resnet50_fpn(pretrained=True).to(device)
2  model.eval()
```

执行结果：可以看到模型结构，请观察最后三个区块，分别是 FeaturePyramidNetwork(FPN)、RegionProposalNetwork(RPN) 及 RoIHeads，它们就是 Faster R-CNN 精华之处，FPN 详细说明可参考 *Feature Pyramid Networks for Object Detection* [24]。

```
(fpn): FeaturePyramidNetwork(
  (inner_blocks): ModuleList(
    (0): Conv2d(256, 256, kernel_size=(1, 1), stride=(1, 1))
    (1): Conv2d(512, 256, kernel_size=(1, 1), stride=(1, 1))
    (2): Conv2d(1024, 256, kernel_size=(1, 1), stride=(1, 1))
    (3): Conv2d(2048, 256, kernel_size=(1, 1), stride=(1, 1))
  )
  (layer_blocks): ModuleList(
    (0): Conv2d(256, 256, kernel_size=(3, 3), stride=(1, 1), padding=(1, 1))
    (1): Conv2d(256, 256, kernel_size=(3, 3), stride=(1, 1), padding=(1, 1))
    (2): Conv2d(256, 256, kernel_size=(3, 3), stride=(1, 1), padding=(1, 1))
    (3): Conv2d(256, 256, kernel_size=(3, 3), stride=(1, 1), padding=(1, 1))
  )
  (extra_blocks): LastLevelMaxPool()
)
(rpn): RegionProposalNetwork(
  (anchor_generator): AnchorGenerator()
  (head): RPNHead(
    (conv): Conv2d(256, 256, kernel_size=(3, 3), stride=(1, 1), padding=(1, 1))
    (cls_logits): Conv2d(256, 3, kernel_size=(1, 1), stride=(1, 1))
    (bbox_pred): Conv2d(256, 12, kernel_size=(1, 1), stride=(1, 1))
  )
)
(roi_heads): RoIHeads(
  (box_roi_pool): MultiScaleRoIAlign(featmap_names=['0', '1', '2', '3'], output_size=
  (box_head): TwoMLPHead(
```

(3) 确定 COCO 数据集类别：COCO 与 Pascal VOC 是对象侦测常用的测试数据集，其中 COCO 有 80 个类别，Pascal VOC 有 20 个类别。

```
1  COCO_INSTANCE_CATEGORY_NAMES = [
2      '__background__', 'person', 'bicycle', 'car', 'motorcycle', 'airplane', 'bus',
3      'train', 'truck', 'boat', 'traffic light', 'fire hydrant', 'N/A', 'stop sign',
4      'parking meter', 'bench', 'bird', 'cat', 'dog', 'horse', 'sheep', 'cow',
5      'elephant', 'bear', 'zebra', 'giraffe', 'N/A', 'backpack', 'umbrella', 'N/A', 'N/A',
6      'handbag', 'tie', 'suitcase', 'frisbee', 'skis', 'snowboard', 'sports ball',
7      'kite', 'baseball bat', 'baseball glove', 'skateboard', 'surfboard', 'tennis racket',
8      'bottle', 'N/A', 'wine glass', 'cup', 'fork', 'knife', 'spoon', 'bowl',
9      'banana', 'apple', 'sandwich', 'orange', 'broccoli', 'carrot', 'hot dog', 'pizza',
10     'donut', 'cake', 'chair', 'couch', 'potted plant', 'bed', 'N/A', 'dining table',
11     'N/A', 'N/A', 'toilet', 'N/A', 'tv', 'laptop', 'mouse', 'remote', 'keyboard', 'cell phone',
12     'microwave', 'oven', 'toaster', 'sink', 'refrigerator', 'N/A', 'book',
13     'clock', 'vase', 'scissors', 'teddy bear', 'hair drier', 'toothbrush'
14  ]
15
16  len(COCO_INSTANCE_CATEGORY_NAMES)
```

执行结果：共 91 类，其中有多类是 NA，原作者在后续的论文中将之删除。

(4) 定义预测函数：这里是关键程序代码，通过预先训练好的模型预测，即可得到预测类别、定界框 (Bounding Box) 及分数，可设定分数的门槛值，将过低分数的定界框删除 ( 定界框即侦测到对象的所在范围 )。

```
1  def get_prediction(img_path, threshold):
2      # 读取图文件
3      img = Image.open(img_path)
4      # 预测
5      transform = T.Compose([T.ToTensor()])
6      img = transform(img)
7      pred = model([img])
8      # 取得预测类别、定界框(bounding box)及分数
9      pred_class = [COCO_INSTANCE_CATEGORY_NAMES[i] \
10                   for i in list(pred[0]['labels'].numpy())]
11     pred_boxes = [[(int(i[0]), int(i[1])), (int(i[2]), int(i[3]))] \
12                   for i in list(pred[0]['boxes'].detach().numpy())]
13     pred_score = list(pred[0]['scores'].detach().numpy())
14
15     # 筛选超过门槛值的框
16     pred_t = [pred_score.index(x) for x in pred_score if x>threshold][-1]
17     pred_boxes = pred_boxes[:pred_t+1]
18     pred_class = pred_class[:pred_t+1]
19     return pred_boxes, pred_class
```

(5) 定义对象侦测的 API：调用上述函数，依结果将图像画出定界框。

```
1  def object_detection_api(img_path, threshold=0.5, rect_th=3, text_size=2, text_th=2):
2      # 预测
3      boxes, pred_cls = get_prediction(img_path, threshold)
4
5      # 画框
6      img = cv2.imread(img_path)
7      img = cv2.cvtColor(img, cv2.COLOR_BGR2RGB)
8      for i in range(len(boxes)):
9          cv2.rectangle(img, boxes[i][0], boxes[i][1],color=(0,255,0), thickness=rect_th)
10         cv2.putText(img,pred_cls[i], (boxes[i][0][0], boxes[i][0][1]-10),
11                     cv2.FONT_HERSHEY_SIMPLEX,
12                     text_size, (0,255,0),thickness=text_th)
13     plt.figure(figsize=(20,30))
14     plt.imshow(img)
15     plt.xticks([])
16     plt.yticks([])
17     plt.show()
```

(6) 测试：调用上述函数。

```
2  object_detection_api('./images_Object_Detection/people.jpg', threshold=0.8)
```

执行结果：可侦测到每一个人，包括重叠的人物。

可测试其他图文件试试看，准确率相当高。

除了 COCO 类别外，读者一定很想训练自定义的类别，*Train your own object detector with Faster-RCNN & PyTorch*[21]一文有说明如何重新训练模型，笔者不在此说明，而会在后续使用 YOLO 模型训练。

Facebook 还提供一个更完整的 Detectron 套件，目前已开发至第 2 版 (Detectron 2)。只能安装在 Linux、macOS 环境，Windows 使用者可以在 Google Colaboratory 上进行测试。

**范例7. 使用Detectron2套件进行对象侦测。**

程序：【**Detectron2 Tutorial.ipynb**】，需在 Google Colaboratory 上执行，请上传程序至 Google 云端硬盘，接着再双击文件即可。记得要在 "运行时间" 选取 GPU。

详细说明请参阅 *Getting Started with Detectron2* [25]。开启范例文件：https://colab.research.google.com/drive/16jcaJoc6bCFAQ96jDe2HwtXj7BMD_-m5。

(1) 确认 PyTorch、GCC 安装完成，且 PyTorch 版本须为 1.7 或以上。
(2) 安装 Detectron2 套件。
(3) 自 Model Zoo 下载 Detectron2 预先训练的模型。
(4) 预测。

```
cfg = get_cfg()
# add project-specific config (e.g., TensorMask) here if you're not running a model in detectron2's core library
cfg.merge_from_file(model_zoo.get_config_file("COCO-InstanceSegmentation/mask_rcnn_R_50_FPN_3x.yaml"))
cfg.MODEL.ROI_HEADS.SCORE_THRESH_TEST = 0.5  # set threshold for this model
# Find a model from detectron2's model zoo. You can use the https://dl.fbaipublicfiles... url as well
cfg.MODEL.WEIGHTS = model_zoo.get_checkpoint_url("COCO-InstanceSegmentation/mask_rcnn_R_50_FPN_3x.yaml")
predictor = DefaultPredictor(cfg)
outputs = predictor(im)
```

(5) 显示对象侦测结果。

扫码看彩图

执行结果：效果超好，就连背景中旁观的人群都可以被正确侦测。

(6) 上传斑马照片。

```
# Upload the results
from google.colab import files
files.upload()
```

(7) 读取文件，进行对象侦测。

```
# 读取文件，进行物件侦测
im = cv2.imread("./zebra.jpg")
cv2_imshow(im)
predictor = DefaultPredictor(cfg)
outputs = predictor(im)
outputs
```

执行结果：侦测到三个定界框

[ 46.5412, 94.6141, 234.9006, 258.9107],[180.8245, 86.8508, 418.6142, 261.7740], [342.8438, 103.8605, 563.8304, 266.2300] 和三个信赖度 [0.9992, 0.9986, 0.9983]，概率都相当高。

(8) 显示对象侦测结果。

```
v = Visualizer(im[:, :, ::-1], MetadataCatalog.get(cfg.DATASETS.TRAIN[0]), scale=1.2)
out = v.draw_instance_predictions(outputs["instances"].to("cpu"))
cv2_imshow(out.get_image()[:, :, ::-1])
```

执行结果。

这个套件真的超强，除了成功抓到所有对象之外，更是做到了实例分割 (Instance Segmentation)，扫描到的对象不仅有定界框，还有准确的屏蔽 (Mask)。

文件后面还示范了以下功能：

(1) 使用自定义的数据集，侦测自己感兴趣的对象。在 Google Colaboratory 上训练只需要几分钟的时间就可以完成。

(2) 人体骨架的侦测。

(3) 全景视频的对象侦测。

## 8-6 YOLO 算法简介

由于 R-CNN 属两阶段 (Two Stage) 的算法，第一阶段先利用区域推荐找出候选区域，第二阶段才是进行对象侦测，所以在侦测速度上始终是一个瓶颈，难以满足实时侦测的要求，后来有学者提出了一阶段 (Single Shot) 的算法，主要区分为：YOLO 及 SSD。

R-CNN 经过一连串的改良后，对象侦测的速度比较如表 8.1，最新版速度比原版加快了 250 倍。

表 8.1　R-CNN 各算法之对象侦测的速度

|  | R-CNN | Fast R-CNN | Faster R-CNN |
|---|---|---|---|
| Test Time per Image | 50s | 2s | 0.2s |
| Speed Up | 1x | 25x | 250x |

看似很好了，然而 YOLO 发明人 Joseph Redmon 在 2016 年的 CVPR 研讨会 (You Only Look Once: Unified, Real-Time Object Detection) 中有两张投影片非常有意思，一辆轿车平均车身长约 8 英尺 (1ft=0.3048m)，假如使用 Faster R-CNN 侦测下一个路况的话，车子早已行驶了 12 英尺。相对的，如果使用 YOLO 侦测下一个路况，车子则只行驶了 2 英尺。安全性是否会提高许多？相信答案已不言而喻，非常有说服力，如图 8.18、图 8.19 所示。

图 8.18　使用 Faster R-CNN 算法对象侦测的速度　　图 8.19　使用 YOLO 算法对象侦测的速度

YOLO(You Only Look Once) 是现在最优秀的对象侦测算法，于 2016 年由 Joseph Redmon 提出，他本人已开发至第 3 版，但因某些原因离开此研究领域，其他学者接手，至 2022 年已开发到第 7 版了，如图 8.20 所示。

- V1：2016 年 5 月，Joseph Redmon
  You Only Look Once: Unified, Real-Time Object Detection
- V2：2017 年 12 月，Joseph Redmon
  YOLO9000: Better, Faster, Stronger
- V3：2018 年 4 月，Joseph Redmon
  YOLOv3: An Incremental Improvement
- V4：2020 年 4 月，Alexey Bochkovskiy
  YOLOv4: Optimal Speed and Accuracy of Object Detection
- V5：2020 年 6 月，Glenn Jocher
  PyTorch based version of YOLOv5
- V6：2022 年 6 月，美团
- V7：2022 年 7 月，王建尧
  YOLOv7: Trainable bag-of-freebies sets new state-of-the-art for real-time object detectors

图 8.20　YOLO 版本演进

YOLO 各版本的平均准确度 (mAP) 与速度的比较，如图 8.21~ 图 8.23 所示。

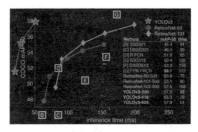

图 8.21　YOLO 版本 v1~v3 的比较

图片来源：YOLO 官网 [26]

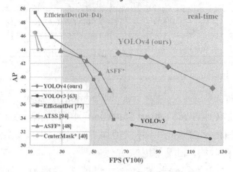

图 8.22　YOLO 版本 v4、v3 的比较

图片来源：*YOLOv4: Optimal Speed and Accuracy of Object Detection* [27]

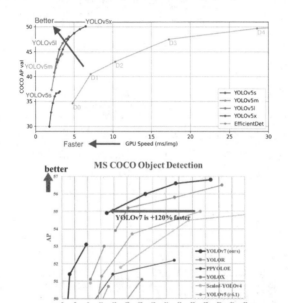

图 8.23　YOLO 版本 v5、v7 的各模型比较

图片来源：YOLO5、YOLO7 GitHub[28]

YOLO 的快速是牺牲部分准确率所换来的，它的流程如图 8.24 所示。

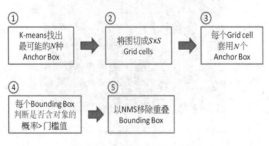

图 8.24　YOLO 的处理流程

(1) 放弃区域推荐，以集群算法 K-Means 从训练数据中找出最常见的 $N$ 种尺寸的候选框 (Anchor Box)。YOLO v5 采用遗传算法 (Genetic algorithm) 生成候选框。

(2) 直接将图像划分成 $(s, s)$ 个网格 (Grid)：每个网格只检查多种不同尺寸的 Anchor Box 是否含有对象。

(3) 输入 CNN 模型，计算每个候选框含有对象的概率。

(4) 同时计算每一个网格可能含有各种对象的概率，假设每一网格最多只能含一个物体。

(5) 合并步骤 (3)、(4) 的信息，并找出合格的候选区域。

(6) 以 NMS 移除重叠定界框 (Bounding Box)。

观察图 8.25，有助于 YOLO 的理解。

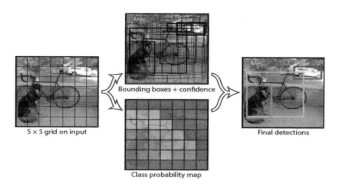

图 8.25　YOLO 处理流程的示意图

图片来源：*You Only Look Once: Unified, Real-Time Object Detection*[29]

YOLO 为追求快速，程序代码采用 C/CUDA 开发，称为 Darknet 架构，可以在 Darknet 执行各种版本的 YOLO 模型，本书不剖析源代码，只聚焦如下重点。

(1) 环境设置。

(2) 范例应用。

(3) 自定义数据集。

# 8-7　YOLO v5 测试

PyTorch 直接支持 YOLO v5，它又分为多种尺寸，大模型准确率 (mAP) 高，但文件大，加载速度慢；小模型准确率稍低，但文件小，加载速度快，适合手机使用。具体如图 8.26 所示。

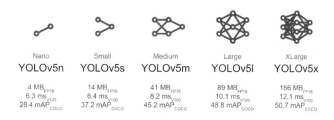

图 8.26　YOLO v5 模型比较

图片来源：PyTorch 官网 YOLO v5 [30]

**范例8. 使用YOLO v5进行对象侦测。**

程序：【**08_07_YOLO5.ipynb**】。

(1) 安装：须安装许多套件，官网直接提供 requirements.txt，指令如下：

pip install -qr https://raw.githubusercontent.com/ultralytics/yolov5/master/requirements.txt

大部分都是 Anaconda 已安装的套件，很快就可以执行完成。

(2) 载入套件。

```
1  import torch
```

(3) 加载小的模型。

```
1  model = torch.hub.load('ultralytics/yolov5', 'yolov5s', pretrained=True).to(device)
```

执行结果：使用 GPU，模型共 213 层。

```
Using cache found in C:\Users\mikec/.cache\torch\hub\ultralytics_yolov5_master
YOLOv5  2021-11-4 torch 1.10.0+cu113 CUDA:0 (NVIDIA GeForce GTX 1050 Ti, 4096MiB)

Fusing layers...
Model Summary: 213 layers, 7225885 parameters, 0 gradients
Adding AutoShape...
```

(4) 预测：可一次输入多个图文件，批处理。

```
1  # 批次处理
2  imgs = ['https://ultralytics.com/images/zidane.jpg',
3          './images_Object_Detection/car.jpg']
4
5  # 预测
6  results = model(imgs)
7
8  # 输出结果
9  results.print()
```

执行结果：显示侦测到的对象类别及个数，还有处理速度。

```
image 1/2: 720x1280 2 persons, 1 tie
image 2/2: 2139x3500 10 cars
Speed: 665.5ms pre-process, 86.1ms inference, 161.3ms NMS per image at shape (2, 3, 416, 640)
```

(5) 预测结果存储：预设会存入 .\runs\detect\exp 目录内。

```
1  results.save()
```

(6) 显示结果：会以默认图片文件编辑工具显示，如不喜欢，可直接开启上述目录观看。

```
1  results.show()
```

执行结果。

(7) 显示定界框、预测概率及类别。

```
1  results.xyxy[0]
```

执行结果：第一张图的 3 个对象，含左上角 / 右下角坐标、预测概率及类别。

```
tensor([[7.54500e+02, 3.92500e+01, 1.14200e+03, 7.13000e+02, 8.76465e-01, 0.00000e+00],
        [1.24000e+02, 1.99875e+02, 9.58000e+02, 7.10500e+02, 5.62500e-01, 0.00000e+00],
        [4.40500e+02, 4.27500e+02, 5.00500e+02, 7.16500e+02, 5.20020e-01, 2.70000e+01]], device='cuda:0')
```

(8) 以表格显示定界框、预测概率及类别，非常贴心。

```
1  results.pandas().xyxy[0]
```

执行结果。

|   | xmin | ymin | xmax | ymax | confidence | class | name |
|---|---|---|---|---|---|---|---|
| 0 | 752.00 | 45.75 | 1148.0 | 716.0 | 0.875977 | 0 | person |
| 1 | 100.25 | 201.50 | 1001.0 | 718.5 | 0.572266 | 0 | person |
| 2 | 438.50 | 422.25 | 510.0 | 720.0 | 0.525879 | 27 | tie |

使用非常简单，加载模型、侦测、显示结果三个步骤。

## 8-8 YOLO v5 模型训练

YOLO v5 之前均依赖 Darknet 架构，须安装 C++ 开发环境，设置比较复杂，训练也要很久，YOLO v5 抛弃 Darknet 架构，直接使用 PyTorch，单一指令就可以进行模型训练，非常方便，只要到官网[30]下载程序，再执行下列指令：

python train.py --img 640 --batch 16 --epochs 5 --data coco128.yaml --weights yolov5s.pt

笔者依 *Custom Object Detection Training using YOLOv5*[31] 一文训练 7 个类别，约 30 分钟就搞定了，之后侦测图像或影片均没有问题。

## 8-9 YOLO v7 测试

YOLO v7 2022 年 7 月推出，在网络上有两个版本：

(1) 王建尧博士版本：继 v4 后，他又改良算法，在 2022 年 7 月发表 YOLO v7 论文 (https://arxiv.org/abs/2207.02696)，并推出对应的程序代码 (https://github.com/WongKinYiu/yolov7)。

(2) JinTian 版本：结合 detectron v2，可轻易做出实例分割 (Instance Segmentation)，程序代码位于 https://github.com/jinfagang/yolov7_d2。

王建尧博士的版本实测步骤如下：

(1) 下载程序 git clone：https://github.com/WongKinYiu/yolov7。

(2) 下载权重文件至 yolov7 目录，网址为 https://github.com/WongKinYiu/yolov7/releases/download/v0.1/yolov7.ptcd yolov7。

(3) 输入以下指令测试，输入为 horses.jpg，侦测的结果在 runs\detect\exp[N] 目录，[N] 为执行的次数：python detect.py --weights yolov7.pt --conf 0.25 --img-size 640 --source inference/images/horses.jpg --view-img。

(4) 执行结果如下：

horses.jpg 及侦测结果：侦测到 5 只马，确实比 YOLO v4 准确。

(5) 同一个程序也可以侦测影片 (test.mp4),使用 GPU,影片播放会更顺畅:
python detect.py --weights yolov7.pt --source ./test.mp4 --view-img。

笔者把 detect.py 简化为 Yolov7_detect.ipynb,说明如下。

(1) 载入套件。

```
1  import time
2  from pathlib import Path
3  import cv2
4  import torch
5  import torch.backends.cudnn as cudnn
6  from numpy import random
7  from models.experimental import attempt_load
8  from utils.datasets import LoadStreams, LoadImages
9  from utils.general import check_img_size, check_requirements, check_imshow, \
10     non_max_suppression, apply_classifier, \
11     scale_coords, xyxy2xywh, strip_optimizer, set_logging, increment_path
12 from utils.plots import plot_one_box
13 from utils.torch_utils import select_device, load_classifier, \
14                               time_synchronized, TracedModel
```

(2) 侦测 GPU。

```
1  device = torch.device("cuda" if torch.cuda.is_available() else "cpu")
```

(3) 加载 YOLO 预先训练好的模型。

```
1  # Load model
2  weights = 'yolov7.pt'
3  model = attempt_load(weights, map_location=device)  # load FP32 model
4  stride = int(model.stride.max())  # model stride
```

(4) 对象侦测函数。

```
1  def detect(source, img_size=640, conf_thres=0.25, save_img=False):
2      dataset = LoadImages(source, img_size=img_size)
3
4      # Get names and colors
5      names = model.module.names if hasattr(model, 'module') else model.names
6      colors = [[random.randint(0, 255) for _ in range(3)] for _ in names]
7
8      # Run inference
9      if device.type != 'cpu':
10         model(torch.zeros(1, 3, img_size, img_size).to(device).type_as(
11             next(model.parameters())))  # run once
12     old_img_w = old_img_h = img_size
13     old_img_b = 1
14
15     t0 = time.time()
16     for path, img, im0s, vid_cap in dataset:
17         img = torch.from_numpy(img).to(device)
18         img = img.float()  # uint8 to fp16/32
19         img /= 255.0  # 0 - 255 to 0.0 - 1.0
20         if img.ndimension() == 3:
21             img = img.unsqueeze(0)
22
23         # Inference
24         t1 = time_synchronized()
25         pred = model(img)[0]
26         t2 = time_synchronized()
27
28         # Apply NMS
29         pred = non_max_suppression(pred, conf_thres)
30         t3 = time_synchronized()
```

```
32        # Process detections
33        for i, det in enumerate(pred):  # detections per image
34            p, s, im0, frame = path, '', im0s, getattr(dataset, 'frame', 0)
35
36            p = Path(p)  # to Path
37            gn = torch.tensor(im0.shape)[[1, 0, 1, 0]]  # normalization gain whwh
38            if len(det):
39                # Rescale boxes from img_size to im0 size
40                det[:, :4] = scale_coords(img.shape[2:], det[:, :4], im0.shape).round()
41
42                # Print results
43                for c in det[:, -1].unique():
44                    n = (det[:, -1] == c).sum()  # detections per class
45                    s += f"{n} {names[int(c)]}{'s' * (n > 1)}, "  # add to string
46
47                # Write results
48                for *xyxy, conf, cls in reversed(det):
49                    # Add bbox to image
50                    label = f'{names[int(cls)]} {conf:.2f}'
51                    plot_one_box(xyxy, im0, label=label, color=colors[int(cls)], line_thickness=1)
52                    print(label)
53
54            cv2.imshow(str(p), im0)
55            cv2.waitKey(0)
56            cv2.destroyAllWindows()
57
58            # Print time (inference + NMS)
59            print(f'{s}Done. ({(1E3 * (t2 - t1)):.1f}ms) Inference, ({(1E3 * (t3 - t2)):.1f}ms) NMS')
60        print(f'Done. ({(time.time() - t0):.3f}s)')
61        return
```

(5) 执行侦测。

```
1  detect('./inference/images/horses.jpg')
```

执行结果：除了显示图片外，还有各对象类别与概率。

```
horse 0.69
horse 0.80
horse 0.86
horse 0.94
horse 0.96
5 horses, Done. (82.0ms) Inference, (1.0ms) NMS
Done. (5.605s)
```

YOLO 特别强调速度，因此，要实际应用于项目，应采用 C/C++ 设置辨识的模块，可参照 YOLO v4(https://github.com/AlexeyAB/darknet)，设置方式可参阅笔者的文章《YOLO v4 设置心得——Windows 环境》(https://ithelp.ithome.com.tw/articles/10231508) 及《YOLO v4 模型训练实操》(https://ithelp.ithome.com.tw/articles/10282549)。

## 8-10 YOLO 模型训练

YOLO 默认模型采用 COCO[32] 或 Pascal VOC 数据集[33]，假若要侦测的对象不在这些类别当中，则需自行训练模型，步骤如下：

(1) 准备数据集：若只是要测试处理影像，不想制作数据集的话，可直接下载 COCO 数据集，内含影像与标注文件 (Annotation)，接着遵循 YOLO 步骤实操，但可能要训练好多天，后续笔者使用 Open Images Dataset，可以选择部分类别，缩短测试时间。

(2) 使用标记工具软件，例如 LabelImg[34]，产生 YOLO 格式的标注文件。LabelImg 安装步骤如下：

① conda install pyqt=5

② .conda install -c anaconda lxml

③ .pyrcc5 -o libs/resources.py resources.qrc

④ 执行 LabelImg，如图 8.27：python labelImg.py

图 8.27　LabelImg 标记工具

(3) 模型训练：训练非常耗时，须耐心等候。

**范例9. 使用自定义数据集训练YOLO模型。**内容参考*Create your own dataset for YOLOv4 object detection in 5 minutes*[35]及*YOLO4 GitHub*[36]这两篇文章的做法。

(1) 下载数据前置处理程序 git clone：https://github.com/theAIGuysCode/OIDv4_ToolKit.git。

(2) 在 OIDv4_ToolKit 目录开启终端机 (cmd)，并安装相关套件：pip install -r requirements.txt。

(3) 至 Open Images Dataset 网站 (https://storage.googleapis.com/openimages/web/index.html ) 下载训练数据，它包含 350 种类别可应用在实例分割 (Instance Segmentation) 上，我们只选取三种类别来测试，避免训练太久。在 OIDv4_ToolKit 目录，执行下列指令，下载训练数据：python main.py downloader --classes Balloon Person Dog --type_csv train --limit 200。

注意，出现"missing files"错误信息时，请输入 y。

(4) 执行下列指令，下载测试数据：python main.py downloader --classes Balloon Person Dog --type_csv test --limit 200。

注意，出现 missing files 错误信息时，请输入 y。

(5) 建立一个 data.yaml 文件，放在 OID/Dataset 目录内，内容如下：

train: ./OID/Dataset/train

val: ./OID/Dataset/test

nc: 3

names: ['Balloon', 'Dog', 'Person']

(6) 执行下列指令，产生 YOLO 标注文件 (Annotation)，即每个图像文件都会有一个同名的标注文件 (*.txt)：

python convert_annotations.py

标注文件的内容为：

< 类别 ID > < 标注框中心点 *X* 坐标 > < 标注框中心点 *Y* 坐标 > < 标注框宽度 > < 标注框高度 >。

(7) 移除 OID\Dataset\train、OID\Dataset\test 子目录下的 Label 目录，包括：

OID\Dataset\train\Balloon\Label；

OID\Dataset\train\Dog\Label；
OID\Dataset\train\Person\Label；
OID\Dataset\test\Balloon\Label；
OID\Dataset\test\Dog\Label；
OID\Dataset\test\Person\Label。

(8) 复制 OID 目录至 yolov7 目录下。

(9) 开启终端机 (cmd)，执行模型训练：

python train.py --batch 4 --cfg cfg/training/yolov7.yaml --img 640 --epochs 55 --data ./OID/Dataset/data.yaml --weights " --name yolov7 --hyp data/hyp.scratch.p5.yaml --device 0

由于笔者的机器 GPU 内存只有 4GB，官网批量 (Batch) 使用 32，会发生内存不足的错误，故改为 4。

结果还是发生内存不足的错误，因此，最后还是搬到 Goggle Colaboratory 上执行，可参阅 YOLOv7_Training.ipynb。

训练时间大约 85 分钟。

每次训练均会产生一个新目录，例如读者训练 3 次，在 runs/train 目录下分别产生以下子目录：yolov7、yolov72、yolov73。

可将子目录内的最佳模型 best.pt 备份下来。

(10) 自 test 目录下或网络任取一文件测试，执行下列指令：

python detect.py --weights ./runs/train/yolov7/weights/best.pt --conf 0.03 --source ./OID/Dataset/test/Balloon/76e41712939b97f2.jpg

效果不是很好，如图 8.28 所示，虽然有捕捉到人和气球，不过概率均偏低，可能与训练数据不足、训练执行周期有关，另外，训练指令的相关参数也可以进行调校。

图 8.28　YOLO 模型测试结果

以上只是笔者简单的试验，文件目录过大，无法放入 GitHub，请读者见谅。上述训练的步骤，在实际项目执行时，应该尚有一些改善空间，但最重要的还是要弄到一台高档的 GPU 机器，用金钱换时间。

另外，笔者也使用《RoboFlow 的海洋水生馆数据集》[37] 进行 YOLO 模型训练，有兴趣的读者可参阅《YOLO v7 实测》[38]。

## 8-11 SSD 算法

Single Shot MultiBox Detector(SSD) 算法与 YOLO 齐名，也属于一阶段的算法，在速度上比 R-CNN 系列算法快，而在准确率 (mAP) 上比 YOLO v1 高，如表 8.2 所示。但后来随着 YOLO 不断的升级改良，SSD 的网络热度变小了。

表 8.2 R-CNN、YOLO、SSD 比较表，数据源：SSD 官网 [39]

| System | VOC2007 test mAP | FPS (Titan X) | Number of Boxes | Input resolution |
|---|---|---|---|---|
| Faster R-CNN (VGG16) | 73.2 | 7 | ~6000 | ~1000 x 600 |
| YOLO (customized) | 63.4 | 45 | 98 | 448 x 448 |
| SSD300* (VGG16) | 77.2 | 46 | 8732 | 300 x 300 |
| SSD512* (VGG16) | 79.8 | 19 | 24564 | 512 x 512 |

SSD 比较特别的地方是它采用 VGG 模型，并且在中间使用多个卷积层撷取特征图 (Feature Map)，同时进行预测，如图 8.29 所示。

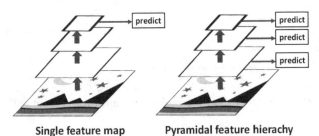

图 8.29 左图为 YOLO 模型，右图为 SSD

详细说明可参阅《一文看尽目标检测算法 SSD 的核心架构与设计思想》[40]。

SSD 也是用 Caffe 架构开发的，SSD 官网 [40] 并未说明在 Windows 操作系统下要如何编译，不过，PyTorch Hub 提供的 SSD300 模型 API 可直接使用。

**范例10. 使用SSD算法进行对象侦测。**

程序：【08_09_SSD.ipynb】。

(1) 加载相关框架。

```
1 import torch
```

(2) 加载模型。

```
1 ssd_model = torch.hub.load('NVIDIA/DeepLearningExamples:torchhub', 'nvidia_ssd').to(device)
2 utils = torch.hub.load('NVIDIA/DeepLearningExamples:torchhub', 'nvidia_ssd_processing_utils')
3 ssd_model.eval()
```

执行结果：可看到 SSD 模型结构非常复杂。

(3) 取得 COCO 类别。

```
1 classes_to_labels = utils.get_coco_object_dictionary()
2 classes_to_labels
```

(4) 下载图像。

```
1  # 下载 3 张图像
2  uris = [
3      'http://images.cocodataset.org/val2017/000000397133.jpg',
4      'http://images.cocodataset.org/val2017/000000037777.jpg',
5      'http://images.cocodataset.org/val2017/000000252219.jpg'
6  ]
7
8  # 转为张量
9  inputs = [utils.prepare_input(uri) for uri in uris]
10 tensor = utils.prepare_tensor(inputs)
```

(5) 预测：SSD 会产生 8732 个候选框，筛选预测概率 > 0.4 的定界框，以删除不准确的候选框。

```
1  # 预测
2  with torch.no_grad():
3      detections_batch = ssd_model(tensor)
4
5  # 筛选预测概率 > 0.4 的定界框
6  results_per_input = utils.decode_results(detections_batch)
7  best_results_per_input = [utils.pick_best(results, 0.40)
8                            for results in results_per_input]
```

(6) 显示结果。

```
1  # 显示结果
2  from matplotlib import pyplot as plt
3  import matplotlib.patches as patches
4
5  for image_idx in range(len(best_results_per_input)):
6      fig, ax = plt.subplots(1)
7      # 显示原图
8      image = inputs[image_idx] / 2 + 0.5
9      ax.imshow(image)
10
11     # 显示侦测结果
12     bboxes, classes, confidences = best_results_per_input[image_idx]
13     for idx in range(len(bboxes)):
14         left, bot, right, top = bboxes[idx]
15         x, y, w, h = [val * 300 for val in \
16                       [left, bot, right - left, top - bot]]
17         rect = patches.Rectangle((x, y), w, h, linewidth=1,
18                                   edgecolor='r', facecolor='none')
19         ax.add_patch(rect)
20         ax.text(x, y, "{} {:.0f}%".format(classes_to_labels[classes[idx] - 1],
21                  confidences[idx]*100), bbox=dict(facecolor='white', alpha=0.5))
22 plt.show()
```

执行结果。

## 8-12　对象侦测的效能衡量指标

对象侦测的效能衡量指标是采用平均精确度均值 (mean Average Precision，mAP)，YOLO 官网展示的图表针对各种模型比较 mAP，如图 8.30 所示。

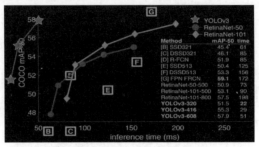

图 8.30　YOLO 与其他模型比较

图片来源：YOLO 官网[26]

第 4 章介绍过的 ROC/AUC 效能衡量指标，是以预测概率为基准，计算各种阈值 (门槛值) 下的真阳率与伪阳率，以伪阳率为 $X$ 轴，真阳率为 $Y$ 轴，绘制出 ROC 曲线。而 mAP 也类似于 ROC/AUC，以 IoU 为基准，计算各种阈值 (门槛值) 下的精确率 (Precision) 与召回率 (Recall)，以召回率为 $X$ 轴，精确率为 $Y$ 轴，绘制出 mAP 曲线。

不过，对象侦测模型通常是多分类，不是二分类，因此，采取计算各个种类的平均精确度，绘制后如图 8.31 左图，通常会调整成图 8.31 右图的粗线，因为，阈值低的精确率一定比阈值高的精确率更好，所以做此调整。

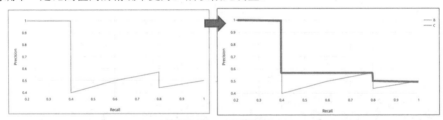

图 8.31　mAP 曲线，左图是实际计算的结果，右图是调整后的结果

## 8-13　总结

本章我们认识了许多对象侦测的算法，包括 HOG、R-CNN、YOLO、SSD，同时也实操了许多范例，如传统的影像金字塔、R-CNN、PyTorch Detectron2、YOLO、TensorFlow Object Detection API，还包含图像和视频侦测，也可自定义数据集训练模型，证明我们的确有能力将对象侦测技术导入项目中使用。

算法各有优劣，Faster R-CNN 虽然速度较慢，但准确度高。尽管 YOLO 早期为了提升执行速度牺牲了准确度，但经过几个版本升级后，准确度也已大幅提高，所以建议读者在实际应用时，还是应该多方尝试，找出最适合的模型，譬如在边缘运算的场域使用轻量模型，不只要求辨识速度快，更要节省内存的使用。

现在许多学者开始研究动态对象侦测，例如姿态 (Pose) 侦测，可用来辨识体育运动姿势是否标准，协助运动员提升成绩，另外还有手势侦测[41]、体感游戏、制作皮影戏[42]等。

# 第 9 章
# 进阶的影像应用

除了对象侦测之外，CNN 还有许多影像方面的应用，例如：
- 语义分割 (Semantic Segmentation)。
- 风格转换 (Style Transfer)。
- 影像标题 (Image Captioning)。
- 姿态辨识 (Pose Detection 或 Action Detection)。
- 生成对抗网络 (GAN) 各种的应用。
- 深度伪造 (Deep Fake)。

本章将继续探讨以上这些应用领域，其中生成对抗网络 (GAN) 的内容较多，会以专章来介绍。

## 9-1 语义分割介绍

对象侦测是以整个对象作为标记 (Label)，而语义分割 (Semantic Segmentation) 则以每个像素 (Pixel) 作为标记 (Label)，区分对象涵盖的区域，如图 9.1 所示。

图 9.1　区分对象涵盖的区域

经过语义分割后产生图 9.2，各对象以不同颜色的像素表示。

图 9.2　经过语义分割的图像

甚至更进一步，进行实例分割 (Instance Segmentation)，相同类别的对象也以不同的颜色表示，如图 9.3 所示。

图 9.3　实例分割

语义分割的应用非常广泛，例如：
(1) 自动驾驶的影像识别。
(2) 医疗诊断：断层扫描 (CT)、核磁共振 (MRI) 的疾病区域标示。
(3) 卫星照片。
(4) 机器人的影像识别。

语义分割的原理是先利用 CNN 进行特征提取 (Feature Extraction)，再运用提取的特征向量来重建影像，如图 9.4 所示。

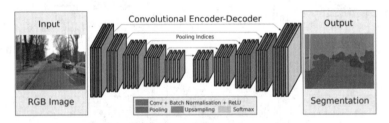

图 9.4　语义分割的示意图

图片来源：*SegNet: A Deep Convolutional Encoder-Decoder Architecture for Image Segmentation* [1]

这种 "原始影像" ➜ "特征提取" ➜ "重建影像" 的做法，称为自动编码器 (AutoEncoder, AE) 架构，许多进阶的算法都以此架构为基础，因此，我们先来探究 AutoEncoder 架构。

## 9-2　自动编码器

自动编码器 (AutoEncoder, AE) 通过特征提取得到训练数据的共同特征，一些噪声会被过滤掉，接着再依据特征向量重建影像，这样就可以达到去噪声 (Denosing) 的目的，此做法也可以扩展到语义分割、风格转换 (Style Transfer)、U-net、生成对抗网络 (GAN) 等各种各样的算法。

AutoEncoder 由 Encoder 与 Decoder 组合而成 ( 如图 9.5 所示 )：

编码器 (Encoder)：即为提取特征的过程，类似于 CNN 模型，但不含最后的分类层 (Dense)。

译码器 (Decoder)：根据提取的特征来重建影像。

图 9.5　自动编码器 (AutoEncoder) 示意图

接下来,我们实操 AutoEncoder,使用 MNIST 数据集,示范如何将噪声去除。

**范例1. 实操AutoEncoder,进行噪声去除。**此范例程序修改自 *Denoising Autoencoder in Pytorch on MNIST dataset*[2]。

用 AutoEncoder 进行噪声去除的相关步骤如图 9.6 所示。

图 9.6　相关步骤

**下列程序代码请参考【09_01_MNIST_Autoencoder.ipynb】。**

(1) 载入相关套件。

```
1  import matplotlib.pyplot as plt
2  import numpy as np
3  import pandas as pd
4  import random
5  import os
6
7  import torch
8  import torchvision
9  from torchvision import transforms
10 from torch.utils.data import DataLoader,random_split
11 from torch import nn
12 import torch.nn.functional as F
13 import torch.optim as optim
14 from sklearn.manifold import TSNE
```

(2) 参数设定。

```
1  PATH_DATASETS = ""  # 预设路径
2  BATCH_SIZE = 256    # 批量
3  device = torch.device("cuda" if torch.cuda.is_available() else "cpu")
4  "cuda" if torch.cuda.is_available() else "cpu"
```

(3) 取得 MNIST 训练数据,只取图像 (X),不需要 Label(Y),因为程序只进行特征提取,不用辨识。

```
1  train_ds = torchvision.datasets.MNIST(PATH_DATASETS, train=True, download=True)
2  test_ds  = torchvision.datasets.MNIST(PATH_DATASETS, train=False, download=True)
```

(4) 显示任意 20 笔 MNIST 图像。

```
1  fig, axs = plt.subplots(4, 5, figsize=(8,8))
2  for ax in axs.flatten():
3      # 随机抽样
4      img, label = random.choice(train_ds)
5      ax.imshow(np.array(img), cmap='gist_gray')
6      ax.set_title('Label: %d' % label)
7      ax.set_xticks([])
8      ax.set_yticks([])
9  plt.tight_layout()
```

执行结果。

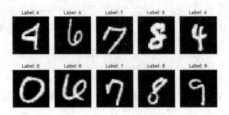

(5) 建立 DataLoader，切割 20% 训练数据作为验证数据，以计算验证数据的损失及准确率。

```
1   # 转为张量
2   train_ds.transform = transforms.ToTensor()
3   test_ds.transform = transforms.ToTensor()
4
5   # 切割20%训练资料作为验证资料
6   m=len(train_ds) # 总笔数
7   train_data, val_data = random_split(train_ds, [int(m-m*0.2), int(m*0.2)])
8
9   train_loader = torch.utils.data.DataLoader(train_data, batch_size=BATCH_SIZE)
10  valid_loader = torch.utils.data.DataLoader(val_data, batch_size=BATCH_SIZE)
11  test_loader = torch.utils.data.DataLoader(test_ds, batch_size=BATCH_SIZE,shuffle=True)
```

(6) 定义 AutoEncoder 模型：含编码器 (Encoder) 及译码器 (Decoder)，编码器使用卷积层提取线条特征向量，译码器以特征向量重建图像。一般编码器只要卷积层提取特征，不需要完全连接层，不过这个范例编码器含有完全连接层，也是可行的。

```
1   class Encoder(nn.Module):
2       def __init__(self, encoded_space_dim,fc2_input_dim):
3           super().__init__()
4
5           # Convolution
6           self.encoder_cnn = nn.Sequential(
7               nn.Conv2d(1, 8, 3, stride=2, padding=1),
8               nn.ReLU(True),
9               nn.Conv2d(8, 16, 3, stride=2, padding=1),
10              nn.BatchNorm2d(16),
11              nn.ReLU(True),
12              nn.Conv2d(16, 32, 3, stride=2, padding=0),
13              nn.ReLU(True)
14          )
15
16          self.flatten = nn.Flatten(start_dim=1)
17
18          self.encoder_lin = nn.Sequential(
19              nn.Linear(3 * 3 * 32, 128),
20              nn.ReLU(True),
21              nn.Linear(128, encoded_space_dim)
22          )
23
24      def forward(self, x):
25          x = self.encoder_cnn(x)
26          x = self.flatten(x)
27          x = self.encoder_lin(x)
28          return x
```

(7) 译码器 (Decoder)：含转置卷积层 (ConvTranspose2d)，也称为反卷积层，把图像放大，详细说明可参考《ConvTranspose2d 原理，深度网络如何进行上采样》[3]，内含动画，它并不是单纯地将一点复制成 $N \times N$ 点，而是会考虑外围的点，类似卷积运算。

```
 1  class Decoder(nn.Module):
 2      def __init__(self, encoded_space_dim,fc2_input_dim):
 3          super().__init__()
 4  
 5          self.decoder_lin = nn.Sequential(
 6              nn.Linear(encoded_space_dim, 128),
 7              nn.ReLU(True),
 8              nn.Linear(128, 3 * 3 * 32),
 9              nn.ReLU(True)
10          )
11  
12          self.unflatten = nn.Unflatten(dim=1, unflattened_size=(32, 3, 3))
13  
14          self.decoder_conv = nn.Sequential(
15              # 反卷积
16              nn.ConvTranspose2d(32, 16, 3, stride=2, output_padding=0),
17              nn.BatchNorm2d(16),
18              nn.ReLU(True),
19              nn.ConvTranspose2d(16, 8, 3, stride=2, padding=1, output_padding=1),
20              nn.BatchNorm2d(8),
21              nn.ReLU(True),
22              nn.ConvTranspose2d(8, 1, 3, stride=2, padding=1, output_padding=1)
23          )
24  
25      def forward(self, x):
26          x = self.decoder_lin(x)
27          x = self.unflatten(x)
28          x = self.decoder_conv(x)
29          x = torch.sigmoid(x)
30          return x
```

(8) 建立模型：根据上述类别建立对象，$d$ 是 encoder 输出个数，也是 decoder 输入个数，可做调整，方便参数调校。

```
1  # 固定随机乱数种子，以利掌握执行结果
2  torch.manual_seed(0)
3  
4  # encoder 输出个数，decoder 输入个数
5  d = 4
6  encoder = Encoder(encoded_space_dim=d,fc2_input_dim=128).to(device)
7  decoder = Decoder(encoded_space_dim=d,fc2_input_dim=128).to(device)
```

(9) 定义损失函数及优化器：可选择其他损失函数及优化器。

```
1  loss_fn = torch.nn.MSELoss()
2  lr= 0.001 # Learning rate
3  
4  params_to_optimize = [
5      {'params': encoder.parameters()},
6      {'params': decoder.parameters()}
7  ]
8  
9  optim = torch.optim.Adam(params_to_optimize, lr=lr)
```

(10) 定义加噪声 (Noise) 的函数：noise_factor 越大，噪声越大。

```
1  def add_noise(inputs,noise_factor=0.3):
2      noise = inputs+torch.randn_like(inputs)*noise_factor
3      noise = torch.clip(noise,0.,1.)
4      return noise
```

(11) 定义训练函数：与一般训练不同的是第 10 行，将数据先加噪声再训练。

```
 1  def train_epoch_den(encoder, decoder, device, dataloader,
 2                      loss_fn, optimizer,noise_factor=0.3):
 3      # 指定为训练阶段
 4      encoder.train()
 5      decoder.train()
 6      train_loss = []
 7      # 训练
 8      for image_batch, _ in dataloader:
 9          # 加杂讯
10          image_noisy = add_noise(image_batch,noise_factor)
```

```
11        image_noisy = image_noisy.to(device)
12        # 编码
13        encoded_data = encoder(image_noisy)
14        # 解码
15        decoded_data = decoder(encoded_data)
16        # 计算损失
17        loss = loss_fn(decoded_data, image_noisy)
18        # 反向传导
19        optimizer.zero_grad()
20        loss.backward()
21        optimizer.step()
22        # print(f'损失：{loss.data}')
23        train_loss.append(loss.detach().cpu().numpy())
24
25    return np.mean(train_loss)
```

(12) 定义测试函数：与一般训练相同，将数据先加噪声再测试，并将译码结果存入 conc_out 变量。

```
1  def test_epoch_den(encoder, decoder, device, dataloader,
2                     loss_fn, noise_factor=0.3):
3      # 指定为评估阶段
4      encoder.eval()
5      decoder.eval()
6      with torch.no_grad(): # No need to track the gradients
7          conc_out = []
8          conc_label = []
9          for image_batch, _ in dataloader:
10             # 加噪声
11             image_noisy = add_noise(image_batch,noise_factor)
12             image_noisy = image_noisy.to(device)
13             # 编码
14             encoded_data = encoder(image_noisy)
15             # 解码
16             decoded_data = decoder(encoded_data)
17             # 输出存入 conc_out 变数
18             conc_out.append(decoded_data.cpu())
19             conc_label.append(image_batch.cpu())
20         # 合并
21         conc_out = torch.cat(conc_out)
22         conc_label = torch.cat(conc_label)
23         # 验证资料的损失
24         val_loss = loss_fn(conc_out, conc_label)
25     return val_loss.data
```

(13) 定义重建图像的函数：显示原图、加噪声的图像及重建的图像，作为比较。

```
1  # fix 中文乱码
2  from matplotlib.font_manager import FontProperties
3  plt.rcParams['font.sans-serif'] = ['Microsoft JhengHei'] # 微软正黑体
4  plt.rcParams['axes.unicode_minus'] = False
5
6  def plot_ae_outputs_den(epoch,encoder,decoder,n=5,noise_factor=0.3):
7      plt.figure(figsize=(10,4.5))
8      for i in range(n):
9          ax = plt.subplot(3,n,i+1)
10         img = test_ds[i][0].unsqueeze(0)
11         image_noisy = add_noise(img,noise_factor)
12         image_noisy = image_noisy.to(device)
13
14         encoder.eval()
15         decoder.eval()
16
17         with torch.no_grad():
18             rec_img = decoder(encoder(image_noisy))
19
20         if epoch == 0:
21             plt.imshow(img.cpu().squeeze().numpy(), cmap='gist_gray')
22             ax.get_xaxis().set_visible(False)
23             ax.get_yaxis().set_visible(False)
24             if i == n//2:
25                 ax.set_title('原图')
```

```
26          ax = plt.subplot(3, n, i + 1 + n)
27          plt.imshow(image_noisy.cpu().squeeze().numpy(), cmap='gist_gray')
28          ax.get_xaxis().set_visible(False)
29          ax.get_yaxis().set_visible(False)
30          if i == n//2:
31              ax.set_title('加噪声')
32
33          if epoch == 0:
34              ax = plt.subplot(3, n, i + 1 + n + n)
35          else:
36              ax = plt.subplot(1, n, i + 1)
37          plt.imshow(rec_img.cpu().squeeze().numpy(), cmap='gist_gray')
38          ax.get_xaxis().set_visible(False)
39          ax.get_yaxis().set_visible(False)
40          if epoch == 0 and i == n//2:
41              ax.set_title('重建图像')
42      plt.subplots_adjust(left=0.1,
43                          bottom=0.1,
44                          right=0.7,
45                          top=0.9,
46                          wspace=0.3,
47                          hspace=0.3)
48      plt.show()
```

(14) 训练。

```
1  noise_factor = 0.3
2  num_epochs = 30
3  history_da={'train_loss':[],'val_loss':[]}
4
5  for epoch in range(num_epochs):
6      # print(f'EPOCH {epoch + 1}/{num_epochs}')
7      # 训练
8      train_loss=train_epoch_den(
9          encoder=encoder,
10         decoder=decoder,
11         device=device,
12         dataloader=train_loader,
13         loss_fn=loss_fn,
14         optimizer=optim,noise_factor=noise_factor)
15     # 验证
16     val_loss = test_epoch_den(
17         encoder=encoder,
18         decoder=decoder,
19         device=device,
20         dataloader=valid_loader,
21         loss_fn=loss_fn,noise_factor=noise_factor)
22     # Print Validation loss
23     history_da['train_loss'].append(train_loss)
24     history_da['val_loss'].append(val_loss)
25     print(f'EPOCH {epoch + 1}/{num_epochs} \t 训练损失：{train_loss:.3f}' +
26           f' \t 验证损失：{val_loss:.3f}')
27     plot_ae_outputs_den(epoch,encoder,decoder,noise_factor=noise_factor)
```

执行结果：随着训练，生成的图像清晰且正确。

第 1 个执行周期如下。

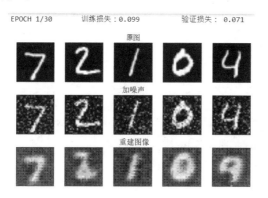

第 30 个执行周期如下：很明显清晰许多。

EPOCH 30/30　　　训练损失：0.050　　　验证损失：0.038

(15) 使用随机数生成图像：输入 4 个数字的向量生成图像，以下程序代码输入由浅色至深色的向量生成图像。

```
def plot_reconstructed(decoder, r0=(-5, 10), r1=(-10, 5), n=10):
    plt.figure(figsize=(20,8.5))
    w = 28
    img = np.zeros((n*w, n*w))
    # 随机数
    for i, y in enumerate(np.linspace(*r1, n)):
        for j, x in enumerate(np.linspace(*r0, n)):
            z = torch.Tensor([[x, y], [x, y]]).reshape(-1,4).to(device)
            # print(z.shape)
            x_hat = decoder(z)
            x_hat = x_hat.reshape(28, 28).to('cpu').detach().numpy()
            img[(n-1-i)*w:(n-1-i+1)*w, j*w:(j+1)*w] = x_hat
    plt.imshow(img, extent=[*r0, *r1], cmap='gist_gray')

plot_reconstructed(decoder, r0=(-1, 1), r1=(-1, 1))
```

执行结果：可生成任意的数字。

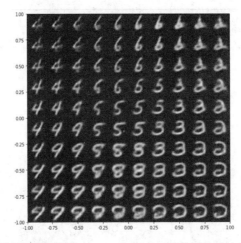

(16) 观察模型中间的潜在因子 (Latent Factor)：编码器的输出设定为 4，故输出 4 个变量。

```
encoded_samples = []
for sample in test_ds:
    img = sample[0].unsqueeze(0).to(device)
    label = sample[1]
    # Encode image
    encoder.eval()
    with torch.no_grad():
        encoded_img = encoder(img)
    # Append to list
    encoded_img = encoded_img.flatten().cpu().numpy()
    encoded_sample = {f"变量 {i}": enc for i, enc in enumerate(encoded_img)}
    encoded_sample['label'] = label
    encoded_samples.append(encoded_sample)

encoded_samples = pd.DataFrame(encoded_samples)
encoded_samples
```

执行结果：测试数据集共 10000 笔数据。

| | Enc. Variable 0 | Enc. Variable 1 | Enc. Variable 2 | Enc. Variable 3 | label |
|---|---|---|---|---|---|
| 0 | -2.708830 | 1.352979 | 0.896718 | -1.138998 | 7 |
| 1 | 0.459103 | 0.874820 | 1.723890 | 0.092751 | 2 |
| 2 | -1.225864 | 2.091084 | 0.537997 | 0.920463 | 1 |
| 3 | 0.379200 | -0.963157 | -0.327787 | 0.176350 | 0 |
| 4 | -0.998168 | 0.381464 | -1.376472 | -0.394563 | 4 |
| ... | ... | ... | ... | ... | ... |
| 9995 | -0.532613 | 0.377049 | 1.552729 | -0.078492 | 2 |
| 9996 | 0.798871 | 0.384637 | 0.918125 | -0.138463 | 3 |
| 9997 | -1.602299 | 0.892798 | -0.702039 | 0.356077 | 4 |
| 9998 | -0.493146 | -0.269544 | -0.146604 | 0.931693 | 5 |
| 9999 | 1.264596 | -0.389804 | -0.610143 | 0.470594 | 6 |

(17) 使用前两个变量为坐标轴绘图。

```
1  import plotly.express as px
2  import plotly.graph_objects as go
3
4  fig = px.scatter(encoded_samples, x='变量 0', y='变量 1',
5                   color=encoded_samples.label.astype(str), opacity=0.7)
6  fig_widget = go.FigureWidget(fig)
7  fig_widget
```

执行结果：测试数据集共 10000 笔数据，阿拉伯数字无法区分得很清楚。

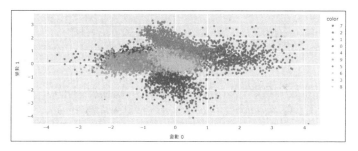

(18) 采用 TSNE 降维，并绘图：将 4 个变量提取成 2 个特征。

```
1  # TSNE 降维
2  tsne = TSNE(n_components=2)
3  tsne_results = tsne.fit_transform(encoded_samples.drop(['label'],axis=1))
4
5  # 绘图
6  fig = px.scatter(tsne_results, x=0, y=1, color=encoded_samples.label.astype(str)
7                   ,labels={'0': 'tsne-变量1', '1': 'tsne-变量2'})
8  fig_widget = go.FigureWidget(fig)
9  fig_widget
```

执行结果：阿拉伯数字区分得很清楚，表示模型效果还不错。

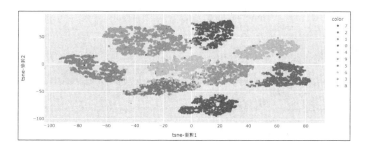

AutoEncoder 属于非监督式学习算法，不需要标记 (Labeling)。另外还有一个 AutoEncoder 的变形 (Variants)，称为 Variational AutoEncoders (VAE)，将编码器输出转为概率分配 ( 通常使用常态分配 )，译码时依据概率分配进行抽样，取得输出，利用此概念去噪声则会更稳健 (Robust)，VAE 常与生成对抗网络 (GAN) 相提并论，可以用来生成影像，如图 9.7 所示。

图 9.7　Variational AutoEncoders (VAE) 的架构

**范例2.** 建立VAE模型，使用MNIST数据集，生成影像。相关步骤如图**9.8**所示。

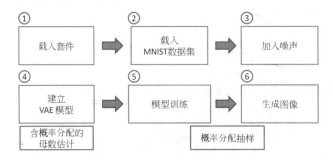

图 9.8　相关步骤

\* VAE 的编码器输出不是特征向量，而是概率分配的 $\mu$ 和 $\log(\delta)$。

本范例程序参考自 *Variational AutoEncoders (VAE) with PyTorch*[4]，并且直接修改自【09_01_MNIST_Autoencoder.ipynb】，大部分的程序代码均不变，故下文仅列出修改的地方。

下列程序代码请参考【09_02_MNIST_VAE.ipynb】。

(1) 定义编码器模型：将输出变成两份数据，分别用来估算常态分配的平均数、变异数。

```python
class Encoder(nn.Module):
    def __init__(self, encoded_space_dim, fc2_input_dim):
        super().__init__()
        
        # Convolution
        self.encoder_cnn = nn.Sequential(
            nn.Conv2d(1, 8, 3, stride=2, padding=1),
            nn.ReLU(True),
            nn.Conv2d(8, 16, 3, stride=2, padding=1),
            nn.BatchNorm2d(16),
            nn.ReLU(True),
            nn.Conv2d(16, 32, 3, stride=2, padding=0),
            nn.ReLU(True)
        )
        
        self.flatten = nn.Flatten(start_dim=1)
        
        self.encoder_lin = nn.Sequential(
            nn.Linear(3 * 3 * 32, 128),
        )
        
        self.encFC1 = nn.Linear(128, encoded_space_dim)
```

```
23          self.encFC2 = nn.Linear(128, encoded_space_dim)
24
25      def forward(self, x):
26          x = self.encoder_cnn(x)
27          x = self.flatten(x)
28          x = self.encoder_lin(x)
29          mu = self.encFC1(x)
30          logVar = self.encFC2(x)
31          return mu, logVar
```

(2) 定义抽样函数：根据平均数 (mu) 和 log 变异数 (log_var) 取随机数。

```
1  def resample(mu, logVar):
2      std = torch.exp(logVar/2)
3      eps = torch.randn_like(std) # N(0, 1) 抽樣
4      return mu + std * eps
```

(3) 译码器不变。

(4) 定义损失函数及优化器：损失函数改用 KL 散度 (Kullback-Leibler Divergence)，它是测量两个分配的差异程度，因为编码器输出的是概率分配，KL 散度通常搭配二分类交叉熵 (Cross Entropy) 作为损失函数。

```
1  # KL divergence
2  def loss_fn(out, imgs, mu, logVar):
3      kl_divergence = 0.5 * torch.sum(1 + logVar - mu.pow(2) - logVar.exp())
4      return F.binary_cross_entropy(out, imgs, size_average=False) - kl_divergence
5
6  lr= 0.001 # Learning rate
7
8  params_to_optimize = [
9      {'params': encoder.parameters()},
10     {'params': decoder.parameters()}
11 ]
12
13 optim = torch.optim.Adam(params_to_optimize, lr=lr)
```

(5) 定义训练函数：在编码器后面调用 resample，自概率分配中随机抽样，再传给译码器 ( 第 13~16 行 )，同时损失函数也调用上述函数 ( 第 18 行 )。

```
1  def train_epoch_den(encoder, decoder, device, dataloader,
2                     loss_fn, optimizer,noise_factor=0.3):
3      # 指定为训练阶段
4      encoder.train()
5      decoder.train()
6      train_loss = []
7      # 训练
8      for image_batch, _ in dataloader:
9          # 加噪声
10         image_noisy = add_noise(image_batch,noise_factor)
11         image_noisy = image_noisy.to(device)
12         # 编码
13         mu, logVar = encoder(image_noisy)
14         encoded_data = resample(mu, logVar)
15         # 解码
16         decoded_data = decoder(encoded_data)
17         # 计算损失
18         loss = loss_fn(decoded_data, image_noisy, mu, logVar)
19
20         # 反向传导
21         optimizer.zero_grad()
22         loss.backward()
23         optimizer.step()
24         # print(f'损失：{loss.data}')
25         train_loss.append(loss.detach().cpu().numpy())
26
27     return np.mean(train_loss)
```

(6) 测试函数：同训练函数方式修改。

```python
def test_epoch_den(encoder, decoder, device, dataloader,
                   loss_fn,noise_factor=0.3):
    # 指定为评估阶段
    encoder.eval()
    decoder.eval()
    val_loss=0.0
    with torch.no_grad(): # No need to track the gradients
        conc_out = []
        conc_label = []
        for image_batch, _ in dataloader:
            # 加噪声
            image_noisy = add_noise(image_batch,noise_factor)
            image_noisy = image_noisy.to(device)
            # 编码
            mu, logVar = encoder(image_noisy)
            encoded_data = resample(mu, logVar)
            # 解码
            decoded_data = decoder(encoded_data)
            # 输出存入 conc_out 变量
            conc_out.append(decoded_data.cpu())
            conc_label.append(image_batch.cpu())
            val_loss += loss_fn(decoded_data.cpu(), image_batch.cpu(), mu, logVar)
        # 合并
        conc_out = torch.cat(conc_out)
        conc_label = torch.cat(conc_label)
        # 验证数据的损失
    return val_loss.data
```

(7) 重建图像的函数：同训练函数方式修改。

```
18          rec_img = decoder(resample(*encoder(image_noisy)))
```

(8) 训练：程序代码不需修改，但要加大训练执行周期，因为随机抽样可能会造成数据不均匀，笔者使用 50 个执行周期，似乎还不够。

执行结果：一开始比 AutoEncoder 糟糕，模糊一片，之后渐渐清晰，但是训练中途，会出现变糟的状况 ( 第 36 执行周期 )，应该是随机抽样，抽到较不好的样本。

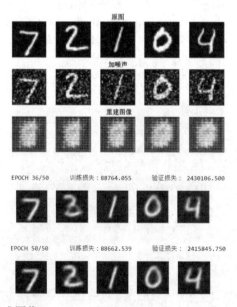

(9) 使用随机数生成图像。

执行结果：比 AutoEncoder 清晰。

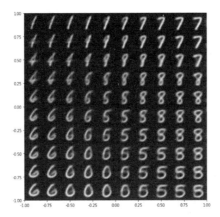

(10) 取前两个变量绘图。

执行结果：所有数字混在一起，无法辨别。

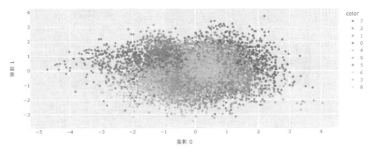

(11) 采用 TSNE 降维，并绘图：将 4 个变量提取成 2 个特征。

执行结果：阿拉伯数字区分得很清楚，表示模型效果还不错。

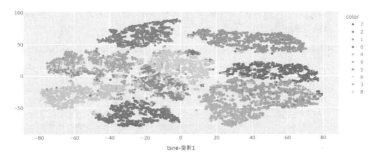

AutoEncoder 模型除了去噪声外，还有以下应用：

(1) 降维 (Dimensionality Reduction)，如上图。

(2) 特征提取 (Feature Extraction)，如编码器。

(3) 影像压缩 (Image Compression)：利用降维，可对影像减色，达到缩小文件的功能。

(4) 影像搜寻 (Image Search)：利用特征提取，以 Cosine Similarity 比对特征。

(5) 异常侦测 (Anomaly Detection)：与影像搜寻相似，但找差异过大的特征。

(6) 遗失值的差补 (Missing Value Imputation)：输入有缺陷的数据，AutoEncoder 可生成完整的数据。

详情可参阅 7 *Applications of Auto-Encoders every Data Scientist should know* 一文 [5]。

## 9-3 语义分割实操

语义分割 (Semantic Segmentation) 或称影像分割 (Image Segmentation) 将每个像素 (Pixel) 作为标记 (Label)，为避免抽样造成像素信息遗失，模型不使用池化层 (Pooling)，学者提出许多算法：

SegNet [1]：全名为影像分割的 Encoder-Decoder 架构 (Deep Convolutional Encoder-Decoder Architecture for Image Segmentation)，使用反卷积放大特征向量，还原图像。

DeepLab [6]：以卷积作用在多种尺寸的图像，得到 Score Map 后，再利用 Score Map 与 Conditional Random Field (CRF) 算法，以内插法 (Interpolate) 的方式还原图像，如图 9.9 所示，详细处理流程可参阅原文。

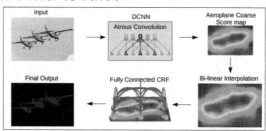

图 9.9　DeepLab 的处理流程

RefineNet [7]：反卷积须占海量存储器，尤其是高分辨率的图像，所以 RefineNet 提出一种节省内存的方法。

PSPnet [8]：使用多种尺寸的池化层，称为金字塔时尚 (Pyramid Fashion)，金字塔掌握影像各个部分的图像数据，利用此金字塔还原图像。

U-Net [9]：广泛应用于生物医学的影像分割，这个模型常被提到，且有许多的变形，所以我们就来认识这个模型。

U-Net 是 AutoEncoder 的变形 (Variant)，由于它的模型结构为 U 形而得名，如图 9.10 所示。

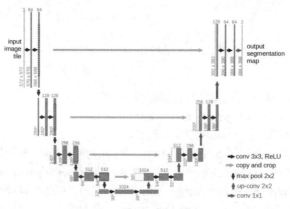

图 9.10　U-Net 模型

图片来源：*U-Net: Convolutional Networks for Biomedical Image Segmentation* [9]

传统 AutoEncoder 的问题点发生在前半段的编码器 (Encoder)，由于它在提取特征的过程中会使输出的尺寸 (Size) 越变越小，接着译码器 (Decoder) 再通过这些变小的特征，重建出一个与原图同样大小的新图像，因此原图的很多信息，如前文所谈的噪声，就没办法传递到译码器了。这个特点应用在去除噪声上是十分恰当的，但假若目标是要侦测异常点 ( 如检测黄斑部病变 ) 的话，那就糟糕了，经过模型过滤后，异常点通通都不见了。

所以，U-Net 在原有编码器与译码器的联系上，增加了一些连接，每一段编码器的输出都与其对面的译码器相连接，使得编码器每一层的信息，都会额外输入到一样尺寸的译码器，如图 9.10 横跨 U 形两侧的中间长箭头，这样在重建的过程中就不会遗失重要信息。

**范例3. 以U-Net实操语义分割，先看看如何建构U-Net模型。程序来自Naoto Usuyama 提供的范例[10]，注意此范例需要较大的GPU内存，笔者PC无法执行，故须在Google Colab上执行。**

语义分割需准备输入图像及屏蔽 (Mask) 图像，屏蔽指出对象所在位置。本范例使用程序产生随机数据，在一个空白图像随意摆放各种形状的对象，之后利用 U-Net 侦测出对象所在位置，并标示不同的颜色。

**下列程序代码请参考【09_03_unet_resnet18.ipynb】。**

(1) 下载源代码。

```
1  import os
2
3  if not os.path.exists("pytorch_unet"):
4      !git clone https://github.com/usuyama/pytorch-unet.git
5
6  %cd pytorch-unet
```

(2) 载入相关套件。

```
1  import matplotlib.pyplot as plt
2  import numpy as np
3  import helper
4  import simulation # simulation.py
```

(3) 测试 simulation.py 生成的图像。

```
1   # 产生3张图像，宽高各为 192，里面有6个随机摆放的图案
2   input_images, target_masks = simulation.generate_random_data(
3                                   192, 192, count=3)
4
5   print("input_images shape and range", input_images.shape,
6         input_images.min(), input_images.max())
7   print("target_masks shape and range", target_masks.shape,
8         target_masks.min(), target_masks.max())
9
10  # 输入图像，改为单色
11  input_images_rgb = [x.astype(np.uint8) for x in input_images]
12
13  # 屏蔽(Mask)图像，使用彩色
14  target_masks_rgb = [helper.masks_to_colorimg(x) for x in target_masks]
15
16  # 显示图像：左边原图为输入，右边屏蔽(Mask)图像为目标
17
18  helper.plot_side_by_side([input_images_rgb, target_masks_rgb])
```

执行结果：产生 3 对图像，宽、高均为 192，里面有 6 个随机摆放的图案，显示图像，左边原图为输入，右边屏蔽 (Mask) 图像为目标 (Target)。

原图像素介于 [0, 255]，屏蔽像素介于 [0, 1]。

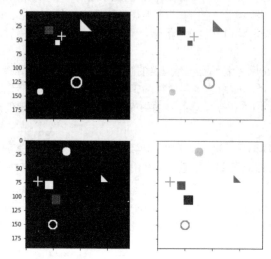

(4) 建立 Dataset。

```python
from torch.utils.data import Dataset, DataLoader
from torchvision import transforms, datasets, models

# 自订数据集，一次回传原图、屏蔽图像各一幅
class SimDataset(Dataset):
    def __init__(self, count, transform=None):
        self.input_images, self.target_masks = \
            simulation.generate_random_data(192, 192, count=count)
        self.transform = transform

    def __len__(self):
        return len(self.input_images)

    def __getitem__(self, idx):
        image = self.input_images[idx]
        mask = self.target_masks[idx]
        if self.transform:
            image = self.transform(image)

        return [image, mask]
```

(5) 建立 DataLoader：产生训练及验证图像各 2000 笔。

```python
# 转换
trans = transforms.Compose([
    transforms.ToTensor(),
    transforms.Normalize([0.485, 0.456, 0.406], [0.229, 0.224, 0.225]) # imagenet
])

# 产生训练及验证图像各2000笔
train_set = SimDataset(2000, transform = trans)
val_set = SimDataset(200, transform = trans)

image_datasets = {
    'train': train_set, 'val': val_set
}

batch_size = 25

dataloaders = {
    'train': DataLoader(train_set, batch_size=batch_size, shuffle=True, num_workers=0),
    'val': DataLoader(val_set, batch_size=batch_size, shuffle=True, num_workers=0)
}
```

执行结果：正确产生 3 对图像无误。

(6) 建立还原转换函数，并测试一批数据。

```
1   import torchvision.utils
2
3   # 还原转换
4   def reverse_transform(inp):
5       inp = inp.numpy().transpose((1, 2, 0))
6       mean = np.array([0.485, 0.456, 0.406])
7       std = np.array([0.229, 0.224, 0.225])
8       inp = std * inp + mean
9       inp = np.clip(inp, 0, 1)
10      inp = (inp * 255).astype(np.uint8)
11
12      return inp
13
14  # 取得一批数据测试
15  inputs, masks = next(iter(dataloaders['train']))
16  print(inputs.shape, masks.shape)
17  plt.imshow(reverse_transform(inputs[3]))
```

(7) 建立 U-Net 模型：使用预先训练的模型 resnet18 作为编码器，在后面建立类似译码器的架构，再将译码器卷积层 (conv) 与对边的编码器神经层相连。由于程序代码过长且重复，这里截取重要程序代码说明。

使用预先训练的模型 resnet18。

```
14      # 载入 resnet18 模型
15      self.base_model = torchvision.models.resnet18(pretrained=True)
16      self.base_layers = list(self.base_model.children())
```

第 59 行建立反卷积层 (Upsample)。

第 61 行找到对面的神经层，并包覆一层卷积层 (Convrelu)，Convrelu 是卷积层加 ReLu Activation Function。

第 63 行将上述两个神经层合并，接到下一层。

其他的程序代码均比照办理。

```
58      # 新增神经层
59      x = self.upsample(x)
60      # 对面的神经层
61      layer2 = self.layer2_1x1(layer2)
62      # 连接新增的神经层及对面的神经层
63      x = torch.cat([x, layer2], dim=1)
64      x = self.conv_up2(x)
```

(8) 定义损失函数：U-Net 使用的损失函数非常特别，采用二分类交叉熵 (Binary Cross Entropy) 加上 Dice Loss，主要是它能克服不平衡的数据集 (Unbalanced Dataset)，并且兼顾精确率 (Precision) 及召回率 (Recall)，因为语义分割主要是要找出屏蔽的位置，而非预测准确率。详细的说明可参阅《语义分割之 Dice Loss 深度分析》[11]。

```
1   from collections import defaultdict
2   import torch.nn.functional as F
3   from loss import dice_loss
4
5   checkpoint_path = "checkpoint.pth"
6
7   # 损失采 binary cross entropy + dice loss
8   def calc_loss(pred, target, metrics, bce_weight=0.5):
9       bce = F.binary_cross_entropy_with_logits(pred, target)
10
11      pred = torch.sigmoid(pred)
12      dice = dice_loss(pred, target)
13
14      loss = bce * bce_weight + dice * (1 - bce_weight)
15
```

```
16      metrics['bce'] += bce.data.cpu().numpy() * target.size(0)
17      metrics['dice'] += dice.data.cpu().numpy() * target.size(0)
18      metrics['loss'] += loss.data.cpu().numpy() * target.size(0)
19
20      return loss
21
22  # 计算效能衡量指标
23  def print_metrics(metrics, epoch_samples, phase):
24      outputs = []
25      for k in metrics.keys():
26          outputs.append(f"{k}: {(metrics[k] / epoch_samples):4f}")
27
28      print(f"{phase}: {', '.join(outputs)}")
```

(9) 建立训练及评估函数：程序代码并无特别之处，与前面的范例类似。

(10) 训练模型：与前面的范例类似。

(11) 预测。

```
1   import math
2
3   # 建立新数据
4   test_dataset = SimDataset(3, transform = trans)
5   test_loader = DataLoader(test_dataset, batch_size=3, shuffle=False, num_workers=0)
6
7   # 取一批数据测试
8   inputs, labels = next(iter(test_loader))
9   inputs = inputs.to(device)
10  labels = labels.to(device)
11  print('inputs.shape', inputs.shape)
12  print('labels.shape', labels.shape)
13
14  # 预测
15  model.eval()
16  pred = model(inputs)
17  pred = torch.sigmoid(pred) # 转为 [0, 1] 之间
18  pred = pred.data.cpu().numpy()
19  print('pred.shape', pred.shape)
20
21  # 原图还原转换
22  input_images_rgb = [reverse_transform(x) for x in inputs.cpu()]
23
24  # 屏蔽
25  target_masks_rgb = [helper.masks_to_colorimg(x) for x in labels.cpu().numpy()]
26
27  # 预测转成图像
28  pred_rgb = [helper.masks_to_colorimg(x) for x in pred]
29
30  ## 左边：原图，中间：屏蔽图像(Target)，右边：预测图像
31  helper.plot_side_by_side([input_images_rgb, target_masks_rgb, pred_rgb])
```

执行结果：下图中间为实际值，右边为预测图像，屏蔽的位置相当准确，当然，这只是示范 U-Net 模型的开发方式，应该以实际案例测试为准。

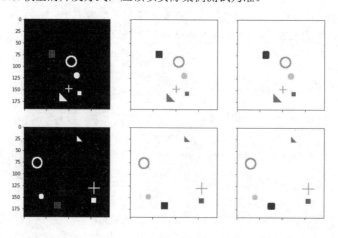

范例4. PyTorch直接支持U-Net预先训练模型，并示范如何应用在脑瘤的诊断上，相关程序代码请参考*U-NET FOR BRAIN MRI*[12]，数据集在*Kaggle Brain MRI segmentation*[13]，文件大小近1GB，有兴趣的读者可下载测试。

**下列程序代码请参考【09_04_brain_segmentation_unet.ipynb】。**

(1) 载入 U-Net 预先训练模型。

```
1  import torch
2  model = torch.hub.load('mateuszbuda/brain-segmentation-pytorch', 'unet',
3      in_channels=3, out_channels=1, init_features=32, pretrained=True)
```

(2) 下载一个范例图片文件。

```
1  import urllib
2
3  url="https://github.com/mateuszbuda/brain-segmentation-pytorch/" + \
4      "raw/master/assets/TCGA_CS_4944.png"
5  filename = "U_Net/TCGA_CS_4944.png"
6  urllib.request.urlretrieve(url, filename)
```

(3) 预测：注意，官网的范例有针对图像进行标准化（第 14 行），但笔者实测结果不佳，故舍弃不用，读者可以解除批注试试看。

```
1  import numpy as np
2  from PIL import Image
3  from torchvision import transforms
4
5  # 开启文件
6  input_image = Image.open(filename)
7
8  # 计算图像的平均值及标准差
9  m, s = np.mean(input_image, axis=(0, 1)), np.std(input_image, axis=(0, 1))
10
11 # 转换
12 preprocess = transforms.Compose([
13     transforms.ToTensor(),
14 # transforms.Normalize(mean=m, std=s),
15 ])
16 input_tensor = preprocess(input_image)
17 input_batch = input_tensor.unsqueeze(0)
18
19 # 如果有GPU，将数据、模型转至 GPU
20 if torch.cuda.is_available():
21     input_batch = input_batch.to('cuda')
22     model = model.to('cuda')
23
24 # 预测
25 with torch.no_grad():
26     output = model(input_batch)
27
28 # 显示有不正常部位的概率
29 print(torch.round(output[0]))
```

执行结果。

```
tensor([[[0., 0., 0.,  ..., 0., 0., 0.],
         [0., 0., 0.,  ..., 0., 0., 0.],
         [0., 0., 0.,  ..., 0., 0., 0.],
         ...,
         [0., 0., 0.,  ..., 0., 0., 0.],
         [0., 0., 0.,  ..., 0., 0., 0.],
         [0., 0., 0.,  ..., 0., 0., 0.]]], device='cuda:0')
```

(4) 比较原图与预测结果。

```
1  import matplotlib.pyplot as plt
2
3  # 原图
4  plt.subplot(1,2,1)
5  plt.imshow(plt.imread(filename))
6  plt.axis('off')
7
8  # 预测结果
9  plt.subplot(1,2,2)
10 plt.imshow(torch.round(output[0]).cpu().numpy().reshape(256, 256), cmap='gray')
11 plt.axis('off')
12 plt.show()
```

执行结果：右图屏蔽位置与原图绿色部位大致吻合。

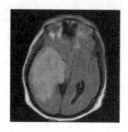

## 9-4 实例分割

上一节的语义分割，同类别的对象只能够以相同颜色呈现，如要做到同类别的对象以不同颜色呈现的话，就需用到实例分割 (Instance Segmentation)。

而实例分割所使用的 Mask R-CNN 算法系由 Facebook AI Research 在 2017 年发布[14]。Mask R-CNN 为 Faster R-CNN 的延伸，不只会框住对象，更能产生不同颜色的屏蔽 (Mask)，如图 9.11 所示。

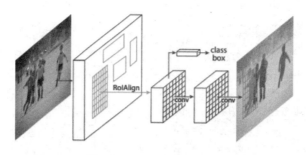

图 9.11　Mask R-CNN 模型

图片来源：Mask R-CNN [14]

除了辨识对象之外，实例分割还有以下延伸的应用。

去背景：侦测到对象后，将对象以外的背景全部去除。

移除特殊的对象：将侦测到的对象移除后，根据周围的颜色填补移除的区域。例如，在观光景点拍照时，最困扰的就是有陌生人一起入镜，这时即可利用此技术将之移除，Photoshop 就有提供类似的功能。

直接来看一个实例，使用 Mask R-CNN 预先设置好的训练模型。

**范例5. 使用Mask R-CNN进行实例分割**。此范例程序修改自 *Mask R-CNN Instance Segmentation with PyTorch*[15]。

**下列程序代码请参考【09_05_Mask_RCNN.ipynb】。**

(1) 载入相关套件。

```
1  from PIL import Image
2  import matplotlib.pyplot as plt
3  import torch
4  import torchvision.transforms as T
5  import torchvision
6  import torch
7  import numpy as np
8  import cv2
9  import random
10 import time
11 import os
```

(2) 定义 COCO 数据集辨识对象名称。

```
1  COCO_INSTANCE_CATEGORY_NAMES = [
2    '__background__', 'person', 'bicycle', 'car', 'motorcycle', 'airplane', 'bus',
3    'train', 'truck', 'boat', 'traffic light', 'fire hydrant', 'N/A', 'stop sign',
4    'parking meter', 'bench', 'bird', 'cat', 'dog', 'horse', 'sheep', 'cow',
5    'elephant', 'bear', 'zebra', 'giraffe', 'N/A', 'backpack', 'umbrella', 'N/A', 'N/A',
6    'handbag', 'tie', 'suitcase', 'frisbee', 'skis', 'snowboard', 'sports ball',
7    'kite', 'baseball bat', 'baseball glove', 'skateboard', 'surfboard', 'tennis racket',
8    'bottle', 'N/A', 'wine glass', 'cup', 'fork', 'knife', 'spoon', 'bowl',
9    'banana', 'apple', 'sandwich', 'orange', 'broccoli', 'carrot', 'hot dog', 'pizza',
10   'donut', 'cake', 'chair', 'couch', 'potted plant', 'bed', 'N/A', 'dining table',
11   'N/A', 'N/A', 'toilet', 'N/A', 'tv', 'laptop', 'mouse', 'remote', 'keyboard', 'cell phone',
12   'microwave', 'oven', 'toaster', 'sink', 'refrigerator', 'N/A', 'book',
13   'clock', 'vase', 'scissors', 'teddy bear', 'hair drier', 'toothbrush'
14 ]
```

(3) 下载 Mask RCNN 预先训练模型。

```
1  # Mask RCNN 预先训练模型
2  model = torchvision.models.detection.maskrcnn_resnet50_fpn(pretrained=True)
3  model
```

执行结果：可以看到完整的模型结构，Faster RCNN 后面加上屏蔽预测层。

```
(mask_roi_pool): MultiScaleRoIAlign(featmap_names=['0', '1', '2', '3'], output_size=(14, 14)
(mask_head): MaskRCNNHeads(
  (mask_fcn1): Conv2d(256, 256, kernel_size=(3, 3), stride=(1, 1), padding=(1, 1))
  (relu1): ReLU(inplace=True)
  (mask_fcn2): Conv2d(256, 256, kernel_size=(3, 3), stride=(1, 1), padding=(1, 1))
  (relu2): ReLU(inplace=True)
  (mask_fcn3): Conv2d(256, 256, kernel_size=(3, 3), stride=(1, 1), padding=(1, 1))
  (relu3): ReLU(inplace=True)
  (mask_fcn4): Conv2d(256, 256, kernel_size=(3, 3), stride=(1, 1), padding=(1, 1))
  (relu4): ReLU(inplace=True)
)
(mask_predictor): MaskRCNNPredictor(
  (conv5_mask): ConvTranspose2d(256, 256, kernel_size=(2, 2), stride=(2, 2))
  (relu): ReLU(inplace=True)
  (mask_fcn_logits): Conv2d(256, 91, kernel_size=(1, 1), stride=(1, 1))
)
```

(4) 定义对象侦测相关函数：包括设定屏蔽的颜色、对象侦测及屏蔽上色、显示结果。

```
1  # 设定屏蔽的颜色
2  def random_colour_masks(image):
3      colours = [[0, 255, 0],[0, 0, 255],[255, 0, 0],[0, 255, 255], \
4                 [255, 255, 0],[255, 0, 255],[80, 70, 180],[250, 80, 190],\
5                 [245, 145, 50],[70, 150, 250],[50, 190, 190]]
6      r = np.zeros_like(image).astype(np.uint8)
7      g = np.zeros_like(image).astype(np.uint8)
8      b = np.zeros_like(image).astype(np.uint8)
9      r[image == 1], g[image == 1], b[image == 1] = colours[random.randrange(0,10)]
10     coloured_mask = np.stack([r, g, b], axis=2)
11     return coloured_mask
```

```
1  # 对象侦测，回传遮罩、边框、类别
2  def get_prediction(img_path, threshold):
3      img = Image.open(img_path)
4      transform = T.Compose([T.ToTensor()])
5      img = transform(img)
6      pred = model([img])
7      pred_score = list(pred[0]['scores'].detach().numpy())
8      pred_t = [pred_score.index(x) for x in pred_score if x>threshold][-1]
9      masks = (pred[0]['masks']>0.5).squeeze().detach().cpu().numpy()
10     pred_class = [COCO_INSTANCE_CATEGORY_NAMES[i] \
11                   for i in list(pred[0]['labels'].numpy())]
12     pred_boxes = [[(int(i[0]), int(i[1])), (int(i[2]), int(i[3]))] \
13                   for i in list(pred[0]['boxes'].detach().numpy())]
14     masks = masks[:pred_t+1]
15     pred_boxes = pred_boxes[:pred_t+1]
16     pred_class = pred_class[:pred_t+1]
17     return masks, pred_boxes, pred_class
```

```
1  # 对象侦测含遮罩上色、显示结果
2  def instance_segmentation_api(img_path, threshold=0.5, rect_th=3,
3                                text_size=2, text_th=2):
4      masks, boxes, pred_cls = get_prediction(img_path, threshold)
5      img = cv2.imread(img_path)
6      img = cv2.cvtColor(img, cv2.COLOR_BGR2RGB)
7      for i in range(len(masks)):
8          rgb_mask = random_colour_masks(masks[i])
9          img = cv2.addWeighted(img, 1, rgb_mask, 0.5, 0)
10         print(boxes[i][0], boxes[i][1])
11         cv2.rectangle(img, boxes[i][0], boxes[i][1],color=(0, 255, 0), \
12                       thickness=rect_th)
13         cv2.putText(img,pred_cls[i], boxes[i][0], cv2.FONT_HERSHEY_SIMPLEX,\
14                     text_size, (0,255,0),thickness=text_th)
15     plt.figure(figsize=(20,30))
16     plt.imshow(img)
17     plt.xticks([])
18     plt.yticks([])
19     plt.show()
```

(5) 流程测试：测试每一个步骤。

```
1  # 显示测试图片文件
2  img = Image.open('./Mask_RCNN/PennFudanPed/PNGImages/FudanPed00001.png')
3  plt.imshow(img)
4  plt.axis('off')
5  plt.show()
```

执行结果：本图来自 *Penn-Fudan Database for Pedestrian Detection and Segmentation*[16]。

(6) 模型预测，并显示模型第一笔预测内容。

```
1  # 模型预测
2  transform = T.Compose([T.ToTensor()])
3  img_tensor = transform(img)
4  
5  model.eval()
6  pred = model([img_tensor])
7  
8  # 显示模型第一笔预测内容
9  pred[0]
```

含边框坐标(Box)、对象类别(Label)、概率(Score)及屏蔽(Mask)，共侦测到3个对象。

```
{'boxes': tensor([[158.9167, 174.7144, 301.3432, 433.7520],
        [418.4989, 165.8801, 535.9410, 490.1069],
        [242.6754, 223.8777, 269.8047, 257.7699]], grad_fn=<StackBackward0>),
 'labels': tensor([ 1,  1, 27]),
 'scores': tensor([0.9998, 0.9996, 0.1162], grad_fn=<IndexBackward0>),
 'masks': tensor([[[[0., 0., 0.,  ..., 0., 0., 0.],
          [0., 0., 0.,  ..., 0., 0., 0.],
          [0., 0., 0.,  ..., 0., 0., 0.],
          ...,
```

(7) 显示屏蔽：一般屏蔽门槛值为 0.5，可做适当调整。

```
1  # 保留屏蔽值>0.5的像素，其他一律为 0
2  masks = (pred[0]['masks']>0.5).squeeze().detach().cpu().numpy()
3
4  # 显示屏蔽
5  plt.imshow(masks[0], cmap='gray')
6  plt.axis('off')
7  plt.show()
```

执行结果。

(8) 屏蔽上色：随机使用不同颜色。

```
1  # 屏蔽上色
2  mask1 = random_colour_masks(masks[0])
3  plt.imshow(mask1)
4  plt.axis('off')
5  plt.show()
```

执行结果。

(9) 原图加屏蔽。

```
1  # 原图加屏蔽
2  blend_img = cv2.addWeighted(np.asarray(img), 0.5, mask1, 0.5, 0)
3  # 第 2 个屏蔽
4  mask2 = random_colour_masks(masks[1])
5  blend_img = cv2.addWeighted(np.asarray(blend_img), 0.5, mask2, 0.5, 0)
6
7  plt.imshow(blend_img)
8  plt.axis('off')
9  plt.show()
```

执行结果：每个屏蔽使用不同的颜色。

(10) 完整 API 测试：结合以上所有步骤成为单一函数 instance_segmentation_api。

```
1  instance_segmentation_api('./Mask_RCNN/people1.jpg', 0.5, rect_th=1,
2                            text_size=1, text_th=1)
```

执行结果：除了屏蔽外还有边框。

(11) 背影及重叠的对象也能侦测到。

```
1  instance_segmentation_api('./Mask_RCNN/people2.jpg', 0.8,
2                            rect_th=1, text_size=1, text_th=1)
```

执行结果。

举一例子说明 Mask-RCNN 实际应用，例如把照片背景模糊化。
(1) 定义函数，侦测所有对象，回传屏蔽。

```
1   # 侦测所有对象, 回传屏蔽
2   def pick_person_mask(img_path, threshold=0.5, rect_th=3, text_size=3, text_th=3):
3       # get the predicted masks and boxes and their corresponding labels
4       masks, boxes, pred_cls = get_prediction(img_path, threshold)
5       # pick the indices belonging to person
6       person_ids = [i for i in range(len(pred_cls)) if pred_cls[i]=="person"]
7       # pick the masks with the person-ids
8       person_masks = masks[person_ids, :, :]
9       # create a single mask out of all the instances and clip them
10      persons_mask = person_masks.sum(axis=0)
11      persons_mask = np.clip(persons_mask, 0,1)
12      return persons_mask
```

(2) 照片背景模糊化。

模型预测，取得人物屏蔽。

把屏蔽 RGB 设为相同值。

使用 OpenCV 的 GaussianBlur 函数将照片模糊化。

照片合成，人物部分采用原图，其他部分使用模糊化的图。

显示原图与生成图，比较背景的处理效果。

```
1   # 读取文件
2   img_path = "./Mask_RCNN/blur.jpg"
3   img = cv2.imread(img_path)
4
5   # 取得人物屏蔽
6   person_mask = pick_person_mask(img_path, threshold=0.5, rect_th=3
7                                  , text_size=3, text_th=3).astype(np.uint8)
8   # 把屏蔽 RGB 设为相同值
9   person_mask = np.repeat(person_mask[:, :, None], 3, axis=2)
10
11  # 照片模糊化
12  img_blur = cv2.GaussianBlur(img, (21, 21), 0)
13
14  # 人物部分采用原图，其他部分使用模糊化的图
15  final_img = np.where(person_mask==1, img, img_blur)
16
17  # fix 中文乱码
18  from matplotlib.font_manager import FontProperties
19  plt.rcParams['font.sans-serif'] = ['Microsoft JhengHei'] # 微软正黑体
20  plt.rcParams['axes.unicode_minus'] = False
21
22  # 显示原图与生成图，比较背景的处理效果
23  plt.figure(figsize=(15,15))
24  plt.subplot(121)
25  plt.title('原图')
26  plt.imshow(img[:,:,::-1])
27  plt.axis('off')
28  plt.subplot(122)
29  plt.title('生成图')
30  plt.imshow(final_img[:,:,::-1])
31  plt.axis('off')
32  plt.show()
```

执行结果：请看程序代码执行结果，人物身旁的落叶全部变模糊了。

关于去背景的应用，网络上还有一个模型 MODNet[17]，去背景效果更好，它提供一个展示程序"MODNet_Image_Matting_Demo_colab.ipynb"，可在 Google Colaboratory 顺利执行，效果如图 9.12 所示，左边为原图，中间为去背景图，右边为屏蔽效果。

图 9.12　效果展示

笔者将程序稍做修改，亦可在本机上执行。展示程序使用 gdown 指定代码 (1mcr7ALciuAsHCpLnrtG_eop5-EYhbCmz) 下载预先训练的模型，这部分不知如何在本机执行，故笔者先在 Colaboratory 执行，下载模型，供本机使用。两个程序及预先训练模型分别收录在本书范例的 MODNet 目录下。

PyTorch 官网提供了另一范例 "TORCHVISION OBJECT DETECTION FINETUNING TUTORIAL"[18]，程序代码执行有错误，且不如上例清楚，但它有一丰富的数据集可供测试。

Facebook AI 试验室另外开发一个套件 Detectron2 [19]，支持更完整、更多的功能，可惜只能安装在 Linux 操作系统上，详情可参阅 Detectron2 官网文件 [20]。

# 9-5　风格转换——人人都可以是毕加索

接着来认识另一个有趣的 AutoEncoder 变形，称为风格转换 (Style Transfer)，把一张照片转换成某一幅画的风格，如图 9.13 所示。读者可以在手机下载 PrismaApp，它能够在拍照后，将照片风格实时转换，内建近二十种的大师画风可供选择，只是转换速度有点慢。

图 9.13　风格转换 (Style Transfer)，原图 + 风格图像 = 生成图像

图片来源：fast-style-transfer GitHub [21]

之前有一则关于美图影像试验室 (MTlab) 的新闻——《催生全球首位 AI 绘图师 Andy，美图抢攻人工智能却面临一大挑战》[22]，该公司号称投资了 1.99 亿元人民币，研发团队超过 60 人，将风格转换速度缩短到 3 秒钟，开发成美图秀秀 App，大受欢迎，之后更趁势推出专属手机，狂销 100 多万台，算得上少数成功的 AI 商业模式。

风格转换算法由 Leon A. Gatys 等学者于 2015 年提出 [23]，主要做法是重新定义损失函数，分为内容损失 (Content Loss) 与风格损失 (Style Loss)，并利用 AutoEncoder 的译码器合成图像，随着训练周期，损失逐渐变小，生成的图像会越接近于原图与风格图的合成。

内容损失函数比较单纯，即原图与生成图像的像素差异平方和，定义如下：

$$J_{\text{content}}(C, G) = \frac{1}{4 \times n_H \times n_W \times n_C} \sum \left(a^{(C)} - a^{(G)}\right)^2$$

其中：

$n_H$、$n_W$ 为原图的宽、高；

$n_C$ 为色彩通道数；

$a^{(C)}$ 为原图的像素；

$a^{(G)}$ 为生成图像的像素。

风格损失函数为该算法的重点，如何量化抽象的画风是一大挑战，Gatys 等学者想到的方法是，先定义 Gram 矩阵 (Matrix)，再利用 Gram 矩阵来定义风格损失。

Gram Matrix：两个特征向量进行点积，代表特征的关联性，显现那些特征是同时出现的，亦即风格。因此，风格损失就是要最小化风格图像与生成图像的 Gram 差异平方和，如下：

$$J_{style}(S,G) = \frac{1}{4 \times n_c^2 \times (n_H \times n_W)^2} \sum_{i=l}^{n_c} \sum_{j=1}^{n_c} (G^{(S)} - G^{(G)})^2$$

其中：

$G^{(S)}$ 为风格图像的 Gram；

$G^{(G)}$ 为生成图像的 Gram。

上式只是单一神经层的风格损失，结合所有神经层的风格损失，定义如下：

$$J_{style}(S,G) = \sum_{l} \lambda^{(l)} J_{style}^{(l)}(S,G)$$

其中：

$\lambda$ 为每一层的权重。

总损失函数：

$$J(G) = \alpha J_{content}(C,G) + \beta J_{style}(S,G)$$

其中：

$\alpha$、$\beta$ 为控制内容与风格的比重，可以控制生成图像要偏重风格的比例。

接下来，我们就来进行实操。

**范例6.** 使用风格转换算法进行图片文件的转换。提醒一下，范例中的内容图即是原图的意思。本范例程序修改自官网提供的范例**NEURAL TRANSFER USING PYTORCH**[24]。

使用风格转换算法进行图文件转换的相关步骤如图 9.14 所示。

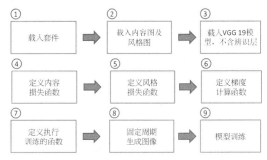

图 9.14 相关步骤

**下列程序代码请参考【09_06_Neural_Style_Transfer.ipynb】。**

(1) 载入相关套件。

```
1  from __future__ import print_function
2  import torch
3  import torch.nn as nn
4  import torch.nn.functional as F
5  import torch.optim as optim
6  from PIL import Image
7  import matplotlib.pyplot as plt
8  import torchvision.transforms as transforms
9  import torchvision.models as models
10 import copy
```

(2) 判断是否使用 GPU。

```
1  device = torch.device("cuda" if torch.cuda.is_available() else "cpu")
```

(3) 定义读取图片文件、显示/转换图像的函数。

```python
# 如果无 GPU 使用较小尺寸的图像
imsize = 512 if torch.cuda.is_available() else 128

# 转换
loader = transforms.Compose([
    transforms.Resize((imsize, imsize)),  # 统一图像尺寸
    transforms.ToTensor()])

# 读取图片文件，转为张量
def image_loader(image_name):
    image = Image.open(image_name)
    image = loader(image).unsqueeze(0)  # 增加一维
    return image.to(device, torch.float)

unloader = transforms.ToPILImage()  # 张量转为 PIL Image 格式

# 显示图像
def imshow(tensor, title=None):
    image = tensor.cpu().clone()  # 复制张量
    image = image.squeeze(0)       # 减少一维
    image = unloader(image)
    plt.axis('off')
    plt.imshow(image)
    if title is not None:
        plt.title(title)
    plt.pause(0.001)  # 显示多张图须停顿，等画面更新
```

(4) 载入内容图片文件(舞者)、风格图片文件(拾穗)。

```python
style_img = image_loader("./StyleTransfer/des_glaneuses.jpg")
content_img = image_loader("./StyleTransfer/dancing.jpg")
print(style_img.shape, content_img.shape)
imshow(style_img, title='Style Image')
imshow(content_img, title='Content Image')
```

执行结果。

(5) 定义内容损失函数：为内容图与生成图特征向量之差的平方和(MSE)，也可以采用上述理论的公式。

```python
class ContentLoss(nn.Module):
    def __init__(self, target,):
        super(ContentLoss, self).__init__()
        # we 'detach' the target content from the tree used
        # to dynamically compute the gradient: this is a stated value,
        # not a variable. Otherwise the forward method of the criterion
        # will throw an error.
        self.target = target.detach()

    def forward(self, input):
        self.loss = F.mse_loss(input, self.target)
        return input
```

(6) 定义风格损失函数：先定义 Gram Matrix 计算函数，再定义风格损失函数。

```python
def gram_matrix(input):
    # a: 批量(=1)
    # b: feature map 数量
    # (c,d): feature maps 维度大小 (N=c*d)
    a, b, c, d = input.size()

    features = input.view(a * b, c * d)  # resise F_XL into \hat F_XL

    G = torch.mm(features, features.t())  # compute the gram product

    return G.div(a * b * c * d)
```

```python
class StyleLoss(nn.Module):
    def __init__(self, target_feature):
        super(StyleLoss, self).__init__()
        self.target = gram_matrix(target_feature).detach()

    def forward(self, input):
        G = gram_matrix(input)
        self.loss = F.mse_loss(G, self.target)
        return input
```

(7) 定义模型：载入 VGG 19 模型。

```python
cnn = models.vgg19(pretrained=True).features.to(device).eval()
```

(8) 定义标准化函数。

```python
cnn_normalization_mean = torch.tensor([0.485, 0.456, 0.406]).to(device)
cnn_normalization_std = torch.tensor([0.229, 0.224, 0.225]).to(device)

# 标准化函数
class Normalization(nn.Module):
    def __init__(self, mean, std):
        super(Normalization, self).__init__()
        # .view the mean and std to make them [C x 1 x 1] so that they can
        # directly work with image Tensor of shape [B x C x H x W].
        # B is batch size. C is number of channels. H is height and W is width.
        self.mean = torch.tensor(mean).view(-1, 1, 1)
        self.std = torch.tensor(std).view(-1, 1, 1)

    def forward(self, img):
        # normalize img
        return (img - self.mean) / self.std
```

(9) 定义内容图和风格图输出的卷积层名称。

```python
# 在下列卷积层后计算损失
content_layers_default = ['conv_4']
style_layers_default = ['conv_1', 'conv_2', 'conv_3', 'conv_4', 'conv_5']
```

(10) 修改模型：只取 VGG19 的卷积层、池化层、ReLU Activation Function 层及 Batch Normalization 层，并计算卷积层后的损失函数。

```python
# 定义卷积层后的损失计算函数
def get_style_model_and_losses(cnn, normalization_mean, normalization_std,
                               style_img, content_img,
                               content_layers=content_layers_default,
                               style_layers=style_layers_default):
    # 标准化
    normalization = Normalization(normalization_mean, normalization_std).to(device)

    # 变量初始化
    content_losses = []
    style_losses = []

    # 模型先加入标准化的神经层
    model = nn.Sequential(normalization)

    i = 0  # increment every time we see a conv
    for layer in cnn.children():
        if isinstance(layer, nn.Conv2d):
```

```
19              i += 1
20              name = f'conv_{i}'
21          elif isinstance(layer, nn.ReLU):
22              name = f'relu_{i}'
23              # inplace=True 效果不佳
24              layer = nn.ReLU(inplace=False)
25          elif isinstance(layer, nn.MaxPool2d):
26              name = f'pool_{i}'
27          elif isinstance(layer, nn.BatchNorm2d):
28              name = f'bn_{i}'
29          else:
30              raise RuntimeError(f'Unrecognized layer: {layer.__class__.__name__}')

32          model.add_module(name, layer)
33
34          if name in content_layers:
35              # add content loss:
36              target = model(content_img).detach()
37              content_loss = ContentLoss(target)
38              model.add_module(f"content_loss_{i}", content_loss)
39              content_losses.append(content_loss)
40
41          if name in style_layers:
42              # add style loss:
43              target_feature = model(style_img).detach()
44              style_loss = StyleLoss(target_feature)
45              model.add_module(f"style_loss_{i}", style_loss)
46              style_losses.append(style_loss)
47
48      # 不加入卷积层后的辨识层
49      for i in range(len(model) - 1, -1, -1):
50          if isinstance(model[i], ContentLoss) or isinstance(model[i], StyleLoss):
51              break
52
53      model = model[:(i + 1)]
54
55      return model, style_losses, content_losses
```

(11) 定义执行训练的函数：以梯度下降法训练模型。

```
1  def get_input_optimizer(input_img):
2      # 设定 input image 要优化
3      optimizer = optim.LBFGS([input_img])
4      return optimizer
5
6  def run_style_transfer(cnn, normalization_mean, normalization_std,
7                         content_img, style_img, input_img, num_steps=300,
8                         style_weight=1000000, content_weight=1):
9      print('Building the style transfer model..')
10     model, style_losses, content_losses = get_style_model_and_losses(cnn,
11         normalization_mean, normalization_std, style_img, content_img)
12
13     # 优化 input image，而不是求权重
14     input_img.requires_grad_(True)
15     model.requires_grad_(False)
16     optimizer = get_input_optimizer(input_img)
17
18     print('优化..')
19     run = [0]
20     while run[0] <= num_steps:
21         def closure():
22             # 限定像素值介于 [0, 1]
23             with torch.no_grad():
24                 input_img.clamp_(0, 1)
25
26             # 计算损失
27             optimizer.zero_grad()
28             model(input_img)
29             style_score = 0
30             content_score = 0
32             for sl in style_losses:
33                 style_score += sl.loss
34             for cl in content_losses:
35                 content_score += cl.loss
36
37             style_score *= style_weight
38             content_score *= content_weight
39
```

```
40          loss = style_score + content_score
41          loss.backward()
42
43          # 显示执行信息
44          run[0] += 1
45          if run[0] % 50 == 0:
46              print("run {}:".format(run))
47              print('Style Loss : {:4f} Content Loss: {:4f}'.format(
48                  style_score.item(), content_score.item()))
49              print()
50
51          return style_score + content_score
52
53      optimizer.step(closure)
54
55  with torch.no_grad():
56      input_img.clamp_(0, 1)
57
58  return input_img
```

(12) 调用上述函数，执行模型训练。

```
1  input_img = content_img.clone()
2  output = run_style_transfer(cnn, cnn_normalization_mean, cnn_normalization_std,
3                              content_img, style_img, input_img)
4
5  plt.ion()
6
7  plt.figure()
8  imshow(output, title='Output Image')
9
10 # sphinx_gallery_thumbnail_number = 4
11 plt.ioff()
12 plt.show()
```

执行结果：执行 300 个步骤。

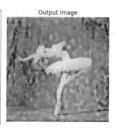

(13) 测试另一组图片文件。

```
1  style_img = image_loader("./StyleTransfer/mirror.jpg")
2  content_img = image_loader("./StyleTransfer/dancing.jpg")
3  print(style_img.shape, content_img.shape)
4  imshow(style_img, title='Style Image')
5  imshow(content_img, title='Content Image')
```

执行结果。

(14) 风格转换。

```
1  input_img = content_img.clone()
2  output = run_style_transfer(cnn, cnn_normalization_mean, cnn_normalization_std,
3                              content_img, style_img, input_img)
4
5  plt.ion()
6
7  plt.figure()
8  imshow(output, title='Output Image')
9
10 # sphinx_gallery_thumbnail_number = 4
11 plt.ioff()
12 plt.show()
```

执行结果。

范例有一些参数可以调整，例如，run_style_transfer 函数的 style_weight、content_weight 参数代表风格图 / 原图影响生成图像的比例，读者可以试试看，也许可以创作出独一无二的画作。

上述程序生成一张图像需要花很长时间，在如今的社交媒体时代，就算这酷炫的效果抓住了大众的眼球，也难以流行，因此在网络上有许多的研究，讨论如何加快算法速度，有兴趣的读者可搜寻 Fast Style Transfer。另外，也有同学问到，如果用同一张风格图，对另一张新的内容图进行风格转换，也要重新训练吗？答案是不一定，这是一个值得研究的课题。开发美图秀秀的公司花了近两亿元人民币，才将速度缩短至 3 秒，可见技术难度颇高，所以，速度绝对是商业模式重要的考虑因素。

风格转换是一个非常有趣的应用，除了转换为名画风格之外，也可将照片卡通化，或是针对脸部美化，凡此种种都值得一试。当然，不只有风格转换算法具有这样的功能，其他像 GAN 或 OpenCV 图像处理也都能做到类似的功能，大家一起天马行空，发挥想象吧！

## 9-6 脸部辨识

脸部辨识 (Facial Recognition) 的应用面非常广泛，国内厂商不论是系统厂商、PC 厂商、NAS 厂商，甚至是电信业者，都已涉猎此领域，推出五花八门的相关产品，目前已有以下应用类型。

智慧保全：结合门禁系统，运用在家庭、学校、员工宿舍、饭店、机场登机检查、出入境比对、黑名单 / 罪犯 / 失踪人口比对等方面。

考勤系统：上下班刷脸取代刷卡。

商店实时监控：实时辨识 VIP 和黑名单客户的进出，进行客户关怀、发送折扣码或记录停留时间，作为商品陈列与改善经营效能的参考依据。

快速结账：以脸部辨识取代刷卡付账。

人流统计：针对有人数容量限制的公共场所，如百货公司、游乐园、体育场馆通过脸部辨识进行人数控管。

情绪分析：辨识脸部情绪，发生意外时能迅速通报救援，或进行满意度调查。

社交软件上传照片的辨识：标注朋友姓名等。

依据技术类别可细分为：

脸部侦测 (Face Detection)：与对象侦测类似，因此运用对象侦测技术即可做到此功能，侦测图像中有哪些脸部及其位置；

脸部特征点检测 (Facial Landmarks Detection)：侦测脸部的特征点，用来比对两张脸是否为同一人；

脸部追踪 (Face Tracking)：在视频中追踪移动中的脸部，可辨识人移动的轨迹。

脸部辨识 (Face Recognition)，分为两种：

脸部识别 (Face Identification)：从 $N$ 个人中找出认识的人；

脸部验证 (Face Verification)：验证脸部特征是否符合特定人，例如，出入境检查旅客是否与其护照上的大头照相符合。

各项脸部辨识技术及支持的套件，如图 9.15 所示。

图 9.15　脸部辨识的技术类别与相关支持套件

接下来，我们就逐一实操这些相关功能。

## 9-6-1　脸部侦测

OpenCV 使用 Haar Cascades 算法进行各种对象的侦测，它会将各种对象的特征记录在 XML 文件，称为级联分类器 (Cascade File)。可在 OpenCV 或 OpenCV-Python 安装目录内找到 (haarcascade_*.xml)，笔者已把相关文件复制到范例程序目录下。

Haar Cascades 技术发展较早，优点是辨识速度快，能够做到实时侦测；缺点则是准确度较差，容易造成伪阳性，即误认脸部特征。它的架构类似卷积，如图 9.16 所示，以各种滤波器 (Filters) 扫描图像，像是眼部比脸颊暗、鼻梁比脸颊亮等。

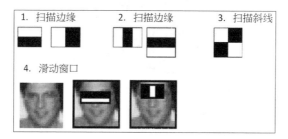

图 9.16　Haar Cascades 以滤波器 (Filters) 扫描图像

**范例1. 使用OpenCV进行脸部侦测(Face Detection)。**

使用 OpenCV 进行脸部侦测的相关步骤如图 9.17 所示。

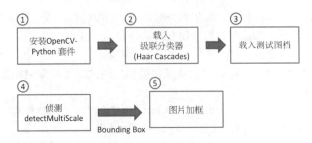

图 9.17 相关步骤

**下列程序代码请参考【09_07_Face Detection_opencv.ipynb】。**

(1) 载入相关套件，包含 OpenCV-Python。

```
1  # 载入相关套件
2  import cv2
3  from cv2 import CascadeClassifier
4  from cv2 import rectangle
5  import matplotlib.pyplot as plt
6  from cv2 import imread
```

(2) 载入脸部的级联分类器 (Face Cascade File)。

```
1  # 载入脸部级联分类器(face cascade file)
2  face_cascade = './cascade_files/haarcascade_frontalface_alt.xml'
3  classifier = cv2.CascadeClassifier(face_cascade)
```

(3) 载入测试图片文件。

```
1   # 载入图片文件
2   image_file = "./images_face/teammates.jpg"
3   image = imread(image_file)
4   
5   # OpenCV 预设为 BGR 色系，转为 RGB 色系
6   im_rgb = cv2.cvtColor(image, cv2.COLOR_BGR2RGB)
7   
8   # 显示图像
9   plt.imshow(im_rgb)
10  plt.axis('off')
11  plt.show()
```

执行结果。

(4) 侦测脸部并显示图像。

```
1  # 侦测脸部
2  bboxes = classifier.detectMultiScale(image)
3  # 脸部加框
4  for box in bboxes:
5      # 取得框的坐标及宽高
6      x, y, width, height = box
7      x2, y2 = x + width, y + height
```

```
 8      # 加白色框
 9      rectangle(im_rgb, (x, y), (x2, y2), (255,255,255), 2)
10
11 # 显示图像
12 plt.imshow(im_rgb)
13 plt.axis('off')
14 plt.show()
```

执行结果：全部人的脸都能被正确侦测到。

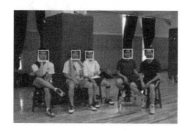

(5) 载入另一图片文件。

```
 1 # 载入图片文件
 2 image_file = "./images_face/classmates.jpg"
 3 image = imread(image_file)
 4
 5 # OpenCV 预设为 BGR 色系，转为 RGB 色系
 6 im_rgb = cv2.cvtColor(image, cv2.COLOR_BGR2RGB)
 7
 8 # 显示图像
 9 plt.imshow(im_rgb)
10 plt.axis('off')
11 plt.show()
```

执行结果：脸部特写。

(6) 侦测脸部并显示图像。

```
 1 # 侦测脸部
 2 bboxes = classifier.detectMultiScale(image)
 3 # 脸部加框
 4 for box in bboxes:
 5     # 取得框的坐标及宽高
 6     x, y, width, height = box
 7     x2, y2 = x + width, y + height
 8     # 加红色框
 9     rectangle(im_rgb, (x, y), (x2, y2), (255,0,0), 5)
10
11 # 显示图像
12 plt.imshow(im_rgb)
13 plt.axis('off')
14 plt.show()
```

执行结果：就算图像中的脸部占据画面较大，还是可以正确侦测到。

(7) 同时载入眼睛与微笑的级联分类器。

```
1  # 载入眼睛级联分类器(eye cascade file)
2  eye_cascade = './cascade_files/haarcascade_eye_tree_eyeglasses.xml'
3  eye_classifier = cv2.CascadeClassifier(eye_cascade)
4
5  # 载入微笑级联分类器(smile cascade file)
6  smile_cascade = './cascade_files/haarcascade_smile.xml'
7  smile_classifier = cv2.CascadeClassifier(smile_cascade)
```

(8) 侦测脸部并显示图像。

```
1  im_rgb_clone = im_rgb.copy()
2  # 侦测脸部
3  bboxes = classifier.detectMultiScale(image)
4  # 脸部加框
5  for box in bboxes:
6      # 取得框的坐标及宽高
7      x, y, width, height = box
8      x2, y2 = x + width, y + height
9      # 加白色框
10     rectangle(im_rgb_clone, (x, y), (x2, y2), (255,0,0), 5)
11
12 # 侦测微笑
13 # scaleFactor=2.5：扫描时每次缩减扫描视窗的尺寸比例。
14 # minNeighbors=20：每一个被选中的视窗至少要有邻近且合格的视窗数
15 bboxes = smile_classifier.detectMultiScale(image, 2.5, 20)
16 #微笑加框
17 for box in bboxes:
18     # 取得框的坐标及宽高
19     x, y, width, height = box
20     x2, y2 = x + width, y + height
21     # 加白色框
22     rectangle(im_rgb_clone, (x, y), (x2, y2), (255,0,0), 5)
23     # break
24
25 # 显示图像
26 plt.imshow(im_rgb_clone)
27 plt.axis('off')
28 plt.show()
```

执行结果：左边人脸的眼睛少抓取了一个，嘴巴误抓取好几个。这是笔者调整 detectMultiScale 参数多次后，所能得到的较佳结果。

(9) 修改程序，以脸部范围侦测眼睛及嘴巴，这样就不会有误判了。

```
1   im_rgb_clone = im_rgb.copy()
2   # 侦测脸部
3   bboxes = classifier.detectMultiScale(image)
4   # 脸部加框
5   for box in bboxes:
6       # 取得框的坐标及宽高
7       x, y, width, height = box
8       x2, y2 = x + width, y + height
9       # 加白色框
10      rectangle(im_rgb_clone, (x, y), (x2, y2), (255,0,0), 5)
11
12      # 侦测眼睛
13      face_box = image[y:y2, x:x2]
14      bboxes_eye = eye_classifier.detectMultiScale(face_box, 1.1, 5)
15      # 加框
16      for box_eye in bboxes_eye:
17          # 取得框的坐标及宽高
18          x, y, width, height = box_eye
19          x2, y2 = x + width, y + height
20          # 加白色框
21          rectangle(im_rgb_clone, (x+box[0], y+box[1]), (x2+box[0], y2+box[1]), (255,0,0), 5)
22
23      # 侦测微笑
24      # scaleFactor=2.5：扫描时每次缩减扫描视窗的尺寸比例。
25      # minNeighbors=20：每一个被选中的视窗至少要有邻近且合格的视窗数
26      bboxes_smile = smile_classifier.detectMultiScale(face_box, 2.5, 20, 0)
27      # 加框
28      for box_smile in bboxes_smile:
29          # 取得框的坐标及宽高
30          x, y, width, height = box_smile
31          x2, y2 = x + width, y + height
32          # 加白色框
33          rectangle(im_rgb_clone, (x+box[0], y+box[1]), (x2+box[0], y2+box[1]), (255,0,0), 5)
34
35  # 显示图像
36  plt.imshow(im_rgb_clone)
37  plt.axis('off')
38  plt.show()
```

执行结果。

detectMultiScale 相关参数的介绍如下：

scaleFactor：设定每次扫描窗口缩小的尺寸比例，设定较小值，会侦测到较多合格的窗口。

minNeighbors：每一个被选中的窗口至少要有邻近且合格的窗口数，设定较大值，会让伪阳性降低，但会使伪阴性提高。

minSize：小于这个设定值，会被过滤掉，格式为 (w, h)。

maxSize：大于这个设定值，会被过滤掉，格式为 (w, h)。

## 9-6-2　MTCNN 算法

Haar Cascades 技术发展较早，使用很简单，但是想要能准确侦测，必须因应图像的色泽、光线、对象大小来调整参数，并不容易。因此，近几年发展改用深度学习算法进行脸部侦测，较知名的算法 MTCNN 系由 Kaipeng Zhang 等学者于 2016 年在 *Joint Face Detection and Alignment using Multi-task Cascaded Convolutional Networks* 发表 [25]。

MTCNN 的架构是运用影像金字塔加上三个神经网络,如图 9.18 所示,四个部分的功能分别为:

影像金字塔 (Image Pyramid):撷取不同尺寸的脸部。
建议网络 (Proposal Network or P-Net):类似区域推荐,找出候选的区域。
强化网络 (Refine Network or R-Net):找出合格框 (Bounding Boxes)。
输出网络 (Output Network or O-Net):找出脸部特征点 (Landmarks)。

乍看之下,会不会觉得有些熟悉?其实 MTCNN 的做法与对象侦测算法 Faster R-CNN 类似。

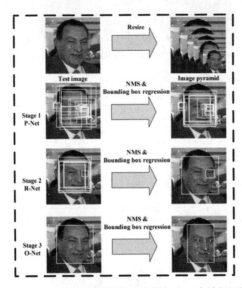

图 9.18　MTCNN 使用影像金字塔加上三个神经网络

原作者使用 Caffe/C 开发[26],许多人将其用 Python 改写,TensorFlow 框架名称为 mtcnn[27],安装指令如下:

pip install mtcnn

PyTorch 框架名称为 facenet-pytorch[28],安装指令如下:

pip install facenet-pytorch

**范例2. 使用MTCNN进行脸部侦测。本范例程序及数据来自官网[27]。**

下列程序代码请参考【09_08_Face Detection_mtcnn.ipynb】。

(1) 载入框架。

```
1  from facenet_pytorch import MTCNN, InceptionResnetV1
2  import torch
3  from torch.utils.data import DataLoader
4  from torchvision import datasets
5  import numpy as np
6  import pandas as pd
7  import os
```

(2) 判断是否使用 GPU。

```
1  device = torch.device("cuda" if torch.cuda.is_available() else "cpu")
```

(3) 载入并显示图片文件。

```
from PIL import Image
import matplotlib.pyplot as plt

image_file = './MTCNN/angelina_jolie/1.jpg'
image = Image.open(image_file)
```

```
# 显示图像
plt.imshow(image)
plt.axis('off')
plt.show()
```

执行结果。

(4) 建立 MTCNN 对象,侦测脸部,MTCNN 参数设定如下:

margin:边框厚度。

min_face_size:最小脸部侦测尺寸。

thresholds:门槛值,须设定三色彩通道。

factor:影像金字塔缩放比例。

post_process:后制处理。

还有许多其他参数,可使用 help(MTCNN) 查看说明。

以下程序均设为默认值。

```
# 建立 MTCNN 物件
mtcnn = MTCNN(
    image_size=160, margin=0, min_face_size=20,
    thresholds=[0.6, 0.7, 0.7], factor=0.709, post_process=True,
    device=device
)
```

(5) 辨识并裁切图片文件,并显示图像:侦测出的图像色彩信道在第一维,须转换至最后一维,部分像素值会超出范围,故须限定像素值范围介于 [0, 1],笔者发现大部分像素值范围散布介于 [-1, 1],故加 1 再除以 2,才不会使图像偏暗。

```
# 辨识
image_cropped = mtcnn(image)
# 色彩通道在第一维,转换至最后一维
image_cropped = torch.permute(image_cropped, (1, 2, 0))
# 限定像素值范围介于 [0, 1]
image_cropped = image_cropped.clamp(-1, 1)
image_cropped = (image_cropped + 1) *.5  # 使像素值介于 [0, 1] 之间
```

```
# 显示图像
plt.imshow(image_cropped)
plt.axis('off')
plt.show()
```

执行结果。

(6) 再看另一个应用，脸部验证，比较多个脸部的相似性，先建立 Inception ResNet 预先训练模型。

```
1  # 建立 inception resnet 预先训练模型
2  resnet = InceptionResnetV1(pretrained='vggface2').eval().to(device)
```

(7) 载入 MTCNN 文件夹下所有影像。

```
1  def collate_fn(x):
2      return x[0]
3
4  dataset = datasets.ImageFolder('./MTCNN')
5  dataset.idx_to_class = {i:c for c, i in dataset.class_to_idx.items()}
6  loader = DataLoader(dataset, collate_fn=collate_fn)
```

(8) 使用 MTCNN 识别脸部，并取得脸部向量。
MTCNN 加参数 return_prob=True，可额外取得脸部识别的概率。

```
1   aligned = []
2   names = []
3   for x, y in loader:
4       x_aligned, prob = mtcnn(x, return_prob=True)
5       if x_aligned is not None:
6           print(f'脸部识别的概率: {prob:8f}')
7           # 取得脸部向量
8           aligned.append(x_aligned)
9           # 取得姓名
10          names.append(dataset.idx_to_class[y])
```

(9) 将识别的脸部转换为嵌入向量。

```
1  aligned = torch.stack(aligned).to(device)
2  embeddings = resnet(aligned).detach().cpu()
```

(10) 比较嵌入向量相似性。

```
1  # 计算夹角
2  dists = [[(e1 - e2).norm().item() for e2 in embeddings] for e1 in embeddings]
3  pd.DataFrame(dists, columns=names, index=names)
```

执行结果：值越小越相似，发现同性较相似。

|  | angelina_jolie | bradley_cooper | kate_siegel | paul_rudd | shea_whigham |
| --- | --- | --- | --- | --- | --- |
| angelina_jolie | 0.000000 | 1.447480 | 0.887728 | 1.429847 | 1.399073 |
| bradley_cooper | 1.447480 | 0.000000 | 1.313749 | 1.013447 | 1.038684 |
| kate_siegel | 0.887728 | 1.313749 | 0.000000 | 1.388377 | 1.379654 |
| paul_rudd | 1.429847 | 1.013447 | 1.388377 | 0.000000 | 1.100503 |
| shea_whigham | 1.399073 | 1.038684 | 1.379654 | 1.100503 | 0.000000 |

再看如何对视频进行脸部追踪，须额外安装一个套件 mmcv：pip install mmcv
(1) 载入套件。

```
1  import mmcv, cv2
2  from PIL import Image, ImageDraw
3  from IPython import display
```

(2) 建立 MTCNN 对象：keep_all=True 会回传所有侦测到的脸部。

```
1  mtcnn = MTCNN(keep_all=True, device=device)
```

(3) 载入视频，并播放视频。

```
1  video_path = './MTCNN/video.mp4'
2  video = mmcv.VideoReader(video_path)
3  frames = [Image.fromarray(cv2.cvtColor(frame, cv2.COLOR_BGR2RGB))
4            for frame in video]
5
6  display.Video(video_path, width=640)
```

执行结果。

(4) 脸部追踪、画框。

```
1   frames_tracked = []
2   for i, frame in enumerate(frames):
3       print('\rTracking frame: {}'.format(i + 1), end='')
4
5       # 脸部追踪
6       boxes, _ = mtcnn.detect(frame)
7
8       # 脸部画框
9       frame_draw = frame.copy()
10      draw = ImageDraw.Draw(frame_draw)
11      for box in boxes:
12          draw.rectangle(box.tolist(), outline=(255, 0, 0), width=6)
13
14      # 存储至 frames_tracked
15      frames_tracked.append(frame_draw.resize((640, 360), Image.BILINEAR))
16  print('\nDone')
```

(5) 播放脸部画框的视频。

```
1  d = display.display(frames_tracked[0], display_id=True)
2  i = 1
3  try:
4      while True:
5          d.update(frames_tracked[i % len(frames_tracked)])
6          i += 1
7  except KeyboardInterrupt:
8      pass
```

执行结果：会连续播放，须按暂停键，才能执行下一格。

(6) 存盘。

```
1  dim = frames_tracked[0].size
2  fourcc = cv2.VideoWriter_fourcc(*'FMP4')
3  video_tracked = cv2.VideoWriter('video_tracked.mp4', fourcc, 25.0, dim)
4  for frame in frames_tracked:
5      video_tracked.write(cv2.cvtColor(np.array(frame), cv2.COLOR_RGB2BGR))
6  video_tracked.release()
```

## 9-6-3 脸部追踪

脸部追踪 (Face Tracking) 可在影片中追踪特定人的脸部，这里使用的套件是 face-recognition，安装指令如下：

pip install face-recognition

注意，face-recognition 是以 dlib 为基础的套件，它使用 C++ 开发，所以要先安装 dlib 套件，在 Windows 作业环境下，必须备妥如下工具设置 dlib：

(1) Microsoft Visual Studio 2019，相关说明请参考笔者的文章《dlib 安装心得 —— Windows 环境》[29]。

(2) CMake for Windows：安装后将 bin 路径 ( 例如 C:\Program Files\CMake\bin) 加入环境变量 Path 中。

(3) 设置 dlib：python setup.py build。

(4) 安装 dlib：python setup.py insatll。

**范例 3. 使用 Face-Recognition 套件进行脸部侦测。**

使用 Face-Recognition 框架进行脸部侦测的相关步骤如图 9.19 所示。

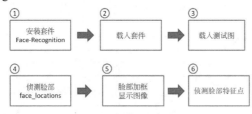

图 9.19　相关步骤

**下列程序代码请参考【09_09_Face_Recognition.ipynb】。**

(1) 载入相关套件，包含 Face-Recognition。

```
1  # 安装套件：pip install face-recognition
2  # 载入相关套件
3  import matplotlib.pyplot as plt
4  from matplotlib.patches import Rectangle, Circle
5  import face_recognition
```

(2) 载入并显示图片文件。

```
1  # 载入图片文件
2  image_file = "./images_face/classmates.jpg"
3  image = plt.imread(image_file)
4
5  # 显示图像
6  plt.imshow(image)
7  plt.axis('off')
8  plt.show()
```

执行结果。

(3) 调用 face_locations 函数侦测脸部。

```
1  # 侦测脸部
2  faces = face_recognition.face_locations(image)
```

(4) 脸部加框，显示图像，注意，框的坐标所代表的方向依次为上 / 左 / 下 / 右 ( 逆时针 )。

```
1   # 脸部加框
2   ax = plt.gca()
3   for result in faces:
4       # 取得框的坐标
5       y1, x1, y2, x2 = result
6       width, height = x2 - x1, y2 - y1
7       # 加红色框
8       rect = Rectangle((x1, y1), width, height, fill=False, color='red')
9       ax.add_patch(rect)
10
11  # 显示图像
12  plt.imshow(image)
13  plt.axis('off')
14  plt.show()
```

执行结果。

(5) 侦测脸部特征点并显示。

```
1   # 侦测脸部特征点并显示
2   from PIL import Image, ImageDraw
3
4   # 载入图片文件
5   image = face_recognition.load_image_file(image_file)
6
7   # 转为 Pillow 图像格式
8   pil_image = Image.fromarray(image)
9
10  # 取得图像绘图对象
11  d = ImageDraw.Draw(pil_image)
12
13  # 侦测脸部特征点
14  face_landmarks_list = face_recognition.face_landmarks(image)
15
16  for face_landmarks in face_landmarks_list:
17      # 显示五官特征点
18      for facial_feature in face_landmarks.keys():
19          print(f"{facial_feature} 特征点: {face_landmarks[facial_feature]}\n")
20
21      # 绘制特征点
22      for facial_feature in face_landmarks.keys():
23          d.line(face_landmarks[facial_feature], width=5, fill='green')
24
25  # 显示图像
26  plt.imshow(pil_image)
27  plt.axis('off')
28  plt.show()
```

执行结果如下：
五官特征点的坐标。

```
chin 特征点: [(958, 485), (968, 525), (982, 562), (999, 598), (1022, 630), (1054, 657), (1092, 677), (1135, 693), (1179, 689), (1220, 670), (1249, 639), (1274, 606), (1291, 567), (1298, 524), (1296, 478), (1291, 433), (1283, 387)]
left_eyebrow 特征点: [(969, 464), (978, 434), (1002, 417), (1032, 413), (1061, 415)]
right_eyebrow 特征点: [(1119, 397), (1142, 373), (1172, 361), (1204, 364), (1228, 382)]
nose_bridge 特征点: [(1098, 440), (1107, 477), (1115, 512), (1124, 548)]
nose_tip 特征点: [(1092, 557), (1112, 562), (1133, 565), (1151, 552), (1167, 538)]
left_eye 特征点: [(1006, 473), (1019, 458), (1038, 454), (1058, 461), (1042, 467), (1024, 472)]
right_eye 特征点: [(1147, 436), (1160, 417), (1179, 409), (1201, 414), (1186, 423), (1167, 430)]
top_lip 特征点: [(1079, 606), (1100, 595), (1121, 586), (1142, 585), (1160, 576), (1186, 570), (1215, 567), (1207, 571), (1164, 585), (1145, 593), (1125, 596), (1088, 605)]
bottom_lip 特征点: [(1215, 567), (1197, 598), (1176, 619), (1155, 628), (1134, 631), (1109, 626), (1079, 606), (1088, 605), (1128, 612), (1149, 610), (1168, 601), (1207, 571)]
```

脸部轮廓画线。

**范例4.** 使用Face-Recognition套件进行视频脸部追踪。程序修改自"face-recognition GitHub"的范例[30]。

使用 Face-Recognition 框架进行视频脸部追踪的相关步骤如图 9.20 所示。

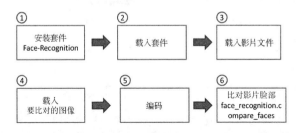

图 9.20  相关步骤

**下列程序代码请参考【09_10_Face_Tracking.ipynb】。**

(1) 载入相关套件。

```
1  # 安装套件: pip install face-recognition
2  # 载入相关套件
3  import matplotlib.pyplot as plt
4  from matplotlib.patches import Rectangle, Circle
5  import face_recognition
6  import cv2
```

(2) 载入影片文件。

```
1  # 载入影片文件
2  input_movie = cv2.VideoCapture("./images_face/short_hamilton_clip.mp4")
3  length = int(input_movie.get(cv2.CAP_PROP_FRAME_COUNT))
4  print(f'影片帧数: {length}')
```

执行结果：影片总帧数为 275。

(3) 指定输出文件名，注意，影片分辨率设为 (640, 360)，故输入的影片不得低于此分辨率，否则输出文件将无法播放。

```
1  # 指定输出文件名
2  fourcc = cv2.VideoWriter_fourcc(*'XVID')
3  # 每秒帧数(fps):29.97，影片解析度(Frame Size):(640, 360)
4  output_movie = cv2.VideoWriter('./images_face/output.avi',
5                                 fourcc, 29.97, (640, 360))
```

(4) 载入要辨识的图像，须先编码 (Encode) 为向量，以方便脸部比对。

```
1   # 载入要辨识的图像
2   image_file = 'lin-manuel-miranda.png'
3   lmm_image = face_recognition.load_image_file("./images_face/"+image_file)
4   # 取得图像编码
5   lmm_face_encoding = face_recognition.face_encodings(lmm_image)[0]
6
7   # obama
8   image_file = 'obama.jpg'
9   obama_image = face_recognition.load_image_file("./images_face/"+image_file)
10  # 取得图像编码
11  obama_face_encoding = face_recognition.face_encodings(obama_image)[0]
12
13  # 设定阵列
14  known_faces = [
15      lmm_face_encoding,
16      obama_face_encoding
17  ]
18
19  # 目标名称
20  face_names = ['lin-manuel-miranda', 'obama']
```

(5) 变数初始化。

```
1  # 变数初始化
2  face_locations = []   # 脸部位置
3  face_encodings = []   # 脸部编码
4  face_names = []       # 脸部名称
5  frame_number = 0      # 帧数
```

(6) 比对脸部并存盘。

```
1   # 侦测脸部并写入输出文件
2   while True:
3       # 读取一帧影像
4       ret, frame = input_movie.read()
5       frame_number += 1
6
7       # 影片播放结束，即跳出循环
8       if not ret:
9           break
10
11      # 将 BGR 色系转为 RGB 色系
12      rgb_frame = frame[:, :, ::-1]
13
14      # 找出脸部位置
15      face_locations = face_recognition.face_locations(rgb_frame)
16      # 编码
17      face_encodings = face_recognition.face_encodings(rgb_frame, face_locations)
18
19      # 比对脸部
20      face_names = []
21      for face_encoding in face_encodings:
22          # 比对脸部编码是否与图片文件符合
23          match = face_recognition.compare_faces(known_faces, face_encoding,
24                                   tolerance=0.50)
25
26          # 找出符合脸部的名称
27          name = None
28          for i in range(len(match)):
29              if match[i] and 0 < i < len(face_names):
30                  name = face_names[i]
```

```
31                break
32
33            face_names.append(name)
34
35    # 输出影片标记脸部位置及名称
36    for (top, right, bottom, left), name in zip(face_locations, face_names):
37        if not name:
38            continue
39
40        # 加框
41        cv2.rectangle(frame, (left, top), (right, bottom), (0, 0, 255), 2)
42
43        # 标记名称
44        cv2.rectangle(frame, (left, bottom - 25), (right, bottom), (0, 0, 255)
45                      , cv2.FILLED)
46        font = cv2.FONT_HERSHEY_DUPLEX
47        cv2.putText(frame, name, (left + 6, bottom - 6), font, 0.5,
48                    (255, 255, 255), 1)
49
50    # 将每一帧影像存储
51    print("Writing frame {} / {}".format(frame_number, length))
52    output_movie.write(frame)
53
54 # 关闭输入文件
55 input_movie.release()
56 # 关闭所有视窗口
57 cv2.destroyAllWindows()
```

执行结果。

```
Writing frame 4 / 275
Writing frame 5 / 275
Writing frame 6 / 275
Writing frame 7 / 275
Writing frame 8 / 275
Writing frame 9 / 275
Writing frame 10 / 275
Writing frame 11 / 275
Writing frame 12 / 275
Writing frame 13 / 275
Writing frame 14 / 275
Writing frame 15 / 275
Writing frame 16 / 275
Writing frame 17 / 275
Writing frame 18 / 275
Writing frame 19 / 275
Writing frame 20 / 275
Writing frame 21 / 275
```

(7) 输出的影片为 images_face/output.avi 文件：观看影片后发现，侦测速度较慢，Obama 并未侦测到，因图片文件是正面照，而图像文件则是侧面的画面。但瑕不掩瑜，大致上仍追踪得到主要影像的动态。

**范例5. 改用WebCam进行脸部实时追踪，程序修改自face-recognition GitHub 的范例**[30]。

**下列程序代码请参考【09_11_Face_Tracking_webcam.ipynb】。**

由于步骤重叠的部分较多，所以只说明与范例 4 有差异的程序代码。

(1) 以读取 WebCam 取代载入影片文件。

```
1 # 指定第一台 WebCam
2 video_capture = cv2.VideoCapture(0)
```

(2) 读取 WebCam 一帧影像：第 4 行。

```
1 # 侦测脸部并即时显示
2 while True:
3     # 读取一帧影像
4     ret, frame = video_capture.read()
```

(3) 侦测脸部的处理均相同，但存盘改成实时显示。

```
46    # 显示每一帧影像
47    cv2.imshow('Video', frame)
```

(4) 按 Q 键即可跳出循环。

```
49    # 按Q键即跳出循环
50    if cv2.waitKey(1) & 0xFF == ord('q'):
51        break
```

原作者也示范一个例子，可在 Raspberry pi 执行，实时进行脸部追踪。

## 9-6-4 脸部特征点侦测

侦测脸部特征点可使用 Face-Recognition、dlib 或者 OpenCV 套件，以上这三种都可侦测到 68 个特征点，如图 9.21 所示。

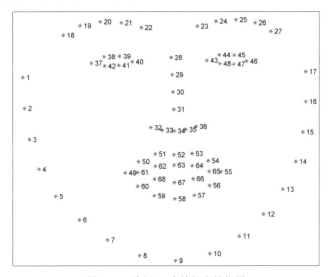

图 9.21　脸部 68 个特征点的位置

讯连科技转投资的玩美移动号称可以侦测脸部特征点达 200 点，相关新闻可参阅《讯连养出 14 亿美元独角兽，玩美移动凭什么赴美 IPO？》[31]。

Face-Recognition 套件侦测脸部特征点已在【09_09_Face_Recognition.ipynb】实操过了，这里不再多作介绍。前面安装的 dlib 套件本身就是包含了机器学习、数值分析、计算器视觉、图像处理等功能的函数库。

**范例5. 使用dlib实操脸部特征点的侦测。**

使用 dlib 实操脸部特征点的侦测的相关步骤如图 9.22 所示。

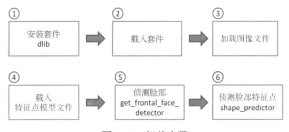

图 9.22　相关步骤

**下列程序代码请参考【09_11_脸部特征点侦测.ipynb】。**

(1) 载入相关套件，imutils 套件是一个简易的图像处理函数库，安装指令如下：
pip install imutils

```
1  # 载入相关套件
2  import dlib
3  import cv2
4  import matplotlib.pyplot as plt
5  from matplotlib.patches import Rectangle, Circle
6  from imutils import face_utils
```

(2) 载入并显示图片文件。

```
1  # 载入图片文件
2  image_file = "./images_face/classmates.jpg"
3  image = plt.imread(image_file)
4
5  # 显示图像
6  plt.imshow(image)
7  plt.axis('off')
8  plt.show()
```

执行结果。

(3) 侦测脸部特征点并显示：

dlib 特征点模型文件为 shape_predictor_68_face_landmarks.dat，可侦测 68 个点，如果只需要侦测 5 个点，可载入 shape_predictor_5_face_landmarks.dat。

dlib.get_frontal_face_detector：侦测脸部。

dlib.shape_predictor：侦测脸部特征点。

```
1  # 载入 dlib 以 HOG 基础的脸部侦测模型
2  model_file = "shape_predictor_68_face_landmarks.dat"
3  detector = dlib.get_frontal_face_detector()
4  predictor = dlib.shape_predictor(model_file)
5
6  # 侦测图像的脸部
7  rects = detector(image)
8
9  print(f'侦测到{len(rects)}张脸部.')
10 # 侦测每张脸的特征点
11 for (i, rect) in enumerate(rects):
12     # 侦测特征点
13     shape = predictor(image, rect)
14
15     # 转为 NumPy 阵列
16     shape = face_utils.shape_to_np(shape)
17
18     # 标示特征点
19     for (x, y) in shape:
20         cv2.circle(image, (x, y), 10, (0, 255, 0), -1)
21
22 # 显示图像
23 plt.imshow(image)
24 plt.axis('off')
25 plt.show()
```

执行结果。

(4) 侦测视频文件也没问题，按 Esc 键即可提前结束。

```
1   # 读取视频文件
2   cap = cv2.VideoCapture('./images_face/hamilton_clip.mp4')
3   while True:
4       # 读取一帧影像
5       _, image = cap.read()
6   
7       # 侦测图像的脸部
8       rects = detector(image)
9       for (i, rect) in enumerate(rects):
10          # 侦测特征点
11          shape = predictor(image, rect)
12          shape = face_utils.shape_to_np(shape)
13  
14          # 标示特征点
15          for (x, y) in shape:
16              cv2.circle(image, (x, y), 2, (0, 255, 0), -1)
17  
18      # 显示影像
19      cv2.imshow("Output", image)
20  
21      k = cv2.waitKey(5) & 0xFF    # 按 Esc 跳离循环
22      if k == 27:
23          break
24  
25  # 关闭输入文件
26  cap.release()
27  # 关闭所有窗口
28  cv2.destroyAllWindows()
```

OpenCV 针对脸部特征点的侦测，提供三种算法：

FacemarkLBF：Shaoqing Ren 等学者于 2014 年发表的 *Face Alignment at 3000 FPS via Regressing Local Binary Features* 中提出 [32]。

FacemarkAAM：Georgios Tzimiropoulos 等学者于 2013 年发表的 *Optimization problems for fast AAM fitting in-the-wild* 中提出 [33]。

FacemarkKamezi：V.Kazemi 和 J. Sullivan 于 2014 年发表的 *One Millisecond Face Alignment with an Ensemble of Regression Trees* 中提出 [34]。

我们分别试验一下，看看有什么差异。

**范例6. 使用OpenCV套件进行脸部特征点侦测。**

使用 OpenCV 框架进行脸部特征点侦测的相关步骤如图 9.23 所示。

图 9.23 相关步骤

下列程序代码请参考【09_12_Landmark_OpenCV.ipynb】。

(1) 载入相关套件：注意，只有 OpenCV 扩充版提供相关 API，所以，须执行下列指令，改安装 OpenCV 扩充版。

卸载：pip uninstall opencv-python opencv-contrib-python

安装套件：pip install opencv-contrib-python

```
1  # 解除安装套件：pip uninstall opencv-python opencv-contrib-python
2  # 安装套件：    pip install opencv-contrib-python
3  # 载入相关套件
4  import cv2
5  import numpy as np
6  from matplotlib import pyplot as plt
```

(2) 载入并显示图片文件：使用 Lena 图像测试。

```
1  # 载入图片文件
2  image_file = "./images_Object_Detection/lena.jpg"
3  image = cv2.imread(image_file)
4
5  # 显示图像
6  image_RGB = cv2.cvtColor(image, cv2.COLOR_BGR2RGB)
7  plt.imshow(image_RGB)
8  plt.axis('off')
9  plt.show()
```

执行结果。

(3) 使用 FacemarkLBF 侦测脸部特征点。

```
1  # 侦测脸部
2  cascade = cv2.CascadeClassifier("./cascade_files/haarcascade_frontalface_alt2.xml")
3  faces = cascade.detectMultiScale(image , 1.5, 5)
4  print("faces", faces)
5
6  # 建立脸部特征点侦测的物件
7  facemark = cv2.cv2.face.createFacemarkLBF()
8  # 训练模型 lbfmodel.yaml 下载自：
9  # https://raw.githubusercontent.com/kurnianggoro/GSOC2017/master/data/lbfmodel.yaml
10 facemark .loadModel("OpenCV/lbfmodel.yaml")
11 # 侦测脸部特征点
12 ok, landmarks1 = facemark.fit(image , faces)
13 print ("landmarks LBF", ok, landmarks1)
```

执行结果：显示脸部和特征点的坐标。

```
faces [[225 205 152 152]]
landmarks LBF True [array([[[201.31314, 268.08807],
        [201.5153 , 293.1106 ],
        [204.91422, 317.07196],
        [210.71988, 340.4278 ],
        [222.97098, 360.37122],
        [240.34521, 375.51422],
        [260.10678, 386.35587],
        [280.64197, 392.04227],
        [298.6573 , 390.89835],
        [311.434  , 384.88406],
        [318.37827, 371.23538],
        [324.82266, 357.113  ],
        [331.87363, 342.1786 ],
        [339.7072 , 327.1501 ],
        [346.04462, 311.9719 ],
        [349.2847 , 296.59448],
        [348.95883, 280.12585],
        [236.43172, 252.06743],
```

(4) 绘制特征点并显示图像。

```
1  # 绘制特征点
2  for p in landmarks1[0][0]:
3      cv2.circle(image, tuple(p.astype(int)), 5, (0, 255, 0), -1)
4
5  # 显示图像
6  image_RGB = cv2.cvtColor(image, cv2.COLOR_BGR2RGB)
7  plt.imshow(image_RGB)
8  plt.axis('off')
9  plt.show()
```

执行结果：很准确，可惜无法侦测到左上角被帽子遮挡的部分。

(5) 改用 FacemarkAAM 来侦测脸部特征点。

```
1  # 建立脸部特征点侦测的图像
2  facemark = cv2.face.createFacemarkAAM()
3  # 训练模型 aam.xml
4  # 下载自：https://github.com/berak/tt/blob/master/aam.xml
5  facemark.loadModel("OpenCV/aam.xml")
6  # 侦测脸部特征点
7  ok, landmarks2 = facemark.fit(image , faces)
8  print ("Landmarks AAM", ok, landmarks2)
```

(6) 绘制特征点并显示图像：过程与前面的程序代码相同。

```
1  # 绘制特征点
2  for p in landmarks2[0][0]:
3      cv2.circle(image, tuple(p.astype(int)), 5, (0, 255, 0), -1)
4
5  # 显示图像
6  image_RGB = cv2.cvtColor(image, cv2.COLOR_BGR2RGB)
7  plt.imshow(image_RGB)
8  plt.axis('off')
9  plt.show()
```

执行结果：左上角反而多出一些错误的特征点。

(7) 换用 FacemarkKamezi 侦测脸部特征点。

```
1  # 建立脸部特征点侦测的图像
2  facemark = cv2.face.createFacemarkKazemi()
3  # 训练模型 face_landmark_model.dat 下载自：
4  # https://github.com/opencv/opencv_3rdparty/tree/contrib_face_alignment_20170818
5  facemark.loadModel("./OpenCV/face_landmark_model.dat")
6  # 侦测脸部特征点
7  ok, landmarks2 = facemark.fit(image , faces)
8  print ("Landmarks Kazemi", ok, landmarks2)
```

(8) 绘制特征点并显示图像：过程与前面的程序代码相同。

```
1  # 绘制特征点
2  for p in landmarks2[0][0]:
3      cv2.circle(image, tuple(p.astype(int)), 5, (0, 255, 0), -1)
4
5  # 显示图像
6  image_RGB = cv2.cvtColor(image, cv2.COLOR_BGR2RGB)
7  plt.imshow(image_RGB)
8  plt.axis('off')
9  plt.show()
```

执行结果：左上角也是多出一些错误的特征点。

## 9-6-5　脸部验证

侦测完脸部特征点后，利用线性代数的法向量比较多张脸的特征点，就能找出哪一张脸最相似，使用 Face-Recognition 或 dlib 套件都可以。

**范例7. 使用Face-Recognition或dlib套件，比对哪一张脸最相似。**

使用 Face-Recognition 进行脸部验证的相关步骤如图 9.24 所示。

图 9.24　相关步骤

**下列程序代码请参考【09_13_Face_Verification.ipynb】。**

(1) 载入相关套件：使用 Face-Recognition 套件。

```
1  # 载入相关套件
2  import face_recognition
3  import numpy as np
4  from matplotlib import pyplot as plt
```

(2) 载入所有要比对的图片文件。

```
1   # 载入图片文件
2   known_image_1 = face_recognition.load_image_file("./images_face/jared_1.jpg")
3   known_image_2 = face_recognition.load_image_file("./images_face/jared_2.jpg")
4   known_image_3 = face_recognition.load_image_file("./images_face/jared_3.jpg")
5   known_image_4 = face_recognition.load_image_file("./images_face/obama.jpg")
6
7   # 标记图片文件名称
8   names = ["jared_1.jpg", "jared_2.jpg", "jared_3.jpg", "obama.jpg"]
9
10  # 显示图像
11  unknown_image = face_recognition.load_image_file("./images_face/jared_4.jpg")
12  plt.imshow(unknown_image)
13  plt.axis('off')
14  plt.show()
```

执行结果。

(3) 图像编码：使用 face_recognition.face_encodings 函数编码。

```
1  # 图像编码
2  known_image_1_encoding = face_recognition.face_encodings(known_image_1)[0]
3  known_image_2_encoding = face_recognition.face_encodings(known_image_2)[0]
4  known_image_3_encoding = face_recognition.face_encodings(known_image_3)[0]
5  known_image_4_encoding = face_recognition.face_encodings(known_image_4)[0]
6  known_encodings = [known_image_1_encoding, known_image_2_encoding,
7                     known_image_3_encoding, known_image_4_encoding]
8  unknown_encoding = face_recognition.face_encodings(unknown_image)[0]
```

(4) 使用 face_recognition.compare_faces 进行比对。

```
1  # 比对
2  results = face_recognition.compare_faces(known_encodings, unknown_encoding)
3  print(results)
```

执行结果：[True, True, True, False]，前三笔符合，完全正确。

(5) 载入相关套件：改用 dlib 套件。

```
1  import dlib
2  import cv2
3  import numpy as np
4  from matplotlib import pyplot as plt
```

(6) 载入模型：包括特征点侦测、编码、脸部侦测。

```
1  # 载入模型
2  pose_predictor_5_point = dlib.shape_predictor("./OpenCV/shape_predictor_5_face_landmarks.dat")
3  face_encoder = dlib.face_recognition_model_v1("./OpenCV/dlib_face_recognition_resnet_model_v1.dat")
4  detector = dlib.get_frontal_face_detector()
```

(7) 定义脸部编码与比对的函数：由于 dlib 无相关现成的函数，必须自行撰写。

```
1  # 找出哪一张脸最相似
2  def compare_faces_ordered(encodings, face_names, encoding_to_check):
3      distances = list(np.linalg.norm(encodings - encoding_to_check, axis=1))
4      return zip(*sorted(zip(distances, face_names)))
5  
6  
7  # 利用线性代数的法向量比较两张脸的特征点
8  def compare_faces(encodings, encoding_to_check):
9      return list(np.linalg.norm(encodings - encoding_to_check, axis=1))
10 
11 # 图像编码
12 def face_encodings(face_image, number_of_times_to_upsample=1, num_jitters=1):
13     # 侦测脸部
14     face_locations = detector(face_image, number_of_times_to_upsample)
15     # 侦测脸部特征点
16     raw_landmarks = [pose_predictor_5_point(face_image, face_location)
17                      for face_location in face_locations]
18     # 编码
19     return [np.array(face_encoder.compute_face_descriptor(face_image,
20                      raw_landmark_set, num_jitters)) for
21                      raw_landmark_set in raw_landmarks]
```

(8) 载入图片文件并显示。

```
1  # 载入图片文件
2  known_image_1 = cv2.imread("./images_face/jared_1.jpg")
3  known_image_2 = cv2.imread("./images_face/jared_2.jpg")
4  known_image_3 = cv2.imread("./images_face/jared_3.jpg")
5  known_image_4 = cv2.imread("./images_face/obama.jpg")
6  unknown_image = cv2.imread("./images_face/jared_4.jpg")
7  names = ["jared_1.jpg", "jared_2.jpg", "jared_3.jpg", "obama.jpg"]
```

(9) 图像编码。

```
1  known_image_1_encoding = face_encodings(known_image_1)[0]
2  known_image_2_encoding = face_encodings(known_image_2)[0]
3  known_image_3_encoding = face_encodings(known_image_3)[0]
4  known_image_4_encoding = face_encodings(known_image_4)[0]
5  known_encodings = [known_image_1_encoding, known_image_2_encoding,
6                     known_image_3_encoding, known_image_4_encoding]
7  unknown_encoding = face_encodings(unknown_image)[0]
```

(10) 比对两张脸。

```
1  # 比对
2  computed_distances = compare_faces(known_encodings, unknown_encoding)
3  computed_distances_ordered, ordered_names = compare_faces_ordered(known_encodings,
4                                                                    names, unknown_encoding)
5  print('比较两张脸的法向量距离：', computed_distances)
6  print('排序：', computed_distances_ordered)
7  print('依相似度排序：', ordered_names)
```

执行结果：显示两张脸的法向量距离，数字越小表示越相似。未知的图像是 Jared，比对结果前三名都是 Jared。

```
比较两张脸的法向量距离： [0.3998327850880958, 0.4104153798439364, 0.3913189516694114, 0.9053701677487068]
排序： (0.3913189516694114, 0.3998327850880958, 0.4104153798439364, 0.9053701677487068)
依相似度排序： ('jared_3.jpg', 'jared_1.jpg', 'jared_2.jpg', 'obama.jpg')
```

## 9-7 光学文字辨识

除了前面介绍的，另外还有很多其他类型的影像应用，例如：

(1) 光学文字辨识 (Optical Character Recognition, OCR)。

(2) 影像修复 (Image Inpainting)：用周围的影像将部分影像进行修复，可用于抹除照片中不喜欢的物件。

(3) 3D 影像的建构与辨识。

利用深度学习开发的影像相关应用系统，也是种类繁多，举例来说：

防疫：是否有戴口罩的侦测、社交距离的计算。

交通：道路拥堵状况的侦测、车速计算、车辆的违规 ( 越线、闯红灯 ) 等。

智能制造：机器人与机器手臂的视觉辅助。

企业运用：考勤、安全监控。

光学文字辨识是把图像中的印刷字辨识为文字，以节省大量的输入时间或抄写错误，可应用于支票号码/金额辨识、车牌辨识 (Automatic Number Plate Recognition, ANPR) 等，但也有人拿来破解登录用的图形码验证 (Captcha)，我们就来看看如何实操 OCR 辨识。

Tesseract OCR 是目前很盛行的 OCR 软件，HP 公司于 2005 年开放源代码 (Open Source)，以 C++ 开发而成，可由源代码设置，或直接安装已设置好的程序，在这里我们采取后者，从 https://github.com/UB-Mannheim/tesseract/wiki 下载最新版 exe 文件，直接执行即可。安装完成后，将安装路径下的 bin 子目录放入环境变量 path 内。若要以 Python 调用 Tesseract OCR，需额外安装 pytesseract 套件，指令如下：

pip install pytesseract

最简单的测试指令如下：

tesseract <图片文件> <辨识结果文件>

例如辨识一张发票文件 ( 如图 9.25 所示 )(./images_ocr/receipt.png) 的指令为：

tesseract ./images_ocr/receipt.png ./images_ocr/result.txt -l eng --psm 6 --dpi 70

其中 -l eng：为辨识英文，--psm 6：指单一区块 (A Single Uniform Block of Text)。相关的参数请参考 Tesseract OCR 官网 [35]，也可以直接执行 tesseract --help-extra，有摘要说明。

图 9.25　发票

图片来源：*A comprehensive guide to OCR with Tesseract, OpenCV and Python* [36]

执行结果：几乎全对，只有特殊符号误判。

**范例8. 以Python调用Tesseract OCR API，辨识中、英文。**

**下列程序代码请参考【09_14_OCR.ipynb】。**

(1) 载入相关套件。

```
1  import cv2
2  import pytesseract
3  import matplotlib.pyplot as plt
```

(2) 载入图片文件并显示。

```
1  # 载入图片文件
2  image = cv2.imread('./images_ocr/receipt.png')
3
4  # 显示图片文件
5  image_RGB = cv2.cvtColor(image, cv2.COLOR_BGR2RGB)
6  plt.figure(figsize=(10,6))
7  plt.imshow(image_RGB)
8  plt.axis('off')
9  plt.show()
```

(3) OCR 辨识：调用 image_to_string 函数。

```
1  # 参数设定
2  custom_config = r'--psm 6'
3  # OCR 辨识
4  print(pytesseract.image_to_string(image, config=custom_config))
```

执行结果：与直接下指令的辨识结果大致相同。

(4) 只辨识数字：参数设定加 outputbase digits。

```
1  # 参数设定，只辨识数字
2  custom_config = r'--psm 6 outputbase digits'
3  # OCR 辨识
4  print(pytesseract.image_to_string(image, config=custom_config))
```

执行结果。

```
0001 122011
4338-
71 2
29.95 19.90
1 3.79
1 4.50
- 28.19
2.50
0 30.69
30.69
```

(5) 只辨识有限字符。

```
1  # 参数设定白名单，只辨识有限字符
2  custom_config = r'-c tessedit_char_whitelist=abcdefghijklmnopqrstuvwxyz --psm 6'
3  # OCR 辨识
4  print(pytesseract.image_to_string(image, config=custom_config))
```

执行结果。

```
elcometoels
heck
erverdeshf
able uests
eefurgrea
efries

udight
ud
ubtotal
alesfax
a
alanceue

hankyouforyourpatronage a
```

(6) 设定黑名单：只辨识有限字符。

```
1  # 参数设定黑名单，只辨识有限字符
2  custom_config = r'-c tessedit_char_blacklist=abcdefghijklmnopqrstuvwxyz --psm 6'
3  # OCR 辨识
4  print(pytesseract.image_to_string(image, config=custom_config))
```

执行结果。

```
W  M]'
C #: 0001 12/20/11
S: J F 4:38 PM
T: 7/1 G: 2
2 B B (€9.95/) 19,90
SIDE: F

1 B L 3.79
1 B 4.50
S-] 28.19
S T 2.50
TOTAL 30.69
BI D 30.69

T é! '
```

(7) 辨识多种文字：先载入并显示图片文件。

```
1  # 载入图片文件
2  image = cv2.imread('./images_ocr/chinese.png')
3
4  # 显示图片文件
5  image_RGB = cv2.cvtColor(image, cv2.COLOR_BGR2RGB)
6  plt.figure(figsize=(10,6))
7  plt.imshow(image_RGB)
8  plt.axis('off')
9  plt.show()
```

- 图片文件如下。

> Tesseract OCR 是目前很盛行的 OCR 软体，HP 公司於 2005 年開放原始程式碼
> (Open Source)，以 C++開發而成的，可由原始程式碼建置安裝，或直接安裝已建
> 置好的程式，我們採取後者，先自 https://github.com/UB-Mannheim/tesseract/wiki
> 下載最新版 exe 檔，直接執行即可。安裝完成後，將安裝路徑下 bin 子目錄放入
> 環境變數 Path 內。若要以 Python 呼叫 Tesseract OCR，需額外安裝 pytesseract 套
> 件，指令如下：
> pip install pytesseract

(8) 辨识多种文字：先自 https://github.com/tesseract-ocr/tessdata_best 下载字库，放入安装目录的 tessdata 子目录内 (C:\Program Files\Tesseract-OCR\tessdata)。

```
1  # 辨识多种文字，中文、日文及英文
2  custom_config = r'-l chi_tra+jpn+eng --psm 6'
3  # OCR 辨识
4  print(pytesseract.image_to_string(image, config=custom_config))
```

执行结果：中文辨识的效果不如预期，即使改用新细明体字体，效果亦不佳。

# 9-8 车牌辨识

车牌辨识 (Automatic Number Plate Recognition, ANPR) 系统已行之有年了，早期用像素逐点辨识，或将数字细线化后，再比对线条，但最近几年改用深度学习进行辨识，它已被应用到许多领域。

机车检验：检验单位的计算机会先进行车牌辨识。

停车场：当车辆进场时，系统会先辨识车牌并记录，要出场时会辨识车牌，自动扣款。

**范例9. 以OpenCV及Tesseract OCR进行车牌辨识。**此范例程序修改自*Car License Plate Recognition using Raspberry Pi and OpenCV*[38]。

以 OpenCV 及 Tesseract OCR 进行车牌辨识的相关步骤如图 9.26 所示。

第 3 篇 | 进阶的影像应用

图 9.26 相关步骤

下列程序代码请参考【09_15_ANPR.ipynb】。

(1) 载入相关套件。

```
1  import cv2
2  import imutils
3  import numpy as np
4  import matplotlib.pyplot as plt
5  import pytesseract
6  from PIL import Image
```

(2) 载入图片文件并显示。

```
1  # 载入图片文件
2  image = cv2.imread('./images_ocr/2.jpg',cv2.IMREAD_COLOR)
3
4  # 显示图片文件
5  image_RGB = cv2.cvtColor(image, cv2.COLOR_BGR2RGB)
6  plt.imshow(image_RGB)
7  plt.axis('off')
8  plt.show()
```

执行结果：此测试图来自原程序。

(3) 直接进行 OCR 辨识车牌号码。

```
1  # 车牌号码 OCR 辨识
2  char_whitelist='ABCDEFGHIJKLMNOPQRSTUVWXYZ1234567890'
3  text = pytesseract.image_to_string(Cropped, config=
4              f'-c tessedit_char_whitelist={char_whitelist} --psm 6 ')
5  print("车牌号码：",text)
```

执行结果：会得到车牌号码以外的许多英数字。

```
车牌号码： W
BA
M YVSS
LS 5 PN 8S
A 7 S
SS N
I PP
ME
AS
4
RR J S
RT 222571
```

| 246 |

(4) 先转为灰度，会比较容易辨识，再提取轮廓。

```
1  # 提取轮廓
2  gray = cv2.cvtColor(image, cv2.COLOR_BGR2GRAY)  # 转为灰度
3  gray = cv2.bilateralFilter(gray, 11, 17, 17)     # 模糊化，去除噪声
4  edged = cv2.Canny(gray, 30, 200)                 # 提取轮廓
5
6  # 显示图片文件
7  plt.imshow(edged, cmap='gray')
8  plt.axis('off')
9  plt.show()
```

执行结果。

(5) 取得等高线区域，并排序，取前 10 个区域。

```
1  # 取得等高线区域，并排序，取前10个区域
2  cnts = cv2.findContours(edged.copy(), cv2.RETR_TREE, cv2.CHAIN_APPROX_SIMPLE)
3  cnts = imutils.grab_contours(cnts)
4  cnts = sorted(cnts, key = cv2.contourArea, reverse = True)[:10]
```

(6) 找第一个含四个点的等高线区域：将等高线区域转为近似多边形，接着寻找四边形的等高线区域。

```
1   # 找第一个含四个点的等高线区域
2   screenCnt = None
3   for i, c in enumerate(cnts):
4       # 计算等高线区域周长
5       peri = cv2.arcLength(c, True)
6       # 转为近似多边形
7       approx = cv2.approxPolyDP(c, 0.018 * peri, True)
8       # 等高线区域维度
9       print(c.shape)
10
11      # 找第一个含四个点的多边形
12      if len(approx) == 4:
13          screenCnt = approx
14          print(i)
15          break
```

(7) 在原图上绘制多边形，框住车牌。

```
1   # 在原图上绘制多边形，框住车牌
2   if screenCnt is None:
3       detected = 0
4       print("No contour detected")
5   else:
6       detected = 1
7
8   if detected == 1:
9       cv2.drawContours(image, [screenCnt], -1, (0, 255, 0), 3)
10      print(f'车牌座标=\n{screenCnt}')
```

(8) 去除车牌以外的图像，找出车牌的上下左右的坐标，计算车牌宽高。

```
1  # 去除车牌以外的图像
2  mask = np.zeros(gray.shape,np.uint8)
3  new_image = cv2.drawContours(mask,[screenCnt],0,255,-1,)
4  new_image = cv2.bitwise_and(image, image, mask=mask)
5
6  # 转为浮点数
7  src_pts = np.array(screenCnt, dtype=np.float32)
8
9  # 找出车牌的上下左右的坐标
10 left = min([x[0][0] for x in src_pts])
11 right = max([x[0][0] for x in src_pts])
12 top = min([x[0][1] for x in src_pts])
13 bottom = max([x[0][1] for x in src_pts])
14
15 # 计算车牌宽高
16 width = right - left
17 height = bottom - top
18 print(f'宽度={width}, 高度={height}')
```

(9) 仿射 (Affine Transformation)，将车牌转为矩形：仿射可将偏斜的梯形转为矩形，笔者发现等高线区域的各点坐标都是以**逆时针排列**，因此，当要找出第一点坐标在哪个方向时，通常它会位于上方或左方，所以不需考虑右下角。

```
1  # 计算仿射(affine transformation)的目标区域坐标，须与撷取的等高线区域座标顺序相同
2  if src_pts[0][0][0] > src_pts[1][0][0] and src_pts[0][0][1] < src_pts[3][0][1]:
3      print('起始点为右上角')
4      dst_pts = np.array([[width, 0], [0, 0], [0, height], [width, height]], dtype=np.float32)
5  elif src_pts[0][0][0] < src_pts[1][0][0] and src_pts[0][0][1] > src_pts[3][0][1]:
6      print('起始点为左下角')
7      dst_pts = np.array([[0, height], [width, height], [width, 0], [0, 0]], dtype=np.float32)
8  else:
9      print('起始点为左上角')
10     dst_pts = np.array([[0, 0], [0, height], [width, height], [width, 0]], dtype=np.float32)
11
12 # 仿射
13 M = cv2.getPerspectiveTransform(src_pts, dst_pts)
14 Cropped = cv2.warpPerspective(gray, M, (int(width), int(height)))
```

(10) 车牌号码 OCR 辨识：限定车牌号码只有大写字母及数字。

```
1  # 车牌号码 OCR 辨识
2  char_whitelist='ABCDEFGHIJKLMNOPQRSTUVWXYZ1234567890'
3  text = pytesseract.image_to_string(Cropped, config=
4          f'-c tessedit_char_whitelist={char_whitelist} --psm 6 ')
5  print("车牌号码：",text)
```

执行结果：HR26BR9044，完全正确。

(11) 显示原图和车牌。

```
1  # 显示原图及车牌
2  cv2.imshow('Orignal image',image)
3  cv2.imshow('Cropped image',Cropped)
4
5  # 车牌存储
6  cv2.imwrite('Cropped.jpg', Cropped)
7
8  # 按 Enter 键结束
9  cv2.waitKey(0)
10
11 # 关闭所有窗口
12 cv2.destroyAllWindows()
```

执行结果。

车牌。

(12) 再使用 images_ocr/1.jpg 测试，车牌为 NAX·6683，辨识为 NAY·6683，X 误认为 Y，有可能是车牌的字体不同，可使用车牌字体供 Tesseract OCR 使用。

车牌。

另外，笔者试验发现，若镜头拉远或拉近，造成车牌过大或过小的话，都有可能辨识错误，所以，实际进行时，镜头最好与车牌距离固定，这样会比较容易辨识。假如图像的画面太杂乱，取到的车牌区域也有可能是错的，而这问题相对容易处理，当 OCR 辨识不到字或者字数不足时，就再找其他的等高线区域，即可解决。

从这个范例可以得知，通常一个实际的案例，并不会像内建的数据集一样，可以直接套用，常常都需要进行前置处理，如灰度化、提取轮廓、找等高线区域、仿射等，数据清理 (Data Clean) 完才可输入模型加以训练，而且为了适应环境变化，这些工作还必须反复进行。所以有人统计，光是收集数据、整理数据、特征工程等工作就占项目 85% 的时间，只有把最繁杂的工作处理好，才是项目成功的关键，这与参加 Kaggle 竞赛是截然不同的感受。

## 9-9　卷积神经网络的缺点

CNN 的应用领域那么多元，相当实用，但是它仍存在一些缺陷。

(1) 卷积不管特征在图像的哪个位置，只针对局部窗口进行特征辨识，因此，如图 9.27 所示两张图，辨识结果是相同的，这种现象称为位置无差异性 (Position Invariant)。

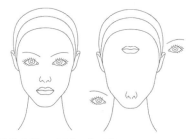

图 9.27　左图是正常的人脸，右图五官移位，两者对 CNN 来说是无差异的

图片来源：Disadvantages of CNN models [39]

(2) 图像中的对象如果经过旋转或倾斜，CNN 就无法辨识了，如图 9.28 所示。

图 9.28　右图为左图旋转近 180°的样子，如此 CNN 就辨识不了

图片来源：Disadvantages of CNN models [39]

(3) 图像坐标转换，人眼可以辨识不同的对象特征，但对于 CNN 来说却难以理解，如图 9.29 所示。

图 9.29　右图为左图上下颠倒，人眼可以看出年轻人与老年人，然而 CNN 就很难理解

图片来源：Disadvantages of CNN models [39]

因此，Geoffrey Hinton 等学者就提出了胶囊算法 (Capsules)，用来改良 CNN 的缺点，有兴趣的读者可以进一步研究 Capsules。

# 第 10 章
# 生成对抗网络

水能载舟，亦能覆舟，AI 虽然给人类带来了许多便利，但也造成不小的危害。近几年泛滥的深度伪造 (Deepfake) 就是一例，它利用 AI 技术伪造政治人物与明星的视频，效果能够做到真假难辨，一旦在网络上散播，就会造成莫大的灾难。深度伪造的基础算法就是生成对抗网络 (Generative Adversarial Network, GAN)，本章就来介绍这一课题。

Facebook 人工智能研究院 Yann LeCun 在接受 Quora 专访时说："GAN 与其变形是近十年最有趣的想法 (This, and the variations that are now being proposed is the most interesting idea in the last 10 years in ML, in my opinion)。"这句话造成 GAN 一炮而红，其作者 Ian Goodfellow 也成为各界竞相邀请演讲的对象。

另外，2018 年 10 月纽约佳士得艺术拍卖会，也卖出第一幅以 GAN 算法绘制的肖像画，如图 10.1，最后得标价为 432 500 美元。有趣的是，画作右下角还列出 GAN 的损失函数，相关报道可参见《全球首次！AI 创作肖像画 10 月佳士得拍卖》[1] 及 Is Artificial Intelligence Set To Become Art's Next Medium?[2]。

图 10.1　Edmond de Belamy 肖像画
图片来源：佳士得网站 [3]

此后有人统计每 28 分钟就有一篇与 GAN 相关的论文发表。

## 10-1　生成对抗网络介绍

关于生成对抗网络有一个很生动的比喻：它是由两个神经网络所组成，一个网络扮演伪钞制造者 (Counterfeiter)，不断制造假钞，另一个网络则扮演警察，不断从伪造者那边拿到假钞，并判断真假，然后，伪造者就根据警察判断结果的回馈不停改良，直到最后假钞变得真假难辨，这就是 GAN 的概念。

"伪钞制造者"称为生成模型 (Generative model)，"警察"则是判别模型

(Discriminative model)，简单的架构如图 10.2 所示。

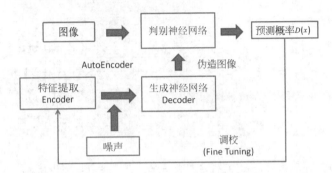

图 10.2　生成对抗网络 (Generative Adversarial Network, GAN) 的架构

处理流程如下：

(1) 先训练判别神经网络：从训练数据中抽取样本，输入判别神经网络，期望预测概率 $D(x) \fallingdotseq 1$，相反地，判断来自生成网络的伪造图片，期望预测概率 $D(G(z)) \fallingdotseq 0$。

(2) 训练生成网络：刚开始以常态分配或均匀分配产生噪声 ($z$)，喂入生成神经网络，生成伪造图片。

(3) 通过判别网络的反向传导 (Backpropagation)，更新生成网络的权重，亦即改良伪造图片的准确度 ( 技术 )，反复训练，直到产生精准的图片为止，如图 10.3 所示。

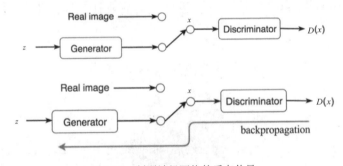

图 10.3　判别神经网络的反向传导

GAN 根据以上流程重新定义损失函数。

(1) 判别神经网络的损失函数：前半段为真实数据的判别，后半段为伪造数据的判别。

$$\max_D V(D) = \mathbb{E}_{x \sim p_{\text{data}}(x)}[\log D(x)] + \mathbb{E}_{z \sim p_z(z)}[\log(1 - D(G(z)))]$$

$\underbrace{\qquad\qquad\qquad}_{\text{recognize real images better}}$ $\underbrace{\qquad\qquad\qquad}_{\text{recognize generated images better}}$

其中：

x 为训练数据，故预测概率 $D(x)$ 越大越好；

E 为为期望值，因为训练数据并不完全相同，故预测概率有高有低；

z 为伪造数据，预测概率 $D(G(z))$ 越小越好，调整为 $1-D(G(z))$，变成越大越好。

两者相加当然是越大越好；

取 Log：它并不会影响最大化求解，通常概率相乘会造成多次方，不容易求解，故取 Log，变成一次方函数。

(2) 生成神经网络的损失函数：即判别神经网络损失函数的右边多项式，生成神经网络期望伪造数据被分类为真的概率越大越好，故差距越小越好。

$$\min_G V(G) = \mathbb{E}_{z \sim p_z(z)}[\log(1 - D(G(z)))]$$

(3) 两个网络损失函数合而为一的表示法如下：

$$\min_G \max_D V(D, G) = \mathbb{E}_{x \sim p_{data}}[\log D(x)] + \mathbb{E}_{z \sim p_z}[\log(1 - D(G(z)))]$$

因为函数左边的多项式与生成神经网络的损失函数无关，故加上亦无碍。

整个算法的伪码如图 10.4 所示，使用小批量梯度下降法，最小化损失函数。

```
Algorithm 1 Minibatch stochastic gradient descent training of generative adversarial nets. The number of steps to apply to the discriminator, k, is a hyperparameter. We used k = 1, the least expensive option, in our experiments.
for number of training iterations do
    for k steps do
        • Sample minibatch of m noise samples {z⁽¹⁾, ..., z⁽ᵐ⁾} from noise prior p_g(z).
        • Sample minibatch of m examples {x⁽¹⁾, ..., x⁽ᵐ⁾} from data generating distribution p_data(x).
        • Update the discriminator by ascending its stochastic gradient:
            ∇_{θ_d} (1/m) Σᵢ₌₁ᵐ [log D(x⁽ⁱ⁾) + log(1 − D(G(z⁽ⁱ⁾)))].
    end for
    • Sample minibatch of m noise samples {z⁽¹⁾, ..., z⁽ᵐ⁾} from noise prior p_g(z).
    • Update the generator by descending its stochastic gradient:
        ∇_{θ_g} (1/m) Σᵢ₌₁ᵐ log(1 − D(G(z⁽ⁱ⁾))).
end for
The gradient-based updates can use any standard gradient-based learning rule. We used momentum in our experiments.
```

图 10.4 GAN 算法的伪码

生成网络希望生成出来的图片越来越逼真，能通过判别网络的检验，而判别网络则希望将生成网络所制造的图片都判定为假数据，两者目标相反，互相对抗，故称为生成对抗网络。

## 10-2 生成对抗网络种类

GAN 不是只有一种模型，其变形非常多，可以参阅 *The GAN Zoo*[4]，有上百种模型，其功能各有不同。

CGAN：参阅 *Pose Guided Person Image Generation*[5]，可生成不同的姿势，如图 10.5 所示。

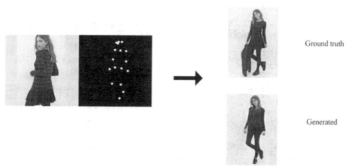

图 10.5 CGAN 算法的姿势生成

ACGAN：参阅 *Towards the Automatic Anime Characters Creation with Generative Adversarial Networks*[6]，可生成不同的动漫人物，如图 10.6 所示。

图 10.6　ACGAN 算法，从左边的动漫角色，生成为右边的新角色

作者有附一个展示的网站 MakeGirlsMoe[7]，可利用不同的模型与参数，生成各种动漫人物，如图 10.7 所示。

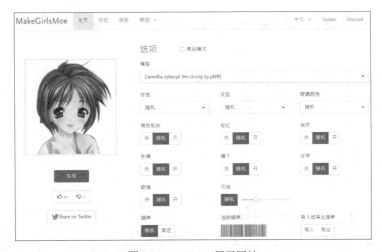

图 10.7　ACGAN 展示网站

CycleGAN：风格转换，作者也有附一个展示的网站 MIL WebDNN[8]，可选择不同的风格图，生成各种风格的照片或视频，如图 10.8 所示。

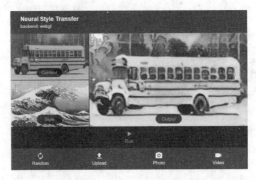

图 10.8　CycleGAN 的展示网站

StarGAN：参阅 *StarGAN: Unified Generative Adversarial Networks for Multi-Domain Image-to-Image Translation*[9]，生成不同的脸部表情，转换肤色、发色或是性别，如图 10.9 所示，程序代码在 StarGAN GitHub[10]。

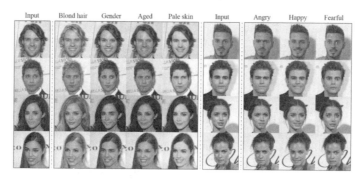

图 10.9　StarGAN 展示，将左边的脸转换肤色、发色、性别或表情

SRGAN：可以生成高分辨率的图像，如图 10.10 所示，参阅 *Photo-Realistic Single Image Super-Resolution Using a Generative Adversarial Network*[11]。

图 10.10　SRGAN 展示，由左而右从低分辨率的图像生成为高分辨率的图像

StyleGAN2：功能与语义分割 (Image Segmentation) 相反，它是从语义分割图渲染成实景图，如图 10.11 所示，参阅 *Analyzing and Improving the Image Quality of StyleGAN*[12]，程序在 Github stylegan2[13]。

图 10.11　StyleGAN2 展示，从右边的图像生成为左边的图像

限于篇幅，仅介绍一小部分的算法，读者如有兴趣可参阅 GAN — Some cool applications of GAN[14] 一文，里面有更多种算法的介绍。

只要修改原创者 GAN 的损失函数，即可产生不同的神奇效果，所以，根据 The GAN Zoo[4] 的统计，GAN 相关的论文数量呈现爆炸性成长，如图 10.12 所示。

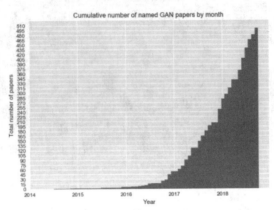

图 10.12　与 GAN 有关的论文数量呈现爆炸性成长

# 10-3　DCGAN

先来实操 DCGAN(Deep Convolutional Generative Adversarial Network) 算法。

**范例1. 以MNIST数据实操DCGAN算法，产生手写阿拉伯数字。此范例程序修改自DCGAN-MNIST-pytorch**[15]。

以 MNIST 数据实操 DCGAN 算法,产生手写阿拉伯数字的相关步骤如图 10.13 所示。

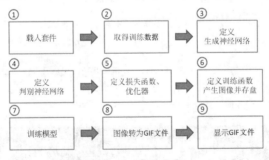

图 10.13　相关步骤

下列程序代码请参考【10_01_DCGAN_MNIST.ipynb】，训练数据集为 MNIST。

(1) 载入相关套件。

```
1  import os
2  import torch
3  from torch import nn
4  from torch.nn import functional as F
5  from torch.utils.data import DataLoader, random_split
6  from torchmetrics import Accuracy
7  from torchvision import transforms
8  from torchvision.datasets import MNIST
9  from torchvision import utils as vutils
```

(2) 设定参数。

```
1  PATH_DATASETS = "" # 预设路径
2  BATCH_SIZE = 64  # 批量
3  device = torch.device("cuda" if torch.cuda.is_available() else "cpu")
4  "cuda" if torch.cuda.is_available() else "cpu"
```

(3) 取得训练数据，转为 Dataset、Data loader。

```
1  # 转换
2  transform=transforms.Compose([
3      transforms.Resize(28),
4      transforms.ToTensor(),
5      transforms.Normalize((0.5,), (0.5,)),
6  ])
7
8  # 下载 MNIST 手写阿拉伯数字 训练数据
9  dataset = MNIST(PATH_DATASETS, train=True, download=True,
10                 transform=transform)
11 dataloader = torch.utils.data.DataLoader(dataset
12                 , batch_size=BATCH_SIZE, shuffle=True)
13
14 # 训练数据的维度
15 print(train_ds.data.shape)
```

(4) 定义神经网络参数：噪声维度为生成神经网络输入的规格。

```
1  nz = 100   # 生成神经网络噪声维度
2  ngf = 64   # 生成神经网络滤波器个数
3  ndf = 64   # 判别神经网络滤波器个数
```

(5) 定义神经网络权重初始值：依据 DCGAN 论文，卷积层权重初始值必须是 $N(0.0, 0.02)$ 的随机数，Batch Normalization 层则为 $N(1.0, 0.02)$。

```
1  def weights_init(m):
2      classname = m.__class__.__name__
3      if classname.find('Conv') != -1:
4          m.weight.data.normal_(0.0, 0.02) # 卷积层权重初始值
5      elif classname.find('BatchNorm') != -1:
6          m.weight.data.normal_(1.0, 0.02) # Batch Normalization 层权重初始值
7          m.bias.data.fill_(0)
```

(6) 定义生成神经网络：

use_bias=False：训练不产生偏差项，因为要生成的影像尽量是像素所构成的。

Conv2DTranspose：反卷积层，进行上采样 (Up Sampling)，由小图插补为大图，strides=(2, 2) 表示宽高各增大 2 倍。

最后产生宽高为 (28,28) 的单色向量。

```
1  class Generator(nn.Module):
2      def __init__(self, nc=1, nz=100, ngf=64):
3          super(Generator, self).__init__()
4          self.main = nn.Sequential(
5              # input is Z, going into a convolution
6              nn.ConvTranspose2d(nz, ngf * 8, 4, 1, 0, bias=False),
7              nn.BatchNorm2d(ngf * 8),
8              nn.ReLU(True),
9              # state size. (ngf*8) x 4 x 4
10             nn.ConvTranspose2d(ngf * 8, ngf * 4, 4, 2, 1, bias=False),
11             nn.BatchNorm2d(ngf * 4),
12             nn.ReLU(True),
13             # state size. (ngf*4) x 8 x 8
14             nn.ConvTranspose2d(ngf * 4, ngf * 2, 4, 2, 1, bias=False),
15             nn.BatchNorm2d(ngf * 2),
16             nn.ReLU(True),
17             # state size. (ngf*2) x 16 x 16
18             nn.ConvTranspose2d(ngf * 2, ngf, 4, 2, 1, bias=False),
19             nn.BatchNorm2d(ngf),
20             nn.ReLU(True),
```

```
21            nn.ConvTranspose2d(ngf, nc, kernel_size=1,
22                               stride=1, padding=2, bias=False),
23            nn.Tanh()
24        )
25
26    def forward(self, input):
27        output = self.main(input)
28        return output
29
30 netG = Generator().to(device)
31 netG.apply(weights_init)
```

(7) 定义判别神经网络：类似一般的 CNN 判别模型，但要去除池化层，避免信息损失。
LeakyReLU：避免数据为 0。

```
1  class Discriminator(nn.Module):
2      def __init__(self, nc=1, ndf=64):
3          super(Discriminator, self).__init__()
4          self.main = nn.Sequential(
5              # input is (nc) x 64 x 64
6              nn.Conv2d(nc, ndf, 4, 2, 1, bias=False),
7              nn.LeakyReLU(0.2, inplace=True),
8              # state size. (ndf) x 32 x 32
9              nn.Conv2d(ndf, ndf * 2, 4, 2, 1, bias=False),
10             nn.BatchNorm2d(ndf * 2),
11             nn.LeakyReLU(0.2, inplace=True),
12             # state size. (ndf*2) x 16 x 16
13             nn.Conv2d(ndf * 2, ndf * 4, 4, 2, 1, bias=False),
14             nn.BatchNorm2d(ndf * 4),
15             nn.LeakyReLU(0.2, inplace=True),
16             # state size. (ndf*4) x 8 x 8
17             nn.Conv2d(ndf * 4, 1, 4, 2, 1, bias=False),
18             nn.Sigmoid()
19         )
20
21     def forward(self, input):
22         output = self.main(input)
23         return output.view(-1, 1).squeeze(1)
24
25 netD = Discriminator().to(device)
26 netD.apply(weights_init)
```

(8) 定义损失函数为二分类交叉熵 (BinaryCrossentropy)，优化器为 Adam。

判别神经网络的损失函数为真实影像加上生成影像的损失函数和，因为判别神经网络会同时接收真实影像和生成影像。

```
1  # 设定损失函数
2  criterion = nn.BCELoss()
3
4  # 设定优化器(optimizer)
5  optimizerD = torch.optim.Adam(netD.parameters(), lr=0.0002, betas=(0.5, 0.999))
6  optimizerG = torch.optim.Adam(netG.parameters(), lr=0.0002, betas=(0.5, 0.999))
```

(9) 进行模型训练：训练了 25 个周期，训练时间需 2 个小时以上。

```
1  fixed_noise = torch.randn(64, nz, 1, 1, device=device)
2  real_label = 1.0
3  fake_label = 0.0
4  niter = 25
5  # 模型训练
6  for epoch in range(niter):
7      for i, data in enumerate(dataloader, 0):
8          ############################################################
9          # (1) 判别神经网路: maximize log(D(x)) + log(1 - D(G(z)))
10         ############################################################
11         # 训练真实数据
12         netD.zero_grad()
13         real_cpu = data[0].to(device)
14         batch_size = real_cpu.size(0)
15         label = torch.full((batch_size,), real_label, device=device)
16
17         output = netD(real_cpu)
```

```
18          errD_real = criterion(output, label)
19          errD_real.backward()
20          D_x = output.mean().item()
21
22          # 训练假数据
23          noise = torch.randn(batch_size, nz, 1, 1, device=device)
24          fake = netG(noise)
25          label.fill_(fake_label)
26          output = netD(fake.detach())
27          errD_fake = criterion(output, label)
28          errD_fake.backward()
29          D_G_z1 = output.mean().item()
30          errD = errD_real + errD_fake
31          optimizerD.step()
32
33          ############################################################
34          # (2) 判别神经网路: maximize log(D(G(z)))
35          ############################################################
36          netG.zero_grad()
37          label.fill_(real_label)
38          output = netD(fake)
39          errG = criterion(output, label)
40          errG.backward()
41          D_G_z2 = output.mean().item()
42          optimizerG.step()
43          if i % 200 == 0:
44              print('[%d/%d][%d/%d] Loss_D: %.4f Loss_G: %.4f D(x): %.4f D(G(z)): %.4f / %.4f'
45                    % (epoch+1, niter, i, len(dataloader),
46                       errD.item(), errG.item(), D_x, D_G_z1, D_G_z2))
47              vutils.save_image(real_cpu,'gan_output/real_samples.png' ,normalize=True)
48              fake = netG(fixed_noise)
49              vutils.save_image(fake.detach(),'gan_output/fake_samples_epoch_%03d.png'
50                                % (epoch), normalize=True)
51      torch.save(netG.state_dict(), 'gan_weights/netG_epoch_%d.pth' % (epoch))
52      torch.save(netD.state_dict(), 'gan_weights/netD_epoch_%d.pth' % (epoch))
```

执行结果：可以观察两个网络的损失变化。

执行过程中会将每一周期的权重存盘。

(10) 新数据预测。

```
1  import matplotlib.pyplot as plt
2
3  batch_size = 25
4  latent_size = 100
5
6  fixed_noise = torch.randn(batch_size, latent_size, 1, 1).to(device)
7  fake_images = netG(fixed_noise)
8  fake_images_np = fake_images.cpu().detach().numpy()
9  fake_images_np = fake_images_np.reshape(fake_images_np.shape[0], 28, 28)
10 R, C = 5, 5
11 for i in range(batch_size):
12     plt.subplot(R, C, i + 1)
13     plt.axis('off')
14     plt.imshow(fake_images_np[i], cmap='gray')
15 plt.show()
```

执行结果：输出的数字已可正常识别。

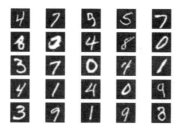

(11) 将训练过程中的存盘图像转为 GIF 文件：需先安装 imageio 套件，以利产生 GIF 动画。

!pip install -q imageio

(12) 产生 GIF 文件。

```python
1  import imageio
2  import glob
3  
4  # 产生 GIF 文件
5  anim_file = './gan_output/dcgan.gif'
6  with imageio.get_writer(anim_file, mode='I') as writer:
7      filenames = glob.glob('./gan_output/fake_samples*.png')
8      filenames = sorted(filenames)
9      for filename in filenames:
10         image = imageio.imread(filename)
11         writer.append_data(image)
```

(13) 显示 GIF 文件。

执行结果：注意，GIF 文件会不断循环播放，也可以打开文件管理选择 dcgan.gif 检视。

或是比较 fake_samples_epoch_000.png、fake_samples_epoch_024.png 的差异，可以看出训练的成效。

接着以名人脸部数据集，生成近似真实的图像，程序修改自 PyTorch 官网 DCGAN Tutorial[16]。

**范例2. 以名人脸部数据集实操DCGAN算法。此范例程序逻辑与【10_01_DCGAN_MNIST.ipynb】几乎相同，只是图像尺寸不同，本数据集为彩色，相关的模型定义要随之改变。**

下列程序代码请参考【10_02_DCGAN_Face.ipynb】，训练数据集为名人脸部。

(1) 载入相关套件。

```python
1  import os
2  import torch
3  from torch import nn
4  from torch.nn import functional as F
5  from torch.utils.data import DataLoader, random_split
6  from torchmetrics import Accuracy
7  from torchvision import transforms
8  from torchvision import datasets
9  from torchvision import utils as vutils
10 import matplotlib.pyplot as plt
11 import numpy as np
```

(2) 设定参数。

```python
1  PATH_DATASETS = ""  # 预设路径
2  BATCH_SIZE = 128    # 批量
3  image_size = 64
4  device = torch.device("cuda" if torch.cuda.is_available() else "cpu")
5  "cuda" if torch.cuda.is_available() else "cpu"
```

(3) 定义神经网络参数。

```
1  nz = 100    # 生成神经网络噪声维度
2  ngf = 64    # 生成神经网络滤波器个数
3  ndf = 64    # 判别神经网络滤波器个数
4  nc = 3      # 颜色通道
```

(4) 自 https://drive.google.com/uc?id=1O7m1010EJjLE5QxLZiM9Fpjs7Oj6e684 下载 img_align_celeba.zip，文件大小约 1.3GB，解压缩至 celeba_gan 目录，产生数据集(dataset)，图像缩放为 (64, 64)。

```
1  # 转换
2  transform=transforms.Compose([
3      transforms.Resize(image_size),
4      transforms.CenterCrop(image_size),
5      transforms.ToTensor(),
6      transforms.Normalize((0.5, 0.5, 0.5), (0.5, 0.5, 0.5)),
7  ])
8
9  # 训练数据
10 dataset = datasets.ImageFolder(root='celeba_gan',
11                                transform=transform)
12 dataloader = torch.utils.data.DataLoader(dataset
13                  , batch_size=BATCH_SIZE, shuffle=True)
14
15 # 显示图片文件
16 real_batch = next(iter(dataloader))
17 plt.figure(figsize=(8,8))
18 plt.axis("off")
19 plt.title("Training Images")
20 plt.imshow(np.transpose(vutils.make_grid(real_batch[0].to(device)[:64]
21                  , padding=2, normalize=True).cpu(),(1,2,0)));
```

执行结果。

(5) 定义神经网络权重初始值：与上例相同。

```
1  def weights_init(m):
2      classname = m.__class__.__name__
3      if classname.find('Conv') != -1:
4          m.weight.data.normal_(0.0, 0.02) # 卷积层权重初始值
5      elif classname.find('BatchNorm') != -1:
6          m.weight.data.normal_(1.0, 0.02) # Batch Normalization 层权重初始值
7          m.bias.data.fill_(0)
```

(6) 定义生成神经网络：结构稍有差异。

```python
class Generator(nn.Module):
    def __init__(self, nc=3, nz=100, ngf=ngf):
        super(Generator, self).__init__()
        self.main = nn.Sequential(
            # input is Z, going into a convolution
            nn.ConvTranspose2d( nz, ngf * 8, 4, 1, 0, bias=False),
            nn.BatchNorm2d(ngf * 8),
            nn.ReLU(True),
            # state size. (ngf*8) x 4 x 4
            nn.ConvTranspose2d(ngf * 8, ngf * 4, 4, 2, 1, bias=False),
            nn.BatchNorm2d(ngf * 4),
            nn.ReLU(True),
            # state size. (ngf*4) x 8 x 8
            nn.ConvTranspose2d( ngf * 4, ngf * 2, 4, 2, 1, bias=False),
            nn.BatchNorm2d(ngf * 2),
            nn.ReLU(True),
            # state size. (ngf*2) x 16 x 16
            nn.ConvTranspose2d( ngf * 2, ngf, 4, 2, 1, bias=False),
            nn.BatchNorm2d(ngf),
            nn.ReLU(True),
            # state size. (ngf) x 32 x 32
            nn.ConvTranspose2d( ngf, nc, 4, 2, 1, bias=False),
            nn.Tanh()
        )

    def forward(self, input):
        output = self.main(input)
        return output

netG = Generator().to(device)
netG.apply(weights_init)
```

(7) 定义判别神经网络：结构稍有差异。

```python
class Discriminator(nn.Module):
    def __init__(self, nc=3, ndf=ndf):
        super(Discriminator, self).__init__()
        self.main = nn.Sequential(
            # input is (nc) x 64 x 64
            nn.Conv2d(nc, ndf, 4, 2, 1, bias=False),
            nn.LeakyReLU(0.2, inplace=True),
            # state size. (ndf) x 32 x 32
            nn.Conv2d(ndf, ndf * 2, 4, 2, 1, bias=False),
            nn.BatchNorm2d(ndf * 2),
            nn.LeakyReLU(0.2, inplace=True),
            # state size. (ndf*2) x 16 x 16
            nn.Conv2d(ndf * 2, ndf * 4, 4, 2, 1, bias=False),
            nn.BatchNorm2d(ndf * 4),
            nn.LeakyReLU(0.2, inplace=True),
            # state size. (ndf*4) x 8 x 8
            nn.Conv2d(ndf * 4, ndf * 8, 4, 2, 1, bias=False),
            nn.BatchNorm2d(ndf * 8),
            nn.LeakyReLU(0.2, inplace=True),
            # state size. (ndf*8) x 4 x 4
            nn.Conv2d(ndf * 8, 1, 4, 1, 0, bias=False),
            nn.Sigmoid()
        )

    def forward(self, input):
        output = self.main(input)
        return output.view(-1, 1).squeeze(1)

netD = Discriminator().to(device)
netD.apply(weights_init)
```

(8) 设定损失函数、优化器 (optimizer)：与上例相同。

```python
# 设定损失函数
criterion = nn.BCELoss()

# 设定优化器(optimizer)
optimizerD = torch.optim.Adam(netD.parameters(), lr=0.0002, betas=(0.5, 0.999))
optimizerG = torch.optim.Adam(netG.parameters(), lr=0.0002, betas=(0.5, 0.999))
```

(9) 进行模型训练：因训练数据更多，只训练 10 周期，程序逻辑与上例相同，不再复制。

执行结果：笔者的计算机执行 1 周期就需花 2~3 个小时，如果单纯使用 CPU，那可以先去睡个觉，隔天再来看结果。若为实际的项目就会有完成时间的压力，这时候 GPU 性能就显得格外重要。

比较第 1 个及第 10 个周期结果，图像质量差异非常明显。

执行 10 个周期的结果：虽然有改善，但仍然不符合预期。

(10) 新数据预测：使用生成模型产生图像，因为是彩色图像，与上例的处理略有差异。

```
1  batch_size = 25
2  latent_size = 100
3
4  fixed_noise = torch.randn(batch_size, latent_size, 1, 1).to(device)
5  # 产生图像，clamp 使像素值介于 [-1, 1] 之间
6  fake_images = netG(fixed_noise).clamp(min=-1, max=1)
7  fake_images_np = fake_images.cpu()
8  fake_images_np = fake_images_np.reshape(-1, 3, image_size, image_size)
9  fake_images_np = torch.permute(fake_images_np, (0, 2, 3, 1)).detach().numpy()
10 fake_images_np = (fake_images_np + 1) *.5   # 使像素值介于 [0, 1] 之间
11 R, C = 5, 5
12 plt.figure(figsize=(8, 8))
13 for i in range(batch_size):
14     plt.subplot(R, C, i + 1)
15     plt.axis('off')
16     plt.imshow(fake_images_np[i])
17 plt.show();
```

执行结果：图像质量差异很大，少数图像质量不错。

## 10-4 Progressive GAN

Progressive GAN 也称为 Progressive Growing GAN 或 PGAN，它是 NVIDIA 2017 年发表的一篇文章 *Progressive Growing of GANs for Improved Quality, Stability, and Variation*[17] 中所提到的算法，它可以生成高画质且稳定的图像，小图像通过层层的神经层不断扩大，直到所要求的尺寸为止，大部分是针对人脸的生成，如图 10.14 所示。

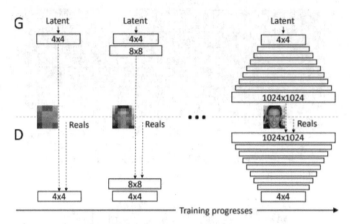

图 10.14　Progressive GAN 的示意图，要生成的图像尺寸越大，神经层就增加越多

图片来源：*Progressive Growing of GANs for Improved Quality, Stability, and Variation*[17]

它厉害的地方是算法生成的尺寸可以大于训练数据集的任何图像，这称为超分辨率 (Super Resolution)。网络架构如图 10.15 所示。

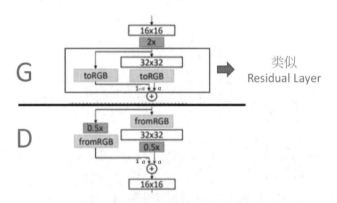

图 10.15　Progressive GAN 的网络架构

生成网络 (G) 使用类似 Residual 神经层，一边输入原图像，另一边输入反卷积层，使用权重 α，确定两个输入层的比例。判别网络 (D) 与生成网络 (G) 做反向操作，进行辨识。使用名人脸部数据集进行模型训练，根据论文估计，使用 8 颗 Tesla V100 GPU，需要训练 4 天左右，才可以得到不错的效果。读到这里，对我们普通用户来说简直是晴天霹雳，根本不用玩了。还好，PyTorch Hub 提供预先训练好的模型，我们马上来测试一下吧。

**范例3.** 再拿名人脸部数据集来实操 **Progressive GAN** 算法。此范例程序来自 **High-quality image generation of fashion, celebrity faces**[18]。

下列程序代码请参考【10_03_PGAN_Face.ipynb】，训练数据集为名人脸部。

(1) 载入相关套件。

```
1  import torch
2  import torchvision
3  import matplotlib.pyplot as plt
```

(2) 载入预先训练好的 PGAN 模型。

```
1  use_gpu = True if torch.cuda.is_available() else False
2
3  # trained on high-quality celebrity faces "celebA" dataset
4  # this model outputs 512 x 512 pixel images
5  model = torch.hub.load('facebookresearch/pytorch_GAN_zoo:hub',
6                         'PGAN', model_name='celebAHQ-512',
7                         pretrained=True, useGPU=use_gpu)
```

输出图像尺寸为 512 × 512，另外还支持 256 × 256 尺寸。

(3) 调用 model.test 产生图像。

```
1  # 产生图像个数
2  num_images = 4
3  # 产生噪声资料
4  noise, _ = model.buildNoiseData(num_images)
5  # 产生图像
6  with torch.no_grad():
7      generated_images = model.test(noise)
8
9  # clamp 使像素值介于 [0, 1] 之间
10 grid = torchvision.utils.make_grid(generated_images.clamp(min=-1, max=1),
11                                    scale_each=True, normalize=True)
12 # permute 设定色彩通道在最后一维
13 plt.imshow(grid.permute(1, 2, 0).cpu().numpy())
14 plt.axis('off');
```

执行结果：效果不是很好，模型训练周期仍不足。

# 10-5 Conditional GAN

DCGAN 生成的图片是随机的，以 MNIST 数据集而言，生成图像的确是数字，但无法控制要生成哪一个数字。Conditional GAN 增加了一个条件 (Condition)，即目标变量 $Y$，用来控制生成的数字。

Conditional GAN 也称为 cGAN，它是 Mehdi Mirz 等学者于 2014 年发表的一篇文章 *Conditional Generative Adversarial Nets*[19] 中所提出的算法，它修改 GAN 损失函数如下：

$$\min_G \max_D V(D,G) = \mathbb{E}_{x \sim p_{\text{data}}(x)}[\log D(x)] + \mathbb{E}_{z \sim p_z(z)}[\log(1 - D(G(z)))]$$

将单纯的 $D(x)$ 改为条件概率 $D(x|y)$。

**范例4.** 以 **Fashion MNIST** 数据集实操 **Conditional GAN** 算法。此范例程序修改自 *Kaggle PyTorch Conditional GAN*[20]。

以 Fashion MNIST 数据集实操 Conditional GAN 算法的相关步骤如图 10.16 所示。

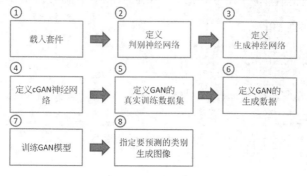

图 10.16　相关步骤

下列程序代码请参考【**10_04_CGAN_FashionMNIST.ipynb**】。

(1) 载入相关套件。

```
1  import torch
2  import torch.nn as nn
3  import pandas as pd
4  import numpy as np
5  from torchvision import transforms
6  from torch.utils.data import Dataset, DataLoader
7  from PIL import Image
8  from torch import autograd
9  from torch.autograd import Variable
10 from torchvision.utils import make_grid
11 import matplotlib.pyplot as plt
12 from torchvision.datasets import FashionMNIST
```

(2) 设定参数。

```
1  BATCH_SIZE = 64    # 批量
2  device = torch.device("cuda" if torch.cuda.is_available() else "cpu")
```

(3) 载入 FashionMNIST 数据。

```
1  # 转换
2  transform=transforms.Compose([
3      transforms.ToTensor(),
4      transforms.Normalize((0.5,), (0.5,)),
5  ])
6
7  dataset = FashionMNIST('', train=True, download=True,
8                  transform=transform)
9  data_loader = torch.utils.data.DataLoader(dataset
10                 , batch_size=BATCH_SIZE, shuffle=True)
```

(4) 定义生成神经网络：输入的标注(Label)需转为 One-hot encoding，成为 10 个变量。

```
1  class Generator(nn.Module):
2      def __init__(self):
3          super().__init__()
4
5          # 设定嵌入层，作为 Label 的输入
6          self.label_emb = nn.Embedding(10, 10)
7
8          self.model = nn.Sequential(
9              nn.Linear(110, 256),
10             nn.LeakyReLU(0.2, inplace=True),
11             nn.Linear(256, 512),
12             nn.LeakyReLU(0.2, inplace=True),
13             nn.Linear(512, 1024),
```

```
14              nn.LeakyReLU(0.2, inplace=True),
15              nn.Linear(1024, 784),
16              nn.Tanh()
17          )
18
19      def forward(self, z, labels):
20          z = z.view(z.size(0), 100)
21          c = self.label_emb(labels)
22          x = torch.cat([z, c], 1)   # 合并输入
23          out = self.model(x)
24          return out.view(x.size(0), 28, 28)
```

(5) 定义判别神经网络：输入的标注 (Label) 处理方式与生成神经网络相同。

```
1   class Discriminator(nn.Module):
2       def __init__(self):
3           super().__init__()
4
5           # 设定嵌入层，作为 Label 的输入
6           self.label_emb = nn.Embedding(10, 10)
7
8           self.model = nn.Sequential(
9               nn.Linear(794, 1024),
10              nn.LeakyReLU(0.2, inplace=True),
11              nn.Dropout(0.3),
12              nn.Linear(1024, 512),
13              nn.LeakyReLU(0.2, inplace=True),
14              nn.Dropout(0.3),
15              nn.Linear(512, 256),
16              nn.LeakyReLU(0.2, inplace=True),
17              nn.Dropout(0.3),
18              nn.Linear(256, 1),
19              nn.Sigmoid()
20          )
21
22      def forward(self, x, labels):
23          x = x.view(x.size(0), 784)
24          c = self.label_emb(labels)
25          x = torch.cat([x, c], 1)   # 合并输入
26          out = self.model(x)
27          return out.squeeze()
```

(6) 建立模型。

```
1   generator = Generator().to(device)
2   discriminator = Discriminator().to(device)
```

(7) 设定损失函数、优化器 (optimizer)。

```
1   criterion = nn.BCELoss()
2   d_optimizer = torch.optim.Adam(discriminator.parameters(), lr=1e-4)
3   g_optimizer = torch.optim.Adam(generator.parameters(), lr=1e-4)
```

(8) 定义生成网络训练函数。

```
1   def generator_train_step(batch_size, discriminator, generator,
2                            g_optimizer, criterion):
3       g_optimizer.zero_grad()
4       z = Variable(torch.randn(batch_size, 100)).to(device)
5       fake_labels = Variable(torch.LongTensor(np.random.randint(0, 10,
6                              batch_size))).to(device)   # 随机乱数 [1, 10]
7       fake_images = generator(z, fake_labels)
8       validity = discriminator(fake_images, fake_labels)
9       g_loss = criterion(validity, Variable(torch.ones(batch_size)).to(device))
10      g_loss.backward()
11      g_optimizer.step()
12      return g_loss.data.item()
```

(9) 定义判别网络训练函数：同时训练真实及伪造影像。

```python
def discriminator_train_step(batch_size, discriminator, generator
                            , d_optimizer, criterion, real_images, labels):
    d_optimizer.zero_grad()

    # 训练真实影像
    real_validity = discriminator(real_images, labels)
    real_loss = criterion(real_validity, Variable(torch.ones(
                batch_size)).to(device))

    # 训练伪造影像
    z = Variable(torch.randn(batch_size, 100)).to(device)
    fake_labels = Variable(torch.LongTensor(np.random.randint(
                0, 10, batch_size))).to(device)  # 随机乱数 [1, 10]
    fake_images = generator(z, fake_labels)
    fake_validity = discriminator(fake_images, fake_labels)
    fake_loss = criterion(fake_validity, Variable(torch.zeros(
                batch_size)).to(device))

    d_loss = real_loss + fake_loss
    d_loss.backward()
    d_optimizer.step()
    return d_loss.data.item()
```

(10) 训练模型。

```python
num_epochs = 30
n_critic = 5
display_step = 300
for epoch in range(num_epochs):
    print('Starting epoch {}...'.format(epoch))
    for i, (images, labels) in enumerate(data_loader):
        real_images = Variable(images).to(device)
        labels = Variable(labels).to(device)
        generator.train()
        batch_size = real_images.size(0)
        d_loss = discriminator_train_step(len(real_images), discriminator,
                                          generator, d_optimizer, criterion,
                                          real_images, labels)

        g_loss = generator_train_step(batch_size, discriminator,
                                      generator, g_optimizer, criterion)

    generator.eval()
    print('g_loss: {}, d_loss: {}'.format(g_loss, d_loss))
    z = Variable(torch.randn(9, 100)).to(device)
    labels = Variable(torch.LongTensor(np.arange(9))).to(device)
    sample_images = generator(z, labels).unsqueeze(1).data.cpu()
    grid = make_grid(sample_images, nrow=3, normalize=True)\
           .permute(1,2,0).numpy()
    plt.imshow(grid)
    plt.axis('off')
    plt.show()
```

执行结果 ( 模型训练需要很长时间，可以先去做其他事，再回来看结果 )：

左图是 epoch 0，中间是 epoch 10，右图是 epoch 29，也是最终结果。可看出生成过程中影像逐渐改善。

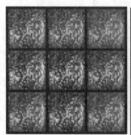
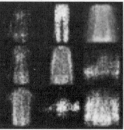

(11) 生成新数据、显示图像：每一栏均为同类对象，程序可指定要生成的对象类别。

```
1  # 标注名称
2  label_names = ['T-Shirt', 'Trouser', 'Pullover', 'Dress', 'Coat'
3                 , 'Sandal', 'Shirt', 'Sneaker', 'Bag', 'Ankle boot']
4  z = Variable(torch.randn(100, 100)).to(device)
5  labels = Variable(torch.LongTensor([i for _ in range(10) for i
6                       in range(10)])).to(device)
7
8  # 生成图像
9  sample_images = generator(z, labels).unsqueeze(1).data.cpu()
10 grid = make_grid(sample_images, nrow=10, normalize=True)\
11                  .permute(1,2,0).numpy()
12
13 # 显示图像
14 fig, ax = plt.subplots(figsize=(15,15))
15 ax.imshow(grid)
16 plt.yticks([])
17 plt.xticks(np.arange(15, 300, 30), label_names,
18            rotation=45, fontsize=20);
```

执行结果。

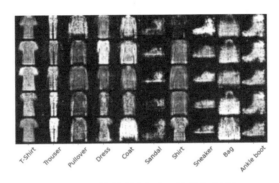

上面是运用 Conditional GAN 很简单的例子，只是把标记 (Label) 一并当作 X，输入模型中训练，作为条件 (Condition) 或限制 (Constraint)。另外还有很多延伸的作法，例如 ColorGAN，它把前置处理的轮廓图作为条件，与噪声一并当作 X，就可以生成与原图相似的图像，并且可以为灰度图上色，如图 10.17 所示。相关细节可参阅 *Colorization Using ConvNet and GAN*[21] 或 *End-to-End Conditional GAN-based Architectures for Image Colourisation*[22]。

图 10.17　ColorGAN

图片来源："Colorization Using ConvNet and GAN"[21]

# 10-6　Pix2Pix

Pix2Pix 为 Conditional GAN 算法的应用，出自 Phillip Isola 等学者在 2016 年发表的 *Image-to-Image Translation with Conditional Adversarial Networks*[23]，它能够将影像进行像素的转换，故称为 Pix2Pix，可应用于：

(1) 将语义分割的街景图转换为真实图像，如图 10.18 所示。
(2) 将语义分割的建筑外观转换为真实图像。
(3) 将卫星照转换为地图，如图 10.19 所示。
(4) 将白天图像转换为夜晚图像，如图 10.20 所示。
(5) 将轮廓图转换为实物图像，如图 10.21 所示。

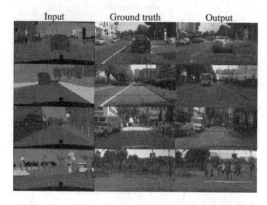

图 10.18　将语义分割的街景图转换为真实图像

以下图片均来自 *Image-to-Image Translation with Conditional Adversarial Networks*[23]

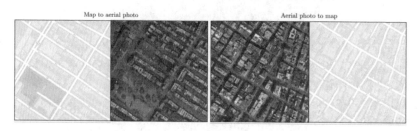

图 10.19　将卫星照片转换为地图，反之亦可

图 10.20　将白天图像转换为夜晚图像

图 10.21　将轮廓图转换为实物图像

生成网络采用 U-net 结构，引进了 Skip-connect 的技巧，即每一层反卷积层的输入都是前一层的输出加与该层对称的卷积层的输出，译码时可从对称的编码器得到对应的信息，使得生成的图像保有原图像的特征。

判别网络额外考虑输入图像的判别，将真实图像、生成图像与输入图像合而为一，作为判别网络的输入，进行辨识。原生的 GAN 在预测像素时，是以真实数据对应的单一像素进行辨识，然而 Pix2Pix 则引用 PatchGAN 的思维，利用卷积将图像切成多个较小的区域，每个像素与对应的区域进行辨识 (Softmax)，计算最大可能的输出。PatchGAN 可参见 *Image-to-Image Translation with Conditional Adversarial Networks*[23]一文。

**范例5. 以CMP Facade Database数据集实操Pix2Pix GAN算法。此范例程序修改自Kaggle Pix2Pix PyTorch[24]。**

CMP Facade Database[25] 共有 12 类的建筑物局部外形，如外观 (façade)、造型 (molding)、屋檐 (cornice)、柱子 (pillar)、窗户 (window)、门 (door) 等。数据集可以 http://efrosgans.eecs.berkeley.edu/pix2pix/datasets/facades.tar.gz 下载。

**下列程序代码请参考【10_05_Pix2Pix.ipynb】。**

(1) 载入相关套件。

```
1  import numpy as np
2  import matplotlib.pyplot as plt
3  import os, time, pickle, json
4  from glob import glob
5  from PIL import Image
6  import cv2
7  from typing import List, Tuple, Dict
8  from statistics import mean
9  from tqdm import tqdm
10
11 import torch
12 import torch.nn as nn
13 from torchvision import transforms
14 from torchvision.utils import save_image
15 from torch.utils.data import DataLoader
```

(2) 载入数据：含训练 (train) 及验证 (val) 数据。

```
1  MEAN = (0.5, 0.5, 0.5,)
2  STD = (0.5, 0.5, 0.5,)
3  RESIZE = 64
4
5  def read_path(filepath) -> List[str]:
6      root_path = "./datasets/facades"
7      path = os.path.join(root_path, filepath)
8      dataset = []
9      for p in glob(path+"/"+"*.jpg"):
10         dataset.append(p)
11     return dataset
12
13
14 class Transform():
15     def __init__(self, resize=RESIZE, mean=MEAN, std=STD):
16         self.data_transform = transforms.Compose([
17             transforms.Resize((resize, resize)),
18             transforms.ToTensor(),
19             transforms.Normalize(mean, std)
20         ])
21
22     def __call__(self, img: Image.Image):
23         return self.data_transform(img)
```

```
26  class Dataset(object):
27      def __init__(self, files: List[str]):
28          self.files = files
29          self.trasformer = Transform()
30
31      def _separate(self, img) -> Tuple[Image.Image, Image.Image]:
32          img = np.array(img, dtype=np.uint8)
33          h, w, _ = img.shape
34          w = int(w/2)
35          return Image.fromarray(img[:, w:, :]), Image.fromarray(img[:, :w, :])
36
37      def __getitem__(self, idx: int) -> Tuple[torch.Tensor, torch.Tensor]:
38          img = Image.open(self.files[idx])
39          input, output = self._separate(img)
40          input_tensor = self.trasformer(input)
41          output_tensor = self.trasformer(output)
42          return input_tensor, output_tensor
43
44      def __len__(self):
45          return len(self.files)
46
47  train = read_path("train")
48  val = read_path("val")
49  train_ds = Dataset(train)
50  val_ds = Dataset(val)
```

(3) 定义图像处理的函数。

```
1   # 使像素值介于 [0, 1] 之间
2   def clamp_image(img: torch.Tensor):
3       img = ((img.clamp(min=-1, max=1)+1)/2).permute(1, 2, 0)
4       return img
5
6   # 显示两个图像
7   def show_img_sample(img: torch.Tensor, img1: torch.Tensor):
8       fig, axes = plt.subplots(1, 2, figsize=(10, 6))
9       ax = axes.ravel()
10      ax[0].imshow(clamp_image(img))
11      ax[0].set_xticks([])
12      ax[0].set_yticks([])
13      ax[0].set_title("label image", c="g")
14      ax[1].imshow(clamp_image(img1))
15      ax[1].set_xticks([])
16      ax[1].set_yticks([])
17      ax[1].set_title("input image", c="g")
18      plt.subplots_adjust(wspace=0, hspace=0)
19      plt.show()
```

(4) 显示两个图像。

```
1   show_img_sample(train_ds.__getitem__(1)[0], train_ds.__getitem__(1)[1])
```

执行结果：左为语义分割图 (Label)，右为实景图 (Photo)。

(5) 建立 Data Loader。

```
1   BATCH_SIZE = 16
2   device = "cuda" if torch.cuda.is_available() else "cpu"
3   torch.manual_seed(0)
4   np.random.seed(0)
5
6   train_dl = DataLoader(train_ds, batch_size=BATCH_SIZE, shuffle=True, drop_last=True)
7   val_dl = DataLoader(val_ds, batch_size=BATCH_SIZE, shuffle=False, drop_last=False)
```

## 第 10 章 | 生成对抗网络

(6) 定义生成网络训练函数：U-Net 结构，多点连接。

```python
class Generator(nn.Module):
    def __init__(self):
        super(Generator, self).__init__()
        self.enc1 = self.conv2Relu(3, 32, 5)
        self.enc2 = self.conv2Relu(32, 64, pool_size=4)
        self.enc3 = self.conv2Relu(64, 128, pool_size=2)
        self.enc4 = self.conv2Relu(128, 256, pool_size=2)

        self.dec1 = self.deconv2Relu(256, 128, pool_size=2)
        self.dec2 = self.deconv2Relu(128+128, 64, pool_size=2)
        self.dec3 = self.deconv2Relu(64+64, 32, pool_size=4)
        self.dec4 = nn.Sequential(
            nn.Conv2d(32+32, 3, 5, padding=2),
            nn.Tanh()
        )

    def conv2Relu(self, in_c, out_c, kernel_size=3, pool_size=None):
        layer = []
        if pool_size:
            # Down width and height
            layer.append(nn.AvgPool2d(pool_size))
        # Up channel size
        layer.append(nn.Conv2d(in_c, out_c, kernel_size,
                               padding=(kernel_size-1)//2))
        layer.append(nn.LeakyReLU(0.2, inplace=True))
        layer.append(nn.BatchNorm2d(out_c))
        layer.append(nn.ReLU(inplace=True))
        return nn.Sequential(*layer)

    def deconv2Relu(self, in_c, out_c, kernel_size=3,
                     stride=1, pool_size=None):
        layer = []
        if pool_size:
            # Up width and height
            layer.append(nn.UpsamplingNearest2d(
                scale_factor=pool_size))
        # Down channel size
        layer.append(nn.Conv2d(in_c, out_c, kernel_size,
                               stride, padding=1))
        layer.append(nn.BatchNorm2d(out_c))
        layer.append(nn.ReLU(inplace=True))
        return nn.Sequential(*layer)

    def forward(self, x):
        x1 = self.enc1(x)
        x2 = self.enc2(x1)
        x3 = self.enc3(x2)
        x4 = self.enc4(x3)  # (b, 256, 4, 4)

        out = self.dec1(x4)
        # concat channel
        out = self.dec2(torch.cat((out, x3), dim=1))
        out = self.dec3(torch.cat((out, x2), dim=1))
        out = self.dec4(torch.cat((out, x1), dim=1))
        return out  # (b, 3, 64, 64)
```

(7) 定义判别神经网络。

```python
class Discriminator(nn.Module):
    def __init__(self):
        super(Discriminator, self).__init__()
        self.layer1 = self.conv2relu(6, 16, 5, cnt=1)
        self.layer2 = self.conv2relu(16, 32, pool_size=4)
        self.layer3 = self.conv2relu(32, 64, pool_size=2)
        self.layer4 = self.conv2relu(64, 128, pool_size=2)
        self.layer5 = self.conv2relu(128, 256, pool_size=2)
        self.layer6 = nn.Conv2d(256, 1, kernel_size=1)

    def conv2relu(self, in_c, out_c, kernel_size=3, pool_size=None, cnt=2):
        layer = []
        for i in range(cnt):
            if i == 0 and pool_size != None:
                # Down width and height
                layer.append(nn.AvgPool2d(pool_size))
            # Down channel size
```

```
18              layer.append(nn.Conv2d(in_c if i == 0 else out_c,
19                                     out_c,
20                                     kernel_size,
21                                     padding=(kernel_size-1)//2))
22              layer.append(nn.BatchNorm2d(out_c))
23              layer.append(nn.LeakyReLU(0.2, inplace=True))
24          return nn.Sequential(*layer)
25
26      def forward(self, x, x1):
27          x = torch.cat((x, x1), dim=1)
28          out = self.layer5(self.layer4(self.layer3(self.layer2(self.layer1(x)))))
29          return self.layer6(out) # (b, 1, 2, 2)
```

(8) 定义训练函数。

```
1   def train_fn(train_dl, G, D, criterion_bce, criterion_mae,
2                optimizer_g, optimizer_d):
3       G.train()
4       D.train()
5       LAMBDA = 100.0
6       total_loss_g, total_loss_d = [], []
7       for i, (input_img, real_img) in enumerate(tqdm(train_dl)):
8           input_img = input_img.to(device)
9           real_img = real_img.to(device)
10
11          real_label = torch.ones(input_img.size()[0], 1, 2, 2)
12          fake_label = torch.zeros(input_img.size()[0], 1, 2, 2)
13          # 生成网络训练
14          fake_img = G(input_img)
15          fake_img_ = fake_img.detach().cpu()
16          out_fake = D(fake_img, input_img).cpu()
17          loss_g_bce = criterion_bce(out_fake, real_label)
18          loss_g_mae = criterion_mae(fake_img, real_img)
19          loss_g = loss_g_bce + LAMBDA * loss_g_mae
20          total_loss_g.append(loss_g.item())
21
22          optimizer_g.zero_grad()
23          optimizer_d.zero_grad()
24          loss_g.backward(retain_graph=True)
25          optimizer_g.step()
26
27          # 判别网络训练
28          out_real = D(real_img.to(device), input_img.to(device))
29          loss_d_real = criterion_bce(out_real.to(device),
30                                      real_label.to(device))
31          out_fake = D(fake_img_.to(device), input_img)
32          loss_d_fake = criterion_bce(out_fake.to(device),
33                                      fake_label.to(device))
34          loss_d = loss_d_real + loss_d_fake
35          total_loss_d.append(loss_d.item())
36
37          optimizer_g.zero_grad()
38          optimizer_d.zero_grad()
39          loss_d.backward()
40          optimizer_d.step()
41      return mean(total_loss_g), mean(total_loss_d), fake_img.detach().cpu()
42
43  def saving_img(fake_img, e):
44      os.makedirs("generated", exist_ok=True)
45      save_image(fake_img, f"generated/fake{str(e)}.png", range=(-1.0, 1.0)
46                 , normalize=True)
47
48  def saving_logs(result):
49      with open("train.pkl", "wb") as f:
50          pickle.dump([result], f)
51
52  def saving_model(D, G, e):
53      os.makedirs("weight", exist_ok=True)
54      torch.save(G.state_dict(), f"weight/G{str(e+1)}.pth")
55      torch.save(D.state_dict(), f"weight/D{str(e+1)}.pth")
56
57  def show_losses(g, d):
58      fig, axes = plt.subplots(1, 2, figsize=(14,6))
59      ax = axes.ravel()
60      ax[0].plot(np.arange(len(g)).tolist(), g)
61      ax[0].set_title("Generator Loss")
62      ax[1].plot(np.arange(len(d)).tolist(), d)
63      ax[1].set_title("Discriminator Loss")
64      plt.show()
```

(9) 进行训练。

```python
def train_loop(train_dl, G, D, num_epoch, lr=0.0002, betas=(0.5, 0.999)):
    G.to(device)
    D.to(device)
    optimizer_g = torch.optim.Adam(G.parameters(), lr=lr, betas=betas)
    optimizer_d = torch.optim.Adam(D.parameters(), lr=lr, betas=betas)
    criterion_mae = nn.L1Loss()
    criterion_bce = nn.BCEWithLogitsLoss()
    total_loss_d, total_loss_g = [], []
    result = {}

    for e in range(num_epoch):
        loss_g, loss_d, fake_img = train_fn(train_dl, G, D, criterion_bce
                                , criterion_mae, optimizer_g, optimizer_d)
        total_loss_d.append(loss_d)
        total_loss_g.append(loss_g)
        saving_img(fake_img, e+1)

        if e%10 == 0:
            saving_model(D, G, e)
    try:
        result["G"] = total_loss_d
        result["D"] = total_loss_g
        saving_logs(result)
        show_losses(total_loss_g, total_loss_d)
        saving_model(D, G, e)
        print("successfully save model")
    finally:
        return G, D

G = Generator()
D = Discriminator()
EPOCH = 10
trained_G, trained_D = train_loop(train_dl, G, D, EPOCH)
```

(10) 定义生成数据的相关函数。

```python
def load_model(name):
    G = Generator()
    G.load_state_dict(torch.load(f"weight/G{name}.pth",
                        map_location={"cuda": "cpu"}))
    G.eval()
    return G.to(device)

def train_show_img(name, G):
    root = "generated"
    fig, axes = plt.subplots(int(name), 1, figsize=(12, 18))
    ax = axes.ravel()
    for i in range(int(name)):
        filename = os.path.join(root, f"fake{str(i+1)}.png")
        ax[i].imshow(Image.open(filename))
        ax[i].set_xticks([])
        ax[i].set_yticks([])

def de_norm(img):
    img_ = img.mul(torch.FloatTensor(STD).view(3, 1, 1))
    img_ = img_.add(torch.FloatTensor(MEAN).view(3, 1, 1)).detach()
    # img_ = ((img_.clamp(min=-1, max=1)+1)/2).permute(1, 2, 0)
    img_ = img_.clamp(min=-1, max=1).permute(1, 2, 0)
    return img_.numpy()

def evaluate(val_dl, name, G):
    with torch.no_grad():
        fig, axes = plt.subplots(6, 8, figsize=(12, 12))
        ax = axes.ravel()
        for input_img, real_img in tqdm(val_dl):
            input_img = input_img.to(device)
            real_img = real_img.to(device)

            fake_img = G(input_img)
            batch_size = input_img.size()[0]
            batch_size_2 = batch_size * 2

            for i in range(batch_size):
                ax[i].imshow(de_norm(input_img[i].cpu()))
                ax[i+batch_size].imshow(de_norm(real_img[i].cpu()))
                ax[i+batch_size_2].imshow(de_norm(fake_img[i].cpu()))
```

```
41          ax[i].set_xticks([])
42          ax[i].set_yticks([])
43          ax[i+batch_size].set_xticks([])
44          ax[i+batch_size].set_yticks([])
45          ax[i+batch_size_2].set_xticks([])
46          ax[i+batch_size_2].set_yticks([])
47          if i == 0:
48              ax[i].set_ylabel("Input Image", c="g")
49              ax[i+batch_size].set_ylabel("Real Image", c="g")
50              ax[i+batch_size_2].set_ylabel("Generated Image", c="r")
51          plt.subplots_adjust(wspace=0, hspace=0)
52          break
```

(11) 测试：生成 5 批新数据。

```
1  train_show_img(5, trained_G)
```

执行结果。

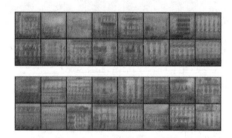

(12) 评估。

```
1  evaluate(val_dl, 5, trained_G)
```

执行结果：第 1~2 排为输入图像；第 3~4 排为真实图像；第 5~6 排为生成图像。

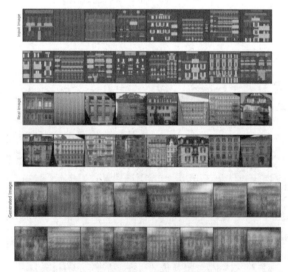

另外可以参阅 Jun-Yan Zhu 开发的项目 pytorch-CycleGAN-and-pix2pix[65]，除了训练的程序，也可以使用预先训练的模型测试，如下指令可以生成 Pix2Pix 图像：

python test.py --dataroot ./datasets/facades --name facades_pix2pix --model pix2pix --direction BtoA

执行结果。右图为生成的 Pix2Pix 图像，亦即输入简单线条的 1_real_A.png 及风格图 1_real_B.png，会生成具有该风格及线条的 1_fake_B.png。

1_real_B.png　　1_real_A.png　　1_fake_B.png

## 10-7　CycleGAN

前面 GAN 算法处理的都是成对转换数据，CycleGAN 则是针对非成对的数据生成图像。成对的意思是一张原始图像对应一张目标图像，如图 10.22 右方表示多对多的数据，也就是给予不同的场域 (Domain)，原始图像就可以合成指定场景的图像。

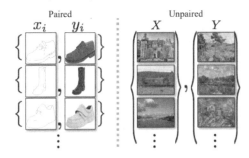

图 10.22　成对的数据 ( 左方 ) 对比非成对的数据 ( 右方 )

以下图片来源均来自 *Unpaired Image-to-Image Translation using Cycle-Consistent Adversarial Networks*[27]

CycleGAN 或称 Cycle-Consistent GAN，也是 Jun-Yan Zhu 等学者于 2017 年发表的一篇文章 *Unpaired Image-to-Image Translation using Cycle-Consistent Adversarial Networks*[27] 中提出的算法，概念如图 10.23 所示。

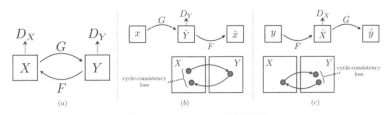

图 10.23　CycleGAN 网络结构

图 10.23(a)：有两个生成网络，$G$ 网络将图像由 $X$ 场域 (Domain) 生成 $Y$ 场域，$F$ 网络则是相反功能，由 $Y$ 场域 (Domain) 生成 $X$ 场域的图像。

图 10.23(b)：引进 cycle consistency losses 概念，可以做到 $x \to G(x) \to F(G(x)) \approx x$，即 $x$ 经过 $G$、$F$ 转换，可得到近似 $x$ 的图像，称为 Forward cycle-consistency loss。

图 10.23(c)：从另一场域 $y$ 开始，也可以做到 $y \to F(y) \to G(F(y)) \approx y$，称为 Backward cycle-consistency loss。

整个模型类似两个 GAN 网络的组合，具备循环机制，因此，损失函数为：

$$L(G, F, DX, DY) = LGAN(G, DY, X, Y) + LGAN(F, DX, Y, X) + \lambda Lcyc(G, F)$$

其中：

$$\mathcal{L}_{\text{cyc}}(G, F) = \mathbb{E}_{x \sim p_{\text{data}}(x)}[\|F(G(x)) - x\|_1] \\ + \mathbb{E}_{y \sim p_{\text{data}}(y)}[\|G(F(y)) - y\|_1]$$

λ 控制 G、F 损失函数的相对重要性。

这种机制可应用到影像增强 (Photo Enhancement)、影像彩色化 (Image Colorization)、风格转换 (Style Transfer) 等功能，如图 10.24 所示，能为一般的马匹加上斑马纹。

图 10.24　CycleGAN 的功能展示

**范例6. 以horse2zebra数据集实操CycleGAN算法。**

由于笔者计算机 GPU 内存只有 4GB，执行本程序会报内存不足错误，故移至 Google Colaboratoy 上执行。数据集请从以下网址下载：https://people.eecs.berkeley.edu/~taesung_park/CycleGAN/datasets/horse2zebra.zip。

**下列程序代码请参考【10_06_ CycleGAN.ipynb】。**

(1) 载入相关套件。

```
1  import os
2  import numpy as np
3  import glob
4  import time
5  import PIL.Image as Image
6  from tqdm.notebook import tqdm
7  from itertools import chain
8  from collections import OrderedDict
9  import random
10
11 import torch
12 import torch.nn as nn
13 import torchvision.utils as vutils
14 import torchvision.datasets as dset
15 import torch.nn.functional as F
16 from torch.utils.data import DataLoader
17 import torchvision.transforms as transforms
18 import matplotlib.pylab as plt
19 import ipywidgets
20 from IPython import display
```

(2) 上传数据集：将数据下载至本机后，再上传至 Colab 虚拟机，也可以直接使用 gdown 指令下载至虚拟机。

```
1  from google.colab import files
2  files.upload()
```

(3) 解压缩，并重新命名目录，以利 ImageFolder 建立 dataset。

```
1  !unzip ./horse2zebra.zip
```

```
1  import shutil, sys
2  shutil.move("./horse2zebra/trainA", "./horses_train/A")
3  shutil.move("./horse2zebra/trainB", "./zebra_train/B")
4  shutil.move("./horse2zebra/testA", "./horses_test/A")
5  shutil.move("./horse2zebra/testB", "./zebra_test/B")
```

(4) 为 4 个子目录建立 dataset、dataloader，以下仅列一段，其他子目录均类似。

```
1   bs = 5
2   workers = 2
3   image_size = (256,256)
4   dataroot = './horses_train/'
5   dataset_horses_train = dset.ImageFolder(root=dataroot,
6                                transform=transforms.Compose([
7                                    transforms.Resize(image_size),
8                                    transforms.CenterCrop(image_size),
9                                    transforms.ToTensor(),
10                                   transforms.Normalize((0, 0, 0), (1, 1, 1)),
11                               ]))
12  dataloader_train_horses = torch.utils.data.DataLoader(dataset_horses_train,
13                               batch_size=bs, shuffle=True, num_workers=workers)
14  real_batch = next(iter(dataloader_train_horses))
15  print(real_batch[0].shape)
16  plt.figure(figsize=(8,8))
17  plt.axis("off")
18  plt.title("Training Images")
19  plt.imshow(np.transpose(vutils.make_grid(real_batch[0].to(device)[:10],
20                               padding=2, normalize=True).cpu(),(1,2,0)))
```

(5) 定义绘制 4 个影像的函数：真实的马、生成的斑马、真实的斑马、生成的马，程序分两个模型，以 G_A2B 将马变成斑马，以 G_B2A 将斑马变成马。

```
1   def plot_images_test(dataloader_test_horses, dataloader_zebra_test):
2       batch_a_test = next(iter(dataloader_test_horses))[0].to(device)
3       real_a_test = batch_a_test.cpu().detach()
4       # 将马变成斑马
5       fake_b_test = G_A2B(batch_a_test ).cpu().detach()
6
7       plt.figure(figsize=(10,10))
8       plt.imshow(np.transpose(vutils.make_grid((real_a_test[:4]+1)/2,
9                               padding=2, normalize=True).cpu(),(1,2,0)))
10      plt.axis("off")
11      plt.title("Real horses")
12      plt.show()
13
14      plt.figure(figsize=(10,10))
15      plt.imshow(np.transpose(vutils.make_grid((fake_b_test[:4]+1)/2,
16                              padding=2, normalize=True).cpu(),(1,2,0)))
17      plt.axis("off")
18      plt.title("Fake zebras")
19      plt.show()
20
21      batch_b_test = next(iter(dataloader_zebra_test))[0].to(device)
22      real_b_test = batch_b_test.cpu().detach()
23      # 将斑马变成马
24      fake_a_test = G_B2A(batch_b_test ).cpu().detach()
25
26      plt.figure(figsize=(10,10))
27      plt.imshow(np.transpose(vutils.make_grid((real_b_test[:4]+1)/2,
28                              padding=2, normalize=True).cpu(),(1,2,0)))
29      plt.axis("off")
30      plt.title("Real zebras")
31      plt.show()
32
33      plt.figure(figsize=(10,10))
34      plt.imshow(np.transpose(vutils.make_grid((fake_a_test[:4]+1)/2,
35                              padding=2, normalize=True).cpu(),(1,2,0)))
36      plt.axis("off")
37      plt.title("Fake horses")
38      plt.show()
```

(6) 定义绘制 8 个影像的函数：额外绘制 Identity horses、Identity zebras、Recovered horses、Recovered zebras，逻辑与上一函数类似，不在此列出程序代码，请直接参阅范例。Identity 表示图像经过 A2B、B2A 两次转换会保持不变。Recovered 则为 B2A、A2B 两次转换的结果。

(7) 定义模型存盘与载入的函数。

```python
def save_models(G_A2B, G_B2A, D_A, D_B, name):
    torch.save(G_A2B, "/content/gdrive/My Drive/model_proj3/"+name+"_G_A2B.pt")
    torch.save(G_B2A, "/content/gdrive/My Drive/model_proj3/"+name+"_G_B2A.pt")
    torch.save(D_A, "/content/gdrive/My Drive/model_proj3/"+name+"_D_A.pt")
    torch.save(D_B, "/content/gdrive/My Drive/model_proj3/"+name+"_D_B.pt")

def load_models(name):
    G_A2B=torch.load("/content/gdrive/My Drive/model_proj3/"+name+"_G_A2B.pt")
    G_B2A=torch.load("/content/gdrive/My Drive/model_proj3/"+name+"_G_B2A.pt")
    D_A=torch.load("/content/gdrive/My Drive/model_proj3/"+name+"_D_A.pt")
    D_B=torch.load("/content/gdrive/My Drive/model_proj3/"+name+"_D_B.pt")
    return G_A2B, G_B2A, D_A, D_B

#save_models(G_A2B, G_B2A, D_A, D_B, "test")
#G_A2B, G_B2A, D_A, D_B= load_models("test")
```

(8) 定义生成网络训练函数。

```python
norm_layer = nn.InstanceNorm2d
class ResBlock(nn.Module):
    def __init__(self, f):
        super(ResBlock, self).__init__()
        self.conv = nn.Sequential(nn.Conv2d(f, f, 3, 1, 1), norm_layer(f), nn.ReLU(),
                                  nn.Conv2d(f, f, 3, 1, 1))
        self.norm = norm_layer(f)
    def forward(self, x):
        return F.relu(self.norm(self.conv(x)+x))

class Generator(nn.Module):
    def __init__(self, f=64, blocks=9):
        super(Generator, self).__init__()
        layers = [nn.ReflectionPad2d(3),
                  nn.Conv2d(  3,   f, 7, 1, 0), norm_layer(  f), nn.ReLU(True),
                  nn.Conv2d(  f, 2*f, 3, 2, 1), norm_layer(2*f), nn.ReLU(True),
                  nn.Conv2d(2*f, 4*f, 3, 2, 1), norm_layer(4*f), nn.ReLU(True)]
        for i in range(int(blocks)):
            layers.append(ResBlock(4*f))
        layers.extend([
                  nn.ConvTranspose2d(4*f, 4*2*f, 3, 1, 1), nn.PixelShuffle(2),
                                  norm_layer(2*f), nn.ReLU(True),
                  nn.ConvTranspose2d(2*f,   4*f, 3, 1, 1), nn.PixelShuffle(2),
                                  norm_layer(  f), nn.ReLU(True),
                  nn.ReflectionPad2d(3), nn.Conv2d(f, 3, 7, 1, 0),
                  nn.Tanh()])
        self.conv = nn.Sequential(*layers)

    def forward(self, x):
        return self.conv(x)
```

(9) 定义判别网络训练函数。

```python
nc=3
ndf=64
class Discriminator(nn.Module):
    def __init__(self):
        super(Discriminator, self).__init__()
        self.main = nn.Sequential(
            # input is (nc) x 128 x 128
            nn.Conv2d(nc,ndf,4,2,1, bias=False),
            nn.LeakyReLU(0.2, inplace=True),
            # state size. (ndf) x 64 x 64
            nn.Conv2d(ndf,ndf*2,4,2,1, bias=False),
            nn.InstanceNorm2d(ndf * 2),
            nn.LeakyReLU(0.2, inplace=True),
            # state size. (ndf*2) x 32 x 32
            nn.Conv2d(ndf*2, ndf * 4, 4, 2, 1, bias=False),
            nn.InstanceNorm2d(ndf * 4),
            nn.LeakyReLU(0.2, inplace=True),
            # state size. (ndf*4) x 16 x 16
            nn.Conv2d(ndf*4,ndf*8,4,1,1),
            nn.InstanceNorm2d(ndf*8),
            nn.LeakyReLU(0.2, inplace=True),
            # state size. (ndf*8) x 15 x 15
            nn.Conv2d(ndf*8,1,4,1,1)
            # state size. 1 x 14 x 14
        )

    def forward(self, input):
        return self.main(input)
```

## 第 10 章 | 生成对抗网络

(10) 定义判别网络及生成网络的损失函数。

```
1  def LSGAN_D(real, fake):
2      return (torch.mean((real - 1)**2) + torch.mean(fake**2))
3
4  def LSGAN_G(fake):
5      return torch.mean((fake - 1)**2)
```

(11) 建立判别及生成网络。

```
1   import itertools
2
3   G_A2B = Generator().to(device)
4   G_B2A = Generator().to(device)
5   D_A = Discriminator().to(device)
6   D_B = Discriminator().to(device)
7
8   # Initialize Loss function
9   criterion_Im = torch.nn.L1Loss()
10
11  # Learning rate for optimizers
12  lr = 0.0002
13
14  # Optimizers
15  # optimizer_G = torch.optim.Adam(itertools.chain(G_A2B.parameters(), G_B2A.parameters()),
16  #                                lr=lr, betas=(0.5, 0.999))
17  optimizer_G_A2B = torch.optim.Adam(G_A2B.parameters(), lr=lr, betas=(0.5, 0.999))
18  optimizer_G_B2A = torch.optim.Adam(G_B2A.parameters(), lr=lr, betas=(0.5, 0.999))
19  optimizer_D_A = torch.optim.Adam(D_A.parameters(), lr=lr, betas=(0.5, 0.999))
20  optimizer_D_B = torch.optim.Adam(D_B.parameters(), lr=lr, betas=(0.5, 0.999))
```

(12) 训练模型：程序代码过长，请直接参阅范例。

(13) 存盘：存储至 CycleGAN 目录。

```
1  if not os.path.exists('./CycleGAN'):
2      os.makedirs('./CycleGAN')
3  # save last check pointing
4  torch.save(G_A2B.state_dict(), f"./CycleGAN/netG_A2B.pth")
5  torch.save(G_B2A.state_dict(), f"./CycleGAN/netG_B2A.pth")
6  torch.save(D_A.state_dict(), f"./CycleGAN/netD_A.pth")
7  torch.save(D_B.state_dict(), f"./CycleGAN/netD_B.pth")
```

(14) 压缩相关模型，并下载模型文件。

```
1  # 压缩相关模型
2  !zip ./model.zip ./CycleGAN/*.*
```

```
1  from google.colab import files
2  files.download('./model.zip')
```

笔者执行程序需要 3 个小时，约完成 7 个执行周期，可以先睡个觉，起床后再看结果。

执行结果：以下仅只截取 2 幅图，左侧为原图，右侧为预测的图像，效果比训练样本差，应该是因为训练的执行周期不足，原文作者执行了 200 个周期，如果也执行 200 个周期，时间太长。

倘若只想测试 CycleGAN 效果，也可以参考 Lornatang_CycleGAN-PyTorch[67]，依照说明，下载程序、数据及预先训练模型后，执行下列指令：

python test_image.py --file assets/horse.png --model-name weights/horse2zebra/netG_A2B.pth --cuda

执行结果会产生 result.png，比较如下，左图为原图 (assets/horse.png)，右图为生成的图像 (result.png)。

再试另一个文件，执行下列指令：

python test_image.py --file assets/apple.png --model-name weights/horse2zebra/netG_A2B.pth --cuda

苹果也加上斑马纹了，这是因为我们使用 horse2zebra 模型，要下载其他模型可参考 weights\download_weights.sh。

# 10-8　GAN 挑战

这一章我们认识了多种不同的 GAN 算法，由于大部分是由同一组学者发表的，因此可以看到演化的路径。原生 GAN 加上条件后，变成 Conditional GAN，再将生成的网络改为对称的 U-Net 后，就变成 Pix2Pix GAN，接着再设定两个 Pix2Pix 循环的网络，就衍生出 Cycle GAN。除此之外，许多算法也会修改损失函数的定义，产生各种意想不到的效果，如表 10.1 所示。本书介绍的算法只是沧海一粟，更多的内容可参考李宏毅老师的 PPT Introduction of Generative Adversarial Network (GAN)[29]。

表 10.1　GAN 各种算法的损失函数表格来源：tensorflow-generative-model-collections[30]

| Name | Paper Link | Value Function |
|---|---|---|
| GAN | Arxiv | $L_D^{GAN} = E[\log(D(x))] + E[\log(1 - D(G(z)))]$<br>$L_G^{GAN} = E[\log(D(G(z)))]$ |
| LSGAN | Arxiv | $L_D^{LSGAN} = E[(D(x) - 1)^2] + E[D(G(z))^2]$<br>$L_G^{LSGAN} = E[(D(G(z)) - 1)^2]$ |

续表

| Name | Paper Link | Value Function |
|------|-----------|----------------|
| WGAN | Arxiv | $L_D^{WGAN} = E[D(x)] - E[D(G(z))]$<br>$L_G^{WGAN} = E[D(G(z))]$<br>$W_D \leftarrow clip\_by\_value(W_D, -0.01, 0.01)$ |
| WGAN_GP | Arxiv | $L_D^{WGAN\_GP} = L_D^{WGAN} + \lambda E[(|\nabla D(\alpha x - (1-\alpha G(z)))| - 1)^2]$<br>$L_G^{WGAN\_GP} = L_G^{WGAN}$ |
| DRAGAN | Arxiv | $L_D^{DRAGAN} = L_D^{GAN} + \lambda E[(|\nabla D(\alpha x - (1-\alpha x_p))| - 1)^2]$<br>$L_G^{DRAGAN} = L_G^{GAN}$ |
| CGAN | Arxiv | $L_D^{CGAN} = E[\log(D(x,c))] + E[\log(1 - D(G(z),c))]$<br>$L_G^{CGAN} = E[\log(D(G(z),c))]$ |
| infoGAN | Arxiv | $L_{D,Q}^{infoGAN} = L_D^{GAN} - \lambda L_I(c,c')$<br>$L_G^{infoGAN} = L_G^{GAN} - \lambda L_I(c,c')$ |
| ACGAN | Arxiv | $L_{D,Q}^{ACGAN} = L_D^{GAN} + E[P(class=c|x)] + E[P(class=c|G(z))]$<br>$L_G^{ACGAN} = L_G^{GAN} + E[P(class=c|G(z))]$ |
| EBGAN | Arxiv | $L_D^{EBGAN} = D_{AE}(x) + \max(0, m - D_{AE}(G(z)))$<br>$L_G^{EBGAN} = D_{AE}(G(z)) + \lambda \cdot PT$ |
| BEGAN | Arxiv | $L_D^{BEGAN} = D_{AE}(x) - k_t D_{AE}(G(z))$<br>$L_G^{BEGAN} = D_{AE}(G(z))$<br>$k_{t+1} = k_t + \lambda(\gamma D_{AE}(x) - D_{AE}(G(z)))$ |

另外，GAN 不光能应用在图像上，还可以结合自然语言处理 (NLP)、强化学习 (RL) 等技术，扩大应用范围，像是高解析图像生成、虚拟人物的生成、数据压缩、文字转语音 (Text To Speech, TTS)、医疗、天文、物理、游戏等，可以参阅 *Tutorial on Deep Generative Models*[31] 一文。

纵使 GAN 应用广泛，但仍然存在一些挑战：

(1) 生成的图像模糊。因为神经网络是根据训练数据求取回归，类似求取每个样本在不同范围的平均值，所以生成的图像会是相似点的平均，导致图像模糊。必须有大量的训练数据，加上相当多的训练周期，才能产生画质较佳的图像。另外，GAN 对超参数特别敏感，包括学习率、滤波器 (Filter) 尺寸，初始值设定得不好，造成生成的数据过差时，判别网络就会判定为伪，到最后生成网络只能一直产生少数类别的数据。

(2) 梯度消失 (Vanishing Gradient)。当生成的数据过差时，判别网络判定为真的概率接近 0，梯度会变得非常小，因此就无法提供良好的梯度来改善生成器，造成生成器梯度消失。发生这种情形时可以使用 leaky ReLU activation function、简化判别网络结构或增加训练周期加以改善。

(3) 模式崩溃 (Mode Collapse)。是指生成器生成的内容过于雷同，缺少变化。如果训练数据的类别不止一种，生成网络则会为了让判别网络辨识的准确率提高，而专注在比较擅长的类别，导致生成的类别缺乏多样性。以制造伪钞来举例，假设钞票分别有 100 元、500 元与 1000 元面额，若伪钞制造者比较善于制作 500 元面额的钞票，模型可能就会全部都制作 500 元面额的伪钞。

(4) 执行训练时间过久。这是最大的问题，反复试验的时候，假如没有高配置的硬件支持，每次调整参数都要好几天，再多的耐心也会消磨殆尽。

## 10-9　深度伪造

深度伪造 (Deepfake) 是目前较新的技术，也是一个 AI 危害人类社会的明显例子。深度伪造大部分是在视频中换脸，由于人在说话时头部会自然摆动，有各种角度的特写，因此，必须要收集特定人 360°的脸部图像，才能让算法成功置换。从网络媒体中收集名人的各种影像是最容易的方式，所以，网络上流传最多的是伪造名人的影片，如政治人物、明星等。

深度伪造的技术基础来自 GAN，类似于前面介绍的 Cycle GAN，架构如图 10.25 所示，也能结合脸部辨识的功能，在抓到脸部特征点 (Landmark) 后，进行原始脸部与要置换脸部的互换。

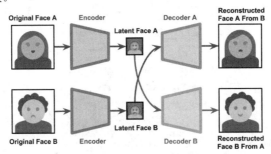

图 10.25　深度伪造的架构示意图

图片来源：*Understanding the Technology Behind DeepFakes*[35]

Aayush Bansal 等学者在 2018 年发表了 *Recycle-GAN: Unsupervised Video Retargeting*[36]，Recycle GAN 是扩充 Cycle GAN 的算法，它的损失函数额外加上了时间同步的相关性 (Temporal Coherence)，如下：

$$L_\tau(P_X) = \sum_t \|x_{t+1} - P_X(x_{1:t})\|^2,$$

其中 temporal predictor $P_X$，$x_{1:t} = (x_1 \ldots x_t)$

类似时间序列 (Time Series)，$t+1$ 时间点的图像应该是 1 至 $t$ 时间点的图像的延续，如图 10.26 所示。因此，生成网络的损失函数如下：

$$L_r(G_X, G_Y, P_Y) = \sum_t \|x_{t+1} - G_X(P_Y(G_Y(x_{1:t})))\|^2$$

其中：$G_y(x_i)$ 是将 $x_i$ 转成 $y_i$ 的生成网络。

John Oliver to Stephen Colbert

图 10.26　视频是连续的变化，因此 *t*+1 时间点的图像应该是 1 至 *t* 时间点的图像的延续

图片来源：*Recycle-GAN: Unsupervised Video Retargeting*[36]

Recycle GAN 算法的演进如图 10.27 所示。

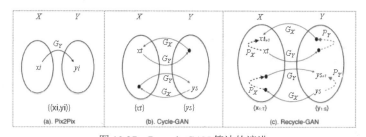

图 10.27　Recycle GAN 算法的演进

图片来源：*Recycle-GAN: Unsupervised Video Retargeting*[36]

图 10.27(a) Pix2Pix 是成对 (Paired data) 转换。

图 10.27(b) CycleGAN 是循环转换，使用成对的网络架构。

图 10.27(c) RecycleGAN 加上时间同步的相关性 (*Px*、*Py*)。

因此整体的 RecycleGAN 的损失函数，包括以下部分：

$$\min_{G,P}\max_{D} L_{rg}(G,P,D) = L_g(G_X,D_X) + L_g(G_Y,D_Y) + \\ \lambda_{rx}L_r(G_X,G_Y,P_Y) + \lambda_{ry}L_r(G_Y,G_X,P_X) + \lambda_{\tau x}L_\tau(P_X) + \lambda_{\tau y}L_\tau(P_Y)$$

另外，还有 Face2Face、嘴型同步技术 (Lip-syncing technology) 等算法，有兴趣的读者可以参阅 Jonathan Hui 的 *Detect AI-generated Images & Deepfakes*[37] 系列文章，里面有大量的图片展示，十分有趣。

Deepfake 的实操可参阅 *DeepFakes in 5 minutes*[38] 一文，它介绍如何利用 DeepFaceLab 套件，在很短的时间内制作出深度伪造的影片，源代码在 DeepFaceLab GitHub[39]，网页附有一个视频 Mini tutorial 说明，只要按步骤执行脚本 (Scripts)，就可以顺利完成影片，不过，它比 GAN 需要更高配置的硬件，笔者就不再测试了。

由于 Deepfake 造成了严重的假新闻灾难，许多学者及企业纷纷提出反制的方法来辨识真假，简单的像是 *Detect AI-generated Images & Deepfakes (Part 1)*[40] 一文所述，可以从脸部边缘是否模糊、是否有随机的噪声以及脸部是否对称等细节来辨别，当然也有大公司推出可辨识影片真假的工具，例如微软的 *Microsoft Video Authenticator*，可参阅 ITHome 相关的报道[41]。

不管是 Deepfake 还是反制的算法，未来发展都值得关注，这起事件也让科学家留意到科学的发展必须兼顾伦理与道德。

# 第 4 篇 自然语言处理

自然语言处理 (Natural Language Processing, NLP) 顾名思义，就是希望计算机能像人类一样，看懂文字或听懂人话，理解语义，并能给予适当的回答，常见的应用是聊天机器人。

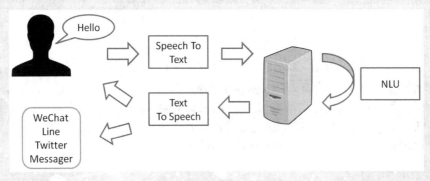

聊天机器人概念示意图

一个简单的计算机与人类的对话，所涵盖的技术包括：

(1) 当人对计算机说话，计算机会先把人的话转成文字，称为语音识别 (Speech recognition) 或语音转文字 (Speech To Text, STT)。

(2) 接着计算机对文字进行解析，了解意图，称为自然语言理解 (Natural Language Understanding, NLU)。

(3) 之后计算机依据对话回答，有两种表达方式：

①以文字回复：从语料库或常用问答 (FAQ) 中找出一段要回复的文字，称为文本生成 (Text Generation)。

②以声音回复：将回复文字转为语音，称为语音合成 (Speech Synthesize) 或文字转语音 (Text To Speech, TTS)。

整个过程看似容易，实则充满了各种挑战，接下来我们把相关技术仔细演练一遍吧！

# 第 11 章
# 自然语言处理的介绍

自然语言处理的发展非常早，大约从 1950 年就开始了，当年英国计算机科学家图灵 (Alan Mathison Turing) 已有先见之明，提出图灵测试 (Turing Test)，目的是测试计算机能否表现出像人类一样的智慧，时至今日，许多聊天机器人如 Siri、小冰等的问世，才算得上真正启动了 NLP 的热潮，即便目前依然无法媲美人类的智慧，但相关技术仍然有许多方面的应用：

(1) 文本分类 (Text classification)。
(2) 信息检索 (Information retrieval)。
(3) 文字校对 (Text proofing)。
(4) 自然语言生成 (Natural language generation)。
(5) 问答系统 (Question answering)。
(6) 机器翻译 (Machine translation)。
(7) 自动摘要 (Automatic summarization)。
(8) 情绪分析 (Sentiment analysis)。
(9) 语音识别 (Speech recognition)。
(10) 音乐方面的应用，比如曲风分类、自动编曲、声音模仿等。

## 11-1 词袋与 TF-IDF

人类的语言具高度暧昧性，一句话可能有多重的意思或隐喻，而计算机当前还无法真正理解语言或文字的意义，因此，现阶段的做法与影像的处理方式类似，先将语音和文字转换成向量，再对向量进行分析或使用深度学习建模，相关研究的进展非常快，这一节我们从最简单的方法开始说起。

词袋 (Bag of Words, BOW) 是把一篇文章进行词汇整理，然后统计每个词汇出现的次数，接着经由前几名的词汇猜测出全文大意，如图 11.1 所示。

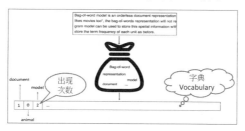

图 11.1　词袋

运行过程如图 11.2 所示。

图 11.2 词袋运行过程

(1) 分词 (Tokenization)：将整篇文章中的每个词汇切开，整理成生字表或字典 (Vocabulary)。英文较单纯，以空白或句点隔开，中文较复杂，须以特殊方式处理。

(2) 前置处理 (Preprocessing)：将词汇做词形还原、转换成小写等。词形还原是动词转为原形、复数转为单数等，避免因为词态不同，词汇统计出现分歧。

(3) 去除停用词 (Stop Word)：be 动词、助动词、代名词、介系词、冠词等不具特殊意义的词汇称为停用词 (Stop Word)，将它们剔除，否则统计结果都是这些词汇出现最多次。

(4) 词汇出现次数统计：计算每个词汇在文章中出现的次数，并由高至低排列。

**范例1.** 以 BOW 实操自动摘要。

**下列程序代码请参考【11_01_BOW.ipynb】。**

(1) 加载相关套件。

```
1  import collections
```

(2) 设定停用词：这里直接设定停用词，许多套件有整理常用的停用词，例如 NLTK、spaCy。

```
2  stop_words=['\n', 'or', 'are', 'they', 'i', 'some', 'by', '-',
3              'even', 'the', 'to', 'a', 'and', 'of', 'in', 'on', 'for',
4              'that', 'with', 'is', 'as', 'could', 'its', 'this', 'other',
5              'an', 'have', 'more', 'at', 'don't', 'can', 'only', 'most']
```

(3) 读取文本文件 news.txt，统计词汇出现的次数，相关数据来自 *South Korea's Convenience Store Culture*[1] 一文。

```
1  # 读取文字文件news.txt，统计字词出现次数
2  maxlen=1000        # 生字表最大个数
3
4  # 生字表的集合
5  word_freqs = collections.Counter()
6  with open('./NLP_data/news.txt','r+', encoding='UTF-8') as f:
7      for line in f:
8          # 转小写、分词
9          words = line.lower().split(' ')
10         # 统计字词出现次数
11         if len(words) > maxlen:
12             maxlen = len(words)
13         for word in words:
14             if not (word in stop_words):
15                 word_freqs[word] += 1
16
17 print(word_freqs.most_common(20))
```

执行结果。

[[('stores', 15), ('convenience', 14), ('korean', 6), ('these', 6), ('one', 6), ('it's', 6), ('from', 5), ('my', 5), ('you', 5), ('their', 5), ('just', 5), ('has', 5), ('new', 4), ('do', 4), ('also', 4), ('which', 4), ('find', 4), ('would', 4), ('like', 4), ('up', 4)]]

前 3 名分别为：

stores：15 次。

convenience：14 次。

korean：6 次。

因此可以猜测整篇文章应该是在讨论"韩国便利商店"(Korea Convenience Store)，结果与标题契合。

BOW 方法十分简单，效果也相当不错，不过它有个缺点，有些词汇不是停用词，也经常出现，但与文章主旨无相关性，譬如上文的 only、most，对猜测全文大意没有帮助，所以，学者提出改良的算法——TF-IDF(Term Frequency - Inverse Document Frequency)，它会针对跨文件常出现的词汇给予较低的分数，例如 only 在每一个文件都出现的话，TF-IDF 对它的评分就相对较低，因此，TF-IDF 的公式定义如下：

$$tf-idf = tf \times idf$$

其中：

tf ( 词频，Term Frequency)：考虑词汇出现在跨文件的次数，分母为在所有文件中出现的次数，分子为在目前文件中出现的次数。

$$tf_{i,j} = \frac{n_{i,j}}{\sum_k n_{k,j}}$$

idf ( 逆向文件频率，Inverse Document Frequency)：考虑词汇出现的文件数，即使单一文件出现特定词汇多次，也只视为 1 次，分子为总文件数，分母为词汇出现的文件数，加 1 是避免分母为 0。

$$idf_{i,j} = \log \frac{|D|}{1+|D_{t_i}|}$$

除了以上的定义，TF-IDF 还有一些变形的公式，可参阅维基百科关于 TF-IDF 的说明[2]。

除了猜测全文大意之外，TF-IDF 也可以应用到文本分类 (Text Classification) 或问题与答案的配对。

**范例2. 以TF-IDF实操问答配对。**

**下列程序代码请参考【11_02_TFIDF.ipynb】。**

(1) 加载相关套件。

```
1  from sklearn.feature_extraction.text import CountVectorizer
2  from sklearn.feature_extraction.text import TfidfTransformer
3  import numpy as np
```

(2) 设定输入数据：最后一句为问题，其他的例句为回答。

```
1  # 例句：最后一句为问题，其他为回答
2  corpus = [
3      'This is the first document.',
4      'This is the second second document.',
5      'And the third one.',
6      'Is this the first document?',
7  ]
```

(3) 将例句转换为词频矩阵，计算各个词汇出现的次数。

```
1  # 将例句转换为词频矩阵，计算各个字词出现的次数。
2  vectorizer = CountVectorizer()
3  X = vectorizer.fit_transform(corpus)
4
5  # 生字表
6  word = vectorizer.get_feature_names()
7  print ("Vocabulary:", word)
```

执行结果。

```
Vocabulary : ['and', 'document', 'first', 'is', 'one', 'second', 'the', 'third', 'this']
```

(4) 查看 4 句话的 BOW。

```
1  print ("BOW=\n", X.toarray())
```

执行结果。

```
BOW=
[[0 1 1 1 0 0 1 0 1]
 [0 1 0 1 0 2 1 0 1]
 [1 0 0 0 1 0 1 1 0]
 [0 1 1 1 0 0 1 0 1]]
```

(5) TF-IDF 转换：将例句转换为 TF-IDF 向量。

```
1  transformer = TfidfTransformer()
2  tfidf = transformer.fit_transform(X)
3  print ("TF-IDF=\n", np.around(tfidf.toarray(), 4))
```

执行结果：每一个元素均介于 [0, 1]，为了显示整齐，取四舍五入，实际运算并不需要。

```
TF-IDF=
[[0.     0.4388 0.542  0.4388 0.     0.     0.3587 0.     0.4388]
 [0.     0.2723 0.     0.2723 0.     0.8532 0.2226 0.     0.2723]
 [0.5528 0.     0.     0.     0.5528 0.     0.2885 0.5528 0.    ]
 [0.     0.4388 0.542  0.4388 0.     0.     0.3587 0.     0.4388]]
```

(6) 比较最后一句与其他例句的相似度：以 cosine_similarity 比较向量的夹角，越接近 1，表示越相似。

```
1  from sklearn.metrics.pairwise import cosine_similarity
2  print (cosine_similarity(tfidf[-1], tfidf[:-1], dense_output=False))
```

执行结果：第一个例句与最后的问句最相似，结果与文意相符合。

```
  (0, 2)    0.1034849000930086
  (0, 1)    0.43830038447620107
  (0, 0)    1.0
```

## 11-2　词汇前置处理

传统上，我们会使用 NLTK(Natural Language Toolkit) 套件来进行词汇的前置处理，它具备非常多的功能，并内含超过 50 个语料库 (Corpora) 可供测试，只可惜它没有支持中文处理，比较新的 spaCy 套件支持多国语言。这里先示范如何运用 NLTK 做一般词汇的前置处理，之后再介绍可以处理中文数据的套件。

NLTK 分为程序和数据两个部分。

(1) 安装 NLTK 程序：pip install nltk。

(2) 安装 NLTK 数据：先执行 Python，再执行 import nltk; nltk.download()，出现如图 11.3 所示页面，包括套件与相关语料库，可下载必要的项目。

# 第 11 章 | 自然语言处理的介绍

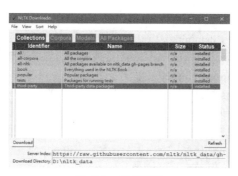

图 11.3　相关页面

由于文件众多，下载时间久，如需安装至第二台计算机，可直接复制下载目录即可，NLTK 加载语料库时，会自动检查所有硬盘的 \nltk_data 目录。

**范例1. 使用NLTK进行词汇的前置处理。**

**下列程序代码请参考【11_03_ 词汇前置处理 .ipynb】。**

(1) 加载相关套件。

```
1  import nltk
```

(2) 输入测试的文章段落如下。

```
1  text="Today is a great day. It is even better than yesterday." + \
2      " And yesterday was the best day ever."
```

(3) 将测试的文章段落分割成字句。

```
1  nltk.sent_tokenize(text)
```

执行结果：分割成三句。

```
['Today is a great day.',
 'It is even better than yesterday.',
 'And yesterday was the best day ever.']
```

(4) 分词 (Tokenize)。

```
1  nltk.word_tokenize(text)
```

执行结果。

```
['Today',
 'is',
 'a',
 'great',
 'day',
 '.',
 'It',
 'is',
 'even',
 'better',
 'than',
 'yesterday',
 '.',
 'And',
 'yesterday',
 'was',
 'the',
 'best',
 'day',
 'ever',
 '.']
```

(5) 词形还原有两种方法：

依字根做词形还原 (Stemming)：速度快，但不一定正确。

依字典规则做词形还原 (Lemmatization)：速度慢，但准确率高。

(6) 字根词形还原 (Stemming)：根据一般文法规则，不管字的含义，直接进行字根词形还原，比如将 keeps 删去 s，crashing 删去 ing，这都正确，但 his 直接删去 s，就会发生错误。

```
1  # 字根词形还原(Stemming)
2  text = 'My system keeps crashing his crashed yesterday, ours crashes daily'
3  ps = nltk.porter.PorterStemmer()
4  ' '.join([ps.stem(word) for word in text.split()])
```

执行结果：his → hi，daily → daili。

```
'My system keep crash hi crash yesterday, our crash daili'
```

(7) 依字典规则的词形还原 (Lemmatization)：查询字典，依单词的不同进行词形还原，如此 his、daily 均不会改变。

```
1  # 依字典规则的词形还原(Lemmatization)
2  text = 'My system keeps crashing his crashed yesterday, ours crashes daily'
3  lem = nltk.WordNetLemmatizer()
4  ' '.join([lem.lemmatize(word) for word in text.split()])
```

执行结果：完全正确。

```
'My system keep crashing his crashed yesterday, ours crash daily'
```

(8) 分词后剔除停用词 (Stopwords)：nltk.corpus.stopwords.words('english') 提供常用的停用词，另外标点符号也可以列入停用词。

```
1   # 标点符号(Punctuation)
2   import string
3   print('标点符号:', string.punctuation)
4
5   # 测试文章段落
6   text="Today is a great day. It is even better than yesterday." + \
7        " And yesterday was the best day ever."
8   # 读取停用词
9   stopword_list = set(nltk.corpus.stopwords.words('english')
10                      + list(string.punctuation))
11
12  # 移除停用词(Removing Stopwords)
13  def remove_stopwords(text, is_lower_case=False):
14      if is_lower_case:
15          text = text.lower()
16      tokens = nltk.word_tokenize(text)
17      tokens = [token.strip() for token in tokens]
18      filtered_tokens = [token for token in tokens if token not in stopword_list]
19      filtered_text = ' '.join(filtered_tokens)
20      return filtered_text, filtered_tokens
21
22  filtered_text, filtered_tokens = remove_stopwords(text)
23  filtered_text
```

执行结果。

标点符号：!"#$%&'()*+,-./:;<=>?@[\]^_`{|}~

(9) 进行 BOW 统计。

```
1  # 测试文章段落
2  with open('./NLP_data/news.txt','r+', encoding='UTF-8') as f:
3      text = f.read()
4
5  filtered_text, filtered_tokens = remove_stopwords(text, True)
6
7  import collections
8  # 生字表的集合
9  word_freqs = collections.Counter()
10 for word in filtered_tokens:
11     word_freqs[word] += 1
12 print(word_freqs.most_common(20))
```

执行结果：同样可以抓到文章大意是"韩国便利商店"。

```
[(''', 35), ('stores', 15), ('convenience', 14), ('one', 8), ('—', 8), ('even', 8), ('seoul', 8), ('city', 7), ('korea', 6),
('korean', 6), ('cities', 6), ('people', 5), ('summer', 4), ('new', 4), ('also', 4), ('find', 4), ('store', 4), ('would', 4),
('like', 4), ('average', 4)]
```

(10) 改用正规表达式 (Regular Expression)：上段程序还是有标点符号未剔除，正规表达式可完全剔除停用词。

```
1  # 剔除停用词(Removing Stopwords)
2  lem = nltk.WordNetLemmatizer()
3  def remove_stopwords_regex(text, is_lower_case=False):
4      if is_lower_case:
5          text = text.lower()
6      tokenizer = nltk.tokenize.RegexpTokenizer(r'\w+') # 筛选文数字(Alphanumeric)
7      tokens = tokenizer.tokenize(text)
8      tokens = [lem.lemmatize(token.strip()) for token in tokens] # 词形还原
9      filtered_tokens = [token for token in tokens if token not in stopword_list]
10     filtered_text = ' '.join(filtered_tokens)
11     return filtered_text, filtered_tokens
12
13 filtered_text, filtered_tokens = remove_stopwords_regex(text, True)
14 word_freqs = collections.Counter()
15 for word in filtered_tokens:
16     word_freqs[word] += 1
17 print(word_freqs.most_common(20))
```

(11) 找出相似词 (Synonyms)：WordNet 语料库内含相似词、相反词与简短说明。

```
1  synonyms = nltk.corpus.wordnet.synsets('love')
2  synonyms
```

执行结果：列出前 10 名，是以例句显示，故许多单词均相同。

```
[Synset('love.n.01'),
 Synset('love.n.02'),
 Synset('beloved.n.01'),
 Synset('love.n.04'),
 Synset('love.n.05'),
 Synset('sexual_love.n.02'),
 Synset('love.v.01'),
 Synset('love.v.02'),
 Synset('love.v.03'),
 Synset('sleep_together.v.01')]
```

(12) 显示相似词说明。

```
1  # 单词说明
2  synonyms[0].definition()
```

执行结果：列出第一个相似词的单词说明。

```
'a strong positive emotion of regard and affection'
```

(13) 显示相似词的例句。

```
1  # 单词的例句
2  synonyms[0].examples()
```

执行结果：列出第一个相似词的例句。

```
['his love for his work', 'children need a lot of love']
```

(14) 找出相反词 (Antonyms)：须先调用 lemmas 进行词形还原，再调用 antonyms。

```
1  antonyms=[]
2  for syn in nltk.corpus.wordnet.synsets('ugly'):
3      for l in syn.lemmas():
4          if l.antonyms():
5              antonyms.append(l.antonyms()[0].name())
6  antonyms
```

执行结果：ugly ➜ beautiful。

(15) 分析词性标签 (POS Tagging)：依照句子结构，显示每个单词的词性。

```
1  text='I am a human being, capable of doing terrible things'
2  sentences=nltk.sent_tokenize(text)
3  for sent in sentences:
4      print(nltk.pos_tag(nltk.word_tokenize(sent)))
```

执行结果。

```
[('I', 'PRP'), ('am', 'VBP'), ('a', 'DT'), ('human', 'JJ'), ('being', 'VBG'), (',', ','), ('capable', 'JJ'), ('of', 'IN'), ('doing', 'VBG'), ('terrible', 'JJ'), ('things', 'NNS')]
```

词性标签 (POS Tagging) 列表如下：

CC (Coordinating Conjunction)：并列连词。

CD (Cardinal Digit)：基数。

DT (Determiner)：量词。

EX (Existential)：存在地，例如 There。

FW (Foreign Word)：外来语。

IN Preposition/Subordinating Conjunction：介词。

JJ Adjective：形容词。

JJR Adjective, Comparative：比较级形容词。

JJS Adjective, Superlative：最高级形容词。

LS (List Marker)：列表标记。

MD (Modal)：情态动词。

NN Noun, Singular：名词单数。

NNS Noun Plural：名词复数。

NNP Proper Noun, Singular：专有名词单数。

NNPS Proper Noun, Plural：专有名词复数。

PDT (Predeterminer)：放在量词的前面，例如 both、a lot of。

POS (Possessive Ending)：所有格，例如 parent's。

PRP (Personal Pronoun)：代名词，例如 I, he, she。

PRP$ Possessive Pronoun：所有格代名词，例如 my, his, hers。

RB Adverb：副词，例如 very, silently。

RBR Adverb, Comparative：比较级副词，例如 better。

RBS Adverb, Superlative：最高级副词，例如 best。
RP Particle：助词，例如 give up。
TO to：例如 go 'to' the store。
UH Interjection：感叹词，例如 errrrrrrrm。
VB Verb, Base Form：动词，例如 take。
VBD Verb, Past Tense：动词过去时，例如 took。
VBG Verb, Gerund/Present Participle：动词进行时，例如 taking。
VBN Verb, Past Participle：动词过去分词，例如 taken。
VBP Verb, Sing Present, non-3d：动词现在时单数，例如 take。
VBZ Verb, 3rd person sing. present：动词现在时复数，例如 takes。
WDT wh-determiner：疑问代名词，例如 which。
WP wh-pronoun who, what：疑问代名词。
WP$ possessive wh-pronoun：疑问代名词所有格，例如 whose。
WRB wh-abverb：疑问副词，例如 where, when。

spaCy 套件提供更强大的分析功能，但由于内容涉及词向量 (Word2Vec)，所以我们留待后续章节再讨论。

# 11-3 词向量

BOW 和 TF-IDF 都只着重于词汇出现在文件中的次数，未考虑语言/文字有上下文的关联，比如，"这间房屋有四扇？"，从上文大概可以推测出最后一个词汇是"窗户"。又譬如，我说喜欢吃辣，那我会点"麻婆豆腐"还是"家常豆腐"呢？相信听到"吃辣"，应该都会猜是"麻婆豆腐"。另一方面，一个语系的单词数有限，中文有几万个字，我们是否也可以比照影像辨识，对所有的单词建构预先训练的模型，之后是否就可以实现转换学习 (Transfer Learning)？

针对上下文的关联，Google 研发团队 Tomas Mikolov 等人于 2013 年提出"词向量"(Word2Vec)，他们收集 1000 亿个单词 (Word) 加以训练，将每个单词改用上下文表达，然后转换为向量，而这就是"词嵌入"(Word Embedding) 的概念，与 TF-IDF 输出是稀疏向量不同，词嵌入的输出是一个稠密的样本空间。

词向量有两种做法 ( 如图 11.4 所示 )：

连续 BOW(Continuous Bag-of-Words, CBOW)：以单词的上下文预测单词。
Continuous Skip-gram Model：刚好相反，以单词预测上下文。

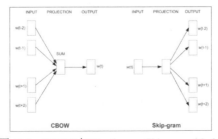

图 11.4　CBOW 与 Continuous Skip-gram Model

图片来源：*Exploiting Similarities among Languages for Machine Translation* [3]

CBOW 算法就是一个深度学习模型，如图 11.5 所示。

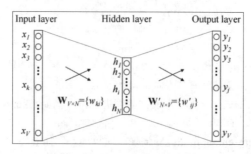

图 11.5　CBOW 的网络结构

图片来源：*An Intuitive Understanding of Word Embeddings: From Count Vectors to Word2Vec* [4]

以单词的上下文为输入，以预测的单词为目标，如同 2-gram 的模型，例句为 "Hey, this is sample corpus using only one context word."，使用 One-hot encoding，输出见表 11.1。

表 11.1　2-gram 与 One-hot encoding

| Input | Output | | Hey | This | is | sample | corpus | using | only | one | context | word |
|---|---|---|---|---|---|---|---|---|---|---|---|---|
| Hey | this | Datapoint 1 | 1 | 0 | 0 | 0 | 0 | 0 | 0 | 0 | 0 | 0 |
| this | hey | Datapoint 2 | 0 | 1 | 0 | 0 | 0 | 0 | 0 | 0 | 0 | 0 |
| is | this | Datapoint 3 | 0 | 0 | 1 | 0 | 0 | 0 | 0 | 0 | 0 | 0 |
| is | sample | Datapoint 4 | 0 | 0 | 1 | 0 | 0 | 0 | 0 | 0 | 0 | 0 |
| sample | is | Datapoint 5 | 0 | 0 | 0 | 1 | 0 | 0 | 0 | 0 | 0 | 0 |
| sample | corpus | Datapoint 6 | 0 | 0 | 0 | 1 | 0 | 0 | 0 | 0 | 0 | 0 |
| corpus | sample | Datapoint 7 | 0 | 0 | 0 | 0 | 1 | 0 | 0 | 0 | 0 | 0 |
| corpus | using | Datapoint 8 | 0 | 0 | 0 | 0 | 1 | 0 | 0 | 0 | 0 | 0 |
| using | corpus | Datapoint 9 | 0 | 0 | 0 | 0 | 0 | 1 | 0 | 0 | 0 | 0 |
| using | only | Datapoint 10 | 0 | 0 | 0 | 0 | 0 | 1 | 0 | 0 | 0 | 0 |
| only | using | Datapoint 11 | 0 | 0 | 0 | 0 | 0 | 0 | 1 | 0 | 0 | 0 |
| only | one | Datapoint 12 | 0 | 0 | 0 | 0 | 0 | 0 | 1 | 0 | 0 | 0 |
| one | only | Datapoint 13 | 0 | 0 | 0 | 0 | 0 | 0 | 0 | 1 | 0 | 0 |
| one | context | Datapoint 14 | 0 | 0 | 0 | 0 | 0 | 0 | 0 | 1 | 0 | 0 |
| context | one | Datapoint 15 | 0 | 0 | 0 | 0 | 0 | 0 | 0 | 0 | 1 | 0 |
| context | word | Datapoint 16 | 0 | 0 | 0 | 0 | 0 | 0 | 0 | 0 | 1 | 0 |
| word | context | Datapoint 17 | 0 | 0 | 0 | 0 | 0 | 0 | 0 | 0 | 0 | 1 |

2-gram 是每次取两个单词，然后逐步向右滑动一个单词，输出如图 11.6 所示。

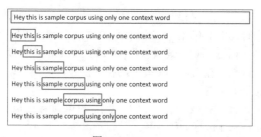

图 11.6　2-gram

接着以第一个单词 One-hot encoding 为输入，第二个单词为预测目标，最后模型预测的是各单词的概率。这是一个简略的说明，当然，实际的模型不会这么简单，还会额外考虑以下情况：

(1) 不只考虑上一个单词，会将上下文各 *n* 个单词都纳入考虑。

(2) 依此类推，输出是 1000 亿个单词的概率，模型应该无法承担如此多的类别，因此改用所谓的"负样本抽样"(Negative Sub-sampling)，只推论输出入是否为上下文，例如 orange, juice，从 P(juice|orange) 改为预测 P(1|<orange, juice>)，亦即从多类别 (1000 亿个) 模型转换成二分类 ( 真 / 假 ) 模型。

CBOW 的优点如下：

(1) 简单，而且比传统确定性模型 (Deterministic methods) 的效能较好。

(2) 对比相关矩阵，CBOW 对内存消耗节省很多。

CBOW 的缺点如下：

(1) 像是 Apple 可能是指水果，但也可能是在指公司名称，遇到这样的情况，CBOW 会取平均值，造成失准，故 CBOW 无法处理一词多义。

(2) CBOW 因为输出高达 1000 亿个单词概率，所以优化求解的收敛十分困难。

因此，后续发展出了 Skip-gram 模型，颠倒输出与输入，改由单词预测上下文，我们再以同样的句子来举例，如表 11.2 所示。

表 11.2　Skip-gram 模型的输出与输入

| Input | Output(Context1) | Output(Context2) |
|---|---|---|
| Hey | this | \<padding\> |
| this | Hey | is |
| is | this | sample |
| sample | is | corpus |
| corpus | sample | corpus |
| using | corpus | only |
| only | using | one |
| one | only | context |
| context | one | word |
| word | context | \<padding\> |

Skip-gram 的优点如下：

(1) 一个单词可以预测多个上下文，解决了一词多义的问题。

(2) 结合负样本抽样 (Negative Sub-sampling) 技术，效能比其他模型佳。负样本可以是任意单词的排列组合，如果要把所有负样本放入训练数据中，数量可能过于庞大，而且会造成不平衡数据 (Imbalanced Data)，因此采用负样本抽样的方法。接下来直接以 Gensim 实操 Skip-gram 模型训练，细节就不介绍，有兴趣的读者可参阅 *NLP 102: Negative Sampling and GloVe*[5]。

我们可以利用预先训练的模型来试验一下，Gensim 和 spaCy 套件均提供 Word2Vec 模型。

先用以下语句安装 Gensim 套件：

pip install gensim

**范例1. 运用Gensim进行相似性比较。**

运用 Gensim 进行相似性比较的相关步骤如图 11.7 所示。

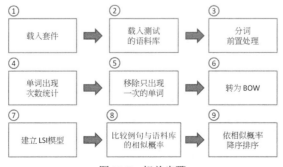

图 11.7　相关步骤

下列程序代码请参考【11_04_gensim_ 相似性比较 .ipynb】。

(1) 加载相关套件。

```
1  import pprint    # 较美观的列印函数
2  import gensim
3  from collections import defaultdict
4  from gensim import corpora
```

(2) 测试的语料库 (Corpus) 如下。

```
1  documents = [
2      "Human machine interface for lab abc computer applications",
3      "A survey of user opinion of computer system response time",
4      "The EPS user interface management system",
5      "System and human system engineering testing of EPS",
6      "Relation of user perceived response time to error measurement",
7      "The generation of random binary unordered trees",
8      "The intersection graph of paths in trees",
9      "Graph minors IV Widths of trees and well quasi ordering",
10     "Graph minors A survey",
11 ]
```

(3) 分词、前置处理。

```
1  # 任意设定一些停用词
2  stoplist = set('for a of the and to in'.split())
3
4  # 分词，转小写
5  texts = [
6      [word for word in document.lower().split() if word not in stoplist]
7      for document in documents
8  ]
9  texts
```

执行结果。

```
[['human', 'machine', 'interface', 'lab', 'abc', 'computer', 'applications'],
 ['survey', 'user', 'opinion', 'computer', 'system', 'response', 'time'],
 ['eps', 'user', 'interface', 'management', 'system'],
 ['system', 'human', 'system', 'engineering', 'testing', 'eps'],
 ['relation', 'user', 'perceived', 'response', 'time', 'error', 'measurement'],
 ['generation', 'random', 'binary', 'unordered', 'trees'],
 ['intersection', 'graph', 'paths', 'trees'],
 ['graph', 'minors', 'iv', 'widths', 'trees', 'well', 'quasi', 'ordering'],
 ['graph', 'minors', 'survey']]
```

(4) 单词出现的次数统计。

```
1  frequency = defaultdict(int)
2  for text in texts:
3      for token in text:
4          frequency[token] += 1
5  frequency
```

执行结果：显示每个单词出现的次数。

```
defaultdict(int,
    {'human': 2,
     'machine': 1,
     'interface': 2,
     'lab': 1,
     'abc': 1,
     'computer': 2,
     'applications': 1,
     'survey': 2,
     'user': 3,
     'opinion': 1,
     'system': 4,
     'response': 2,
     'time': 2,
     'eps': 2,
     'management': 1,
     'engineering': 1,
     'testing': 1,
     'relation': 1,
     'perceived': 1,
     'error': 1,
     'measurement': 1,
     'generation': 1,
     'random': 1,
```

(5) 移除只出现一次的单词：仅专注在较常出现的关键词。

```
1  texts = [
2      [token for token in text if frequency[token] > 1]
3      for text in texts
4  ]
5  texts
```

执行结果：每句筛选的结果。

```
[['human', 'interface', 'computer'],
 ['survey', 'user', 'computer', 'system', 'response', 'time'],
 ['eps', 'user', 'interface', 'system'],
 ['system', 'human', 'system', 'eps'],
 ['user', 'response', 'time'],
 ['trees'],
 ['graph', 'trees'],
 ['graph', 'minors', 'trees'],
 ['graph', 'minors', 'survey']]
```

(6) 转为 BOW。

```
1  # 转为字典
2  dictionary = corpora.Dictionary(texts)
3
4  # 转为 BOW
5  corpus = [dictionary.doc2bow(text) for text in texts]
6  corpus
```

执行结果。

```
[[(0, 1), (1, 1), (2, 1)],
 [(0, 1), (3, 1), (4, 1), (5, 1), (6, 1), (7, 1)],
 [(2, 1), (5, 1), (7, 1), (8, 1)],
 [(1, 1), (5, 2), (8, 1)],
 [(3, 1), (6, 1), (7, 1)],
 [(9, 1)],
 [(9, 1), (10, 1)],
 [(9, 1), (10, 1), (11, 1)],
 [(4, 1), (10, 1), (11, 1)]]
```

(7) 建立 LSI (Latent semantic indexing) 模型：可指定议题的个数，每一项议题皆由所有单词加权组合而成。

```
1  # 建立 LSI (Latent semantic indexing) 模型
2  from gensim import models
3
4  # num_topics=2：取二维，即两个议题
5  lsi = models.LsiModel(corpus, id2word=dictionary, num_topics=2)
6
7  # 两个议题的 LSI 公式
8  lsi.print_topics(2)
```

执行结果：两项议题的公式。

```
[(0,
  '0.644*"system" + 0.404*"user" + 0.301*"eps" + 0.265*"time" + 0.265*"response" + 0.240*"computer" + 0.221*"human" + 0.206*"survey" + 0.198*"interface" + 0.036*"graph"'),
 (1,
  '0.623*"graph" + 0.490*"trees" + 0.451*"minors" + 0.274*"survey" + -0.167*"system" + -0.141*"eps" + -0.113*"human" + 0.107*"response" + 0.107*"time" + -0.072*"interface"')]
```

(8) 测试 LSI (Latent semantic indexing) 模型。

```
1  # 例句
2  doc = "Human computer interaction"
3
4  # 测试 LSI (Latent semantic indexing) 模型
5  vec_bow = dictionary.doc2bow(doc.lower().split())
6  vec_lsi = lsi[vec_bow]
7  print(vec_lsi)
```

执行结果：将例句代入到两项议题公式中，计算 LSI 值，结果比较接近第一项议题。

[(0, 0.4618210045327157), (1, -0.07002766527900067)]

(9) 比较例句与文章段落内每一个句子的相似概率。

```
1  from gensim import similarities
2
3  # 比较例句与语料库的相似性索引
4  index = similarities.MatrixSimilarity(lsi[corpus])
5
6  # 比较例句与语料库的相似概率
7  sims = index[vec_lsi]
8
9  # 显示语料库的索引值及相似概率
10 print(list(enumerate(sims)))
```

执行结果：将例句代入到两项议题公式中，计算 LSI 值。

```
[(0, 0.998093), (1, 0.93748635), (2, 0.9984453), (3, 0.98658866), (4, 0.90755945), (5, -0.12416792), (6, -0.1063926), (7, -0.09879464), (8, 0.05004177)]
```

(10) 按照概率进行降序排序。

```
1  sims = sorted(enumerate(sims), key=lambda item: -item[1])
2  for doc_position, doc_score in sims:
3      print(doc_score, documents[doc_position])
```

执行结果：前两句概率最大，依语义判断结果正确无误。

```
0.9984453 The EPS user interface management system
0.998093 Human machine interface for lab abc computer applications
0.98658866 System and human system engineering testing of EPS
0.93748635 A survey of user opinion of computer system response time
0.90755945 Relation of user perceived response time to error measurement
0.05004177 Graph minors A survey
-0.09879464 Graph minors IV Widths of trees and well quasi ordering
-0.1063926 The intersection graph of paths in trees
-0.12416792 The generation of random binary unordered trees
```

Srijith Rajamohan 提供一段简单的程序代码，以自定义的数据集实操 Word2Vec，包括 CBOW、Skip-Gram，虽然不含 hierarchical softmax、negative sampling 等算法，但也有助于更深入了解 Word2Vec，详细说明可参阅 *Word2Vec in Pytorch - Continuous Bag of Words and Skipgrams*[6]。

**范例2. Word2Vec实操。** 笔者将部分程序代码删除，以利概念的理解，读者若要观看所有的程序代码可参阅原文。

下列程序代码请参考【**11_05_Word2Vec_Pytorch.ipynb**】，数据集为 **nlp_data/word2vec_test.txt**。

(1) 先实操 CBOW，加载相关套件。

```
1  import torch
2  import torch.nn as nn
3  import torch.nn.functional as F
4  import torch.optim as optim
5  import numpy as np
6  import urllib.request
7  from nltk.tokenize import RegexpTokenizer
8  from nltk.corpus import stopwords
9  from nltk import word_tokenize
10 import sklearn
11 from sklearn.cluster import KMeans
12 from sklearn.metrics.pairwise import euclidean_distances
```

(2) 参数设定。

以上文 3 个单词预测目前的单词，CBOW 应该也要考虑下文，不过这是简单练习，就不修正了。

嵌入层输出为每个单词转化的向量维度，可设任意值，第 12 章会有详细的说明。

```
1  torch.manual_seed(1)    # 固定乱数种子
2  CONTEXT_SIZE = 3        # 上下文个数
3  EMBEDDING_DIM = 10      # 嵌入层输出维度
```

(3) 定义文字处理函数。

```
1  # 以值(value)找键值(key)
2  def get_key(word_id):
3      for key,val in word_to_ix.items():
4          if(val == word_id):
5              return key
6      return ''
7
8  # 分词及前置处理
9  def read_data(file_path, remove_stopwords = False):
10     tokenizer = RegexpTokenizer(r'\w+')
11     if file_path.lower().startswith('http'):
12         data = urllib.request.urlopen(file_path)
13         data = data.read().decode('utf8')
14     else:
15         data = open(file_path, encoding='utf8').read()
16     tokenized_data = word_tokenize(data)
17     if remove_stopwords:
18         stop_words = set(stopwords.words('english'))
19     else:
20         stop_words = set([])
21     stop_words.update(['.',',',':',';','(',')','#','--','...','"'])
22     cleaned_words = [ i for i in tokenized_data if i not in stop_words ]
23     return(cleaned_words)
```

(4) 读取文件，作为测试的文本 (Text)：可读取本机或网络文件，也可以使用 NLTK 内建的语料库 (Corpus)。

```
1  test_sentence = read_data('./nlp_data/word2vec_test.txt')
2
3  # 或读取其他文件
4  #test_sentence = 'https://www.gutenberg.org/files/57884/57884-0.txt')
```

(5) 进行 N-grams 处理：取得单词的上文。

```
1  ngrams = []
2  for i in range(len(test_sentence) - CONTEXT_SIZE):
3      tup = [test_sentence[j] for j in np.arange(i , i + CONTEXT_SIZE) ]
4      ngrams.append((tup,test_sentence[i + CONTEXT_SIZE]))
5
6  print(ngrams[0], ngrams[1])
```

执行结果：文本为 Empathy for the poor may not come easily to people…，处理后前两笔数据为 (['Empathy', 'for', 'the'], 'poor') (['for', 'the', 'poor'], 'may')，即 poor 上文为 Empathy for the，may 上文为 for the poor。

(6) 词汇表设定并建立字典，以利于由单词取得它的代码。

```
1  # 取得词汇表(vocabulary)
2  vocab = set(test_sentence)
3  print("单词个数：",len(vocab))
4
5  # 建立字典，以单词取得代码
6  word_to_ix = {word: i for i, word in enumerate(vocab)}
```

(7) 建立 CBOW 模型：依次为 embeddings、linear、relu、linear、log_softmax 神经层，embeddings 为嵌入层，会将输入转换为实数的向量空间。

```python
class CBOWModeler(nn.Module):
    def __init__(self, vocab_size, embedding_dim, context_size):
        super(CBOWModeler, self).__init__()
        self.embeddings = nn.Embedding(vocab_size, embedding_dim)
        self.linear1 = nn.Linear(context_size * embedding_dim, 128)
        self.linear2 = nn.Linear(128, vocab_size)

    def forward(self, inputs):
        # embeds -> linear -> relu -> linear -> log_softmax
        embeds = self.embeddings(inputs).view((1, -1))
        out1 = F.relu(self.linear1(embeds))
        out2 = self.linear2(out1)
        log_probs = F.log_softmax(out2, dim=1)
        return log_probs

    def predict(self, input):
        # 以上下文预测
        context_idxs = torch.LongTensor([word_to_ix[w] for w in input])
        res = self.forward(context_idxs)
        res_arg = torch.argmax(res)
        res_val, res_ind = res.sort(descending=True)
        res_val = res_val[0][:3]   # 前3个预测值
        res_ind = res_ind[0][:3]   # 前3个预测索引值
        for arg in zip(res_val,res_ind):
            print([(key,val,arg[0]) for key,val in word_to_ix.items()
                                     if val == arg[1]])
```

(8) 模型训练：输入单词及上文的编码，进行梯度下降法的训练。

```python
losses = []
loss_function = nn.NLLLoss()
model = CBOWModeler(len(vocab), EMBEDDING_DIM, CONTEXT_SIZE)
optimizer = optim.SGD(model.parameters(), lr=0.001)

for epoch in range(400):
    total_loss = 0
    for context, target in ngrams:
        # 以单字取得代码
        context_idxs = torch.LongTensor([word_to_ix[w] for w in context])

        # 梯度下降
        model.zero_grad()
        log_probs = model(context_idxs)
        loss = loss_function(log_probs, torch.LongTensor([word_to_ix[target]]))
        loss.backward()
        optimizer.step()
        total_loss += loss.item()
    losses.append(total_loss)
```

(9) 模型预测：输入上文，即 3 个单词，预测下一个单词。

```python
model.predict(['of','all','human'])
```

执行结果：上文为 of all human，预测下一个单词前 3 名为 afflictions、it、neither，以 of all human 搜寻文本，果然下一个单词为 afflictions。

(10) 接着进行 Skip-gram，以相同文本进行 N-grams 处理。

```python
ngrams = []
for i in range(len(test_sentence) - CONTEXT_SIZE):
    tup = [test_sentence[j] for j in np.arange(i + 1 , i + CONTEXT_SIZE + 1) ]
    ngrams.append((test_sentence[i],tup))
print(ngrams[0], ngrams[1])
```

执行结果：与 CBOW 相反，单词 Empathy 的下文为 for the poor。

(11) 建立 Skip-Gram 模型：神经层与 CBOW 类似。

```
1  class SkipgramModeler(nn.Module):
2      def __init__(self, vocab_size, embedding_dim, context_size):
3          super(SkipgramModeler, self).__init__()
4          self.embeddings = nn.Embedding(vocab_size, embedding_dim)
5          self.linear1 = nn.Linear(embedding_dim, 128)
6          self.linear2 = nn.Linear(128, context_size * vocab_size)
7          #self.parameters['context_size'] = context_size
8  
9      def forward(self, inputs):
10         # embeds -> linear -> relu -> linear -> log_softmax
11         embeds = self.embeddings(inputs).view((1, -1))
12         out1 = F.relu(self.linear1(embeds))
13         out2 = self.linear2(out1)
14         log_probs = F.log_softmax(out2, dim=1).view(CONTEXT_SIZE,-1)
15         return log_probs
16  
17     def predict(self,input):
18         context_idxs = torch.LongTensor([word_to_ix[input]])
19         res = self.forward(context_idxs)
20         res_arg = torch.argmax(res)
21         res_val, res_ind = res.sort(descending=True)
22         indices = [res_ind[i][0] for i in np.arange(0,3)]
23         for arg in indices:
24             print([(key, val) for key,val in word_to_ix.items()
25                 if val == arg ])
```

(12) 模型训练：神经层与 CBOW 类似。

```
1  losses = []
2  loss_function = nn.NLLLoss()
3  model = SkipgramModeler(len(vocab), EMBEDDING_DIM, CONTEXT_SIZE)
4  optimizer = optim.SGD(model.parameters(), lr=0.001)
5  
6  # Freeze embedding layer
7  #model.freeze_layer('embeddings')
8  
9  for epoch in range(550):
10     total_loss = 0
11     # model.predict('psychologically')
12  
13     for context, target in ngrams:
14         context_idxs = torch.LongTensor([word_to_ix[context]])
15         model.zero_grad()
16         log_probs = model(context_idxs)
17         target_list = torch.LongTensor([word_to_ix[w] for w in target])
18         loss = loss_function(log_probs, target_list)
19         loss.backward()
20         optimizer.step()
21         total_loss += loss.item()
22     losses.append(total_loss)
```

(13) 模型预测：输入 1 个单词，预测下文。

```
1  model.predict('psychologically')
```

执行结果：单词 psychologically 的下文为 and physically incapacitating，以 psychologically 搜寻文本，下文果然吻合。

这个范例是让我们体验一下 Word2Vec 的实操，如果要正规地训练 Word2Vec，可使用 Gensim 套件。Gensim 不仅提供 Word2Vec 预先训练模型，也支持自定义数据训练的功能，预先训练模型可提供一般内容的推论，但如果内容属于特殊领域，则应该自行训练模型会比较恰当，Gensim Word2Vec 的用法请参考 Gensim 官网 Word2Vec 说明文件[7]。

## 范例3. 运用Gensim进行Word2Vec训练与测试。

**下列程序代码请参考【11_06_gensim_Word2Vec.ipynb】。**

(1) 加载相关套件。

```
1  import gzip
2  import gensim
```

(2) 以 Gensim 进行简单测试：把 Gensim 内建的语料库 common_texts 作为训练数据，并且对 "hello" "world" "michael" 三个单词进行训练，产生词向量。

```
1  from gensim.test.utils import common_texts
2
3  # size：词向量的大小，window：考虑上下文各自的长度
4  # min_count：单词至少出现的次数，workers：执行绪个数
5  model_simple = gensim.models.Word2Vec(sentences=common_texts, window=1,
6                                         min_count=1, workers=4)
7  # 传回 有效的字数及总处理字数
8  model_simple.train([["hello", "world", "michael"]], total_examples=1, epochs=2)
```

执行结果：回传两个值 (0, 6)，包括所有执行周期的有效字数与总处理字数，其中前者为内部处理的逻辑，不太理解，后者数字为 6=3 个单词 × 2 个执行周期。

train() 的参数有很多，可参阅 Gensim 官网 Word2Vec 说明文件，这里仅摘录此范例所用到的参数。

sentences：训练数据。
size：产生的词向量大小。
window：考虑上下文各自的长度。
min_count：单词至少出现的次数。
workers：线程的个数。

(3) 另一个例子。

```
1  sentences = [["cat", "say", "meow"], ["dog", "say", "woof"]]
2
3  model_simple = gensim.models.Word2Vec(min_count=1)
4  model_simple.build_vocab(sentences)  # 建立生字表(vocabulary)
5  model_simple.train(sentences, total_examples=model_simple.corpus_count
6                     , epochs=model_simple.epochs)
```

执行结果：回传 (1, 30)，其中 30=6 个单词 × 5 个执行周期。

(4) 实例测试：载入 OpinRank 语料库，文章内容是关于车辆与旅馆的评论。

```
1  # 载入 OpinRank 语料库：关于车辆与旅馆的评论
2  data_file="./Word2Vec/reviews_data.txt.gz"
3
4  with gzip.open (data_file, 'rb') as f:
5      for i,line in enumerate (f):
6          print(line)
7          break
```

执行结果。

b"Oct 12 2009 \tNice trendy hotel location not too bad.\tI stayed in this hotel for one night. As this is a fairly new place some of the taxi drivers did not know where it was and/or did not want to drive there. Once I have eventually arrived at the hotel, I was very pleasantly surprised with the decor of the lobby/ground floor area. It was very stylish and modern. I found the reception's staff geeting me with 'Aloha' a bit out of place, but I guess they are briefed to say that to keep up the coroporate image.As I have a Starwood Preferred Guest member, I was given a small gift upon-check in. It was only a couple of fridge magnets in a gift box, but nevertheless a nice gesture.My room was nice and roomy, there are tea and coffee facilities in each room and you get two complimentary bottles of water plus some toiletries by 'bliss'.The location is not great. It is at the last metro stop and you then need to take a taxi, but if you are not planning on going to see the historic sites in Beijing, then you will be ok.I chose to have some breakfast in the hotel, which was really tasty and there was a good selection of dishes. There are a couple of computers to use in the communal area, as well as a pool table. There is also a small swimming pool and a gym area.I would definitely stay in this hotel again, but only if I did not plan to travel to central Beijing, as it can take a long time. The location is ok if you plan to do a lot of shopping, as there is a big shopping centre just few minutes away from the hotel and there are plenty of eating options around, including restaurants that serve a dog meat!\t\r\n"

(5) 读取 OpinRank 语料库，并进行前置处理，如分词。

```
1   # 读取 OpinRank 语料库，并做前置处理
2   def read_input(input_file):
3       with gzip.open (input_file, 'rb') as f:
4           for i, line in enumerate (f):
5               # 前置处理
6               yield gensim.utils.simple_preprocess(line)
7
8   # 载入 OpinRank 语料库，分词
9   documents = list(read_input(data_file))
10  documents
```

执行结果：为一个 List。

```
[['oct',
  'nice',
  'trendy',
  'hotel',
  'location',
  'not',
  'too',
  'bad',
  'stayed',
  'in',
  'this',
  'hotel',
  'for',
  'one',
  'night',
  'as',
  'this',
```

(6) Word2Vec 模型训练：约需 10 分钟。

```
1   # Word2Vec 模型训练，约10分钟
2   model = gensim.models.Word2Vec(documents, size=150, window=10,
3                                  min_count=2, workers=10)
4   model.train(documents,total_examples=len(documents),epochs=10)
```

执行结果：(303,484,226, 415,193,580)，处理达数亿个单词。

接下来进行各种测试。

(7) 测试"dirty"的相似词。

```
1   # 测试'fdirty'相似词
2   w1 = "dirty"
3   model.wv.most_similar(positive=w1) # positive：相似词
```

执行结果：显示 10 个最相似的单词。

```
[('filthy', 0.8602699041366577),
 ('stained', 0.7798251509666443),
 ('dusty', 0.7683317065238953),
 ('unclean', 0.7638086676597595),
 ('grubby', 0.757234513759613),
 ('smelly', 0.7431163787841797),
 ('dingy', 0.7304496169090271),
 ('disgusting', 0.7111263275146484),
 ('soiled', 0.7099645733833313),
 ('mouldy', 0.706375241279602)]
```

(8) 测试"france"的相似词：topn 可指定列出前 *n* 名。

```
1   # 测试'france'相似词
2   w1 = ["france"]
3   model.wv.most_similar (positive=w1, topn=6) # topn：只列出前 n 名
```

执行结果：显示 6 个最相似的单词。

```
[('germany', 0.6627413034439087),
 ('canada', 0.6545147895812988),
 ('spain', 0.644172728061676),
 ('england', 0.6122641563415527),
 ('mexico', 0.6106705665588379),
 ('rome', 0.6044377684593201)]
```

(9) 同时测试多个词汇："床、床单、枕头"的相似词与"长椅"的相反词。

```
1  # 测试'床、床单、枕头'相似词及'长椅'相反词
2  w1 = ["bed",'sheet','pillow']
3  w2 = ['couch']
4  model.wv.most_similar (positive=w1, negative=w2, topn=10)  # negative : 相反词
```

执行结果：显示 10 个最适合的单词。

```
[('duvet', 0.7157680988311768),
 ('blanket', 0.7036269903182983),
 ('mattress', 0.7003698348999023),
 ('quilt', 0.7003640532493591),
 ('matress', 0.6967926621437073),
 ('pillowcase', 0.665346086025238),
 ('sheets', 0.6376352310180664),
 ('pillows', 0.6317484378814697),
 ('comforter', 0.6119856834411621),
 ('foam', 0.6095048785209656)]
```

(10) 比较两个词汇的相似概率。

```
1  model.wv.similarity(w1="dirty",w2="smelly")
```

执行结果：相似概率为 0.7431163。

(11) 挑选出较不相似的词汇。

```
1  model.wv.doesnt_match(["cat","dog","france"])
```

执行结果：france。

(12) 接着测试加载预先训练模型，有两种方式：程序直接下载或者手动下载后再读取文件。

程序直接下载预先训练的模型。

```
1  import gensim.downloader as api
2  wv = api.load('word2vec-google-news-300')
```

手动下载后加载，预先训练模型的下载网址为：https://drive.google.com/file/d/0B7XkCwpI5KDYNlNUTTlSS21pQmM/edit。

```
1  from gensim.models import KeyedVectors
2
3  # 每个词向量有 300 个元素
4  model = KeyedVectors.load_word2vec_format(
5      './Word2Vec/GoogleNews-vectors-negative300.bin', binary=True)
```

接下来进行各种测试。

(13) 取得 dog 的词向量。

```
1  model['dog']
```

执行结果：共有 300 个元素。

```
array([ 5.12695312e-02, -2.23388672e-02, -1.72851562e-01,  1.61132812e-01,
       -8.44726562e-02,  5.73730469e-02,  5.85937500e-02, -8.25195312e-02,
       -1.53808594e-02, -6.34765625e-02,  1.79687500e-01, -4.23828125e-01,
       -2.25830078e-02, -1.66015625e-01, -2.51464844e-01,  1.07421875e-01,
       -1.99218750e-01,  1.59179688e-01, -1.87500000e-01, -1.20117188e-01,
        1.55273438e-01, -9.91210938e-02,  1.42578125e-01, -1.64062500e-01,
       -8.93554688e-02,  2.00195312e-01, -1.49414062e-01, -2.03125000e-01,
        3.28125000e-01,  2.44140625e-01, -9.71679688e-02, -8.20312500e-02,
       -3.63769531e-02, -8.59375000e-02, -9.86328125e-02,  7.78198242e-03,
       -1.34277344e-02,  5.27343750e-02,  1.48437500e-01,  3.33984375e-01,
```

(14) 测试 woman, king 的相似词和 man 的相反词。

```
1  model.most_similar(positive=['woman', 'king'], negative=['man'])
```

执行结果：这就是有名的 king - man + woman = queen。

```
[('queen', 0.7118192911148071),
 ('monarch', 0.6189674139022827),
 ('princess', 0.5902431011199951),
 ('crown_prince', 0.5499460697174072),
 ('prince', 0.5377321243286133),
 ('kings', 0.5236844420433044),
 ('Queen_Consort', 0.5235945582389832),
 ('queens', 0.518113374710083),
 ('sultan', 0.5098593235015869),
 ('monarchy', 0.5087411999702454)]
```

(15) 挑选出较不相似的词汇。

```
1  model.doesnt_match("breakfast cereal dinner lunch".split())
```

执行结果：cereal( 麦片 ) 与三餐较不相似。

(16) 比较两词相似概率。

```
1  model.similarity('woman', 'man')
```

执行结果：概率为 0.76640123，'woman', 'man' 是相似的。

由上面测试可以知道，对于一般的文字判断，使用预先训练模型都相当准确，但是，如果要判断特殊领域的相关内容，效果可能就会打折。举例来说，Kaggle 上有一个很有趣的数据集"辛普生对话"(Dialogue Lines of The Simpsons)，是有关辛普生家庭的卡通剧情问答，像是询问剧中人物 Bart 与 Nelson 的相似度，结果只有 0.5，这是由于在卡通里面他们虽然是朋友，但不是很亲近，假如使用预先训练模型来推论问题的话，答案应该就不会如此精确，除此之外，还有很多例子，读者有空不妨测试看看此范例程序"Gensim Word2Vec Tutorial"[8]。

之前都是比较单词的相似度，然而更常见的需求是对语句 (Sentence) 的比对，譬如常见问答集 (FAQ) 或是对话机器人，系统会先比对问题的相似度，再将答案回复给使用者，Gensim 支持 Doc2Vec 算法，可进行语句相似度比较，程序代码如下。

(1) 笔者从 Starbucks 官网抓取了一段 FAQ 的标题当作测试语料库。

```
1  import numpy as np
2  import nltk
3  import gensim
4  from gensim.models import Word2Vec
5  from gensim.models.doc2vec import Doc2Vec, TaggedDocument
6  from sklearn.metrics.pairwise import cosine_similarity
7
8  # 测试语料
9  f = open('./FAQ/starbucks_faq.txt', 'r', encoding='utf8')
```

```
10  corpus = f.readlines()
11  # print(corpus)
12
13  # 参数设定
14  MAX_WORDS_A_LINE = 30  # 每行最多字数
15
16  # 标点符号(Punctuation)
17  import string
18  print('标点符号:', string.punctuation)
19
20  # 读取停用词
21  stopword_list = set(nltk.corpus.stopwords.words('english')
22                      + list(string.punctuation) + ['\n'])
```

(2) 训练 Doc2Vec 模型。

```
1   # 分词函数
2   def tokenize(text, stopwords, max_len = MAX_WORDS_A_LINE):
3       return [token for token in gensim.utils.simple_preprocess(text
4                       , max_len=max_len) if token not in stopwords]
5
6   # 分词
7   document_tokens=[] # 整理后的字词
8   for line in corpus:
9       document_tokens.append(tokenize(line, stopword_list))
10
11  # 设定为 Gensim 标签文件格式
12  tagged_corpus = [TaggedDocument(doc, [i]) for i, doc in
13                   enumerate(document_tokens)]
14
15  # 训练 Doc2Vec 模型
16  model_d2v = Doc2Vec(tagged_corpus, vector_size=MAX_WORDS_A_LINE, epochs=200)
17  model_d2v.train(tagged_corpus, total_examples=model_d2v.corpus_count,
18                  epochs=model_d2v.epochs)
```

(3) 比较语句的相似度。

```
1   # 测试
2   questions = []
3   for i in range(len(document_tokens)):
4       questions.append(model_d2v.infer_vector(document_tokens[i]))
5   questions = np.array(questions)
6   # print(questions.shape)
7
8   # 测试语句
9   # text = "find allergen information"
10  text = "mobile pay"
11  filtered_tokens = tokenize(text, stopword_list)
12  # print(filtered_tokens)
13
14  # 比较语句相似度
15  similarity = cosine_similarity(model_d2v.infer_vector(
16      filtered_tokens).reshape(1, -1), questions, dense_output=False)
17
18  # 选出前 10 名
19  top_n = np.argsort(np.array(similarity[0]))[::-1][:10]
20  print(f'前 10 名 index:{top_n}\n')
21  for i in top_n:
22      print(round(similarity[0][i], 4), corpus[i].rstrip('\n'))
```

执行结果：以"mobile pay"(手机支付)寻找前 10 名相似的语句，结果还不错。读者可再试试其他语句，笔者测试其他的结果并不理想，后面改用 BERT 模型时，准确率会提升许多。

另外 TensorBoard 还提供一个词嵌入的可视化工具 Embedding Projector[9]，可以观察单词间的距离，支持 3D 的向量空间，读者可以按如下步骤操作：

①在右方的搜寻字段输入单词后，系统就会显示候选字。

②选择其中一个候选字，接着系统会显示相似字，且利用各种算法 (PCA、T-SNE、UMAP) 来降维，以 3D 接口显示单词间的距离。

③单击"Isolate 101 points"：只显示距离最近的 101 个单词。

(4) 也可以修改词嵌入的模型：Word2Vec All、Word2Vec 10K、GNMT( 全球语言神经机器翻译 ) 等，如图 11.8 所示。

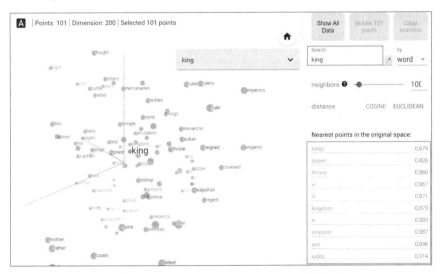

图 11.8　TensorFlow Embedding Projector 可视化工具

# 11-4　GloVe 模型

GloVe(Global Vectors) 是由斯坦福大学 Jeffrey Pennington 等学者于 2014 提出的另一套词嵌入模型，与 Word2Vec 齐名，他们认为 Word2Vec 并未考虑全局的概率分配，只以移动窗口内的词汇为样本，没有掌握全文的信息，因此，提出了词汇共现矩阵 (word-word cooccurrence matrix)，考虑词汇同时出现的概率，解决 Word2Vec 只看局部的缺陷以及 BOW 稀疏向量空间的问题，详细内容可参阅 *GloVe: Global Vectors for Word Representation*[10]。

GloVe 有 4 个预先训练好的模型：

glove.42B.300d.zip [11]：430 亿词汇，300 维向量，占 1.75 GB 的文件。

glove.840B.300d.zip [12]：8400 亿词汇，300 维向量，占 2.03 GB 的文件。

glove.6B.300d.zip [13]：60 亿词汇，300 维向量，占 822 MB 的文件。

glove.twitter.27B.zip [14]：270 亿词汇，200 维向量，占 1.42 GB 的文件。

GloVe 词向量模型文件的格式十分简单，每列是一个单词，每个字段以空格隔开，第一栏为单词，第二栏以后为该单词的词向量。所以，通常把模型文件读入后，转为字典 (dict) 的数据类型，以利查询。

**范例 4. GloVe 测试**。

**下列程序代码请参考【11_07_GloVe.ipynb】**。

(1) 载入 GloVe 词向量文件 glove.6B.300d.txt。

```
1  # 载入相关套件
2  import numpy as np
3
4  # 载入GloVe词向量文件 glove.6B.300d.txt
5  embeddings_dict = {}
6  with open("./glove/glove.6B.300d.txt", 'r', encoding="utf-8") as f:
7      for line in f:
8          values = line.split()
9          word = values[0]
10         vector = np.asarray(values[1:], "float32")
11         embeddings_dict[word] = vector
```

(2) 取得 GloVe 的词向量：任选一个单词 (love) 测试，取得 GloVe 的词向量。

```
1  # 随意测试一个单词(love)，取得 GloVe 的词向量
2  embeddings_dict['love']
```

部分执行结果。

```
array([-4.5205e-01, -3.3122e-01, -6.3607e-02,  2.8325e-02, -2.1372e-01,
        1.6839e-01, -1.7186e-02,  4.7309e-02, -5.2355e-02, -9.8706e-01,
        5.3762e-01, -2.6893e-01, -5.4294e-01,  7.2487e-02,  6.6193e-02,
       -2.1814e-01, -1.2113e-01, -2.8832e-01,  4.8161e-01,  6.9185e-01,
       -2.0022e-01,  1.0082e+00, -1.1865e-01,  5.8710e-01,  1.8482e-01,
        4.5799e-02, -1.7836e-02, -3.3952e-01,  2.9314e-01, -1.9951e-01,
       -1.8930e-01,  4.3267e-01, -6.3181e-01, -2.9510e-01, -1.0547e+00,
        1.8231e-01, -4.5040e-01, -2.7800e-01, -1.4021e-01,  3.6785e-02,
        2.6487e-01, -6.6712e-01, -1.5204e-01, -3.5001e-01,  4.0864e-01,
       -7.3615e-02,  6.7630e-01,  1.8274e-01, -4.1660e-02,  1.5014e-01,
        2.5216e-01, -1.0109e-01,  3.1915e-02, -1.1298e-01, -4.0147e-01,
        1.7274e-01,  1.8497e-01,  2.4456e-01,  6.8777e-01, -2.7019e-01,
        8.0728e-01, -5.8296e-02,  4.0550e-01,  3.9893e-01, -9.1688e-02,
       -5.2080e-01,  2.4570e-01,  6.3001e-02,  2.1421e-01,  3.3197e-01,
       -3.4299e-01, -4.8735e-01,  2.2264e-02,  2.7862e-01,  2.3881e-01,
```

(3) 指定以欧几里得 (Euclidean) 距离计算相似性：找出最相似的 10 个单词。

```
1  # 以欧几里得(Euclidean)距离计算相似性
2  from scipy.spatial.distance import euclidean
3
4  def find_closest_embeddings(embedding):
5      return sorted(embeddings_dict.keys(),
6                    key=lambda word: euclidean(embeddings_dict[word], embedding))
7
8  print(find_closest_embeddings(embeddings_dict["king"])[1:10])
```

执行结果：大部分与 "king" 的意义相似。

'queen', 'monarch', 'prince', 'kingdom', 'reign', 'ii', 'iii', 'brother', 'crown'

(4) 任选 100 个单词，并以散布图观察单词的相似度。

```
1  # 任意选 100 个单词
2  words = list(embeddings_dict.keys())[100:200]
3  # print(words)
4
5  from sklearn.manifold import TSNE
6  import matplotlib.pyplot as plt
7
8  # 以 T-SNE 降维至二个特征
9  tsne = TSNE(n_components=2)
10 vectors = [embeddings_dict[word] for word in words]
11 Y = tsne.fit_transform(vectors)
12
13 # 绘制散布图，观察单词相似度
14 plt.figure(figsize=(12, 8))
15 plt.axis('off')
16 plt.scatter(Y[:, 0], Y[:, 1])
17 for label, x, y in zip(words, Y[:, 0], Y[:, 1]):
18     plt.annotate(label, xy=(x, y), xytext=(0, 0), textcoords="offset points")
```

执行结果：每次的执行结果均不相同，可以看到相似词都集中在局部区域。

## 11-5 中文处理

前面介绍的都是英文语料，中文是否也可以比照办理呢？答案是肯定的，NLP 所有做法都有考虑非英语系的支持。Jieba 套件提供中文分词的功能，而 spaCy 套件则有支持中文语料的模型，现在我们就来介绍这两个套件的用法。

Jieba 的主要功能包括：

(1) 分词 (Tokenization)。

(2) 关键词提取 (Keyword Extraction)。

(3) 词性标注 (POS)。

Jieba 安装指令如下：

(1) pip install jieba。

(2) 默认为简体字典，若要使用繁体须自 https://github.com/APCLab/jieba-tw/tree/master/jieba 下载繁体字典，直接覆盖安装目录的文件 dict.txt，也可以于程序中使用 set_dictionary() 设定繁体字典。

**范例5. 以Jieba套件进行中文分词。**

**下列程序代码请参考【11_08_ 中文 _NLP.ipynb】。**

(1) 简体字分词：包含三种模式。

全模式 (Full Mode)：显示所有可能的词组。

精确模式：只显示最有可能的词组，此为默认模式。

搜索引擎模式：使用隐马尔可夫模型 (HMM)。

```python
1  # 载入相关套件
2  import numpy as np
3  import jieba
4
5  # 分词
6  text = "小明硕士毕业于中国科学院计算所，后在日本京都大学深造"
7  # cut_all=True：全模式
8  seg_list = jieba.cut(text, cut_all=True)
9  print("全模式: " + "/ ".join(seg_list))
10
11 # cut_all=False：精确模式
12 seg_list = jieba.cut(text, cut_all=False)
13 print("精确模式: " + "/ ".join(seg_list))
14
15 # cut_for_search：搜索引擎模式
16 seg_list = jieba.cut_for_search(text)
17 print('搜索引擎模式: ', ', '.join(seg_list))
```

执行结果。

全模式：小/ 明/ 硕士/ 毕业/ 于/ 中国/ 中国科学院/ 科学/ 科学院/ 学院/ 计算/ 计算所/ ,/ 后/ 在/ 日本/ 日本京都大学/ 京都/ 京都大学/ 大学/ 深造
精确模式：小明/ 硕士/ 毕业/ 于/ 中国科学院/ 计算所/ ,/ 后/ 在/ 日本京都大学/ 深造
搜索引擎模式：小明, 硕士, 毕业, 于, 中国, 科学, 学院, 科学院, 中国科学院, 计算, 计算所, ,, 后, 在, 日本, 京都, 大学, 日本京都大学, 深造

(2) 繁体字分词：先调用 set_dictionary()，设定繁体字典 dict.txt。

```python
1  # 设定繁体字典
2  jieba.set_dictionary('./jieba/dict.txt')
3
4  # 分词
5  text = "新竹的交通大學在新竹的大學路上"
6
7  # cut_all=True：全模式
8  seg_list = jieba.cut(text, cut_all=True)
9  print("全模式: " + "/ ".join(seg_list))
10
11 # cut_all=False：精确模式
12 seg_list = jieba.cut(text, cut_all=False)
13 print("精确模式: " + "/ ".join(seg_list))
14
15 # cut_for_search：搜索引擎模式
16 seg_list = jieba.cut_for_search(text)
17 print('搜索引擎模式: ', ', '.join(seg_list))
```

执行结果。

全模式：新竹/ 的/ 交通/ 交通大/ 大學/ 在/ 新竹/ 的/ 大學/ 大學路/ 學路/ 路上
精确模式：新竹/ 的/ 交通/ 大學/ 在/ 新竹/ 的/ 大學路/ 上
搜索引擎模式： 新竹, 的, 交通, 大學, 在, 新竹, 的, 大學, 學路, 大學路, 上

(3) 分词后，显示词汇的位置。

```python
1  text = "新竹的交通大学在新竹的大学路上"
2  result = jieba.tokenize(text)
3  print("单字\t开始位置\t结束位置")
4  for tk in result:
5      print(f"{tk[0]}\t{tk[1]:-2d}\t{tk[2]:-2d}")
```

执行结果。

| 单字 | 开始位置 | 结束位置 |
|---|---|---|
| 新竹 | 0 | 2 |
| 的 | 2 | 3 |
| 交通 | 3 | 5 |
| 大学 | 5 | 7 |
| 在 | 7 | 8 |
| 新竹 | 8 | 10 |
| 的 | 10 | 11 |
| 大学路 | 11 | 14 |
| 上 | 14 | 15 |

(4) 加词：假如词汇不在默认的字典中，可使用 add_word() 将词汇加入字典中，各行各业的专门术语都可以利用此方式加入，不必直接修改 dict.txt。

```
1   # 测试语句
2   text = "张惠妹在演唱会演唱三天三夜"
3
4   # 加词前的分词
5   seg_list = jieba.cut(text, cut_all=False)
6   print("加词前的分词: " + "/ ".join(seg_list))
7
8   # 加词
9   jieba.add_word('三天三夜')
10
11  seg_list = jieba.cut(text, cut_all=False)
12  print("加词后的分词: " + "/ ".join(seg_list))
```

执行结果：原本"三天三夜"分为两个词"三天三""夜"，加词后，分词就正确了。

加词前的分词: 张/ 惠妹/ 在/ 演唱/ 会/ 演唱/ 三天三/ 夜
加词后的分词: 张/ 惠妹/ 在/ 演唱/ 会/ 演唱/ 三天三夜

(5) 取得词性 (POS) 标注：调用 posseg.cut 函数，可使用 POSTokenizer 自定义分词器。

```
1   # 设定简体字典
2   jieba.set_dictionary('jieba/dict_org.txt')
3
4   # 测试语句
5   text = "张惠妹在演唱会演唱三天三夜"
6
7   # 词性(POS) 标注
8   words = jieba.posseg.cut(text)
9   for word, flag in words:
10      print(f'{word} {flag}')
```

执行结果。

张惠妹 N
在 P
演唱会 N
演唱 Vt
三天三夜 x

词性代码表可参阅《汇整中文与英文的词性标注代号》[16] 一文，内文有完整的说明与范例。

# 11-6　spaCy 套件

spaCy 套件支持超过 64 种语言，不只有 Wod2Vec 词向量模型，也支持 BERT 预先训练的模型，主要的功能见表 11.3。

表 11.3　spaCy 框架主要功能

| 项次 | 功能 | 说明 |
| --- | --- | --- |
| 1 | 分词 (Tokenization) | 词汇切割 |
| 2 | 词性标签 (POS Tagging) | 分析语句中每个单词的词性 |
| 3 | 文法解析 (Dependency Parsing) | 依文法解析单词的相依性 |
| 4 | 词性还原 (Lemmatization) | 还原成词汇的原形 |

续表

| 项次 | 功能 | 说明 |
| --- | --- | --- |
| 5 | 语句切割<br>(Sentence Boundary Detection) | 将文章段落切割成多个语句 |
| 6 | 命名实体识别<br>(Named Entity Recognition) | 识别语句中的命名实体，例如人名、地点、机构名称等 |
| 7 | 实体链接<br>(Entity Linking) | 根据知识图谱链接实体 |
| 8 | 相似性比较 (Similarity) | 单词或语句的相似性比较 |
| 9 | 文本分类<br>(Text Classification) | 对文章或语句进行分类 |
| 10 | 语义标注<br>(Rule-based Matching) | 类似 Regular expression，依据语义找出词汇的顺序 |
| 11 | 模型训练 (Training) | |
| 12 | 模型存盘 (Serialization) | |

(1) 可利用 spaCy 网页[17]的菜单产生安装指令，产生的指令如下：
- pip install spacy。
- 支持 GPU，须配合 CUDA 版本：pip install -U spacy[cuda111]。
  - cuda111：为 cuda v11.1 版。

(2) 下载词向量模型，spaCy 称为 pipeline，指令如下：
- 英文：python -m spacy download en_core_web_sm。
- 中文：python -m spacy download zh_core_web_sm。
- 其他语系可参考『spaCy Quickstart 网页』[18]。

(3) 词向量模型分成大型 (lg)、中型 (md)、小型 (sm)。

(4) 中文分词有三个选项，可在组态文件 (config.cfg) 选择：
char：默认选项。
jieba：使用 Jieba 套件分词。
pkuseg：支持多领域分词，可参阅 pkuseg GitHub[19]，依照文件说明，pkuseg 的各项功能 (Precision、Recall、F1) 比 jieba 要好。

spaCy 相关功能的展示，可参考 spaCy 官网 *spaCy 101: Everything you need to know*[20] 的说明，以下就依照该文测试相关的功能。

**范例6. spaCy相关功能测试。**

**下列程序代码请参考【11_09_spaCy_test.ipynb】。**

(1) 加载相关套件。

```
1  import spacy
```

(2) 加载小型词向量模型。

```
1  nlp = spacy.load("en_core_web_sm")
```

(3) 分词及取得词性标签 (POS Tagging)。
token 的属性可参阅 spaCy Token[21]。
词性标签表则请参考 glossary.py GitHub[22]。

```
1  # 分词及取得词性标签(POS Tagging)
2  doc = nlp("Apple is looking at buying U.K. startup for $1 billion")
3  for token in doc:
4      print(token.text, token.pos_, token.dep_)
```

执行结果。

```
Apple PROPN nsubj
is AUX aux
looking VERB ROOT
at ADP prep
buying VERB pcomp
U.K. PROPN dobj
startup NOUN advcl
for ADP prep
$ SYM quantmod
1 NUM compound
billion NUM pobj
```

(4) 取得词性标签 (POS Tagging) 详细信息。

```
1  doc = nlp("Apple is looking at buying U.K. startup for $1 billion")
2  for token in doc:
3      print(token.text, token.pos_, token.dep_)
```

执行结果。

```
Apple Apple PROPN NNP nsubj Xxxxx True False
is be AUX VBZ aux xx True True
looking look VERB VBG ROOT xxxx True False
at at ADP IN prep xx True True
buying buy VERB VBG pcomp xxxx True False
U.K. U.K. PROPN NNP dobj X.X. False False
startup startup NOUN NN advcl xxxx True False
for for ADP IN prep xxx True True
$ $ SYM $ quantmod $ False False
1 1 NUM CD compound d False False
billion billion NUM CD pobj xxxx True False
```

(5) 以 displaCy Visualizer 显示语义分析图，displacy.serve 的参数请参阅 displaCy visualizer 的说明文件[23]。

```
1  from spacy import displacy
2
3  displacy.render(doc, style="dep")
```

执行结果如下，如果使用在 py 文件内，可执行 display. serve，它会是一个网站，须使用网页浏览 http://127.0.0.1:5000。箭头表示依存关系，例如 looking 的主词是 Apple，buying 的受词是 UK。

(6) 以 displaCy visualizer 标示命名实体 (Named Entity)。

```
1  text = "When Sebastian Thrun started working on self-driving cars " + \
2          "at Google in 2007, few people outside of the company took him seriously."
3
4  doc = nlp(text)
5  # style="ent" : 实体
6  displacy.render(doc, style="ent")
```

执行结果：Sebastian Thrun 是一个人名 (Person)，2007 是日期 (Date)。

(7) 繁体中文分词。

```
1  nlp = spacy.load("zh_core_web_sm")
2  doc = nlp("清華大學位於北京")
3  for token in doc:
4      print(token.text, token.pos_, token.dep_)
```

执行结果：spaCy 中文不分简 / 繁体，相关功能以简体为主，故大学被切割成两个词，结果不太正确，建议实际执行时可以先用简体分词后，再转回繁体。

```
清華 NOUN compound:nn
大 ADJ amod
學位 NOUN nsubj
於 ADP case
北京 PROPN ROOT
```

(8) 简体中文分词。

```
1  nlp = spacy.load("zh_core_web_sm")
2  doc = nlp("清华大学位于北京")
3  for token in doc:
4      print(token.text, token.pos_, token.dep_)
```

执行结果。

```
清华 PROPN compound:nn
大学 NOUN nsubj
位于 VERB ROOT
北京 PROPN dobj
```

(9) 显示中文语义分析图。

```
1  from spacy import displacy
2
3  displacy.render(doc, style="dep")
```

执行结果。

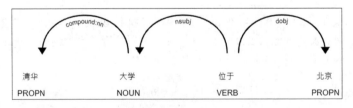

(10) 分词，并判断是否在字典中 (Out of Vocabulary, OOV)。

```
1  nlp = spacy.load("en_core_web_md")
2  tokens = nlp("dog cat banana afskfsd")
3
4  for token in tokens:
5      print(token.text, token.has_vector, token.vector_norm, token.is_oov)
```

执行结果：afskfsd 不在字典中，注意，必须使用中型 (md) 以上的模型，小型 (sm) 会出现错误。

```
dog True 7.0336733 False
cat True 6.6808186 False
banana True 6.700014 False
afskfsd False 0.0 True
```

(11) 相似度比较。

```
1  nlp = spacy.load("en_core_web_md")
2
3  # 测试两语句
4  doc1 = nlp("I like salty fries and hamburgers.")
5  doc2 = nlp("Fast food tastes very good.")
6
7  # 两语句的相似度比较
8  print(doc1, "<->", doc2, doc1.similarity(doc2))
9
10 # 关键字的相似度比较
11 french_fries = doc1[2:4]
12 burgers = doc1[5]
13 print(french_fries, "<->", burgers, french_fries.similarity(burgers))
```

执行结果。

```
I like salty fries and hamburgers. <-> Fast food tastes very good. 0.7799485853415737
salty fries <-> hamburgers 0.7304624
```

# 第 12 章
# 自然语言处理的算法

第 11 章我们认识了自然语言处理的前置处理和词向量应用，接下来将探讨自然语言处理相关的深度学习算法。

自然语言的推断 (Inference) 不仅需要考虑语文上下文的关联，还要考虑人类特殊的能力——记忆力。譬如，我们从小就学习历史，讲到治水，第一个可能会想到治水的人就是大禹，这就是记忆力的影响，就算时间再久远，都会深印在脑中。因此，NLP 相关的深度学习算法要能够提升预测准确率，模型就必须额外添加上下文关联与记忆力的功能。

我们会依照循环神经网络发展的轨迹依次说明，从简单的 RNN、LSTM、注意力机制 (Attention) 到 Transformer 等算法，包括目前最强大的 BERT 模型。

## 12-1 循环神经网络

一般神经网络以回归为基础，以特征 (x) 预测目标 (y)，但 NLP 的特征并不互相独立，它们有上下文的关联，因此，循环神经网络 (Recurrent Neural Network, RNN) 就像自回归 (Auto-regression) 模型一样，会考虑同一层前面的神经元影响。可以用数学式表示两者的差异 ( 如图 12.1 所示 )。

回归：$y=Wx+b$

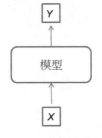

图 12.1　回归的示意图

RNN：
$h_t = W * h_{t-1} + U * x_t + b$
$y = V * h_t$

其中，$W$、$U$、$V$ 都是权重，$h$ 为隐藏层的输出。

可以看到时间点 $t$ 的 $h$ 会受到前一时间点的 $h_{t-1}$ 影响，如图 12.2 所示。

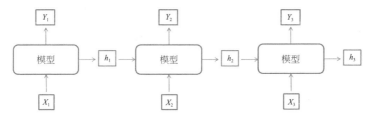

图 12.2　RNN 的示意图

由于每一个时间点的模型都类似,因此又可简化为如图 12.3 所示的循环网络,这不仅有助于理解,在开发时也可简化为递归结构。

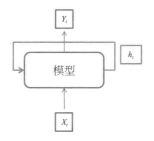

图 12.3　RNN 循环

归纳上述说明,一般神经网络假设同一层的神经元是互相独立的,而 RNN 则将同一层的前一个神经元也视为输入。

PyTorch 直接支持嵌入层 (Embedding layer)、RNN 神经层以及相关的文字处理的辅助函数,以上均含在 torchtext 模块中,下面让我们通过实操逐步了解各函数用法。

torchtext 需额外安装,指令如下:

pip install torchtext

**范例 1. 简单的 RNN 模型测试。**

下列程序代码请参考【12_01_RNN_test.ipynb】。

(1) 加载相关框架:以下先对嵌入层、RNN 神经层进行简单测试,以了解它们的输入 / 输出规格及参数的设定。

```
1  import torch
2  import torch.nn as nn
3  import torch.nn.functional as F
4  import torch.optim as optim
5  import torchtext
```

(2) 嵌入层:在自然语言处理时,通常会在 RNN 神经层前先插入一个嵌入层,将输入转换成二维矩阵,即每个单词以一维向量表示,第二维为语句长度。之前使用 BOW 时,在词汇表 (Vocabulary) 很大的情况下,转换后会造成稀疏矩阵,即矩阵中的元素大部分为 0,不仅浪费内存空间,还会影响计算的效能,改用嵌入层可以将输入转换为稠密的向量空间 (dense vector)。嵌入层的参数如下:

num_embeddings:词汇表的单词个数。

embedding_dim:输出向量的元素个数。

**padding_idx**：指定索引值的权重不参与梯度下降，为固定值，通常是指不包含在词汇表中的单词，一般是插入词汇表最前面，即索引值为 0，若不填此参数，表示输入语句不会有词汇表外的单词。

**freeze**：是否冻结嵌入层，若先利用 Word2Vec/GloVe 等预先训练模型转换为词向量，嵌入层就不用参与训练，此参数即可设为 True，或之后下指令 embeds.weight.requires_grad = False。

其他参数请参阅 PyTorch 官网嵌入层说明[1]。

(3) 嵌入层测试：输入两笔数据，内含值 (0~5) 为在词汇表中的索引值，故 nn.Embedding 第一个参数为 6，表示词汇表含 6 个单词。

```
1  x = torch.LongTensor([[0,1,2], [3,4,5]])
2  embeds = nn.Embedding(6, 5)
3  print(embeds(x))
```

执行结果：nn.Embedding 第二个参数为 5，表示每个单词以 5 个实数表示。

```
tensor([[[ 0.6531, -1.9722, -0.6393,  0.9719, -0.5552],
         [ 1.7436,  0.6179,  0.5530, -0.0325,  0.9319],
         [ 0.0876, -0.2328,  2.6156, -0.7486, -1.1053]],

        [[-0.6745, -0.8500, -0.7149, -1.9410, -0.1172],
         [ 0.4337, -2.5339,  0.5160,  0.1252, -1.2865],
         [-0.7444,  0.9612,  0.7005, -0.3367, -0.9618]]],
       grad_fn=<EmbeddingBackward0>)
```

(4) 显示嵌入层的起始权重：与其他的神经层一样，起始权重都是取随机数，嵌入层起始权重默认为标准常态分配 N(0, 1)。

```
1  embeds.weight
```

执行结果：6×5 矩阵。

```
tensor([[-0.2854,  0.4994, -1.2292,  0.0285, -1.4484],
        [-0.4937,  0.7987, -0.5471,  1.5526,  1.3826],
        [-0.1778, -1.9945,  0.3916,  0.7550,  0.2322],
        [ 0.2465,  0.7877,  0.3312,  0.5031, -0.3601],
        [ 0.9708,  0.3138, -0.4496,  1.8550,  0.6466],
        [ 2.1568,  0.5826, -1.4558,  0.1674,  1.6133]])
```

输出是依照索引值查询上表而来的。若模型经过训练，权重会不断地更新，输出也会随之改变。

(5) 输入改为 1~6：nn.Embedding 第一个参数须改为 7，即词汇表应含 0~6，共 7 个单词，因为输入最大索引值为 6。

```
1  x = torch.LongTensor([[1,2,3], [4,5,6]])
2  embeds = nn.Embedding(7, 5)
3  print(embeds(x))
```

(6) 以英文单词输入：需先利用字典 (word_to_ix)，将单词转为索引值，再输入至嵌入层。

```
1  # 测试资料
2  word_to_ix = {"hello": 0, "world": 1}
3  # 词汇表(vocabulary)含2个单词，转换为5维的向量
4  embeds = nn.Embedding(2, 5)
5  # 测试 hello
6  lookup_tensor = torch.LongTensor([word_to_ix["hello"]])
7  hello_embed = embeds(lookup_tensor)
8  print(hello_embed)
```

执行结果：
[[0.2347, 0.0490, 0.1800, 0.6384, 0.4259]]

(7) RNN 层测试：RNN 神经层通常会接在嵌入层后面，也可以单独使用，参数如下。

input_size：特征个数。

hidden_size：隐藏层的神经元个数 $H_{out}$。

num_layers：RNN 神经层的层数，层数大于 1，称为堆栈 (Stacked) RNN，比 TensorFlow 方便很多，TensorFlow 需层层设定，每一层的相关参数必须正确设定才能运作。

dropout：若大于 0，则 RNN 后会加上 dropout 层，此参数为它抛弃神经元的比例。

bidirectional：RNN 预设只考虑上文，若要同时考虑上下文，可设为 True。

RNN 输入的维度可以是二维 ($L$、$H_{in}$) 或三维 ($L$、$N$、$H_{in}$)，后者含批量。

$L$ 是序列长度 (sequence length)，即字符串长度。

$H_{in}$ 即特征个数 (input_size)。

$N$ 是批量 (batch size)。

batch_first：若是 True，输入维度须为 ($N$、$L$、$H_{in}$)，若是二维 ($L$、$H_{in}$)，则无影响，参数默认值为 False。

RNN 输出有两个，若输入维度是二维，以下的 $N$ 也会去掉：

输出 (Output)：维度为 ($L$、$N$、$H_{out}$)，包含最后一层的输出特征，**注意，采双向 (bidirectional=True) 时，输出个数会有 2 倍，接在后面的神经层输入参数设定要相符，才不会出错。**

隐藏层状态 (hidden state)：维度为 ($num\_layers$、$N$、$H_{out}$)，包含每一层最后的隐藏层状态，**注意，采双向 (bidirectional=True) 时，维度为 ($num\_layers \times 2$、$N$、$H_{out}$)，接在后面的神经层输入参数设定要相符，才不会出错。**

- RNN 两种输出的区别如图 12.4 所示，例如输入为"TAKE"有 4 个字母，经 RNN 处理后，输出 (Output) 为每个字母预测的结果，隐藏层状态 (hidden state) 为每一神经层最后的隐藏层状态。如果只要最后的结果，通常会取 hn[-1]。

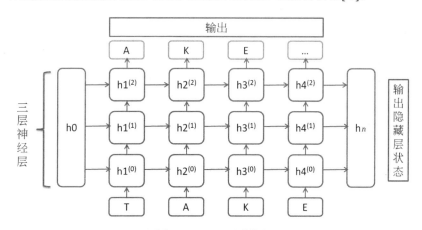

图 12.4　RNN 两种输出

- 其他参数请参阅 PyTorch 官网 RNN 层说明 [2]。

(8) 测试：输入为二维 ($L$、$H_{in}$)。

```
1  # 测试资料
2  input = torch.randn(5, 10)
3  # 建立 RNN 物件
4  rnn = nn.RNN(10, 20, 2)
5  # RNN 处理
6  output, hn = rnn(input)
7  # 显示输出及隐藏层的维度
8  print(output.shape, hn.shape)
```

执行结果：输出维度为 [5, 20] ($L$、$H_{out}$)，隐藏层维度为 [2, 20] ($num\_layers$、$H_{out}$)。

(9) 测试：输入为三维 ($L$、$N$、$H_{in}$)。

```
1  # 测试资料
2  input = torch.randn(5, 4, 10)
3  # 建立 RNN 物件
4  rnn = nn.RNN(10, 20, 2)
5  # RNN 处理
6  output, hn = rnn(input)
7  # 显示输出及隐藏层的维度
8  print(output.shape, hn.shape)
```

执行结果：输出维度为 [5, 4, 20] ($L$、$N$、$H_{out}$)，隐藏层维度为 [2, 4, 20] ($num\_layers$、$N$、$H_{out}$)。

(10) RNN 的输入可以有初始的隐藏层状态 ($h_0$)，$h_0$ 最后一维需等于 $H_{out}$。

```
1   # 测试资料
2   input = torch.randn(5, 3, 10)
3   # 建立 RNN 物件
4   rnn = nn.RNN(10, 20, 2)
5   # 隐藏层的输入
6   h0 = torch.randn(2, 3, 20)
7   # RNN 处理
8   output, hn = rnn(input, h0)
9   # 显示输出及隐藏层的维度
10  print(output.shape, hn.shape)
```

执行结果：输出维度为 [5, 3, 20]，隐藏层维度为 [2, 3, 20]。

接着介绍 PyTorch 前置处理功能，除了 NLTK、spaCy 等套件外，PyTorch 也提供简单的前置处理功能。

(11) 分词。

```
1  from torchtext.data.utils import get_tokenizer
2  
3  tokenizer = get_tokenizer('basic_english')
4  
5  text = 'Could have done better.'
6  tokenizer(text)
```

执行结果：['could', 'have', 'done', 'better', '.']。

(12) 词汇表处理：PyTorch 词汇表对象，可提供词汇表建立、单词与索引值互转等功能。

```
1   from torchtext.vocab import vocab
2   from collections import Counter, OrderedDict
3   
4   # BOW 统计
5   counter = Counter(tokenizer(text))
6   # 依出现次数降幂排列
7   sorted_by_freq_tuples = sorted(counter.items(),
8                                  key=lambda x: x[1], reverse=True)
9   # 建立词汇字典
10  ordered_dict = OrderedDict(sorted_by_freq_tuples)
11  
12  # 建立词汇表物件，并加一个未知单字(unknown)的索引值
13  vocab_object = torchtext.vocab.vocab(ordered_dict, specials=["<unk>"])
14  # 设定词汇表预设值为未知单字(unknown)的索引值
15  vocab_object.set_default_index(vocab_object["<unk>"])
16  
17  # 测试
18  vocab_object['done']
```

(13) 取得词汇表的所有单词。

```
1  vocab_object.get_itos()
```

执行结果：['<unk>', 'could', 'have', 'done', 'better', '.']。

(14) 取得词汇表的单词个数。

```
1  vocab_object.__len__()
```

执行结果：6。

(15) 数据转换函数：去除标点符号、建立词汇表对象。

```
1   import string
2
3   def create_vocabulary(text_list):
4       # 取得标点符号
5       stopwords = list(string.punctuation)
6
7       # 去除标点符号
8       clean_text_list = []
9       clean_tokens_list = []
10      for text in text_list:
11          tokens = tokenizer(text)
12          clean_tokens = []
13          for w in tokens:
14              if w not in stopwords:
15                  clean_tokens.append(w)
16          clean_tokens_list += clean_tokens
17          clean_text_list.append(' '.join(clean_tokens))
18
19      # 建立词汇表对象
20      counter = Counter(clean_tokens_list)
21      sorted_by_freq_tuples = sorted(counter.items(),
22                                     key=lambda x: x[1], reverse=True)
23      ordered_dict = OrderedDict(sorted_by_freq_tuples)
24      vocab_object = torchtext.vocab.vocab(ordered_dict, specials=["<unk>"])
25      vocab_object.set_default_index(vocab_object["<unk>"])
26
27      # 将输入字串转为索引值：自词汇表物件查询索引值
28      clean_index_list = []
29      for clean_tokens_list in clean_text_list:
30          clean_index_list.append(
31              vocab_object.lookup_indices(clean_tokens_list.split(' ')))
32
33      # 输出 词汇表对象、去除标点符号的字串阵列、字串阵列的索引值
34      return vocab_object, clean_text_list, clean_index_list
```

整合以上功能，实操一个简单的案例，说明相关的处理步骤，如图 12.5 所示。

图 12.5　相关步骤

(1) 建立词汇表：整理输入语句，截长补短，使语句长度一致。

```
1   maxlen = 4        # 语句最大字数
2   # 测试数据
3   docs = ['Well done!',
4           'Good work',
5           'Great effort',
6           'nice work',
7           'Excellent!',
8           'Weak',
9           'Poor effort!',
10          'not good',
```

```
11            'poor work',
12            'Could have done better']
13
14 vocab_object, clean_text_list, clean_index_list = create_vocabulary(docs)
15
16 # 若字串过长，删除多余单词
17 clean_index_list = torchtext.functional.truncate(clean_index_list, maxlen)
18
19 # 若字串长度不足，后面补 0
20 while len(clean_index_list[0]) < maxlen:
21     clean_index_list[0] += [0]
22 torchtext.functional.to_tensor(clean_index_list, 0) # 0:不足补0
```

执行结果。

```
tensor([[ 6,  2,  0,  0],
        [ 3,  1,  0,  0],
        [ 7,  4,  0,  0],
        [ 8,  1,  0,  0],
        [ 9,  0,  0,  0],
        [10,  0,  0,  0],
        [ 5,  4,  0,  0],
        [11,  3,  0,  0],
        [ 5,  1,  0,  0],
        [12, 13,  2, 14]])
```

(2) 嵌入层转换。

```
1 embeds = nn.Embedding(vocab_object.__len__(), 5)
2 X = torchtext.functional.to_tensor(clean_index_list, 0) # 0:不足补0
3 embed_output = embeds(X)
4 print(embed_output.shape)
```

执行结果：[10, 4, 5]。

(3) 再接完全连接层 (Linear)，进行分类预测，正面情绪为 1，负面情绪为 0。

嵌入层输出为二维或三维，而完全连接层输入为一维，故须使用 reshape 转换成一维。

```
 1 class RecurrentNet(nn.Module):
 2     def __init__(self, vocab_size, embed_dim, num_class):
 3         super().__init__()
 4         self.embedding = nn.Embedding(vocab_size, embed_dim)
 5         self.fc = nn.Linear(embed_dim * maxlen, num_class) # 要乘以 maxlen
 6         self.embed_dim = embed_dim
 7         self.init_weights()
 8
 9     def init_weights(self):
10         initrange = 0.5
11         self.embedding.weight.data.uniform_(-initrange, initrange)
12         self.fc.weight.data.uniform_(-initrange, initrange)
13         self.fc.bias.data.zero_()
14
15     def forward(self, text):
16         embedded = self.embedding(text)
17         out = embedded.reshape(embedded.size(0), -1) # 转换成1维
18         return self.fc(out)
19
20 model = RecurrentNet(vocab_object.__len__(), 10, 1)
```

以上模型也可使用 nn.EmbeddingBag：EmbeddingBag，会将词向量平均，也可以设定加总 (mode="sum") 或最大值 (mode="max")，会将二维转换成一维。

```
1 class RecurrentNet(nn.Module):
2     def __init__(self, vocab_size, embed_dim, num_class):
3         super().__init__()
4         self.embedding = nn.EmbeddingBag(vocab_size, embed_dim)
5         self.fc = nn.Linear(embed_dim, num_class)
6         self.embed_dim = embed_dim
7         self.init_weights()
8
```

```
 9    def init_weights(self):
10        initrange = 0.5
11        self.embedding.weight.data.uniform_(-initrange, initrange)
12        self.fc.weight.data.uniform_(-initrange, initrange)
13        self.fc.bias.data.zero_()
14
15    def forward(self, text):
16        embedded = self.embedding(text)
17        return self.fc(embedded)
18
19 model = RecurrentNet(vocab_object.__len__(), 10, 1)
```

(4) 模型训练。

```
 1 # 定义 10 个语句的正面(1)或负面(0)的情绪
 2 y = torch.FloatTensor([1,1,1,1,1,0,0,0,0,0])
 3 X = torchtext.functional.to_tensor(clean_index_list, 0) # 0:不足补0
 4
 5 # 指定优化器、损失函数
 6 criterion = torch.nn.MSELoss()
 7 optimizer = torch.optim.Adam(model.parameters())
 8
 9 # 模型训练
10 for epoch in range(1000):
11     outputs = model.forward(X) #forward pass
12     optimizer.zero_grad()
13     loss = criterion(outputs.reshape(-1), y)
14     loss.backward()
15     optimizer.step()
16     if epoch % 100 == 0:
17         #print(outputs.shape)
18         print(f"Epoch: {epoch}, loss: {loss.item():1.5f}")
```

执行结果：经过训练后，观察损失逐渐降低。

```
Epoch: 0, loss: 0.48356
Epoch: 100, loss: 0.17130
Epoch: 200, loss: 0.07601
Epoch: 300, loss: 0.03124
Epoch: 400, loss: 0.00997
Epoch: 500, loss: 0.00296
Epoch: 600, loss: 0.00108
Epoch: 700, loss: 0.00049
Epoch: 800, loss: 0.00025
Epoch: 900, loss: 0.00013
```

(5) 训练数据预测。

```
1 model.eval()
2 model(X)
```

执行结果：概率在 0.5 以上为正面，反之为负面，前 5 句为正面，后 5 句为负面，结果与真实答案相符。

```
tensor([[ 1.0000e+00],
        [ 9.8935e-01],
        [ 1.0092e+00],
        [ 9.9873e-01],
        [ 9.9942e-01],
        [-2.1182e-03],
        [-1.3330e-02],
        [ 5.5699e-03],
        [ 1.5421e-02],
        [-1.7509e-06]], grad_fn=<AddmmBackward0>)
```

(6) 测试数据预测。

```
1   # 测试数据
2   test_docs = ['great effort', 'well done',
3               'poor effort']
4   
5   # 转成数值
6   clean_index_list = []
7   for text in test_docs:
8       clean_index_list.append(vocab_object.lookup_indices(text.split(' ')))
9       while len(clean_index_list[0]) < maxlen:
10          clean_index_list[0] += [0]
11  
12  clean_index_list = torchtext.functional.truncate(clean_index_list, maxlen)
13  X = torchtext.functional.to_tensor(clean_index_list, 0) # 0:不足补0
14  model(X)
```

执行结果：判断正确。

```
tensor([[ 1.0004e+00],
        [ 1.0005e+00],
        [-8.5166e-04]], grad_fn=<AddmmBackward0>)
```

以上是使用 PyTorch 内建的嵌入层转换，如果使用预先训练好的词向量转换是否更方便且更准确呢？毕竟词向量的训练样本较齐全，且输出更高的维度。PyTorch 支持 GloVe、FastText 及 CharNGram 三种词向量，各有多种模型及维度，整理如下，也可参阅 PyTorch 源代码[3]：

- charngram.100d
- fasttext.en.300d
- fasttext.simple.300d
- glove.42B.300d
- glove.840B.300d
- glove.twitter.27B.25d
- glove.twitter.27B.50d
- glove.twitter.27B.100d
- glove.twitter.27B.200d
- glove.6B.50d
- glove.6B.100d
- glove.6B.200d
- glove.6B.300d

以下我们就以 GloVe 为例进行测试。

(1) 读取 GloVe 50 维的词向量，转换为 GloVe 50 维的词向量。

```
1   # https://pytorch.org/text/stable/vocab.html#glove
2   examples = ['great']
3   vec = torchtext.vocab.GloVe(name='6B', dim=50)
4   ret = vec.get_vecs_by_tokens(examples, lower_case_backup=True)
5   ret
```

执行结果。

```
tensor([[-0.0266,  1.3357, -1.0280, -0.3729,  0.5201, -0.1270, -0.3543,  0.3782,
         -0.2972,  0.0939, -0.0341,  0.9296, -0.1402, -0.6330,  0.0208, -0.2153,
          0.9692,  0.4765, -1.0039, -0.2401, -0.3632, -0.0048, -0.5148, -0.4626,
          1.2447, -1.8316, -1.5581, -0.3747,  0.5336,  0.2088,  3.2209,  0.6455,
          0.3744, -0.1766, -0.0242,  0.3379, -0.4190,  0.4008, -0.1145,  0.0512,
         -0.1521,  0.2986, -0.4405,  0.1109, -0.2463,  0.6625, -0.2695, -0.4966,
         -0.4162, -0.2549]])
```

(2) 显示词向量大小。

```
1  vec.vectors.size()
```

执行结果：(400000, 50) 表示此模型含 40 万个单词，每一单词以 50 维向量表示。

(3) 查询单词的词向量索引值。

```
1  vec.stoi['great']
```

(4) 建立模型：Embedding 不需要训练，直接设定嵌入层权重，详细说明可参阅 *How to use Pre-trained Word Embeddings in PyTorch*[4]。

```
 1  class RecurrentNet(nn.Module):
 2      def __init__(self, weights_matrix, num_embeddings, embedding_dim, num_class):
 3          super().__init__()
 4          self.embedding = nn.EmbeddingBag(num_embeddings, embedding_dim)
 5          # 设定嵌入层权重
 6          self.embedding.load_state_dict({'weight': weights_matrix})
 7          self.fc = nn.Linear(embedding_dim, num_class)
 8  
 9      def forward(self, text):
10          embedded = self.embedding(text)
11          return self.fc(embedded)
```

(5) 测试数据转换。

```
 1  docs = ['Well done!',
 2          'Good work',
 3          'Great effort',
 4          'nice work',
 5          'Excellent!',
 6          'Weak',
 7          'Poor effort!',
 8          'not good',
 9          'poor work',
10          'Could have done better']
11  
12  # 将词汇表转为词向量
13  clean_text_list = []
14  clean_tokens_list = []
15  for i, text in enumerate(docs):
16      tokens = tokenizer(text.lower())
17      clean_tokens = []
18      for w in tokens:
19          if w not in stopwords:
20              clean_tokens.append(w)
21      clean_tokens_list += clean_tokens
22      clean_text_list.append(clean_tokens)
23      tokens_vec = vec.get_vecs_by_tokens(clean_tokens)
24  vocab_list = list(set(clean_tokens_list))
25  weights_matrix = vec.get_vecs_by_tokens(vocab_list)
```

(6) 定义 10 个语句的正面 (1) 或负面 (0) 的情绪，并将 10 个语句转换为词汇表索引值。

```
1  # 定义 10 个语句的正面(1)或负面(0)的情绪
2  y = torch.FloatTensor([1,1,1,1,1,0,0,0,0,0])
3  X = torch.LongTensor(np.zeros((len(docs), maxlen)))
4  for i, item in enumerate(clean_text_list):
5      for j, token in enumerate(item):
6          if token in vocab_list:
7              X[i, j] = vocab_list.index(token)
8  X
```

执行结果。

```
tensor([[ 9,  6,  0,  0],
        [10,  4,  0,  0],
        [13, 12,  0,  0],
        [ 0,  4,  0,  0],
        [ 7,  0,  0,  0],
        [ 8,  0,  0,  0],
        [ 2, 12,  0,  0],
        [ 5, 10,  0,  0],
        [ 2,  4,  0,  0],
        [ 3, 11,  6,  1]])
```

(7) 模型训练：将词汇表的词向量 (weights_matrix) 输入模型，设定为嵌入层权重。

```python
1  # 建立模型物件
2  model = RecurrentNet(torch.FloatTensor(weights_matrix), len(vocab_list), 50, 1)
3
4  # 指定优化器、损失函数
5  criterion = torch.nn.MSELoss()
6  optimizer = torch.optim.Adam(model.parameters())
7
8  # 模型训练
9  for epoch in range(1000):
10     outputs = model.forward(X) #forward pass
11     optimizer.zero_grad()
12     loss = criterion(outputs.reshape(-1), y)
13     loss.backward()
14     optimizer.step()
15     if epoch % 100 == 0:
16         #print(outputs.shape)
17         print(f"Epoch: {epoch}, loss: {loss.item():1.5f}")
```

(8) 观察训练数据的预测结果。

```python
1  model.eval()
2  model(X)
```

执行结果：完全正确。

```
tensor([[ 1.0002e+00],
        [ 1.0006e+00],
        [ 1.0018e+00],
        [ 9.9592e-01],
        [ 1.0006e+00],
        [ 7.7283e-04],
        [-2.3356e-03],
        [-3.8705e-04],
        [ 3.5572e-03],
        [-4.7453e-05]], grad_fn=<AddmmBackward0>)
```

(9) 观察测试数据的预测结果。

```python
1  # 测试数据
2  test_docs = ['great effort', 'well done',
3              'poor effort']
4
5  # 转成数值
6  X = torch.LongTensor(np.zeros((len(test_docs), maxlen)))
7  clean_text_list = []
8  for i, text in enumerate(test_docs):
9      tokens = tokenizer(text.lower())
10     clean_tokens = []
11     for w in tokens:
12         if w not in stopwords:
13             clean_tokens.append(w)
14     clean_text_list.append(clean_tokens)
15
16 for i, item in enumerate(clean_text_list):
17     for j, token in enumerate(item):
18         if token in vocab_list:
19             X[i, j] = vocab_list.index(token)
20
21 # 预测
22 model.eval()
23 model(X)
```

执行结果：完全正确。

以上方式并不能预测训练数据以外的单词，为了改善此缺点，以下将 GloVe 所有词向量设定为嵌入层权重。

(1) 建立模型。

```python
1  class RecurrentNet2(nn.Module):
2      def __init__(self, vec, embedding_dim, num_class):
3          super().__init__()
4          # 将整个词向量设定为嵌入层权重，且嵌入层设为不训练
5          self.embedding = nn.EmbeddingBag.from_pretrained(vec, freeze=True)
6          self.fc = nn.Linear(embedding_dim, num_class)
7  
8      def forward(self, text):
9          embedded = self.embedding(text)
10         return self.fc(embedded)
11 
12 model = RecurrentNet2(vec.vectors, vec.dim, 1)
```

(2) 将训练数据转换为 GloVe 词向量索引值。

```python
1  # 测试数据
2  docs = ['Well done!',
3          'Good work',
4          'Great effort',
5          'nice work',
6          'Excellent!',
7          'Weak',
8          'Poor effort!',
9          'not good',
10         'poor work',
11         'Could have done better']
12 
13 # 转成数值
14 X = torch.LongTensor(np.zeros((len(docs), maxlen)))
15 
16 for i, text in enumerate(docs):
17     tokens = tokenizer(text.lower())
18     clean_tokens = []
19     j=0
20     for w in tokens:
21         if w not in stopwords:
22             # 转成词向量索引值
23             X[i, j] = vec.stoi[w]
24             j+=1
25 X
```

(3) 模型训练。

```python
1  # 指定优化器、损失函数
2  criterion = torch.nn.MSELoss()
3  optimizer = torch.optim.Adam(model.parameters())
4  
5  # 模型训练
6  for epoch in range(1000):
7      outputs = model.forward(X) #forward pass
8      optimizer.zero_grad()
9      loss = criterion(outputs.reshape(-1), y)
10     loss.backward()
11     optimizer.step()
12     if epoch % 100 == 0:
13         #print(outputs.shape)
14         print(f"Epoch: {epoch}, loss: {loss.item():1.5f}")
15 
16 model.eval()
17 model(X)
```

(4) 输入训练数据以外的单词测试。

```python
1  # 测试数据
2  test_docs = ['great job', 'well done',
3               'poor job']
4  
5  # 转成数值
6  X = torch.LongTensor(np.zeros((len(test_docs), maxlen)))
7  for i, text in enumerate(test_docs):
8      tokens = tokenizer(text.lower())
9      clean_tokens = []
10     j=0
11     for w in tokens:
12         if w not in stopwords:
13             X[i, j] = vec.stoi[w]
14             j+=1
15 X
```

执行结果：job 索引值不会为 0。

```
tensor([[353, 664,   0,   0],
        [143, 751,   0,   0],
        [992, 664,   0,   0]])
```

(5) 观察测试数据的预测结果。

```
1  model.eval()
2  model(X)
```

执行结果：与之前结果略为不同，若测试语句较长，效果就能显现。

```
tensor([[ 0.6623],
        [ 0.8730],
        [-0.4088]], grad_fn=<AddmmBackward0>)
```

## 12-2　PyTorch 内建文本数据集

PyTorch 提供非常多内建的文本数据集 (Text Datasets)，可用于文本分类 (Text Classification)、语言模型 (Language Modeling)、机器翻译 (Machine Translation)、序列标注 (Sequence Tagging)、问答集 (Question Answer) 及非监督式学习 (Unsupervised Learning)，详细说明可参阅 *PyTorch TorchText Datasets*[5]。要使用这些数据集，需要额外安装套件，指令如下：

pip install torchdata

TorchData 提供 DataPipe 数据格式，是一种迭代器 (Iterator) 的数据结构，方便逐批读取数据，可轻易和 DataLoader 整合，相关数据可参考 *PyTorch TorchData Tutorial*[6]。

**范例2. 实操新闻的文本分类(Text Classification)**。本范例程序来自 Text classification with the torchtext library[7]。

**下列程序代码请参考【12_02_Text_Classification.ipynb】**。

数据集：AG News 是 Antonio Gulli 新闻语料库的子集合，相关说明可参阅《AG 语料库介绍》[8]。共有 4 类新闻：国际 (World)、运动 (Sports)、商业 (Business) 及科技 (Sci/Tec)。

PyTorch 内建文本数据集的相关步骤如图 12.6 所示。

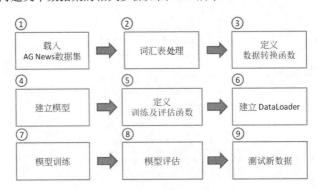

图 12.6　相关步骤

(1) 载入 AG News 数据集。

```
1  import torch
2  from torchtext.datasets import AG_NEWS
3
4  news = AG_NEWS(split='train')
5
6  type(news)
```

执行结果如下,MapperIterDataPipe 是 DataPipe 的衍生类别:
torch.utils.data.datapipes.iter.callable.MapperIterDataPipe

(2) 建立迭代器 (Iterator):逐批读取数据,以节省内存的使用。

```
1  train_iter = iter(AG_NEWS(split='train'))
```

(3) 测试:读取下一笔数据。

```
1  next(train_iter)
```

执行结果:

```
(3,
 "Wall St. Bears Claw Back Into the Black (Reuters) Reuters - Short-sellers, Wall Street's dwindling\\band of ultra-cynics, are seeing green again.")
```

(4) 词汇表处理。

```
1   from torchtext.data.utils import get_tokenizer
2   from torchtext.vocab import build_vocab_from_iterator
3
4   # 分词
5   tokenizer = get_tokenizer('basic_english')
6
7   # 建立 Generator 函数
8   def yield_tokens(data_iter):
9       for _, text in data_iter:
10          yield tokenizer(text)
11
12  # 由 train_iter 建立词汇字典
13  vocab = build_vocab_from_iterator(yield_tokens(train_iter), specials=["<unk>"])
14
15  # 设定预设的索引值
16  vocab.set_default_index(vocab["<unk>"])
```

(5) 判断 GPU 是否存在。

```
1  device = torch.device("cuda" if torch.cuda.is_available() else "cpu")
```

(6) 测试词汇字典。

```
1  # 测试词汇字典,取得单词的索引值
2  vocab(['here', 'is', 'an', 'example'])
```

执行结果:[475, 21, 30, 5297]。

(7) 参数设定。

```
1  EPOCHS = 10 # 训练周期数
2  LR = 5      # 学习率
3  BATCH_SIZE = 64 # 训练批量
4  # 取得标注个数
5  num_class = len(set([label for (label, text) in news]))
6  vocab_size = len(vocab)
7  emsize = 64
```

(8) 定义数据转换函数。

```
1  text_pipeline = lambda x: vocab(tokenizer(x))  # 分词,取得单词的索引值
2  label_pipeline = lambda x: int(x) - 1  # 换成索引值
```

(9) 测试数据转换。

```
1  print(text_pipeline('here is an example'))
2  label_pipeline('10')
```

执行结果：[475, 21, 30, 5297]，与词汇字典测试结果相同。

(10) 建立模型：仅使用嵌入层及完全连接层，嵌入层将文字转换成向量，完全连接层依据向量进行分类。

```
1   from torch import nn
2
3   class TextClassificationModel(nn.Module):
4       def __init__(self, vocab_size, embed_dim, num_class):
5           super().__init__()
6           self.embedding = nn.EmbeddingBag(vocab_size, embed_dim, sparse=True)
7           self.fc = nn.Linear(embed_dim, num_class)
8           self.init_weights()
9
10      def init_weights(self):
11          initrange = 0.5
12          self.embedding.weight.data.uniform_(-initrange, initrange)
13          self.fc.weight.data.uniform_(-initrange, initrange)
14          self.fc.bias.data.zero_()
15
16      def forward(self, text, offsets):
17          embedded = self.embedding(text, offsets)
18          return self.fc(embedded)
19
20  model = TextClassificationModel(vocab_size, emsize, num_class).to(device)
```

(11) 定义训练及评估函数。

```
1   import time
2
3   # 训练函数
4   def train(dataloader):
5       model.train()
6       total_acc, total_count = 0, 0
7       log_interval = 500
8       start_time = time.time()
9
10      for idx, (label, text, offsets) in enumerate(dataloader):
11          optimizer.zero_grad()
12          predicted_label = model(text, offsets)
13          loss = criterion(predicted_label, label)
14          loss.backward()
15          torch.nn.utils.clip_grad_norm_(model.parameters(), 0.1)
16          optimizer.step()
17          total_acc += (predicted_label.argmax(1) == label).sum().item()
18          total_count += label.size(0)
19          if idx % log_interval == 0 and idx > 0:
20              elapsed = time.time() - start_time
21              print('| epoch {:3d} | {:5d}/{:5d} batches '
22                    '| accuracy {:8.3f}'.format(epoch, idx, len(dataloader),
23                                                total_acc/total_count))
24              total_acc, total_count = 0, 0
25              start_time = time.time()
26
27  # 评估函数
28  def evaluate(dataloader):
29      model.eval()
30      total_acc, total_count = 0, 0
31
32      with torch.no_grad():
33          for idx, (label, text, offsets) in enumerate(dataloader):
34              predicted_label = model(text, offsets)
35              loss = criterion(predicted_label, label)
36              total_acc += (predicted_label.argmax(1) == label).sum().item()
37              total_count += label.size(0)
38      return total_acc/total_count
```

(12) 建立 DataLoader：collate_batch 是 DataLoader 的前置处理函数，将特征 (X)、目标 (Y) 整理成所需格式。

注意，使用 EmbeddingBag 时，每笔数据不需要补 0，变成等长，只要在调用

EmbeddingBag 时，第二个参数放入每笔数据的起始位置即可，程序代码中的 offsets 就是在记录这个信息。

```python
from torch.utils.data import DataLoader
from torch.utils.data.dataset import random_split
from torchtext.data.functional import to_map_style_dataset

# 批次处理
def collate_batch(batch):
    label_list, text_list, offsets = [], [], [0]
    for (_label, _text) in batch:
        label_list.append(label_pipeline(_label))
        processed_text = torch.tensor(text_pipeline(_text), dtype=torch.int64)
        text_list.append(processed_text)
        offsets.append(processed_text.size(0)) # 设定每笔资料的起始位置
    label_list = torch.tensor(label_list, dtype=torch.int64)
    offsets = torch.tensor(offsets[:-1]).cumsum(dim=0)  # 每笔资料的起始位置累加
    text_list = torch.cat(text_list)
    return label_list.to(device), text_list.to(device), offsets.to(device)

train_iter, test_iter = AG_NEWS()
# 转换为 DataSet
train_dataset = to_map_style_dataset(train_iter)
test_dataset = to_map_style_dataset(test_iter)
# 资料切割，95% 作为训练资料
num_train = int(len(train_dataset) * 0.95)
split_train_, split_valid_ = \
    random_split(train_dataset, [num_train, len(train_dataset) - num_train])

# 建立 DataLoader
train_dataloader = DataLoader(split_train_, batch_size=BATCH_SIZE,
                              shuffle=True, collate_fn=collate_batch)
valid_dataloader = DataLoader(split_valid_, batch_size=BATCH_SIZE,
                              shuffle=True, collate_fn=collate_batch)
test_dataloader = DataLoader(test_dataset, batch_size=BATCH_SIZE,
                             shuffle=True, collate_fn=collate_batch)
```

(13) 模型训练。

```python
criterion = torch.nn.CrossEntropyLoss()
optimizer = torch.optim.SGD(model.parameters(), lr=LR)
scheduler = torch.optim.lr_scheduler.StepLR(optimizer, 1.0, gamma=0.1)

total_accu = None
for epoch in range(1, EPOCHS + 1):
    epoch_start_time = time.time()
    train(train_dataloader)
    accu_val = evaluate(valid_dataloader)
    if total_accu is not None and total_accu > accu_val:
        scheduler.step()
    else:
        total_accu = accu_val
    print('-' * 59)
    print('| end of epoch {:3d} | time: {:5.2f}s | '
          'valid accuracy {:8.3f} '.format(epoch,
                                            time.time() - epoch_start_time,
                                            accu_val))
    print('-' * 59)
```

执行结果。

```
| epoch   8 |   500/ 1782 batches | accuracy    0.940
| epoch   8 |  1000/ 1782 batches | accuracy    0.938
| epoch   8 |  1500/ 1782 batches | accuracy    0.940
-----------------------------------------------------
| end of epoch   8 | time: 11.22s | valid accuracy    0.916
-----------------------------------------------------
| epoch   9 |   500/ 1782 batches | accuracy    0.938
| epoch   9 |  1000/ 1782 batches | accuracy    0.941
| epoch   9 |  1500/ 1782 batches | accuracy    0.938
-----------------------------------------------------
| end of epoch   9 | time: 11.85s | valid accuracy    0.916
-----------------------------------------------------
| epoch  10 |   500/ 1782 batches | accuracy    0.939
| epoch  10 |  1000/ 1782 batches | accuracy    0.940
| epoch  10 |  1500/ 1782 batches | accuracy    0.939
-----------------------------------------------------
| end of epoch  10 | time: 11.29s | valid accuracy    0.916
-----------------------------------------------------
```

(14) 模型评估。

```
1  print(f'测试资料准确度：{evaluate(test_dataloader):.3f}')
```

执行结果，准确度：0.904。

(15) 测试新数据。

```
1   # 新闻类别
2   ag_news_label = {1: "World",
3                    2: "Sports",
4                    3: "Business",
5                    4: "Sci/Tec"}
6
7   # 预测
8   def predict(text, text_pipeline):
9       with torch.no_grad():
10          text = torch.tensor(text_pipeline(text)).to(device)
11          output = model(text, torch.tensor([0]).to(device))
12          return output.argmax(1).item() + 1
13
14  # 测试数据
15  ex_text_str = "MEMPHIS, Tenn. - Four days ago, Jon Rahm was \
16      enduring the season's worst weather conditions on Sunday at The \
17      Open on his way to a closing 75 at Royal Portrush, which \
18      considering the wind and the rain was a respectable showing. \
19      Thursday's first round at the WGC-FedEx St. Jude Invitational \
20      was another story. With temperatures in the mid-80s and hardly any \
21      wind, the Spaniard was 13 strokes better in a flawless round. \
22      Thanks to his best putting performance on the PGA Tour, Rahm \
23      finished with an 8-under 62 for a three-stroke lead, which \
24      was even more impressive considering he'd never played the \
25      front nine at TPC Southwind."
26
27  print(ag_news_label[predict(ex_text_str, text_pipeline)])
```

执行结果：Sports，属于运动类新闻。

(16) 从美国职业棒球大联盟 (MLB) 摘一段新闻存入文件，进行测试。

```
1  my_test = open('./nlp_data/news.txt', encoding='utf8').read()
2  print(ag_news_label[predict(my_test, text_pipeline)])
```

执行结果：Sports，属于运动类新闻。

上述模型仅使用嵌入层及完全连接层，若要插入 RNN 层，模型建构如下，请参阅程序【12_03_RNN_Text_Classification.ipynb】。

```
1   from torch import nn
2
3   class TextClassificationModel(nn.Module):
4       def __init__(self, vocab_size, embed_dim, num_class):
5           super().__init__()
6           self.embedding = nn.EmbeddingBag(vocab_size, embed_dim, sparse=True)
7           self.rnn = nn.RNN(embed_dim, 32)
8           self.fc = nn.Linear(32, num_class)
9           self.init_weights()
10
11      def init_weights(self):
12          initrange = 0.5
13          self.embedding.weight.data.uniform_(-initrange, initrange)
14          self.fc.weight.data.uniform_(-initrange, initrange)
15          self.fc.bias.data.zero_()
16
17      def forward(self, text, offsets):
18          embedded = self.embedding(text, offsets)
19          rnn_out, h_out = self.rnn(embedded)
20          return self.fc(rnn_out)
21
22  model = TextClassificationModel(vocab_size, emsize, num_class).to(device)
```

在文本分类上，是否使用 RNN 层并不重要，但若是其他任务，例如文本生成 (Text Generation)，要预测下一个单词或句子，RNN 层就很重要了。

## 12-3 长短期记忆网络

简单 (Vanilla) RNN 只考虑上文 ( 上一个神经元 )，如果要同时考虑下文，可以直接设定参数 bidirectional=True 即可，要多个 RNN，也只要设定参数 num_layers=N。

但是简单的 RNN 还是有一个瑕疵，它跟 CNN 一样，为简化模型均假设权值共享 (Shared Weights)，因此，公式由 $h_t$、$h_{t-1}$、$h_{t-2}$…，往前推：

$h_t = W * h_{t-1} + U * x_t + b$

$h_{t-1} = W * h_{t-2} + U * x_{t-1} + b$

➔ $h_t = W * (W * h_{t-2} + U * x_{t-1} + b) + U * x_t + b$

➔ $h_t = (W^2 * h_{t-2} + W * U * x_{t-1} + W * b) + U * x_t + b$

若 $W<1$，则越前面的神经层 $W^n$ 会越小，造成影响力越小，这种现象称为"梯度消失"(Vanishing Gradient)。

反之，若 $W>1$，则越前面的神经层 $W^n$ 会越大，造成梯度爆炸 (Exploding Gradient)，优化求解无法收敛。

梯度消失导致考虑的上文长度有限，因此，Hochreiter 和 Schmidhuber 于 1997 年提出"长短期记忆网络"(Long Short Term Memory Network, LSTM) 算法，额外维护一条记忆网络，希望能让较久远的记忆发挥影响力。图 12.7 比较了 LSTM 与简单 RNN 的差别。

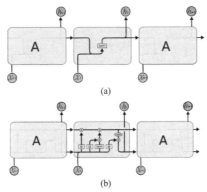

(a)

(b)

图 12.7　RNN 与 LSTM 内部结构的比较，(a) 为 RNN，(b) 为 LSTM

图片来源：Understanding LSTM Networks [9]

我们将图 12.7 的 LSTM 进行拆解，就可以了解 LSTM 的运算机制。

(1) 额外维护一条记忆线 (Cell state)，如图 12.8 所示。

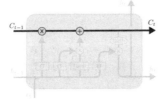

图 12.8　维护记忆线

(2) LSTM 多了四个门 (Gate)，用来维护记忆网络与预测网络，即图 12.7 中的 ⊗、⊕。

(3) 遗忘门 (Forget Gate)：决定之前记忆是否删除，$\sigma$ 为 sigmoid 神经层，输出为 0 时，乘以原记忆，表示删除，反之则为保留记忆，如图 12.9 所示。

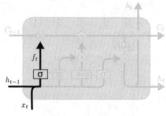

图 12.9　遗忘门

$$f_t = \sigma(w_f \cdot [h_{t-1}, x_t] + b_f)$$

(4) 输入门 (Input Gate)：输入含目前的特征 ($x_t$) 加 **t**-1 时间点的隐藏层 ($h_{t-1}$)，通过 σ(sigmoid)，得到输出 ($i_t$)，而记忆 ($C_t$) 使用 tanh activation function，其值介于 (-1, 1)，两者相乘，再加入到记忆在线，如图 12.10 所示。

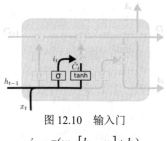

图 12.10　输入门

$$i_t = \sigma(w_i \cdot [h_{t-1}, x_t] + b_i)$$

$$\tilde{C}_t = \tan h(W_C \cdot [h_{t-1}, x_t] + b_C)$$

(5) 更新门 (Update Gate)：更新记忆 ($C_t$)，为之前的记忆加上目前增加的信息，如图 12.11 所示。

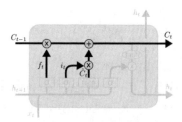

图 12.11　更新门

$$C_t = f_t * C_{t-1} + i_t * \tilde{C}_t$$

(6) 输出门 (Output Gate)：将目前 RNN 神经层的输出，乘以更新的记忆 ($C_t$)，即为 LSTM 输出，如图 12.12 所示。

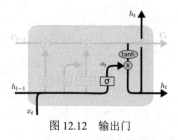

图 12.12　输出门

$$o_t = \sigma(W_o[h_{t-1}, x_t] + b_o)$$
$$h_t = o_t * \tanh(C_t)$$

依照前面的拆解，大概就能知道 LSTM 是怎么保存及使用记忆了，网络上也有人直接用 NumPy 开发 LSTM，可以彻底了解上述运算，不过，既然深度学习框架均已直接定义 LSTM 神经层了，我们就直接拿来用了。

LSTM 神经层与 RNN 参数设定几乎一样，详细说明可参阅《PyTorch LSTM 神经层》[10]，输出会多一个内存状态 (cn)，如图 12.13 所示。

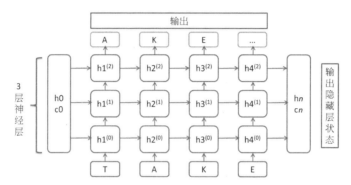

图 12.13　输出隐藏状态

**范例3.** 以LSTM实操情绪分析(Sentiment Analysis)，判别评论属于正面或负面。

下列程序代码请参考【12_05_LSTM_IMDB_Sentiment_Analysis.ipynb】。

数据集：影评数据集 (IMDB movie review)，IMDB(Internet Movie Database) 是收集全球影音信息的网站，它提供的语料库，让各界可以进行相关研究。

(1) 载入 IMDB 数据集：与 AG News 数据集加载指令相似。

```
1  import torch
2  from torchtext.datasets import IMDB
3
4  imdb = IMDB(split='train')
5
6  type(imdb)
```

(2) 取得下一笔数据测试。

```
1  train_iter = iter(IMDB(split='train'))
2
3  data = next(train_iter)
4  data
```

执行结果：第一个字段为正面 (pos) 或负面 (neg) 情绪。

```
('neg',
 'I rented I AM CURIOUS-YELLOW from my video store because of all the controversy that surrounded it when it was first released in 1967. I also heard that at first it was seized by U.S. customs if it ever tried to enter this country, therefore being a fan of films considered "controversial" I really had to see this for myself.<br /><br />The plot is centered around a young Swedish drama student named Lena who wants to learn everything she can about life. In particular she wants to focus her attentions to making some sort of documentary on what the average Swede thought about certain political issues such as the Vietnam War and race issues in the United States. In between asking politicians and ordinary denizens of Stockholm about their opinions on politics, she has sex with her drama teacher, classmates, and married men.<br /><br />What kills me about I AM CURIOUS-YELLOW is that 40 years ago, this was considered pornographic. Really, the sex and nudity scenes are few and far between, even then it\'s not shot like some cheaply made porno. While my countrymen mind find it shocking, in reality sex and nudity are a major staple in Swedish cinema. Even Ingmar Bergman, arguably their answer to good old boy John Ford, had sex scenes in his films.<br /><br />I do commend the filmmakers for the fact that any sex shown in the film is shown for artistic purposes rather than just to shock people and make money to be shown in pornographic theaters in America. I AM CURIOUS-YELLOW is a good film for anyone wanting to study the meat and potatoes (no pun intended) of Swedish cinema. But really, this film doesn\'t have much of a plot.')
```

(3) 之后的程序代码均类似，就不重复说明了，完整内容请参照范例程序，以下仅列出重大差异处。

(4) 建立模型：使用 "Embedding + LSTM + Linear" 神经层，LSTM 采用双向 (bidirectional=True)，会产生两倍的输出，故下一层输入个数要乘以 2。

```python
from torch import nn

class TextClassificationModel(nn.Module):
    def __init__(self, vocab_size, embed_dim, num_class):
        super().__init__()
        self.embedding = nn.EmbeddingBag(vocab_size, embed_dim, sparse=True)
        self.rnn = nn.LSTM(embed_dim, hidden_dim, bidirectional=True)
        self.fc = nn.Linear(hidden_dim * 2, num_class)
        self.init_weights()

    def init_weights(self):
        initrange = 0.5
        self.embedding.weight.data.uniform_(-initrange, initrange)
        self.fc.weight.data.uniform_(-initrange, initrange)
        self.fc.bias.data.zero_()

    def forward(self, text, offsets):
        embedded = self.embedding(text, offsets)
        rnn_out, h_out = self.rnn(embedded)
        return self.fc(rnn_out)

model = TextClassificationModel(vocab_size, emsize, num_class).to(device)
```

(5) 模型评估执行结果为 0.874。笔者改用 RNN 模型执行，结果为 0.827，LSTM 虽然准确率较高，但差异并不明显。RNN 程序为【12_04_RNN_IMDB_Sentiment_Analysis.ipynb】。

# 12-4 自定义数据集

以上均使用 PyTorch 内建数据，如果是自行收集的文件数据，要如何处理呢？仍以 IMDB 数据集为例。但是数据集是外部文件，而非内建数据集，可自 https://ai.stanford.edu/~amaas/data/sentiment/aclImdb_v1.tar.gz 下载，解压缩，可能需要解压缩两次，会先解压缩为 aclImdb_v1.tar，再解压缩为 aclImdb 目录，文件个数非常多，需耐心等候。

**范例4**. 使用自定义数据集，以 LSTM 实操情绪分析(Sentiment Analysis)。程序代码与上例架构类似，仅说明差异之处。

下列程序请参考【12_06_LSTM_Custom_IMDB_Sentiment_Analysis.ipynb】。

(1) 自定义数据集：与【06_05_Data_Augmentation_MNIST.ipynb】的 CustomImageDataset 类似，实操 __init__、__getitem__、__len__ 方法即可，唯一要注意的是第 22 行，以次目录名称作为标注 (Label)，即 Y。

```python
# 数据集所在目录
data_base_path = './aclImdb/'

class ImdbDataset(torch.utils.data.Dataset):
    def __init__(self, mode):
        super(ImdbDataset, self).__init__()
        if mode == "train":
            text_path = [os.path.join(data_base_path, i) for i in ["train/neg", "train/pos"]]
        else:
            text_path = [os.path.join(data_base_path, i) for i in ["test/neg", "test/pos"]]
        # print(text_path)
```

```
13         self.total_file_path_list = []
14         for i in text_path:
15             self.total_file_path_list.extend([os.path.join(i, j) for j in os.listdir(i)])
16         # print(len(self.total_file_path_list))
17
18     def __getitem__(self, idx):
19         cur_path = self.total_file_path_list[idx]
20         cur_filename = os.path.basename(cur_path)
21         # print(cur_path)
22         label = 0 if cur_path.find('/neg') > 0 else 1
23         # text = tokenizer(open(cur_path, encoding="utf-8").read().strip())
24         text = open(cur_path, encoding="utf-8").read().strip()
25         return label, text
26
27     def __len__(self):
28         return len(self.total_file_path_list)
```

(2) 测试 Dataset：取得下一笔数据测试。

```
1  dataset = ImdbDataset(mode="train")
2  print(dataset[0])
```

执行结果：0 代表负面，1 代表正面。

```
(0, "Story of a man who has unnatural feelings for a pig. Starts out with a opening scene that is a terrific example of absurd
comedy. A formal orchestra audience is turned into an insane, violent mob by the crazy chantings of it's singers. Unfortunately
it stays absurd the WHOLE time with no general narrative eventually making it just too off putting. Even those from the era sho
uld be turned off. The cryptic dialogue would make Shakespeare seem easy to a third grader. On a technical level it's better th
an you might think with some good cinematography by future great Vilmos Zsigmond. Future stars Sally Kirkland and Frederic Forr
est can be seen briefly.")
```

(3) 建立 DataLoader：collate_batch 函数不变，第 17 行直接通过 DataSet 转成 DataLoader，无须经过 Iterator。

```
1  from torch.utils.data import DataLoader
2
3  # 批次处理
4  def collate_batch(batch):
5      label_list, text_list, offsets = [], [], [0]
6      for (_label, _text) in batch:
7          label_list.append(label_pipeline(_label))
8          processed_text = torch.tensor(text_pipeline(_text), dtype=torch.int64)
9          text_list.append(processed_text)
10         offsets.append(processed_text.size(0))  # 设定每笔数据的起始位置
11     label_list = torch.tensor(label_list, dtype=torch.int64)
12     offsets = torch.tensor(offsets[:-1]).cumsum(dim=0)  # 单词的索引值累加
13     text_list = torch.cat(text_list)
14     return label_list.to(device), text_list.to(device), offsets.to(device)
15
16 dataloader = DataLoader(dataset, batch_size=BATCH_SIZE, shuffle=True,
17                         collate_fn=collate_batch)
```

(4) 测试 DataLoader。

```
1  # 取得3笔数据
2  for idx,(label,text, offset) in enumerate(dataloader):
3      print("idx : ",idx)
4      print("label:",label)
5      print("text:",text)
6      print("offset:",offset)
7      if idx >= 2:
8          break
```

执行结果：可以观察 offset 的数值，它是每一笔在该批数据的起始位置。

```
idx : 0
label: tensor([1, 0, 0, 0, 1, 1, 0, 1, 0, 0, 0, 1, 0, 0, 0, 1, 1, 0, 0, 0, 0,
        1, 1, 1, 1, 1, 0, 0, 0, 1, 0, 0, 1, 0, 1, 1, 0, 0, 1, 1, 0, 1,
        0, 0, 0, 1, 1, 1, 0, 1, 1, 1, 1, 1, 0, 1, 0, 1])
text: tensor([ 1039,    10,    16,   ...,     7, 10889,   156])
offset: tensor([    0,   275,   700,  1040,  1245,  2053,  2268,  2477,  2615,  2946,
         3038,  3442,  4040,  4182,  4371,  4480,  4610,  4874,  5089,  5278,
         5470,  6123,  6354,  6721,  7220,  7545,  8196,  8290,  8495,  8722,
         8973,  9134,  9346,  9559,  9735,  9872, 10007, 10620, 10754, 11239,
        11926, 12088, 12274, 12468, 12592, 12829, 13017, 13157, 13291, 13464,
        13599, 13745, 14528, 14734, 15103, 15246, 15389, 15624, 15708, 15958,
        16102, 16168, 16863])
idx : 1
label: tensor([0, 0, 1, 0, 0, 1, 1, 0, 1, 0, 0, 0, 0, 0, 1, 0, 1, 1, 0, 0, 0, 1,
        1, 0, 0, 1, 1, 0, 1, 1, 1, 0, 1, 0, 0, 1, 0, 1, 0, 1, 0, 1, 1,
        1, 1, 1, 1, 1, 1, 0, 0, 1, 1, 0, 1, 0, 0, 0, 0])
text: tensor([  59,   12, 1212,  ...,    2,  130,   35])
offset: tensor([    0,   214,   663,   917,  1046,  1294,  1330,  1709,  1973,  2147,
         2306,  2441,  2779,  3046,  3701,  4811,  5177,  5316,  5831,  6122,
```

(5) 后续的程序代码均与上例相同，实测准确率也与上例差不多。

网络上也有一篇文章《PyTorch 实现 IMDB 数据集情感分类》[11]，采用较一般性的处理方法，可以和本程序对照比较。

在网络上 RNN 系列的算法较为普遍地采用 LSTM，如果是较复杂的应用，可以考虑使用堆栈 LSTM (Stacked LSTM)、双向 (bidirectional)，只要设定 nn.LSTM 类别中设定参数 num_layers、bidirectional，就可轻易达成，比 TensorFlow 容易多了，但要注意，模型输出及隐藏层状态的维度均会因设定不同而改变，可参阅前文说明。

# 12-5　时间序列预测

RNN 特性是考虑上下文 (Context) 的影响，恰好与时间序列 (Time Series) 相同，例如气温、股票行情、营收等，数据都有时间的相关性及延续性，接下来看一个实例，使用 RNN 进行时间序列的预测，并借以说明 RNN 并不只能用作文本分类。

**范例5.** 时间序列预测，以LSTM算法预测航空公司的未来营收，包括以下各种模型测试。

(1) 前期数据为 X，当期数据为 Y。
(2) 取前 3 期的数据作为 X，即以 $t$-3、$t$-2、$t$-1 期预测 $t$ 期。
(3) 以 $t$-3 期 ( 单期 ) 预测 $t$ 期。
(4) 将每个周期再分多批训练。
(5) Stacked LSTM：使用多层的 LSTM。

此范例程序修改自 Time Series Prediction with LSTM Recurrent Neural Networks in Python with Keras[12]。

**下列程序代码请参考【12_07_LSTM_Time_Series.ipynb】。**

时间序列预测的相关步骤如图 12.14 所示。

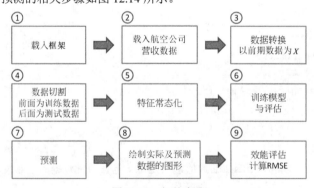

图 12.14　相关步骤

(1) 加载相关框架。

```
1  import torch
2  import torch.nn as nn
3  import torch.nn.functional as F
4  import torch.optim as optim
5  import torchtext
6  import numpy as np
7  import pandas as pd
8  import os
9  import matplotlib.pyplot as plt
```

(2) 判断 GPU 是否存在。

```
1  device = torch.device("cuda" if torch.cuda.is_available() else "cpu")
```

(3) 加载航空公司的营收数据：这份数据年代久远，每月一笔数据，时间是 1949 年 1 月至 1960 年 12 月。

```
1  df = pd.read_csv('./nlp_data/airline-passengers.csv')
2  df.head()
```

(4) 绘图：通过图表可以发现，航空公司的营收除了有淡旺季之分以外，还有逐步上升的成长趋势。

```
1  df2 = df.set_index('Month')
2  df2.plot(legend=None)
3  plt.xticks(rotation=30);
```

执行结果。

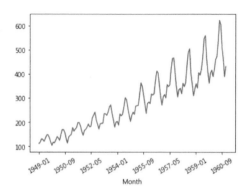

(5) 转换数据：以前期数据为 X，当期数据为 Y，以前期营收预测当期营收。因此训练数据和测试数据不采取随机切割，前面 2/3 为训练数据，后面 1/3 为测试数据，并对特征进行常态化。

```
1   from sklearn.preprocessing import MinMaxScaler
2   look_back = 1 # 以前N期数据为 X，当期数据为 Y
3
4   # 函数：以前N期数据为 X，当前期数据为 Y
5   def create_dataset(data1, look_back):
6       x, y = [], []
7       for i in range(len(data1)-look_back-1):
8           _x = data1[i:(i+look_back)]
9           _y = data1[i+look_back]
10          x.append(_x)
11          y.append(_y)
12
13      return torch.Tensor(np.array(x)), torch.Tensor(np.array(y))
14
15  dataset = df2[['Passengers']].values.astype('float32')
16
17  # X 常态化
18  scaler = MinMaxScaler()
19  dataset = scaler.fit_transform(dataset)
20
21  # 数据分割
22  train_size = int(len(dataset) * 0.67)
23  test_size = len(dataset) - train_size
24  train_data, test_data = dataset[0:train_size,:], dataset[train_size:len(dataset),:]
25
26  trainX, trainY = create_dataset(train_data, look_back)
27  testX, testY = create_dataset(test_data, look_back)
28  dataset.shape, trainY.shape
```

(6) 建立模型：LSTM 串接 Linear，函数可设定多层的 LSTM。

LSTM 的 batch_first=True，是将输出的数据中批量移至第一维。

LSTM 串接 Linear 的数据使用取最后一层的隐藏层状态，并转成二维。也可以取用最后一个输出，请参见程序【12_08_LSTM_Time_Series_Use_Output.ipynb】。

```python
class TimeSeriesModel(nn.Module):
    def __init__(self, look_back, hidden_size=4, num_layers=1):
        super().__init__()
        self.hidden_size = hidden_size
        self.num_layers = num_layers
        self.rnn = nn.LSTM(1, self.hidden_size, num_layers=self.num_layers
                           , batch_first=True)
        self.fc = nn.Linear(self.hidden_size, 1)
        self.init_weights()
    
    def init_weights(self):
        initrange = 0.5
        self.fc.weight.data.uniform_(-initrange, initrange)
        self.fc.bias.data.zero_()
    
    def forward(self, x):
        #print(x.shape)
        # rnn_out, h_out = self.rnn(x)
        h_0 = torch.zeros(self.num_layers, x.size(0), self.hidden_size)
        c_0 = torch.zeros(self.num_layers, x.size(0), self.hidden_size)
        out, (h_out, _) = self.rnn(x, (h_0, c_0))
        #print(h_out.shape)
        
        # 取最后一层的 h，并转成二维
        h_out = h_out[-1].view(-1, self.hidden_size)
        return self.fc(h_out)

model = TimeSeriesModel(look_back, hidden_size=4, num_layers=1).to(device)
```

(7) 模型训练：由于训练数据较少，故训练较多执行周期 (2000)。

```python
num_epochs = 2000
learning_rate = 0.01

def train(trainX, trainY):
    criterion = torch.nn.MSELoss()    # MSE
    optimizer = torch.optim.Adam(model.parameters(), lr=learning_rate)
    
    for epoch in range(num_epochs):
        optimizer.zero_grad()
        outputs = model(trainX)
        if epoch <= 0: print(outputs.shape)
        loss = criterion(outputs, trainY)
        loss.backward()
        optimizer.step()
        if epoch % 100 == 0:
            print(f"Epoch: {epoch}, loss: {loss.item():.5f}")

train(trainX, trainY)
```

执行结果：大约在 600 周期前就收敛了。

```
Epoch: 0, loss: 0.03078
Epoch: 100, loss: 0.00193
Epoch: 200, loss: 0.00191
Epoch: 300, loss: 0.00190
Epoch: 400, loss: 0.00190
Epoch: 500, loss: 0.00190
Epoch: 600, loss: 0.00190
Epoch: 700, loss: 0.00190
Epoch: 800, loss: 0.00190
Epoch: 900, loss: 0.00190
Epoch: 1000, loss: 0.00190
```

(8) 模型评估。

```
1  model.eval()
2  trainPredict = model(trainX).detach().numpy()
3  testPredict = model(testX).detach().numpy()
4  trainPredict.shape
```

(9) 预测后还原常态化,计算训练及测试数据的均方根误差 (RMSE)。

```
1   from sklearn.metrics import mean_squared_error
2   import math
3
4   # 还原常态化的训练及测试数据
5   trainPredict = scaler.inverse_transform(trainPredict)
6   trainY_actual = scaler.inverse_transform(trainY.reshape(-1, 1))
7   testPredict = scaler.inverse_transform(testPredict)
8   testY_actual = scaler.inverse_transform(testY.reshape(-1, 1))
9   print(trainY_actual.shape, trainPredict.shape)
10
11  # 计算 RMSE
12  trainScore = math.sqrt(mean_squared_error(trainY_actual, trainPredict.reshape(-1)))
13  print(f'Train RMSE: {trainScore:.2f}')
14  testScore = math.sqrt(mean_squared_error(testY_actual, testPredict.reshape(-1)))
15  print(f'Test RMSE:  {testScore:.2f}')
```

执行结果:训练数据和测试数据的 RMSE 分别为 22.56、56.09。

(10) 绘制实际数据和预测数据的图表。

```
1   # 训练数据的 X/Y
2   trainPredictPlot = np.empty_like(dataset)
3   trainPredictPlot[:, :] = np.nan
4   trainPredictPlot[1:len(trainPredict)+look_back, :] = trainPredict
5
6   # 测试数据 X/Y
7   testPredictPlot = np.empty_like(dataset)
8   testPredictPlot[:, :] = np.nan
9   testPredictPlot[-testPredict.shape[0]-1:-1, :] = testPredict
10
11  # 绘图
12  plt.plot(scaler.inverse_transform(dataset), label='Actual')
13  plt.plot(trainPredictPlot, label='train predict')
14  plt.plot(testPredictPlot, label='test predict')
15  plt.legend()
16  plt.show()
```

执行结果:请参阅程序,蓝色线条为实际值,橘色线条为训练数据的预测值,绿色线条为测试数据的预测值。预测值与实际值相差不远。

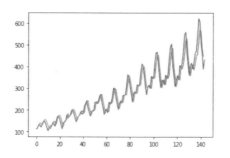

扫码看彩图

(11) Loopback 改为 3:$X$ 由前 1 期改为前 3 期,即以 $t$-3、$t$-2、$t$-1 期预测 $t$ 期。

```
1   # 以前期数据为 X,当前期数据为 Y
2   look_back = 3
3   trainX, trainY = create_dataset(train_data, look_back)
4   testX, testY = create_dataset(test_data, look_back)
5
6   model = TimeSeriesModel(look_back, hidden_size=4, num_layers=1).to(device)
7   train(trainX, trainY)
```

执行结果：需要较长周期收敛。

(12) 模型训练、评估与绘制图表：程序均无须修改。

执行结果：训练数据和测试数据的 RMSE 近似于 1 期预测结果，使用多期预测，有移动平均的效果，预测曲线会比较平缓，比较不容易受到激烈变化的样本点影响。

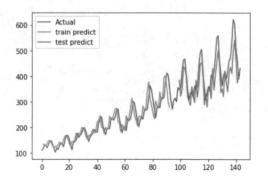

扫码看彩图

(13) 多层 (Stacked)LSTM 预测：将 LSTM 层数设为 3(num_layers=3)。

```
1  # 以前期数据为 X，当前期数据为 Y
2  look_back = 3
3  trainX, trainY = create_dataset(train_data, look_back)
4  testX, testY = create_dataset(test_data, look_back)
5
6  model = TimeSeriesModel(look_back, hidden_size=4, num_layers=3).to(device)
7  train(trainX, trainY)
```

执行结果：训练数据和测试数据的 RMSE 分别为 11.75、122.53，测试 RMSE 比较差，应该是因为数据很单纯，使用太复杂的网络结构反而没有帮助。

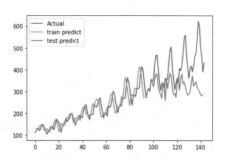

扫码看彩图

## 12-6　Gate Recurrent Unit

Gate Recurrent Unit (GRU) 也是 RNN 变形的算法，由 Kyunghyun Cho 在 2014 年提出的，可参阅 *Empirical Evaluation of Gated Recurrent Neural Networks on Sequence Modeling*[13]，主要就是要改良 LSTM 缺陷。

(1) LSTM 计算过慢，GRU 可改善训练速度。

(2) 简化 LSTM 模型，节省内存的空间。

LSTM 是由遗忘门 (Forget Gate) 与输入门 (Input Gate) 来维护记忆状态 (Cell State)，然而因为这部分太过耗时，所以 GRU 废除记忆状态，直接使用隐藏层输出 ($h_t$)，并且将前述两个门改由更新门(Update Gate)替代，两个模型的架构比较如图12.15所示。

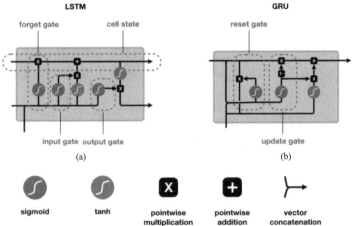

图 12.15　LSTM 与 GRU 内部结构的比较，(a) 为 LSTM，(b) 为 GRU

图片来源：*Illustrated Guide to LSTM's and GRU's: A step by step explanation*[14]

**范例6. 以GRU预测航空公司乘客数。**

**下列程序代码请参考【12_09_GRU_Time_Series.ipynb】。**

(1) 只要将nn.LSTM换为nn.GRU即可，PyTorch的GRU没有记忆线，故须把c删除。

```python
class TimeSeriesModel(nn.Module):
    def __init__(self, look_back, hidden_size=4, num_layers=1):
        super().__init__()
        self.hidden_size = hidden_size
        self.num_layers = num_layers
        self.rnn = nn.GRU(1, self.hidden_size, num_layers=self.num_layers
                          , batch_first=True)
        self.fc = nn.Linear(self.hidden_size, 1)
        self.init_weights()
    
    def init_weights(self):
        initrange = 0.5
        self.fc.weight.data.uniform_(-initrange, initrange)
        self.fc.bias.data.zero_()
    
    def forward(self, x):
        #print(x.shape)
        # rnn_out, h_out = self.rnn(x)
        h_0 = torch.zeros(self.num_layers, x.size(0), self.hidden_size)
        out, h_out = self.rnn(x, h_0)
        #print(h_out.shape)
        
        # 取最后一层的 h，并转成二维
        h_out = h_out[-1].view(-1, self.hidden_size)
        return self.fc(h_out)

model = TimeSeriesModel(look_back, hidden_size=4, num_layers=1).to(device)
```

(2) 其他程序代码均与【12_07_LSTM_Time_Series.ipynb】相同，执行结果也类似。

虽然GRU作者提出效能比较图表，说明GRU的效能比LSTM好，不过，稍微吐槽一下，笔者实际测试的结果，差异并不明显，而且网络上也较少提到GRU，大多仍以LSTM为主流，因此就不详细研究了。

## 12-7 股价预测

由于 LSTM 与时间序列模型类似，网络上有许多文章探讨"以 LSTM 预测股票价格"，我们就来实操一下。

**范例7. 以LSTM算法预测股价**。此范例程序修改自 Predicting stock prices with LSTM[15]，且程序流程与【12_07_LSTM_Time_Series】近似，以下仅说明有重大差异之处。

下列程序代码请参考【12_10_Stock_Forecast.ipynb】。数据集：本范例使用亚马逊企业股票[16]。

(1) 加载测试数据。

```
1  df = pd.read_csv('./nlp_data/AMZN_2006-01-01_to_2018-01-01.csv')
2  df.head()
```

执行结果。

| Date | Open | High | Low | Close | Volume | Name |
|---|---|---|---|---|---|---|
| 2006-01-03 | 47.47 | 47.85 | 46.25 | 47.58 | 7582127 | AMZN |
| 2006-01-04 | 47.48 | 47.73 | 46.69 | 47.25 | 7440914 | AMZN |
| 2006-01-05 | 47.16 | 48.20 | 47.11 | 47.65 | 5417258 | AMZN |
| 2006-01-06 | 47.97 | 48.58 | 47.32 | 47.87 | 6154285 | AMZN |
| 2006-01-09 | 46.55 | 47.10 | 46.40 | 47.08 | 8945056 | AMZN |

(2) 绘图：绘制收盘价线图。

```
1  df2 = df.set_index('Date')
2  df2.Close.plot(legend=None)
3  plt.xticks(rotation=30);
```

执行结果。

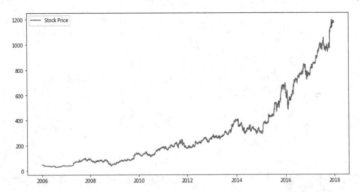

(3) 一次预测 1 天，训练模型，并绘制实际数据和预测数据的图表：与之前的程序代码相同。

执行结果：若一次只预测 1 天，结果看起来还不错，但差异会逐渐变大。

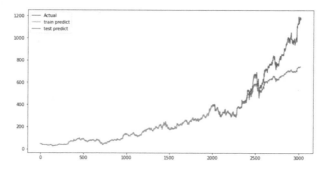

扫码看彩图

模型改用 LSTM 的输出取代隐藏层状态，可提高准确率。

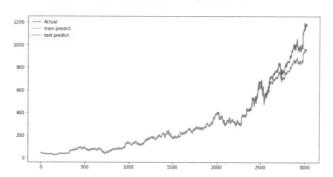

扫码看彩图

以上程序一次只能预测一天，要能一次预测多天，可将目标 (Y) 改为多个变量，例如每笔数据有 10 个 Y，请参看程序"预测多期"段落。

(1) 前置处理函数：多加一个参数 forward_days，为预测天数。

```
# 函数：以前N期数据为 X，当前期数据为 Y
def create_dataset(data1, look_back, forward_days):
    x, y = [], []
    for i in range(len(data1) - look_back - forward_days + 1):
        _x = data1[i:(i+look_back)]
        _y = data1[i+look_back:(i+look_back+forward_days)]
        x.append(_x)
        y.append(_y)

    x, y = np.array(x), np.array(y)
    return torch.Tensor(x), torch.Tensor(y.reshape(y.shape[0], y.shape[1]))
```

(2) 建立数据集。

```
look_back = 10   # 以前10期数据为 X
forward_days = 10   # 预测天数
trainX, trainY = create_dataset(train_data, look_back, forward_days)
testX, testY = create_dataset(test_data, look_back, forward_days)
```

(3) 建立模型：多加一个参数 forward_days。

```
class TimeSeriesModel(nn.Module):
    def __init__(self, look_back, forward_days, hidden_size=4, num_layers=1):
        super().__init__()
        self.hidden_size = hidden_size
        self.num_layers = num_layers
        self.rnn = nn.LSTM(1, self.hidden_size, num_layers=self.num_layers
                           , batch_first=True)
        self.fc = nn.Linear(self.hidden_size, forward_days)
        self.init_weights()
```

```python
11      def init_weights(self):
12          initrange = 0.5
13          self.fc.weight.data.uniform_(-initrange, initrange)
14          self.fc.bias.data.zero_()
15
16      def forward(self, x):
17          #print(x.shape)
18          # rnn_out, h_out = self.rnn(x)
19          h_0 = torch.zeros(self.num_layers, x.size(0), self.hidden_size)
20          c_0 = torch.zeros(self.num_layers, x.size(0), self.hidden_size)
21          out, (h_out, _) = self.rnn(x, (h_0, c_0))
22          #print(h_out.shape)
23
24          # 取最后一层的 h，并转成二维
25          # h_out = h_out[-1].view(-1, self.hidden_size)
26          # return self.fc(h_out)
27          # 取最后一个输出，并转成二维
28          flatten_output = out[:, -1].view(-1, self.hidden_size)
29          return self.fc(flatten_output)
30
31  model = TimeSeriesModel(look_back, forward_days, hidden_size=20,
32                           num_layers=1).to(device)
```

(4) 模型训练与评估：程序代码不变。

执行结果：训练 RMSE= 7.67，测试 RMSE= 29.24，均很理想。

(5) 绘制测试数据的预测值与实际值的比较。

```python
1   plt.figure(figsize=(12,6))
2
3   # 真实数据
4   plt.plot(range(len(dataset)), scaler.inverse_transform(dataset), 'b', label='Actual')
5
6   # 训练数据
7   for i in range(trainPredict.shape[0]):
8       plt.plot(range(i, i+forward_days), trainPredict[i], 'orange')
9
10  # 测试数据
11  for i in range(testPredict.shape[0]):
12      plt.plot(range(i+trainPredict.shape[0], i+trainPredict.shape[0]+forward_days),
13              testPredict[i], 'r')
14  plt.show()
```

执行结果：预测很接近实际值，看起来似乎很好。

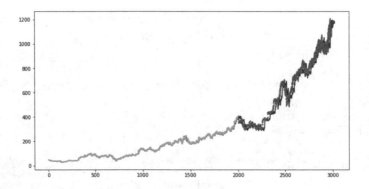

(6) 拉近看：绘制训练数据的前 20 条预测值。

```python
1   n = 20
2   plt.figure(figsize=(12,6))
3   # 真实数据
4   plt.plot(range(forward_days*n), scaler.inverse_transform(dataset[:forward_days*n]), 'b', label='Actual')
5
6   # 训练数据
7   for i in range(n):
8       plt.plot(range(i*forward_days, (i+1)*forward_days), trainPredict[i*forward_days], 'r')
9
10  plt.show()
```

执行结果：预测值完全没有抓到实际值的变化，上图其实只是镜头拉远的效果而已。

由执行结果可以看出，模型无法抓到股价的变动，原因如下：

(1) 股价非稳态 (Non-stationary)：每个时间点的股价平均数与标准偏差都不一致，违反回归的基本假设，例如上图，股价数据有长期下降趋势，并非完全随机跳动。因此，时间序列预测通常会将股价转换为收益率 (Return Rate)，使数据呈现稳态后，再输入模型。

(2) 股价变化非常大，单纯以历史数据预测未来股价，模型过于简化。

(3) 可以进一步使用回测 (Back Testing) 衡量模型的有效性，观察策略是否奏效，简单的回测做法可参考笔者撰写的《算法交易 (Algorithmic Trading) 实操》[17]一文。

虽然试验效果不佳，但 LSTM 与传统的时间序列 (Time Series) 理论相比，LSTM 提供更有弹性的做法，除了股价外，我们可以输入更多的变量，例如各种财务比率、技术指标、总经指标、筹码面指标等，有兴趣的读者不妨试验看看。

## 12-8 注意力机制

RNN 从之前的隐藏层状态取得上文的信息，LSTM 则额外维护一条记忆线 (Cell State)，目的都是希望能借由记忆的方式来提高预测的准确性，但是，两个算法都局限于上文的序列顺序，导致越靠近预测目标的信息，权重越大。实际上当我们在阅读一篇文章时，往往会对文中的标题、人事时地物或强烈的形容词特别注意，这就是所谓的注意力机制 (Attention Mechanism)，对于图像也是如此，如图 12.16 所示，婴儿的脸部与右下方整叠的纸尿裤是注意力热区。

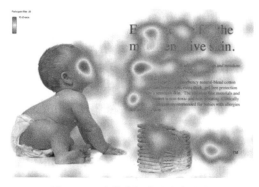

图 12.16　人类的视觉注意力分布

图片来源：《深度学习中的注意力机制 (2017 版 )》[18]

通过注意力机制，额外把重点单词或部位纳入考虑，而不只是上下文。它的概念影响巨大，近几年发展的模型均受到它的启发，例如 Transformer、BERT 等，下面我们就来讲一下它的做法。

机器翻译 (Neural Machine Translation, NMT) 属于文本生成的模型，一般称为序列到序列 (Sequence to Sequence, Seq2Seq) 模型，是 Encoder-Decoder 的变形，结构如图 12.17 所示，其中的 Context Vector 是 Encoder 输出的上下文向量，类似 CNN AutoEncoder 提取的特征向量。

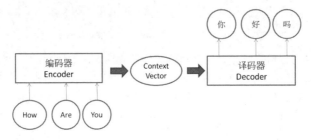

图 12.17　Seq2Seq 模型，How are you 翻译为"你好吗"

也可以应用于对话问答，如图 12.18 所示。

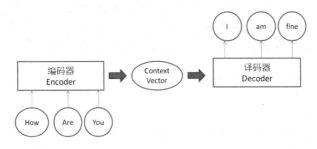

图 12.18　对话问答，问 How are you，回答 I am fine

而注意力机制就是把编码器 (encoder) 的隐藏层输出都乘上一个权重 (weight)，与 Context Vector 混合计算成 Attention Vector，用以预测下一个词汇，这个机制会应用到译码器的每一层，如图 12.19 所示。

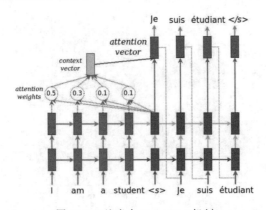

图 12.19　注意力 (Attention) 机制

权重 (weight) 的计算方式有两种，分别由 Luong 和 Bahdanau 提出的乘法与加法的公式，它利用完全连接层 (Dense) 及 Softmax activation function 优化求得整个语句内每个词汇可能的概率，如图 12.20 所示。

$$\alpha_{ts} = \frac{\exp(\text{score}(\boldsymbol{h}_t, \bar{\boldsymbol{h}}_s))}{\sum_{s'=1}^{S} \exp(\text{score}(\boldsymbol{h}_t, \bar{\boldsymbol{h}}_{s'}))} \quad \text{[Attention weights]}$$

$$\boldsymbol{c}_t = \sum_s \alpha_{ts} \bar{\boldsymbol{h}}_s \quad \text{[Context vector]}$$

$$\boldsymbol{a}_t = f(\boldsymbol{c}_t, \boldsymbol{h}_t) = \tanh(\boldsymbol{W}_c[\boldsymbol{c}_t; \boldsymbol{h}_t]) \quad \text{[Attention vector]}$$

$$\text{score}(\boldsymbol{h}_t, \bar{\boldsymbol{h}}_s) = \begin{cases} \boldsymbol{h}_t^\top \boldsymbol{W} \bar{\boldsymbol{h}}_s & \text{[Luong's multiplicative style]} \\ \boldsymbol{v}_a^\top \tanh(\boldsymbol{W}_1 \boldsymbol{h}_t + \boldsymbol{W}_2 \bar{\boldsymbol{h}}_s) & \text{[Bahdanau's additive style]} \end{cases}$$

图 12.20　Attention weight 公式

虚拟程序代码如下：

(1) score = FC(tanh(FC(EO) + FC(H)))，其中：

　　FC：完全连接层。

　　EO：Encoder 输出 (output)。

　　H：所有隐藏层输出。

　　tanh：tanh activation function。

(2) Attention weights = softmax(score, axis = 1)。

(3) Context vector = sum(Attention weights * EO, axis = 1)。

(4) Embedding output = 译码器的 input 经嵌入层 (Embedding layer) 处理后的输出。

(5) Attention Vector = concat(embedding output, context vector)。

推荐《浅谈神经机器翻译 & 用 Transformer 与 TensorFlow 2 英翻中》[19] 一文的流程图动画，有助了解 Seq2Seq 模型的特点，如图 12.21 所示。

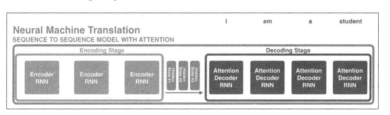

图 12.21　Seq2Seq 模型加上注意力机制

注意力机制的目的就是要建构如图 12.22 所示关联，eating 与 apple 是高度相关，eating 与 green 是低度相关。

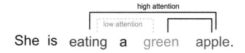

图 12.22　注意力机制范例

图片来源：*Attention? Attention!*[20]

范例**8.** 使用**Seq2Seq**架构，加上注意力**(Attention)**机制，实操神经机器翻译**(NMT)**。此范例修改自**PyTorch**官网所提供的范例Translation with a sequence to sequence network and attention[21]。

**下列程序代码请参考【12_11_seq2seq_translation.ipynb】。**

数据集：自 https://download.pytorch.org/tutorial/data.zip 下载法文/英文对照文件 data.zip，解压缩后取得 eng-fra.txt。

神经机器翻译的相关步骤如图 12.23 所示。

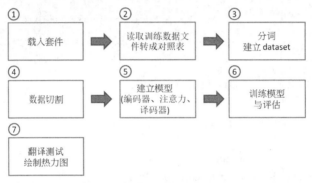

图 12.23　相关步骤

(1) 加载相关套件。

```
1  %matplotlib inline
2  from __future__ import unicode_literals, print_function, division
3  from io import open
4  import unicodedata
5  import string
6  import re
7  import random
8
9  import torch
10 import torch.nn as nn
11 from torch import optim
12 import torch.nn.functional as F
```

(2) 文字的前置处理函数。

```
1  SOS_token = 0   # 字句的开头加一标志
2  EOS_token = 1   # 字句的结尾加一标志
3
4  # 前置处理函数
5  class Lang:
6      def __init__(self, name):
7          self.name = name
8          self.word2index = {}  # 单词转代码的字典
9          self.word2count = {}
10         self.index2word = {0: "SOS", 1: "EOS"}  # 代码转单字的字典
11         self.n_words = 2  # Count SOS and EOS
12
13     def addSentence(self, sentence):
14         for word in sentence.split(' '):
15             self.addWord(word)  # 分词
16
17     def addWord(self, word):  # 建立词汇表
18         if word not in self.word2index:
19             self.word2index[word] = self.n_words
20             self.word2count[word] = 1
21             self.index2word[self.n_words] = word
22             self.n_words += 1
23         else:
24             self.word2count[word] += 1
```

```python
# Unicode 转 ASCII
# https://stackoverflow.com/a/518232/2809427
def unicodeToAscii(s):
    return ''.join(
        c for c in unicodedata.normalize('NFD', s)
        if unicodedata.category(c) != 'Mn'
    )

# 转小写、去除前后的空白、去除标点符号
def normalizeString(s):
    s = unicodeToAscii(s.lower().strip())
    s = re.sub(r"([.!?])", r" \1", s)
    s = re.sub(r"[^a-zA-Z.!?]+", r" ", s)
    return s
```

```python
# 读取文件
def readLangs(lang1, lang2, reverse=False):
    print("Reading lines...")

    # 读取文件,分行
    lines = open(f'./nlp_data/{lang1}-{lang2}.txt', encoding='utf-8').\
        read().strip().split('\n')

    # 每行分栏
    pairs = [[normalizeString(s) for s in l.split('\t')] for l in lines]

    # 法文翻译为英文,或反向
    if reverse:
        pairs = [list(reversed(p)) for p in pairs]
        input_lang = Lang(lang2)
        output_lang = Lang(lang1)
    else:
        input_lang = Lang(lang1)
        output_lang = Lang(lang2)

    return input_lang, output_lang, pairs
```

```python
# 每句只限翻译 10 个单词
MAX_LENGTH = 10

# 特殊缩写字对照表
eng_prefixes = (
    "i am ", "i m ",
    "he is", "he s ",
    "she is", "she s ",
    "you are", "you re ",
    "we are", "we re ",
    "they are", "they re "
)

# 超过 10 个单词的字句删除
def filterPair(p):
    return len(p[0].split(' ')) < MAX_LENGTH and \
        len(p[1].split(' ')) < MAX_LENGTH and \
        p[1].startswith(eng_prefixes)


def filterPairs(pairs):
    return [pair for pair in pairs if filterPair(pair)]
```

```python
# 前置处理:整合以上函数
def prepareData(lang1, lang2, reverse=False):
    input_lang, output_lang, pairs = readLangs(lang1, lang2, reverse)
    print("Read %s sentence pairs" % len(pairs))
    pairs = filterPairs(pairs)
    print("Trimmed to %s sentence pairs" % len(pairs))
    print("Counting words...")
    for pair in pairs:
        input_lang.addSentence(pair[0])
        output_lang.addSentence(pair[1])
    print("Counted words:")
    print(input_lang.name, input_lang.n_words)
    print(output_lang.name, output_lang.n_words)
    return input_lang, output_lang, pairs

# eng-fra.txt 文件读取与处理
input_lang, output_lang, pairs = prepareData('eng', 'fra', True)
# 随机显示一笔
print(random.choice(pairs))
```

执行结果：共 135842 句，超过 10 个单词的字句删除，剩 10599 句，法文词汇表有 4345 个单词，英文词汇表有 2803 个单词，随机显示一笔法文 / 英文对照。

```
Reading lines...
Read 135842 sentence pairs
Trimmed to 10599 sentence pairs
Counting words...
Counted words:
fra 4345
eng 2803
['tu es prudent .', 'you re careful .']
```

(3) 建立 Seq2Seq 模型：分为 Encoder、Decoder 两个网络，先建立 Encoder 模型，含嵌入层、GRU 层，输入法文。

```
 1  class EncoderRNN(nn.Module):
 2      def __init__(self, input_size, hidden_size):
 3          super(EncoderRNN, self).__init__()
 4          self.hidden_size = hidden_size
 5
 6          self.embedding = nn.Embedding(input_size, hidden_size)
 7          self.gru = nn.GRU(hidden_size, hidden_size)
 8
 9      def forward(self, input, hidden):
10          embedded = self.embedding(input).view(1, 1, -1)
11          output = embedded
12          output, hidden = self.gru(output, hidden)
13          return output, hidden
14
15      def initHidden(self):
16          return torch.zeros(1, 1, self.hidden_size, device=device)
```

(4) 建立 Decoder 模型：含嵌入层、GRU 层，使用 Encoder 的输出 (context vector) 作为隐藏层的初始状态 (h0)，结合英文输入。

```
 1  class DecoderRNN(nn.Module):
 2      def __init__(self, hidden_size, output_size):
 3          super(DecoderRNN, self).__init__()
 4          self.hidden_size = hidden_size
 5
 6          self.embedding = nn.Embedding(output_size, hidden_size)
 7          self.gru = nn.GRU(hidden_size, hidden_size)
 8          self.out = nn.Linear(hidden_size, output_size)
 9          self.softmax = nn.LogSoftmax(dim=1)
10
11      def forward(self, input, hidden):
12          output = self.embedding(input).view(1, 1, -1)
13          output = F.relu(output)
14          output, hidden = self.gru(output, hidden)
15          output = self.softmax(self.out(output[0]))
16          return output, hidden
17
18      def initHidden(self):
19          return torch.zeros(1, 1, self.hidden_size, device=device)
```

(5) Attention Decoder：加上注意力的 Decoder 模型，是上面 Decoder 的加强版。

将 Encoder 的输出与英文输入结合，利用完全连接层 (Linear) 计算注意力的权重，模型如图 12.24 所示，其中 attn 是完全连接层。

# 第 12 章 | 自然语言处理的算法

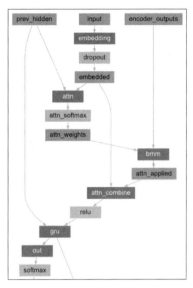

图 12.24 Decoder 模型

```
1  class AttnDecoderRNN(nn.Module):
2      def __init__(self, hidden_size, output_size, dropout_p=0.1, max_length=MAX_LENGTH):
3          super(AttnDecoderRNN, self).__init__()
4          self.hidden_size = hidden_size
5          self.output_size = output_size
6          self.dropout_p = dropout_p
7          self.max_length = max_length
8
9          self.embedding = nn.Embedding(self.output_size, self.hidden_size)
10         self.attn = nn.Linear(self.hidden_size * 2, self.max_length)
11         self.attn_combine = nn.Linear(self.hidden_size * 2, self.hidden_size)
12         self.dropout = nn.Dropout(self.dropout_p)
13         self.gru = nn.GRU(self.hidden_size, self.hidden_size)
14         self.out = nn.Linear(self.hidden_size, self.output_size)

16     def forward(self, input, hidden, encoder_outputs):
17         embedded = self.embedding(input).view(1, 1, -1)
18         embedded = self.dropout(embedded)
19
20         attn_weights = F.softmax(
21             self.attn(torch.cat((embedded[0], hidden[0]), 1)), dim=1)
22         attn_applied = torch.bmm(attn_weights.unsqueeze(0),
23                                  encoder_outputs.unsqueeze(0))
24
25         output = torch.cat((embedded[0], attn_applied[0]), 1)
26         output = self.attn_combine(output).unsqueeze(0)
27
28         output = F.relu(output)
29         output, hidden = self.gru(output, hidden)
30
31         output = F.log_softmax(self.out(output[0]), dim=1)
32         return output, hidden, attn_weights
33
34     def initHidden(self):
35         return torch.zeros(1, 1, self.hidden_size, device=device)
```

(6) 定义张量转换函数：字句与张量互转。

```
1  def indexesFromSentence(lang, sentence):
2      return [lang.word2index[word] for word in sentence.split(' ')]
3
4  def tensorFromSentence(lang, sentence):
5      indexes = indexesFromSentence(lang, sentence)
6      indexes.append(EOS_token)
7      return torch.tensor(indexes, dtype=torch.long, device=device).view(-1, 1)
8
9  def tensorsFromPair(pair):
10     input_tensor = tensorFromSentence(input_lang, pair[0])
11     target_tensor = tensorFromSentence(output_lang, pair[1])
12     return (input_tensor, target_tensor)
```

(7) 定义模型训练函数：程序代码实现有两种模式 "Free-Running" "Teacher Forcing"，Free-Running 就是一般 RNN 的模式，以上一步骤的隐藏层输出 $h_{t-1}$，作为下一步骤的输入，而 Teacher Forcing 则采用上一步骤的实际标注 ($Y$)，作为下一步骤的输入，就像有一位老师随时在旁边指导，这样做的好处是训练时不会偏离正轨过大，但缺点则是易造成过度拟合，如果测试数据与训练数据概率分布不同时，会造成测试准确度不佳。

本范例采用混合模式，以第 1 行程序代码决定采用 Teacher Forcing 的概率，每一周期都采用随机数，决定是否采用 Teacher Forcing（第 27 行）。

第 36 行程序代码采用 Teacher Forcing，第 44 行采用 Free-Running。

```
1   teacher_forcing_ratio = 0.5 # 采用 Teacher Forcing 的概率
2
3   def train(input_tensor, target_tensor, encoder, decoder, encoder_optimizer,
4           decoder_optimizer, criterion, max_length=MAX_LENGTH):
5       encoder_hidden = encoder.initHidden()
6
7       encoder_optimizer.zero_grad()
8       decoder_optimizer.zero_grad()
9
10      input_length = input_tensor.size(0)
11      target_length = target_tensor.size(0)
12
13      encoder_outputs = torch.zeros(max_length, encoder.hidden_size, device=device)
14
15      loss = 0
16
17      for ei in range(input_length):
18          encoder_output, encoder_hidden = encoder(
19              input_tensor[ei], encoder_hidden)
20          encoder_outputs[ei] = encoder_output[0, 0]
21
22      decoder_input = torch.tensor([[SOS_token]], device=device)
23
24      decoder_hidden = encoder_hidden
25
26      # 取随机乱数，决定是否采用 Teacher Forcing
27      use_teacher_forcing = True if random.random() < teacher_forcing_ratio else False
28
29      # Teacher Forcing 模式
30      if use_teacher_forcing:
31          for di in range(target_length):
32              decoder_output, decoder_hidden, decoder_attention = decoder(
33                  decoder_input, decoder_hidden, encoder_outputs)
34              loss += criterion(decoder_output, target_tensor[di])
35              # 以上一步骤的实际标注(Y)，作为下一步骤的输入
36              decoder_input = target_tensor[di]
37
38      else: # Free-Running 模式
39          for di in range(target_length):
40              decoder_output, decoder_hidden, decoder_attention = decoder(
41                  decoder_input, decoder_hidden, encoder_outputs)
42              topv, topi = decoder_output.topk(1)
43              # 以上一步骤的隐藏层输出，作为下一步骤的输入
44              decoder_input = topi.squeeze().detach()
45
46              loss += criterion(decoder_output, target_tensor[di])
47              if decoder_input.item() == EOS_token:
48                  break
49
50      loss.backward()
51
52      encoder_optimizer.step()
53      decoder_optimizer.step()
54
55      return loss.item() / target_length
```

(8) 定义计算运行时间的函数。

```python
import time
import math

# 换算为分钟
def asMinutes(s):
    m = math.floor(s / 60)
    s -= m * 60
    return f'{m}m {s}s'

# 计算执行时间
def timeSince(since, percent):
    now = time.time()
    s = now - since
    es = s / (percent)
    rs = es - s
    return f'{asMinutes(s)} (- {asMinutes(rs)})'
```

(9) 定义训练周期函数：整合以上函数。

```python
def trainIters(encoder, decoder, n_iters, print_every=1000, plot_every=100
               , learning_rate=0.01):
    start = time.time()
    plot_losses = []
    print_loss_total = 0  # 初始化列印的损失值
    plot_loss_total = 0   # 初始化绘制的损失值

    # 定义优化器、损失函数
    encoder_optimizer = optim.SGD(encoder.parameters(), lr=learning_rate)
    decoder_optimizer = optim.SGD(decoder.parameters(), lr=learning_rate)
    training_pairs = [tensorsFromPair(random.choice(pairs))
                      for i in range(n_iters)]
    criterion = nn.NLLLoss()

    # 训练周期
    for iter in range(1, n_iters + 1):
        training_pair = training_pairs[iter - 1]
        input_tensor = training_pair[0]
        target_tensor = training_pair[1]

        loss = train(input_tensor, target_tensor, encoder,
                     decoder, encoder_optimizer, decoder_optimizer, criterion)
        print_loss_total += loss
        plot_loss_total += loss

        if iter % print_every == 0:
            print_loss_avg = print_loss_total / print_every
            print_loss_total = 0
            print(f'{timeSince(start, iter / n_iters)}' +
                  f' ({iter} {iter / n_iters * 100}%) {print_loss_avg:.4f}')

        if iter % plot_every == 0:
            plot_loss_avg = plot_loss_total / plot_every
            plot_losses.append(plot_loss_avg)
            plot_loss_total = 0

    showPlot(plot_losses)
```

(10) 定义绘图函数：使用 plt.switch_backend('agg')，需加 "%matplotlib inline"，图形才会显示。

```python
import matplotlib.pyplot as plt
plt.switch_backend('agg')
import matplotlib.ticker as ticker
import numpy as np

def showPlot(points):
    plt.figure()
    fig, ax = plt.subplots()
    # this locator puts ticks at regular intervals
    loc = ticker.MultipleLocator(base=0.2)
    ax.yaxis.set_major_locator(loc)
    plt.plot(points)
```

(11) 定义模型评估函数。

```python
def evaluate(encoder, decoder, sentence, max_length=MAX_LENGTH):
    with torch.no_grad():
        input_tensor = tensorFromSentence(input_lang, sentence)
        input_length = input_tensor.size()[0]
        encoder_hidden = encoder.initHidden()

        encoder_outputs = torch.zeros(max_length, encoder.hidden_size, device=device)

        for ei in range(input_length):
            encoder_output, encoder_hidden = encoder(input_tensor[ei],
                                                     encoder_hidden)
            encoder_outputs[ei] += encoder_output[0, 0]

        decoder_input = torch.tensor([[SOS_token]], device=device)  # SOS

        decoder_hidden = encoder_hidden

        decoded_words = []
        decoder_attentions = torch.zeros(max_length, max_length)

        for di in range(max_length):
            decoder_output, decoder_hidden, decoder_attention = decoder(
                decoder_input, decoder_hidden, encoder_outputs)
            decoder_attentions[di] = decoder_attention.data
            topv, topi = decoder_output.data.topk(1)
            if topi.item() == EOS_token:
                decoded_words.append('<EOS>')
                break
            else:
                decoded_words.append(output_lang.index2word[topi.item()])

            decoder_input = topi.squeeze().detach()

        return decoded_words, decoder_attentions[:di + 1]
```

(12) 模型训练：笔者使用 GTX 1050 Ti 显卡，训练 75000 周期约 70 分钟。

```python
hidden_size = 256
encoder1 = EncoderRNN(input_lang.n_words, hidden_size).to(device)
attn_decoder1 = AttnDecoderRNN(hidden_size, output_lang.n_words,
                               dropout_p=0.1).to(device)

trainIters(encoder1, attn_decoder1, 75000, print_every=5000)
```

执行结果。

```
4m 28s (- 62m 38s) (5000 6%) 2.8160
9m 1s (- 58m 38s) (10000 13%) 2.2702
13m 36s (- 54m 24s) (15000 20%) 1.9482
18m 11s (- 50m 2s) (20000 26%) 1.7116
23m 10s (- 46m 20s) (25000 33%) 1.5424
28m 9s (- 42m 14s) (30000 40%) 1.3399
33m 15s (- 38m 0s) (35000 46%) 1.2217
38m 4s (- 33m 18s) (40000 53%) 1.0693
42m 45s (- 28m 30s) (45000 60%) 0.9944
47m 24s (- 23m 42s) (50000 66%) 0.8953
52m 8s (- 18m 57s) (55000 73%) 0.8258
56m 52s (- 14m 13s) (60000 80%) 0.7477
61m 35s (- 9m 28s) (65000 86%) 0.6795
66m 14s (- 4m 43s) (70000 93%) 0.6376
70m 52s (- 0m 0s) (75000 100%) 0.5740
```

(13) 任选 10 笔训练数据评估。

```python
def evaluateRandomly(encoder, decoder, n=10):
    for i in range(n):
        pair = random.choice(pairs)
        print('>', pair[0])
        print('=', pair[1])
        output_words, attentions = evaluate(encoder, decoder, pair[0])
        output_sentence = ' '.join(output_words)
        print('<', output_sentence)
        print('')

evaluateRandomly(encoder1, attn_decoder1)
```

执行结果：每一笔含三列，第 1 列为输入值，第 2 列为实际值，第 3 列为预测值，大部分预测均正确。

```
> elles en sont presque la .
= they re almost here .
< they re almost here . <EOS>

> je travaille pour une entreprise de commerce .
= i m working for a trading firm .
< i m a a for a few . <EOS>

> nous sommes des survivantes .
= we re survivors .
< we re survivors . <EOS>

> il est impopulaire pour une raison quelconque .
= he is unpopular for some reason .
< he is unpopular for some reason . <EOS>
```

(14) Attention 可视化：输入一句法文，显示关联图，X 轴为输入，Y 轴为预测值。

```
1  output_words, attentions = evaluate(
2      encoder1, attn_decoder1, "je suis trop froid .")
3  plt.matshow(attentions.numpy())
```

执行结果：可以看出不是呈对角线，也就是每个英文单词依赖的注意力，不是对齐的法文单词。

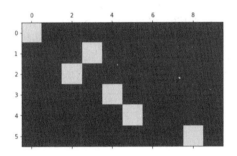

(15) Attention 可视化加强版：把单词显示在坐标轴。

```
1  def showAttention(input_sentence, output_words, attentions):
2      # Set up figure with colorbar
3      fig = plt.figure()
4      ax = fig.add_subplot(111)
5      cax = ax.matshow(attentions.numpy(), cmap='bone')
6      fig.colorbar(cax)
7
8      # Set up axes
9      ax.set_xticklabels([''] + input_sentence.split(' ') +
10                         ['<EOS>'], rotation=90)
11     ax.set_yticklabels([''] + output_words)
12
13     # Show label at every tick
14     ax.xaxis.set_major_locator(ticker.MultipleLocator(1))
15     ax.yaxis.set_major_locator(ticker.MultipleLocator(1))
16
17     plt.show()
18
19 def evaluateAndShowAttention(input_sentence):
20     output_words, attentions = evaluate(
21         encoder1, attn_decoder1, input_sentence)
22     print('input =', input_sentence)
23     print('output =', ' '.join(output_words))
24     showAttention(input_sentence, output_words, attentions)
25
26 # 测试4句法文
27 evaluateAndShowAttention("elle a cinq ans de moins que moi .")
28 evaluateAndShowAttention("elle est trop petit .")
29 evaluateAndShowAttention("je ne crains pas de mourir .")
30 evaluateAndShowAttention("c est un jeune directeur plein de talent .")
```

执行结果：英文的 is too 都依赖法文的 petit 较大。

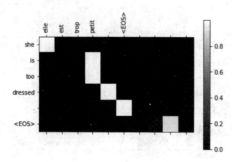

本例还可以进行以下试验：

(1) 改为中翻英，可以自 https://www.manythings.org/anki/ 下载 cmn-eng.zip，内容为简体中文，可使用 MS Word 翻译为繁体中文。

(2) 使用 Word2Vec/GloVe 取代嵌入层的训练。

(3) 使用 Stacked LSTM、改变隐藏层神经元个数，比较运行时间及准确率。

Seq2Seq 模型处理一对一的训练数据，还有下列领域的应用：

(1) 人机界面：指令与执行。

(2) 聊天机器人：聊天与回应。

(3) 常用问答集 (FAQ)：问与答。

除了一对一外，按照输入/输出的个数不同，也可以进行各种形态的应用，如图 12.25 所示。

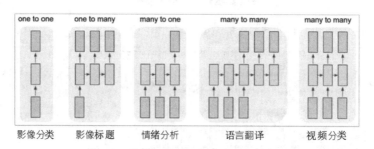

图 12.25　Seq2Seq 模型的各种形态和应用

图片来源：*The Unreasonable Effectiveness of Recurrent Neural Networks*[22]。

(1) 一对一 (One to One)：固定长度的输入 (input) 与输出 (output)，即一般的神经网络模型。例如影像分类，输入一张影像后，预测这张影像所属的类别。

(2) 一对多 (One to Many)：单一输入，多个输出。例如影像标题 (Image Captioning)，输入一个影像后，接着侦测影像内的多个对象，并一一给予标题，这称为"Sequence Output"。

(3) 多对一 (Many to One)：多个输入，单一输出。例如情绪分析 (Sentiment Analysis)，输入一大段话后，判断这段话是正面或负面的情绪表达，这称为"Sequence input"。

(4) 多对多 (Many to Many)：多个输入，多个输出。例如语言翻译 (Machine Translation)，输入一段英文句子后，翻译成中文，这称为"Sequence Input and Sequence Output"。

(5) 另一种多对多 (Many to Many)：多个输入，多个输出同步 (Synchronize)。例如视频分类 (Video Classification)，输入一段影片后，每一帧 (Frame) 都各产生一个标题，这称为"Synced Sequence Input and Output"。

## 12-9 Transformer 架构

Google 的学者 Ashish Vaswani 等人于 2017 年依照 Seq2Seq 模型加上注意力机制，提出了 Transformer 架构，如图 12.26 所示。架构一推出后，立即跃升为 NLP 近年来最强大的算法，而 *Attention Is All You Need*[23] 一文也被公认是必读的文章，各种改良的算法也纷纷出笼，例如 BERT、GPT-2、GPT-3、XLNet、ELMo、T5 等，几乎抢占了 NLP 大部分的版面，接下来我们就来认识一下 Transformer 与其相关的算法。

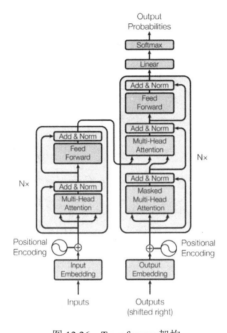

图 12.26　Transformer 架构

图片来源：*Attention Is All You Need*[23]

### 12-9-1　Transformer 原理

RNN/LSTM/GRU 有个最大的缺点，因为要以上文预测目前的目标，必须以序列的方式，依次执行每一个节点的训练，造成执行效能过慢。而 Transformer 为了克服这一问题，提出"自注意力机制"(Self-Attention Mechanism) 及多头 (Multi-Head) 机制，取代模型中的 RNN 神经层，以平行处理的方式计算所有的输出，步骤如下，请同时参考图 12.26。

(1) 首先，输入向量 (Input Vector) 被表征为 $Q$、$K$、$V$ 向量 (Vector)。

$K$：Key Vector，为 Encoder 隐藏层状态的键值，即上下文的词向量。
$V$：Value Vector，为 Encoder 隐藏层状态的输出值。
$Q$：Query Vector，为 Decoder 的前一期输出。

故自注意力机制对应 $Q$、$K$、$V$，共有三种权重，而单纯的注意力机制则只有一种权重——Attention Weight。

利用神经网络优化可以找到三种权重的最佳值，然后以输入向量个别乘以三种权重，即可求得 $Q$、$K$、$V$ 向量。

(2) $Q$、$K$、$V$ 再经过图 12.27 所示的运算即可得到自注意力矩阵。

点积 (Dot product) 运算：$Q \times K$，计算输入向量与上下文词汇的相似度。

特征缩放：$Q$、$K$ 维度开根号，通常 $Q$、$K$ 维度是 64，故 $\sqrt{64} = 8$。

Softmax 运算：将上述结果转为概率。

找出要重视的上下文词汇：以 Value Vector 乘以上述概率，较大值为要重视的上下文词汇。

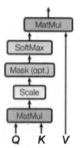

图 12.27　"自注意力机制"运算

图片来源：*Attention Is All You Need*[23]

图 12.27 运算过程以数学公式表达如下，即自注意力矩阵公式：

$$Attention(Q,K,V) = soft\max\left(\frac{QK^T}{\sqrt{d_k}}\right)V$$

其中 $d_k$ 为 Key Vector 维度，通常是 64。

(3) 自注意力机制是多头 (Multi-Head) 的，通常是 8 个头，如图 12.27 所示的机制，经过内积 (Dot product) 运算，串联这 8 个头，如图 12.28 所示。

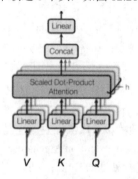

图 12.28　"自注意力机制"多头运算

图片来源：*Attention Is All You Need*[23]

多头自注意力矩阵公式如下：

$$MultiHead(Q,K,V) = concat(head_1 head_2...head_n)W_O$$
$$where, head_i = Attention\left(QW_i^Q, KW_i^K, VW_i^V\right)$$

(4) 最后加上其他的神经层，如图 12.26，就构成了 Transformer 网络架构。

要了解详细的计算过程请参考 *Illustrated: Self-Attention*[24] 一文，它还附有精美的动画，另外，*The Illustrated Transformer*[25] 也值得一读。

总而言之，自注意力机制就是要找出应该关注的上下文词汇，举例来说：

The animal didn't cross the street because it was too tired.

其中的 it 是代表 animal 还是 street？

通过自注意力机制，可以帮我们找出 it 与上下文词汇的关联度，进而判断出 it 所代表的是 animal，如图 12.29 所示。

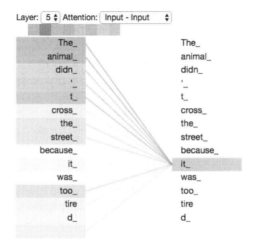

图 12.29　自注意力机制示意图，it 与上下文词汇的关联度

图片来源：*The Illustrated Transformer*[23]

## 12-9-2　Transformer 效能

依 *Self-attention in NLP*[26] 一文中的试验，上述 Transformer 网络在一台 8 颗 NVIDIA P100 GPU 的服务器上运作，大约要 3.5 天才能完成训练，英文 / 德文翻译的准确率 (BLEU) 约 28.4 分，英文 / 法文翻译的准确率约 41.8 分。BLEU(Bilingual Evaluation Understudy) 是专为双语言翻译所设定的效能衡量指标，是根据 n-gram 的相符数目 ( 不考虑顺序 )，乘以对应的权重而得到的分数，详细的计算可参考 *A Gentle Introduction to Calculating the BLEU Score for Text in Python*[27] 一文。

一般做法会分为两阶段：

(1) 基础模型 (Base Model)：利用大量的语料库训练一个基础模型，例如 BERT 使用维基百科上 33 亿单词训练，完成后存储成预先训练模型 (Pre-Trained Model)。

(2) 微调 (Fine tuning)：利用预先训练模型，进行转移学习 (Transfer Learning)，应用到各种任务 (Task)，如图 12.30 所示。

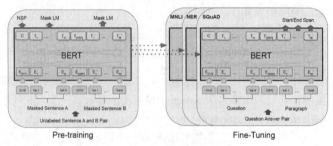

图 12.30　Transformer 架构的转移学习

图片来源：*BERT: Pre-training of Deep Bidirectional Transformers for Language Understanding*[28]

General Language Understanding Evaluation(GLEU) 效能评判的任务如表 12.1 所示。

**表 12.1　General Language Understanding Evaluation 效能评价的任务**

| # | Task | 说明 |
|---|---|---|
| 1 | MNLI | 推理，给定一个前提(Premise)，推断假设(Hypothesis)是否成立，三分类(推理、矛盾或中立) |
| 2 | QQP | 相似性，判断Quora上的两个问题句是否同义 |
| 3 | QNLI | Q&A，判断文本是否包含问题的答案 |
| 4 | STS-B | 相似性，预测两个句子的相似性，包括5个级别 |
| 5 | MRPC | 相似性，判断两个句子是否同义 |
| 6 | RTE | 推理，类似于MNLI，二分类判断 |
| 7 | SWAG | 接续，从四个句子中选择为可能为前句下文 |
| 8 | SST-2 | 情绪分析 |
| 9 | CoLA | 语义判断，判定句子是否语法正确 |
| 10 | SQuAD | Q&A，自文本找出问题的答案 |
| 11 | CoNLL | NER |

两阶段的做法类似之前介绍的影像辨识的预先训练模型，大型研究机构创建各种预先训练模型后，我们就可以花费较少的精力与时间，利用这些模型微调以完成各项特定的任务，如图 12.31 所示。

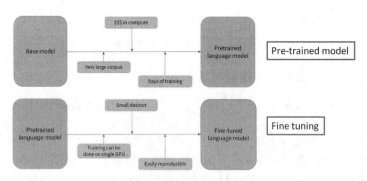

图 12.31　Transformer 两阶段特点

图片来源：*Hugging Face Transformers*[29]

# 12-10　BERT

BERT (Bidirectional Encoder Representations from Transformers) 顾名思义，就是双向的 Transformer，在 2018 年由 Google 的 Jacob Devlin 等学者首度发表，是目前最强大的

Transformer 算法，详情可参阅 *BERT: Pre-training of Deep Bidirectional Transformers for Language Understanding*[28] 一文。

Word2Vec/GloVe 每个单词只以一个多维的词向量表示，但是，一词多义是所有语系共有的现象，例如，Apple 可以表示水果也可以表示苹果公司，Bank 可以表示银行也可以表示岸边。BERT 为解决这个问题，提出上下文相关 (Context Dependent) 的概念，输入改为一整个句子，而不是一个单词，例如：

We go to the river bank. ➡ bank 是岸边。

I need to go to bank to make a deposit. ➡ bank 是银行。

BERT 算法比 Transformer 更复杂，要花更多的时间训练，为什么还要介绍呢？有以下两点原因：

(1) BERT 有各种各样的预先训练模型 (Pre-trained model)，可以进行转移学习。

(2) BERT 支持中文模型。

虽然没办法训练模型，但为了在实务上能灵活运用，我们还是要稍微理解一下 BERT 的运作原理，免得到时候误用，还不知道为什么错，那就很麻烦了。

BERT 使用两个训练策略：

(1) Masked LM (MLM)。

(2) Next Sentence Prediction (NSP)。

## 12-10-1 Masked LM

RNN/LSTM/GRU 都是以序列的方式逐一产生输出，导致训练速度过慢，而 Masked LM (MLM) 则可以克服这个问题，训练数据在输入模型前，有 15% 的词汇先以 [MASK] 符号取代，即所谓的屏蔽 (Mask)，之后算法就试图用未屏蔽的词汇来预测被屏蔽的词汇，类似 CBOW。Masked LM 的架构如图 12.32 所示。

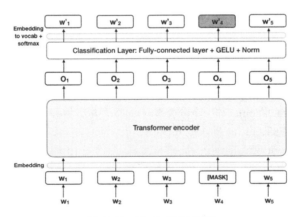

图 12.32　Masked LM 架构

图片来源：*BERT Explained: State of the art language model for NLP*[30]

## 12-10-2 Next Sentence Prediction

Next Sentence Prediction(NSP)( 如图 12.33 所示 ) 训练数据含两个语句，NSP 预测第 2 句是否第 1 句的接续下文。训练时会取样正负样本各占 50%，进行以下的前置处理。

(1) 符号词嵌入 (Token embedding)：[CLS] 插在第 1 句的前面，[SEP] 插在每一句的后面。

(2) 语句词嵌入 (Sentence Embedding)：在每个符号（词汇）上加注它是属于第 1 句或第 2 句。

(3) 位置词嵌入 (Positional Embedding)：在每个符号（词汇）上加注它是在合并语句中的第几个位置。

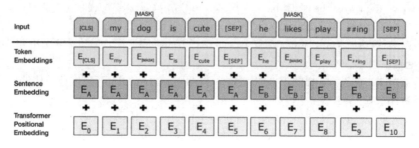

图 12.33　Next Sentence Prediction(NSP)。

图片来源：*BERT Explained: State of the art language model for NLP*[30]

这三种词嵌入就类似于前面"自注意力机制"的 *Q*、*K*、*V*。

BERT 训练时会结合两个算法，目标是最小化两个策略的合并损失函数，注意，MLM、NSP 两项任务都不需要标注数据，可直接由语料库产生训练数据，不需要人工标注，比 ImageNet 轻松很多，这种方式也称为"自我学习"(Self Learning)。语音的训练也可以如法炮制，使用自我学习进行基础模型的训练。

根据 BERT GitHub[31] 说明，模型训练 (Pre-trained model) 在 4~16 个 TPU 的服务器上要训练 4 天的时间，这又是身为平凡小民的笔者不可承受之重，因此，我们还是乖乖地下载预先训练好的模型，然后集中火力在效能微调 (Fine-tuning) 上比较实际。

## 12-10-3　BERT 效能微调

效能微调 (Fine-tuning) 就是根据不同的应用领域，加入各行业别的知识，使 BERT 能更聪明，有以下应用类型。

(1) 分类 (Classification)：加一个分类层 (Dense)，进行情绪分析 (Sentiment Analysis) 的判别，可参考 BERT GitHub 的程序 run_classifier.py。

(2) 问答 (Question Answering)：比如 SQuAD 数据集，输入一个问题后，能够在全文中标示出答案的开头与结束的位置，可参考 BERT GitHub 的程序 run_squad.py。

(3) 命名实体识别 (Named Entity Recognition, NER)：输入一段文字后，可以标注其中的实体，如人名、组织、日期等。

以分类为例，测试处理程序如下：

(1) 自 GLUE(https://gluebenchmark.com/tasks) 下载数据集，较有名的是 Quora Question Pairs，它是科技问答网站的问题配对，辨识问题相似与否。

(2) 下载预先训练模型 BERT-Base(https://storage.googleapis.com/bert_models/2018_10_18/uncased_L-12_H-768_A-12.zip)，解压缩至某一目录，例如 BERT_BASE_DIR 所指向的目录。

(3) 效能微调：下列是 Linux 指令，Windows 作业环境下可直接把变量代入。

```
export BERT_BASE_DIR=/path/to/bert/uncased_L-12_H-768_A-12
export GLUE_DIR=/path/to/glue

python run_classifier.py \
  --task_name=MRPC \
  --do_train=true \
  --do_eval=true \
  --data_dir=$GLUE_DIR/MRPC \
  --vocab_file=$BERT_BASE_DIR/vocab.txt \
  --bert_config_file=$BERT_BASE_DIR/bert_config.json \
  --init_checkpoint=$BERT_BASE_DIR/bert_model.ckpt \
  --max_seq_length=128 \
  --train_batch_size=32 \
  --learning_rate=2e-5 \
  --num_train_epochs=3.0 \
  --output_dir=/tmp/mrpc_output/
```

(4) 得到结果如下。

```
***** Eval results *****
  eval_accuracy = 0.845588
  eval_loss = 0.505248
  global_step = 343
  loss = 0.505248
```

(5) 预测：参数 do_predict=true，输入放在 input/test.tsv，执行结果在 output/test_results.tsv。

```
export BERT_BASE_DIR=/path/to/bert/uncased_L-12_H-768_A-12
export GLUE_DIR=/path/to/glue
export TRAINED_CLASSIFIER=/path/to/fine/tuned/classifier

python run_classifier.py \
  --task_name=MRPC \
  --do_predict=true \
  --data_dir=$GLUE_DIR/MRPC \
  --vocab_file=$BERT_BASE_DIR/vocab.txt \
  --bert_config_file=$BERT_BASE_DIR/bert_config.json \
  --init_checkpoint=$TRAINED_CLASSIFIER \
  --max_seq_length=128 \
  --output_dir=/tmp/mrpc_output/
```

也可以使用 SQuAD 问答 (Question Answering) 数据集，与上述程序类似。

注意，效能微调使用 GPU 时，BERT GitHub 建议为 Titan X 或 GTX 1080，否则容易发生内存不足的情形。看到这里，可能有些读者 ( 包括笔者 ) 脸上有三条线了，还好有一些框架可以让我们直接试验，无须使用上述程序。

BERT 的变形很多，故统称为 "BERTology"，可参阅 A Primer in BERTology: What we know about how BERT works[32]，该论文发表时就有将近 20 多种，每一种模型的大小、速度、效能及适用的任务都有所不同，如图 12.34 所示。

| | Model | Compression | Performance | Speedup | Model | Evaluation |
|---|---|---|---|---|---|---|
| | BERT-base (Devlin et al., 2019) | ×1 | 100% | ×1 | $BERT_{12}$ | All GLUE tasks, SQuAD |
| | BERT-small | ×3.8 | 91% | - | $BERT_4$† | All GLUE tasks |
| Distillation | DistilBERT (Sanh et al., 2019a) | ×1.5 | 90%§ | ×1.6 | $BERT_6$ | All GLUE tasks, SQuAD |
| | $BERT_6$-PKD (Sun et al., 2019a) | ×1.6 | 98% | ×1.9 | $BERT_6$ | No WNLI, CoLA, STS-B; RACE |
| | $BERT_3$-PKD (Sun et al., 2019a) | ×2.4 | 92% | ×3.7 | $BERT_3$ | No WNLI, CoLA, STS-B; RACE |
| | Aguilar et al. (2019), Exp. 3 | ×1.6 | 93% | - | $BERT_6$ | CoLA, MRPC, QQP, RTE |
| | BERT-48 (Zhao et al., 2019) | ×62 | 87% | ×77 | $BERT_{12}$*† | MNLI, MRPC, SST-2 |
| | BERT-192 (Zhao et al., 2019) | ×5.7 | 93% | ×22 | $BERT_{12}$*† | MNLI, MRPC, SST-2 |
| | TinyBERT (Jiao et al., 2019) | ×7.5 | 96% | ×9.4 | $BERT_4$† | No WNLI; SQuAD |
| | MobileBERT (Sun et al., 2020) | ×4.3 | 100% | ×4 | $BERT_{24}$† | No WNLI; SQuAD |
| | PD (Turc et al., 2019) | ×1.6 | 98% | ×2.5‡ | $BERT_6$ | No WNLI, CoLA and STS-B |
| | WaLDORf (Tian et al., 2019) | ×4.4 | 93% | ×9 | $BERT_8$†∥ | SQuAD |
| | MiniLM (Wang et al., 2020b) | ×1.65 | 99% | ×2 | $BERT_6$ | No WNLI, STS-B, $MNLI_{mm}$; SQuAD |
| | MiniBERT(Tsai et al., 2019) | ×6** | 98% | ×27** | $mBERT_3$† | CoNLL-18 POS and morphology |
| | BiLSTM-soft (Tang et al., 2019) | ×110 | 91% | ×434‡ | $BiLSTM_1$ | No MNLI-mm; SQuAD |
| Quanti-zation | Q-BERT-MP (Shen et al., 2019) | ×13 | 98%‡ | - | $BERT_{12}$ | MNLI, SST-2, CoNLL-03, SQuAD |
| | BERT-QAT (Zafrir et al., 2019) | ×4 | 99% | - | $BERT_{12}$ | No WNLI, MNLI; SQuAD |
| | GOBO(Zadeh and Moshovos, 2020) | ×9.8 | 99% | - | $BERT_{12}$ | MNLI |
| Pruning | McCarley et al. (2020), ff2 | ×2.2‡ | 98%‡ | ×1.9‡ | $BERT_{24}$ | SQuAD, Natural Questions |
| | RPP (Guo et al., 2019) | ×1.7‡ | 99%‡ | - | $BERT_{24}$ | No WNLI, STS-B, SQuAD |
| | Soft MvP (Sanh et al., 2020) | ×33 | 94%¶ | - | $BERT_{12}$ | MNLI, QQP, SQuAD |
| | IMP (Chen et al., 2020), rewind 50% | ×1.4–2.5 | 94–100% | - | $BERT_{12}$ | No WNLI; SQuAD |
| Other | ALBERT-base (Lan et al., 2020b) | ×9 | 97% | - | $BERT_{12}$† | MNLI, SST-2 |
| | ALBERT-xxlarge (Lan et al., 2020b) | ×0.47 | 107% | - | $BERT_{12}$† | MNLI, SST-2 |
| | BERT-of-Theseus (Xu et al., 2020) | ×1.6 | 98% | ×1.9 | $BERT_6$ | No WNLI |
| | PoWER-BERT (Goyal et al., 2020) | N/A | 99% | ×2–4.5 | $BERT_{12}$ | No WNLI; RACE |

图 12.34　BERT 家族 (BERTology)

图片来源：*A Primer in BERTology: What we know about how BERT works*[32]

# 12-11　Transformers 框架

Transformers 框架是由 Hugging Face 所开发的，功能十分强大，它支持数十种模型，包括 BERT、GPT、T5、XLNet、XLM 等架构，详情请参阅 Transformers GitHub[33]，其中 BERT 就涵盖了各种变形。

接下来我们就拿 Transformers 这个框架当例子，做一些试验，它包含以下功能：

(1) 情绪分析 (Sentiment analysis)。

(2) 文本生成 (Text generation)：限英文。

(3) 命名实体识别 (Named Entity Recognition, NER)。

(4) 问题回答 (Question Answering)。

(5) 漏字填空 (Filling masked text)。

(6) 文本摘要 (Text Summarization)：提取文章大意。

(7) 翻译 (Translation)。

(8) 特征提取 (Feature extraction)：类似词向量，将文字转换为向量。

除了文字以外，最近 Transformers 也扩展功能至影像 (Image)、语音 (Audio)。

## 12-11-1　Transformers 框架范例

先安装框架，指令如下：

pip install transformers

**范例1. 情绪分析(Sentiment analysis)**。此范例程序修改自Transformers官网的"Quick tour"[34]。

下列程序代码请参考【12_12_Transformers_Sentiment_Analysis.ipynb】。

(1) 加载相关框架。

```
1  from transformers import pipeline
```

(2) 加载模型：BERT 有许多变形，下列指令默认下载 distilbert-base-uncased-finetuned-sst-2-english 模型，SST-2 即"The Stanford Sentiment Treebank"数据集。

若出现"torch_scatter installed with the wrong CUDA version"错误，表示目前 torch-scatter 的版本过新，需安装 2.0.8 版，指令如下：

pip install torch-scatter==2.0.8

```
1  # 载入模型
2  classifier = pipeline('sentiment-analysis')
```

(3) 情绪分析测试。

```
1  # 正面
2  print(classifier('We are very happy to show you the 🤗 transformers library.'))
3
4  # 负面
5  print(classifier('I hate this movie.'))
6
7  # 否定句也可以正确分类
8  print(classifier('the movie is not bad.'))
```

执行结果：非常准确，否定句也可以正确分类，不像之前的 RNN/LSTM/GRU 碰到否定句都无法正确分类。

```
[{'label': 'POSITIVE', 'score': 0.9997795224189758}]
[{'label': 'NEGATIVE', 'score': 0.9996869564056396}]
[{'label': 'POSITIVE', 'score': 0.999536395072937}]
```

(4) 可一次测试多笔。

```
1  results = classifier(["We are very happy.",
2                         "We hope you don't hate it."])
3  for result in results:
4      print(f"label: {result['label']}, with score: {round(result['score'], 4)}")
```

执行结果：非常准确，就连否定句有可能是中性的这点也能够分辨，像是 don't hate 不讨厌，但不意味是喜欢，所以分数只有 0.5。

```
label: POSITIVE, with score: 0.9999
label: NEGATIVE, with score: 0.5309
```

(5) 多语系支持：BERT 支持 100 多种语系，提供 24 种模型，以 BERT-base 的文件名 uncased_L-12_H-768_A-12.zip 为例来说明，L-12：12 层神经层，H-768：768 个隐藏层神经元，A-12：12 个头。注意，这里的多语系是指欧美的语言，并不包括中文。

(6) 西班牙文 (Spanish) 测试：笔者也不懂西班牙文，可使用 Google 翻译。

```
1  # 负面, I hate this movie
2  print(classifier('Odio esta pelicula.'))
3
4  # the movie is not bad.
5  print(classifier('la pelicula no esta mal.'))
```

执行结果：第一句只得 1 颗星，第二句得到 3 颗星。

```
[{'label': '1 star', 'score': 0.4615824222564697}]
[{'label': '3 stars', 'score': 0.6274545788764954}]
```

(7) 法文 (French) 测试。

```
1  # 负面, I hate this movie
2  print(classifier('Je deteste ce film.'))
3
4  # the movie is not bad.
5  print(classifier('le film n\'est pas mal.'))
```

执行结果：与西班牙文测试结果相同。

```
[{'label': '1 star', 'score': 0.631117582321167}]
[{'label': '3 stars', 'score': 0.5710769295692444}]
```

**范例2. 问题回答(Question Answering)**：从一段文章中截取一段文字作为问题的答案。本范例程序修改自Transformers官网Summary of the tasks的Extractive Question Answering [35]。

下列程序代码请参考【12_13_Question_Answering.ipynb】。

(1) 加载相关框架。

```
1  from transformers import pipeline
```

(2) 加载模型：参数须设为question-answering。

```
1  nlp = pipeline("question-answering")
```

(3) 设定训练数据。

```
1  context = r"Extractive Question Answering is the task of extracting an answer " + \
2  "from a text given a question. An example of a question answering " + \
3  "dataset is the SQuAD dataset, which is entirely based on that task. " + \
4  "If you would like to fine-tune a model on a SQuAD task, you may " + \
5  "leverage the examples/question-answering/run_squad.py script."
```

(4) 测试两笔数据。

```
1  result = nlp(question="What is extractive question answering?", context=context)
2  print(f"Answer: '{result['answer']}', score: {round(result['score'], 4)}",
3        f", start: {result['start']}, end: {result['end']}")
4
5  print()
6
7  result = nlp(question="What is a good example of a question answering dataset?",
8               context=context)
9  print(f"Answer: '{result['answer']}', score: {round(result['score'], 4)}",
10       f", start: {result['start']}, end: {result['end']}")
```

执行结果：非常准确，通常是从训练数据中节录一段文字当作回答。

```
Answer: 'the task of extracting an answer from a text given a question', score: 0.6226 , start: 33, end: 94
Answer: 'SQuAD dataset', score: 0.5053 , start: 146, end: 159
```

(5) 结合分词(Tokenizer)：可自定义分词器，断句会比较准确，请参阅 *Using tokenizers from Tokenizers*[36]，下面使用预设的分词器。

(6) 载入分词器(Tokenizer)。

```
1  from transformers import AutoTokenizer, AutoModelForQuestionAnswering
2  import torch
3
4  # 结合分词(Tokenizer)
5  tokenizer = AutoTokenizer.from_pretrained("bert-large-uncased-whole-word-masking-finetuned-squad")
6  model = AutoModelForQuestionAnswering.from_pretrained("bert-large-uncased-whole-word-masking-finetuned-squad")
```

(7) 加载训练数据。

```
1  text = r"""
2  🤗 Transformers (formerly known as pytorch-transformers and pytorch-pretrained-bert) provides general-purpose
3  architectures (BERT, GPT-2, RoBERTa, XLM, DistilBert, XLNet…) for Natural Language Understanding (NLU) and Natural
4  Language Generation (NLG) with over 32+ pretrained models in 100+ languages and deep interoperability between
5  TensorFlow 2.0 and PyTorch.
6  """
```

(8) 设定问题。

```
1  questions = [
2      "How many pretrained models are available in 🤗 Transformers?",
3      "What does 🤗 Transformers provide?",
4      "🤗 Transformers provides interoperability between which frameworks?",
5  ]
```

(9) 推测答案。

```
1  for question in questions:
2      inputs = tokenizer(question, text, add_special_tokens=True, return_tensors="pt")
3      input_ids = inputs["input_ids"].tolist()[0]
4  
5      outputs = model(**inputs)
6      answer_start_scores = outputs.start_logits
7      answer_end_scores = outputs.end_logits
8  
9      # Get the most likely beginning of answer with the argmax of the score
10     answer_start = torch.argmax(answer_start_scores)
11     # Get the most likely end of answer with the argmax of the score
12     answer_end = torch.argmax(answer_end_scores) + 1
13  
14     answer = tokenizer.convert_tokens_to_string(
15         tokenizer.convert_ids_to_tokens(input_ids[answer_start:answer_end])
16     )
17  
18     print(f"Question: {question}")
19     print(f"Answer: {answer}")
```

执行结果：非常准确。

```
Question: How many pretrained models are available in 🤗 Transformers?
Answer: over 32 +

Question: What does 🤗 Transformers provide?
Answer: general - purpose architectures

Question: 🤗 Transformers provides interoperability between which frameworks?
Answer: tensorflow 2. 0 and pytorch
```

**范例3. 漏字填空(Masked Language Modeling)**：遮住一个单词，由上下文猜测该单词。此范例程序修改自Transformers官网Summary of the tasks的Masked Language Modeling [37]。

下列程序代码请参考【12_14_Masked_Language_Modeling.ipynb】。

(1) 加载相关框架。

```
1  from transformers import pipeline
```

(2) 加载模型：参数须设为 fill-mask。

```
1  nlp = pipeline("fill-mask")
```

(3) 测试。

```
1  from pprint import pprint
2  pprint(nlp(f"HuggingFace is creating a {nlp.tokenizer.mask_token} " + \
3         "that the community uses to solve NLP tasks."))
```

执行结果：列出前5名与其分数，框起来的即是猜测的单词。

```
[{'score': 0.17927466332912445,
  'sequence': 'HuggingFace is creating a tool that the community uses to solve '
              'NLP tasks.',
  'token': 3944,
  'token_str': ' tool'},
 {'score': 0.11349395662546158,
  'sequence': 'HuggingFace is creating a framework that the community uses to '
              'solve NLP tasks.',
  'token': 7208,
  'token_str': ' framework'},
 {'score': 0.05243542045354843,
  'sequence': 'HuggingFace is creating a library that the community uses to '
              'solve NLP tasks.',
  'token': 5560,
  'token_str': ' library'},
 {'score': 0.03493538498878479,
  'sequence': 'HuggingFace is creating a database that the community uses to '
              'solve NLP tasks.',
  'token': 8503,
  'token_str': ' database'},
 {'score': 0.02860254235565624,
  'sequence': 'HuggingFace is creating a prototype that the community uses to '
              'solve NLP tasks.',
  'token': 17715,
  'token_str': ' prototype'}]
```

(4) 结合分词 (Tokenizer)。

```
1  # 载入相关套件
2  from transformers import AutoModelForMaskedLM, AutoTokenizer
3  import torch
4
5  # 结合分词器(Tokenizer)
6  tokenizer = AutoTokenizer.from_pretrained("distilbert-base-cased")
7  model = AutoModelForMaskedLM.from_pretrained("distilbert-base-cased")
```

(5) 推测答案。

```
1  sequence = f"Distilled models are smaller than the models they mimic. " + \
2      f"Using them instead of the large versions would help {tokenizer.mask_token} " + \
3      "our carbon footprint."
4  inputs = tokenizer(sequence, return_tensors="pt")
5  mask_token_index = torch.where(inputs["input_ids"] == tokenizer.mask_token_id)[1]
6  token_logits = model(**inputs).logits
7  mask_token_logits = token_logits[0, mask_token_index, :]
8  top_5_tokens = torch.topk(mask_token_logits, 5, dim=1).indices[0].tolist()
9  for token in top_5_tokens:
10     print(sequence.replace(tokenizer.mask_token, tokenizer.decode([token])))
```

执行结果：列出前 5 名与其分数，框起来的即是填上的单词。

```
Distilled models are smaller than the models they mimic. Using them instead of the large versions would help reduce our carbon footprint.
Distilled models are smaller than the models they mimic. Using them instead of the large versions would help increase our carbon footprint.
Distilled models are smaller than the models they mimic. Using them instead of the large versions would help decrease our carbon footprint.
Distilled models are smaller than the models they mimic. Using them instead of the large versions would help offset our carbon footprint.
Distilled models are smaller than the models they mimic. Using them instead of the large versions would help improve our carbon footprint.
```

**范例4. 文本生成(Text Generation)**：输入一段文字，框架会自动生成下一句话，这里使用 GPT-2算法，并非BERT算法，它属于Transformer算法的变形，目前已发展到GPT-3。此范例程序修改自Transformers官网Summary of the tasks的Text Generation [38]。

下列程序代码请参考【12_15_Text_Generation.ipynb】。

(1) 加载相关框架。

```
1  from transformers import pipeline
```

(2) 加载模型：参数须设为 text-generation。

```
1  text_generator = pipeline("text-generation")
```

(3) 测试：max_length=50 表示最大生成字数，do_sample=False 表示不随机产生，反之为 True 时，每次生成的内容都会不同，像是聊天机器人，使用者会期望机器人表达能够有变化，不要每次都回答一样的答案。例如，问"How are you?"，机器人有时候回答"I am fine"，有时候回答"great""not bad"。

```
1  print(text_generator("As far as I am concerned, I will",
2                       max_length=50, do_sample=False))
```

执行结果：每次生成的内容均相同。

```
[{'generated_text': 'As far as I am concerned, I will be the first to admit that I am not a fan of the idea of a "free market." I think that the idea of a free market is a bit of a stretch. I think that the idea'}]
```

(4) 测试：do_sample=True 表示随机产生，每次生成的内容均不同。

```
1  print(text_generator("As far as I am concerned, I will",
2                       max_length=50, do_sample=True))
```

执行结果：每次生成的内容均不相同。

```
[{'generated_text': 'As far as I am concerned, I will not be using the name \'Archer\', even though it\'d make all of me cry!\n\n"I\'ll wait until they leave me, you know, on this little ship, of course,'}]
```

(5) 结合分词 (Tokenizer)：这里使用 XLNet 算法，而非 BERT 算法，它也属于 Transformer 算法的变形。

```
1  # 载入相关套件
2  from transformers import AutoModelForCausalLM, AutoTokenizer
3
4  # 结合分词器(Tokenizer)
5  model = AutoModelForCausalLM.from_pretrained("xlnet-base-cased")
6  tokenizer = AutoTokenizer.from_pretrained("xlnet-base-cased")
```

(6) 短文与提示：针对短文，XLNet 通常要补零 (Padding)，因为它是针对开放式 (Open-Ended) 问题而设计的，但 GPT-2 则不用。

```
1  # 短文
2  PADDING_TEXT = """In 1991, the remains of Russian Tsar Nicholas II and his family
3  (except for Alexei and Maria) are discovered.
4  The voice of Nicholas's young son, Tsarevich Alexei Nikolaevich, narrates the
5  remainder of the story. 1883 Western Siberia,
6  a young Grigori Rasputin is asked by his father and a group of men to perform magic.
7  Rasputin has a vision and denounces one of the men as a horse thief. Although his
8  father initially slaps him for making such an accusation, Rasputin watches as the
9  man is chased outside and beaten. Twenty years later, Rasputin sees a vision of
10 the Virgin Mary, prompting him to become a priest. Rasputin quickly becomes famous,
11 with people, even a bishop, begging for his blessing. <eod> </s> <eos>"""
12
13 # 提示
14 prompt = "Today the weather is really nice and I am planning on "
```

(7) 推测答案。

```
1  inputs = tokenizer(PADDING_TEXT + prompt, add_special_tokens=False,
2                     return_tensors="pt")["input_ids"]
3
4  prompt_length = len(tokenizer.decode(inputs[0]))
5  outputs = model.generate(inputs, max_length=250, do_sample=True,
6                           top_p=0.95, top_k=60)
7  generated = prompt + tokenizer.decode(outputs[0])[prompt_length + 1 :]
8
9  print(generated)
```

执行结果。

```
Today the weather is really nice and I am planning on anning on getting some good photos. I need to take some long-running pictures of the past few weeks and "in the moment."<eop> We are on a beach, right on the coast of Alaska. It is beautiful. It is peaceful. It is very quiet. It is peaceful. I am trying not to be too self-centered. But if the sun doesn
```

范例5. 命名实体识别(Named Entity Recognition, NER)：找出重要的人/地/物。此范例程序修改自Transformers官网Summary of the tasks的Named Entity Recognition[39]。

下列程序代码请参考【12_16_NER.ipynb】，预设使用 CoNLL-2003 NER 数据集

(1) 加载相关框架。

```
1 from transformers import pipeline
```

(2) 加载模型：参数须设为 ner。

```
1 nlp = pipeline("ner")
```

(3) 测试。

```
1 # 测试数据
2 sequence = "Hugging Face Inc. is a company based in New York City. " \
3            "Its headquarters are in DUMBO, therefore very" \
4            "close to the Manhattan Bridge."
5
6 # 推测答案
7 import pandas as pd
8 df = pd.DataFrame(nlp(sequence))
9 df
```

执行结果：显示所有实体(Entity)，word 字段中以 ## 开头的，表示与其前一个词汇结合也是一个实体，例如 ##gging，前一个词汇为 Hu，即表示 Hu、Hugging 均为实体。

entity 字段有下列实体类别：

O：非实体。

B-MISC：杂项实体的开头，接在另一个杂项实体的后面。

I-MISC：杂项实体。

B-PER：人名的开头，接在另一个人名的后面。

I-PER：人名。

B-ORG：组织的开头，接在另一个组织的后面。

I-ORG：组织。

B-LOC：地名的开头，接在另一个地名的后面。

I-LOC：地名。

| | word | score | entity | index | start | end |
|---|---|---|---|---|---|---|
| 0 | Hu | 0.999511 | I-ORG | 1 | 0 | 2 |
| 1 | ##gging | 0.989597 | I-ORG | 2 | 2 | 7 |
| 2 | Face | 0.997970 | I-ORG | 3 | 8 | 12 |
| 3 | Inc | 0.999376 | I-ORG | 4 | 13 | 16 |
| 4 | New | 0.999341 | I-LOC | 11 | 40 | 43 |
| 5 | York | 0.999193 | I-LOC | 12 | 44 | 48 |
| 6 | City | 0.999341 | I-LOC | 13 | 49 | 53 |
| 7 | D | 0.986336 | I-LOC | 19 | 79 | 80 |
| 8 | ##UM | 0.939624 | I-LOC | 20 | 80 | 82 |
| 9 | ##BO | 0.912139 | I-LOC | 21 | 82 | 84 |
| 10 | Manhattan | 0.983919 | I-LOC | 29 | 113 | 122 |
| 11 | Bridge | 0.992424 | I-LOC | 30 | 123 | 129 |

(4) 结合分词 (Tokenizer)。

```
1  # 载入相关框架
2  from transformers import AutoModelForTokenClassification, AutoTokenizer
3  import torch
4
5  # 结合分词器(Tokenizer)
6  model_name = "dbmdz/bert-large-cased-finetuned-conll03-english"
7  model = AutoModelForTokenClassification.from_pretrained(model_name)
8  tokenizer = AutoTokenizer.from_pretrained("bert-base-cased")
```

(5) 测试。

```
1   # NER 类别
2   label_list = [
3       "O",        # 非实体
4       "B-MISC",   # 杂项实体的开头,接在另一杂项实体的后面
5       "I-MISC",   # 杂项实体
6       "B-PER",    # 人名的开头,接在另一人名的后面
7       "I-PER",    # 人名
8       "B-ORG",    # 组织的开头,接在另一组织的后面
9       "I-ORG",    # 组织
10      "B-LOC",    # 地名的开头,接在另一地名的后面
11      "I-LOC"     # 地名
12  ]
13
14  # 测试数据
15  sequence = "Hugging Face Inc. is a company based in New York City. " \
16             "Its headquarters are in DUMBO, therefore very" \
17             "close to the Manhattan Bridge."
18
19  # 推测答案
20  inputs = tokenizer(sequence, return_tensors="pt")
21  tokens = inputs.tokens()
22
23  outputs = model(**inputs).logits
24  predictions = torch.argmax(outputs, dim=2)
25
26  for token, prediction in zip(tokens, predictions[0].numpy()):
27      print((token, model.config.id2label[prediction]))
```

执行结果：代码对照说明可参照第 3~11 行。

```
[('[CLS]', 'O'), ('Hu', 'I-ORG'), ('##gging', 'I-ORG'), ('Face', 'I-ORG'), ('Inc', 'I-ORG'), ('.', 'O'), ('is', 'O'), ('a', 'O'), ('company', 'O'), ('based', 'O'), ('in', 'O'), ('New', 'I-LOC'), ('York', 'I-LOC'), ('City', 'I-LOC'), ('.', 'O'), ('It', 's', 'O'), ('headquarters', 'O'), ('are', 'O'), ('in', 'O'), ('D', 'I-LOC'), ('##UM', 'I-LOC'), ('##BO', 'I-LOC'), (',', 'O'), ('therefore', 'O'), ('very', 'O'), ('##c', 'O'), ('##lose', 'O'), ('to', 'O'), ('the', 'O'), ('Manhattan', 'I-LOC'), ('Bridge', 'I-LOC'), ('.', 'O'), ('[SEP]', 'O')]
```

**范例6. 文本摘要(Text Summarization)**：从篇幅较长的文章中整理出摘要。此范例程序修改自Transformers官网Summary of the tasks的Summarization[40]。

下列程序代码请参考【12_17_Text_Summarization.ipynb】，测试数据集是 CNN 新闻和 Daily Mail 媒体刊登的文章。

(1) 加载相关框架。

```
1  from transformers import pipeline
```

(2) 加载模型：参数须设为 summarization。

```
1  summarizer = pipeline("summarization")
```

(3) 测试。

```
1  # 测试数据
2  ARTICLE = """ New York (CNN) When Liana Barrientos was 23 years old, she got married in Westchester County, New York.
3  A year later, she got married again in Westchester County, but to a different man and without divorcing her first husband.
4  Only 18 days after that marriage, she got hitched yet again. Then, Barrientos declared "I do" five more times, sometimes onl
5  In 2010, she married once more, this time in the Bronx. In an application for a marriage license, she stated it was her "fir
```

```
 6  Barrientos, now 39, is facing two criminal counts of "offering a false instrument for filing in the first degree," referring
 7  2010 marriage license application, according to court documents.
 8  Prosecutors said the marriages were part of an immigration scam.
 9  On Friday, she pleaded not guilty at State Supreme Court in the Bronx, according to her attorney, Christopher Wright, who de
10  After leaving court, Barrientos was arrested and charged with theft of service and criminal trespass for allegedly sneaking
11  Annette Markowski, a police spokeswoman. In total, Barrientos has been married 10 times, with nine of her marriages occurrin
12  All occurred either in Westchester County, Long Island, New Jersey or the Bronx. She is believed to still be married to four
13  Prosecutors said the immigration scam involved some of her husbands, who filed for permanent residence status shortly after
14  Any divorces happened only after such filings were approved. It was unclear whether any of the men will be prosecuted.
15  The case was referred to the Bronx District Attorney\'s Office by Immigration and Customs Enforcement and the Department of
16  Investigation Division. Seven of the men are from so-called "red-flagged" countries, including Egypt, Turkey, Georgia, Pakis
17  Her eighth husband, Rashid Rajput, was deported in 2006 to his native Pakistan after an investigation by the Joint Terrorism
18  If convicted, Barrientos faces up to four years in prison.  Her next court appearance is scheduled for May 18.
19  """
20
21  # 推测答案
22  print(summarizer(ARTICLE, max_length=130, min_length=30, do_sample=False))
```

执行结果：摘要内容还算看得懂。

```
[{'summary_text': ' Liana Barrientos, 39, is charged with two counts of "offering a false instrument for filing in the first de
gree" In total, she has been married 10 times, with nine of her marriages occurring between 1999 and 2002 . At one time, she wa
s married to eight men at once, prosecutors say .'}]
```

(4) 结合 Tokenizer：T5 是 Google Text-To-Text Transfer Transformer 的模型，它提供一个框架可以使用多种的模型、损失函数、超参数，来进行不同的任务 (Tasks)，例如翻译、语义接受度检查、相似度比较、文本摘要等，详细说明可参阅 *Exploring Transfer Learning with T5: the Text-To-Text Transfer Transformer*[41]。

```
 1  # 载入相关框架
 2  from transformers import AutoModelForSeq2SeqLM, AutoTokenizer
 3
 4  model = AutoModelForSeq2SeqLM.from_pretrained("t5-base")
 5
 6  # 结合分词器(Tokenizer)
 7  tokenizer = AutoTokenizer.from_pretrained("t5-base")
 8
 9  # T5 最多限 512 个单字
10  inputs = tokenizer("summarize: " + ARTICLE, return_tensors="pt",
11                     max_length=512, truncation=True)
12  outputs = model.generate(
13         inputs["input_ids"], max_length=150, min_length=40,
14         length_penalty=2.0, num_beams=4, early_stopping=True
15  )
16
17  print(tokenizer.decode(outputs[0]))
```

执行结果：T5 最多只可输入 512 个词汇，故将多余的文字截断，产生的摘要也还可以看得懂。

```
['<pad> prosecutors say the marriages were part of an immigration scam. if convicted, barrientos faces two criminal counts of
"offering a false instrument for filing in the first degree" she has been married 10 times, nine of them between 1999 and 200
2.']
```

**范例7.** 翻译(Translation)功能。此范例程序修改自Transformers官网Summary of the tasks的Translation [42]。

下列程序代码请参考【12_18_Translation.ipynb】，使用 WMT English to German dataset( 英翻德数据集 )。

(1) 加载相关框架。

```
1  from transformers import pipeline
```

(2) 加载模型：参数设为 translation_en_to_de 表示英文翻译为德文。

```
1  translator = pipeline("translation_en_to_de")
```

(3) 测试。

```
1  text = "Hugging Face is a technology company based in New York and Paris"
2  print(translator(text, max_length=40))
```

执行结果。

[{'translation_text': 'Hugging Face ist ein Technologieunternehmen mit Sitz in New York und Paris.'}]

(4) 结合 Tokenizer。

```
1  # 载入相关框架
2  from transformers import AutoModelForSeq2SeqLM, AutoTokenizer
3
4  model = AutoModelForSeq2SeqLM.from_pretrained("t5-base")
5
6  # 结合分词器(Tokenizer)
7  tokenizer = AutoTokenizer.from_pretrained("t5-base")
8  text = "translate English to German: Hugging Face is a " + \
9         "technology company based in New York and Paris"
10 inputs = tokenizer(text, return_tensors="pt")
11 outputs = model.generate(inputs["input_ids"], max_length=40,
12                          num_beams=4, early_stopping=True)
13
14 print(tokenizer.decode(outputs[0]))
```

执行结果。

['<pad> Hugging Face ist ein Technologieunternehmen mit Sitz in New York und Paris.']

## 12-11-2　Transformers 框架效能微调

Transformers 也有提供效能微调的功能，可参阅 Transformers 官网 Training and fine-tuning[43] 的网页说明，我们现在就来练习整个程序。Transformers 效能微调可使用下列三种方式：

(1) TensorFlow v2。

(2) PyTorch。

(3) Transformers 的 Trainer。

**范例 7. 以 Transformers 的 Trainer 对象进行效能微调，此范例程序修改自官网案例[44]。**

**下列程序代码请参考【12_19_Custom_Training.ipynb】。**

Transformers 框架效能微调的相关步骤如图 12.35 所示。

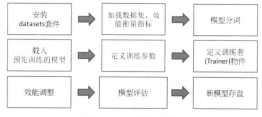

图 12.35　相关步骤

(1) 安装 datasets 套件：GLUE Benchmark(https://gluebenchmark.com/tasks) 包含许多任务 (Task) 与测试数据集。

pip install datasets

(2) 定义 GLUE 所有任务 (Task)。

```
1  GLUE_TASKS = ["cola", "mnli", "mnli-mm", "mrpc", "qnli", "qqp", "rte", "sst2", "stsb", "wnli"]
```

(3) 指定任务为 cola。

```
1  task = "cola"
2  # 预先训练模型
3  model_checkpoint = "distilbert-base-uncased"
4  # 批量
5  batch_size = 16
```

(4) 加载数据集、效能衡量指标：每个数据集有不同的效能衡量指标。

```
1  import datasets
2
3  actual_task = "mnli" if task == "mnli-mm" else task
4  # 载入资料集
5  dataset = datasets.load_dataset("glue", actual_task)
6  # 载入效能衡量指标
7  metric = datasets.load_metric('glue', actual_task)
```

(5) 显示 dataset 数据内容：dataset 数据类型为 DatasetDict，可参考 Transformers 官网的《DatasetDict 说明文件》[45]。

```
1  dataset
```

执行结果：训练数据 8551 笔，验证数据 1043 笔，测试数据 1063 笔。

```
DatasetDict({
    train: Dataset({
        features: ['sentence', 'label', 'idx'],
        num_rows: 8551
    })
    validation: Dataset({
        features: ['sentence', 'label', 'idx'],
        num_rows: 1043
    })
    test: Dataset({
        features: ['sentence', 'label', 'idx'],
        num_rows: 1063
    })
})
```

(6) 显示第一笔内容。

```
1  dataset["train"][0]
```

执行结果：正面 / 负面的情绪分析数据。

```
{'idx': 0,
 'label': 1,
 'sentence': "Our friends won't buy this analysis, let alone the next one we propose."}
```

(7) 定义随机抽取数据函数。

```
1   import random
2   import pandas as pd
3   from IPython.display import display, HTML
4
5   # 随机抽取数据函数
6   def show_random_elements(dataset, num_examples=10):
7       picks = []
8       for _ in range(num_examples):
9           pick = random.randint(0, len(dataset)-1)
10          while pick in picks:
11              pick = random.randint(0, len(dataset)-1)
12          picks.append(pick)
13
14      df = pd.DataFrame(dataset[picks])
15      for column, typ in dataset.features.items():
16          if isinstance(typ, datasets.ClassLabel):
17              df[column] = df[column].transform(lambda i: typ.names[i])
18      display(HTML(df.to_html()))
```

(8) 随机抽取 10 笔数据查看。

```
1  show_random_elements(dataset["train"])
```

执行结果。

|   | idx | label | sentence |
|---|---|---|---|
| 0 | 2722 | unacceptable | A bicycle lent to me. |
| 1 | 6537 | unacceptable | Who did you arrange for to come? |
| 2 | 1451 | unacceptable | The cages which we donated wire for the convicts to build with are strong. |
| 3 | 3119 | acceptable | Cynthia munched on peaches. |
| 4 | 3399 | acceptable | Jackie chased the thief. |
| 5 | 4705 | acceptable | Nina got Bill elected to the committee. |
| 6 | 1942 | unacceptable | Every student who ever goes to Europe ever has enough money. |
| 7 | 5889 | acceptable | Bob gave Steve the syntax assignment. |
| 8 | 4162 | acceptable | This Government have been more transparent in the way they have dealt with public finances than any previous government. |
| 9 | 1261 | acceptable | I know two men behind me. |

(9) 显示效能衡量指标。

```
1  metric
```

执行结果：包含准确率 (Accuracy)、F1、Pearson 关联度 (Correlation)、Spearman 关联度 (Correlation)、Matthew 关联度 (Correlation)。

```
Metric(name: "glue", features: {'predictions': Value(dtype='int64', id=None),
ge: """
Compute GLUE evaluation metric associated to each GLUE dataset.
Args:
    predictions: list of predictions to score.
        Each translation should be tokenized into a list of tokens.
    references: list of lists of references for each translation.
        Each reference should be tokenized into a list of tokens.
Returns: depending on the GLUE subset, one or several of:
    "accuracy": Accuracy
    "f1": F1 score
    "pearson": Pearson Correlation
    "spearmanr": Spearman Correlation
    "matthews_correlation": Matthew Correlation
```

(10) 产生两笔随机数，测试效能衡量指标。

```
1  import numpy as np
2  
3  fake_preds = np.random.randint(0, 2, size=(64,))
4  fake_labels = np.random.randint(0, 2, size=(64,))
5  metric.compute(predictions=fake_preds, references=fake_labels)
```

(11) 模型分词：前置处理以利测试，可取得生词表 (Vocabulary)，设定 use_fast=True 就能够快速处理。

```
1  from transformers import AutoTokenizer
2  
3  # 分词
4  tokenizer = AutoTokenizer.from_pretrained(model_checkpoint, use_fast=True)
```

(12) 测试两笔数据，进行分词。

```
1  tokenizer("Hello, this one sentence!", "And this sentence goes with it.")
```

(13) 定义任务的数据集字段。

```
1   task_to_keys = {
2       "cola": ("sentence", None),
3       "mnli": ("premise", "hypothesis"),
4       "mnli-mm": ("premise", "hypothesis"),
5       "mrpc": ("sentence1", "sentence2"),
6       "qnli": ("question", "sentence"),
7       "qqp": ("question1", "question2"),
8       "rte": ("sentence1", "sentence2"),
9       "sst2": ("sentence", None),
10      "stsb": ("sentence1", "sentence2"),
11      "wnli": ("sentence1", "sentence2"),
12  }
```

(14) 测试第一笔数据。

```
1  sentence1_key, sentence2_key = task_to_keys[task]
2  if sentence2_key is None:
3      print(f"Sentence: {dataset['train'][0][sentence1_key]}")
4  else:
5      print(f"Sentence 1: {dataset['train'][0][sentence1_key]}")
6      print(f"Sentence 2: {dataset['train'][0][sentence2_key]}")
```

执行结果：Our friends won't buy this analysis, let alone the next one we propose.。

(15) 测试 5 笔数据分词。

```
1  def preprocess_function(examples):
2      if sentence2_key is None:
3          return tokenizer(examples[sentence1_key], truncation=True)
4      return tokenizer(examples[sentence1_key], examples[sentence2_key], truncation=True)
5
6  preprocess_function(dataset['train'][:5])
```

执行结果。

```
{'input_ids': [[101, 2256, 2814, 2180, 1005, 1056, 4965, 2023, 4106, 1010, 2292, 2894, 1996, 2279, 2028, 2057, 16599, 1012, 10
2], [101, 2028, 2062, 18404, 2236, 3989, 1998, 1045, 1005, 1049, 3228, 2039, 1012, 102], [101, 2028, 2062, 18404, 2236, 3989, 2
030, 1045, 1005, 1049, 3228, 2039, 1012, 102], [101, 1996, 2062, 2057, 2817, 16025, 1010, 1996, 13675, 16103, 2121, 2027, 2131,
1012, 102], [101, 2154, 2011, 2154, 1996, 8866, 2024, 2893, 14163, 8024, 3771, 1012, 102]], 'attention_mask': [[1, 1, 1, 1, 1,
1, 1, 1, 1, 1, 1, 1, 1, 1, 1, 1, 1, 1, 1], [1, 1, 1, 1, 1, 1, 1, 1, 1, 1, 1, 1, 1, 1], [1, 1, 1, 1, 1, 1, 1, 1, 1, 1, 1, 1, 1,
1], [1, 1, 1, 1, 1, 1, 1, 1, 1, 1, 1, 1, 1, 1], [1, 1, 1, 1, 1, 1, 1, 1, 1, 1, 1, 1, 1]]}
```

(16) 接下来进行效能微调 (Fine tuning)，先加载预先训练的模型。

```
1  from transformers import AutoModelForSequenceClassification, TrainingArguments, Trainer
2
3  # 载入预先训练的模型
4  num_labels = 3 if task.startswith("mnli") else 1 if task=="stsb" else 2
5  model = AutoModelForSequenceClassification.from_pretrained(model_checkpoint, num_labels=num_labels)
```

(17) 定义训练参数：可参阅 Transformers 官网的《TrainingArguments 说明文件》[46]。

```
1  metric_name = "pearson" if task == "stsb" else "matthews_correlation" \
2                           if task == "cola" else "accuracy"
3
4  args = TrainingArguments(
5      "test-glue",
6      evaluation_strategy = "epoch",
7      learning_rate=2e-5,
8      per_device_train_batch_size=batch_size,
9      per_device_eval_batch_size=batch_size,
10     num_train_epochs=5,
11     weight_decay=0.01,
12     load_best_model_at_end=True,
13     metric_for_best_model=metric_name,
14 )
```

(18) 定义效能衡量指标计算的函数。

```
1  def compute_metrics(eval_pred):
2      predictions, labels = eval_pred
3      if task != "stsb":
4          predictions = np.argmax(predictions, axis=1)
5      else:
6          predictions = predictions[:, 0]
7      return metric.compute(predictions=predictions, references=labels)
```

(19) 定义训练者 (Trainer) 对象：参数包含额外增加的训练数据。

```
1  validation_key = "validation_mismatched" if task == "mnli-mm" else \
2                   "validation_matched" if task == "mnli" else "validation"
3
4  trainer = Trainer(
5      model,
6      args,
7      train_dataset=encoded_dataset["train"],
8      eval_dataset=encoded_dataset[validation_key],
9      tokenizer=tokenizer,
10     compute_metrics=compute_metrics
11 )
```

(20) 在预先训练好的模型基础上继续训练，即是效能调整，在笔者的 PC 上至少训练了 20 小时。

```
1  trainer.train()
```

执行结果。

| Epoch | Training Loss | Validation Loss | Matthews Correlation | Runtime | Samples Per Second |
|---|---|---|---|---|---|
| 1 | 0.519900 | 0.484644 | 0.437994 | 301.078100 | 3.464000 |
| 2 | 0.352600 | 0.519489 | 0.505773 | 299.051900 | 3.488000 |
| 3 | 0.231000 | 0.538032 | 0.556475 | 1863.316700 | 0.560000 |
| 4 | 0.180900 | 0.733648 | 0.515271 | 241.590500 | 4.317000 |
| 5 | 0.130700 | 0.787703 | 0.538738 | 242.532000 | 4.300000 |

训练时间统计。

```
TrainOutput(global_step=2675, training_loss=0.27276652897629783, metrics={'train_runtime': 57010.5155, 'train_samples_per_second': 0.047, 'total_flos': 356073036950940.0, 'epoch': 5.0, 'init_mem_cpu_alloc_delta': 757764096, 'init_mem_gpu_alloc_delta': 268953088, 'init_mem_cpu_peaked_delta': 273670144, 'init_mem_gpu_peaked_delta': 0, 'train_mem_cpu_alloc_delta': -935870464, 'train_mem_gpu_alloc_delta': 1077715968, 'train_mem_cpu_peaked_delta': 1757851648, 'train_mem_gpu_peaked_delta': 21298176})
```

(21) 模型评估。

```
1  trainer.evaluate()
```

执行结果。

```
{'eval_loss': 0.5380318760871887,
 'eval_matthews_correlation': 0.5564748164739529,
 'eval_runtime': 229.9053,
 'eval_samples_per_second': 4.537,
 'epoch': 5.0,
 'eval_mem_cpu_alloc_delta': 507904,
 'eval_mem_gpu_alloc_delta': 0,
 'eval_mem_cpu_peaked_delta': 0,
 'eval_mem_gpu_peaked_delta': 20080128}
```

(22) 新模型存盘：未来就能通过 from_pretrained() 加载此效能调整后的模型进行预测。

```
1  trainer.save_model('./cola')
```

(23) 新数据预测。

```
 1  class SimpleDataset:
 2      def __init__(self, tokenized_texts):
 3          self.tokenized_texts = tokenized_texts
 4
 5      def __len__(self):
 6          return len(self.tokenized_texts["input_ids"])
 7
 8      def __getitem__(self, idx):
 9          return {k: v[idx] for k, v in self.tokenized_texts.items()}
10
11  texts = ["Hello, this one sentence!", "And this sentence goes with it."]
12  tokenized_texts = tokenizer(texts, padding=True, truncation=True)
13  new_dataset = SimpleDataset(tokenized_texts)
14  trainer.predict(new_dataset)
```

执行结果：每笔以最大值作为预测结果。

```
PredictionOutput(predictions=array([[-0.55236566,  0.32417056],
       [-1.5994813 ,  1.4773667 ]], dtype=float32), label_ids=None, metrics={'test_runtime': 4.0388, 'test_samples_per_second': 0.495, 'test_mem_cpu_alloc_delta': 20480, 'test_mem_gpu_alloc_delta': 0, 'test_mem_cpu_peaked_delta': 0, 'test_mem_gpu_peaked_delta': 609280})
```

(24) 之后可进行参数调校，这边笔者就不继续往下做了，要不然的话，机器应该会烧坏。

## 12-11-3　Transformers 中文模型

中研院也在 Transformers 架构下开发繁体中文预先训练的模型，称为"CKIP Transformers"[47]，使用说明可参阅官网文件[48]，它提供 ALBERT、BERT、GPT2 模型及分词 (Tokenization)、词性标记 (POS)、命名实体识别 (NER) 等功能。

**范例8. CKIP Transformers测试**，此范例程序修改自官网文件。

**下列程序代码请参考【12_20_CKIP_Transformers.ipynb】。**

(1) 加载相关框架。

```
1  from ckip_transformers import __version__
2  from ckip_transformers.nlp import CkipWordSegmenter, \
3                                   CkipPosTagger, CkipNerChunker
4  import torch
```

(2) 加载模型：包含分词 (Tokenization)、词性标记 (POS)、命名实体识别 (NER)。
device 参数：可指定使用 CPU 或 GPU，设为 -1 代表使用 CPU。
level 参数：level 1 最快，level 3 最精准。

```
1  # 指定 device 以使用 GPU，设为 -1（预设值）代表不使用 GPU
2  device = 0 if torch.cuda.is_available() else -1
3
4  ws_driver = CkipWordSegmenter(level=3, device=device)   # 分词
5  pos_driver = CkipPosTagger(level=3, device=device)      # 词性标记(POS)
6  ner_driver = CkipNerChunker(level=3, device=device)     # 命名实体识别(NER)
```

(3) 测试：任意剪辑两段新闻进行测试。

```
1  text=['''
2  便利商店除了提供微波食品，也有贩卖烤地瓜。一位网友近日在社群网站分享，
3  针对自己在3家超商食用烤地瓜后的看法，并以"甜度"作为评价标准，这则PO文引起许多网友讨论。
4  ''',
5  '''
6  中秋连假兰屿涌入大量游客，但受梅花台风影响，明天（11日）后壁湖往返兰屿，及台东往返兰屿海运全数停航
7  ，东部航务中心请旅客利用今天航班，提前搭乘船班返台，并请注意航班开停航情形。''']
8
9  ws  = ws_driver(text)
10 pos = pos_driver(ws)
11 ner = ner_driver(text)
```

(4) 显示执行结果。

```
1  # 显示分词、词性标记结果
2  def pack_ws_pos_sentece(sentence_ws, sentence_pos):
3      res = []
4      for word_ws, word_pos in zip(sentence_ws, sentence_pos):
5          res.append(f"{word_ws}({word_pos})")
6      return " ".join(res)
7
8  # 显示执行结果
9  for sentence, sentence_ws, sentence_pos, sentence_ner in zip(text, ws, pos, ner):
10     print(sentence)
11     print(pack_ws_pos_sentece(sentence_ws, sentence_pos))
12     for entity in sentence_ner:
13         print(entity)
14     print()
```

第一段执行结果：分词、词性标记都还算正确，包括标点符号。最后两行找到数字 NER。

```
(WHITESPACE)  便利商店(Nc)  除了(P)  提供(VD)  微波(Na)  食品(Na)  ，(COMMACATEGORY)  也(D)  有(V_2)  贩卖(VD)  烤(VC)  地瓜(Na)
。(PERIODCATEGORY)  一(Neu)  位(Nf)  网友(Na)  近日(Nd)  在(P)  社群(Na)  网站(Nc)  分享(VJ)  ，(COMMACATEGORY)
(WHITESPACE)  针对(P)  自己(Nh)  在(P)  3(Neu)  家(Nf)  超商(Nc)  食用(VC)  烤(VC)  地瓜(Na)  后(Ng)  的(DE)  看法(Na)  ，(COMMACAT
EGORY)  并(Cbb)  以(P)  "(Nb)  甜度(Na)  "(FW)  作为(VG)  评价(Na)  标准(Na)  ，(COMMACATEGORY)  这(Nep)  则(Nf)  PO文(FW)  引起(V
C)  许多(Neqa)  网友(Na)  讨论(VE)  。(PERIODCATEGORY)
(WHITESPACE)
NerToken(word='一', ner='CARDINAL', idx=(22, 23))
NerToken(word='3', ner='CARDINAL', idx=(42, 43))
```

第二段执行结果：有标记到许多地名及日期 NER。

```
(WHITESPACE)  中秋(Nd)  连假(Na)  兰屿(Nc)  拥入(VCL)  大量(Neqa)  游客(Na)  ，(COMMACATEGORY)  但(Cbb)  受(P)  梅花(Na)  台风(Na)
影响(VC)  ，(COMMACATEGORY)  明天(Nd)  （(PARENTHESISCATEGORY)  11日(Neu)  ）(PARENTHESISCATEGORY)  后壁湖(Nc)  往返(VCL)  兰屿(N
c)  ，(COMMACATEGORY)  及(Caa)  台东(Nc)  往返(VCL)  兰屿(Nc)  海运(Na)  全数(Neqa)  停航(VH)
(WHITESPACE)  ，(COMMACATEGORY)  东部(Ncd)  航高(Na)  中心(Nc)  谓(VF)  旅客(Na)  利用(VC)  今天(Nd)  航班(Na)  ，(COMMACATEGORY)
提前(VB)  搭乘(VC)  船班(Na)  返台(VA)  ，(COMMACATEGORY)  并(Cbb)  请(VF)  注意(VK)  航班(Na)  开(VC)  停航(VH)  情形(Na)  。(PERI
ODCATEGORY)
NerToken(word='中秋', ner='DATE', idx=(1, 3))
NerToken(word='兰屿', ner='GPE', idx=(5, 7))
NerToken(word='梅花台风', ner='EVENT', idx=(16, 20))
NerToken(word='明天', ner='DATE', idx=(23, 25))
NerToken(word='后壁湖', ner='GPE', idx=(30, 33))
NerToken(word='兰屿', ner='GPE', idx=(35, 37))
NerToken(word='台东', ner='GPE', idx=(39, 41))
NerToken(word='兰屿', ner='GPE', idx=(43, 45))
NerToken(word='今天', ner='DATE', idx=(64, 66))
NerToken(word='台', ner='GPE', idx=(76, 77))
```

### 12-11-4 后续努力

以上只就官方的文件与范例介绍，Transformers 框架的功能越来越强大，要熟悉完整功能，尚待后续努力地试验，BERT 的变形不少，这些变形统称为"BERTology"，预先训练的模型可参阅"Transformers Pretrained models"[49]，有提供轻量型模型，如 ALBERT、TinyBERT，也有提供复杂模型，像是 GPT-3，号称有 1750 亿个参数，更多内容可参阅《AI 趋势周报》第 142 期报导[50]。

Transformer 架构的出现已经完全颠覆了 NLP 的发展，过往的 RNN/LSTM 模型虽然仍然可以拿来应用，但是遵循 Transformer 架构的模型在准确率上确实有明显的优势，因此，推测后续的研究方向应该会逐渐转移到 Transformer 架构上，而且它不只可应用于 NLP，也开始将触角伸向影像辨识领域"Vision Transformer"，由此可见 Transformers 框架日益重要，详情可参阅《AI 趋势周报》第 167 期报导[51]。

## 12-12 总结

本章我们介绍了处理自然语言的相关模型与其演进，包括 RNN、LSTM、GRU、注意力机制、Transformer、BERT 等，同时也实操了许多范例，像是情绪分析 (Sentiment Analysis)、神经机器翻译 (NMT)、语句相似度的比较、问答系统、文本摘要、命名实体识别 (NER)、时间序列 (Time Series) 预测等。相信各位对于 NLP 应用已有基本的认识，若想能灵活应用，还需要找些项目或题目实操。提醒一下，由于目前 Transformer 系列的模型在准确度方面已经超越 RNN/LSTM/GRU，所以如果是项目应用，建议应优先采用 Transformer 模型。

# 第 13 章
# ChatBot

这几年 NLP 的应用范围相当广泛，例如聊天机器人——ChatBot，几乎每一家企业都有这方面的需求，从售前支持 (Pre-sale)、销售 (Sales) 到售后服务 (Post Services) 等方面 ( 如图 13.1 所示 )，用途十分多元，而支持系统功能的技术则涵盖了 NLP、NLU、NLG，既要能解析对话 (NLP)、理解问题 (NLU)，又要能回答得体、幽默、周全 (NLG)，技术范围几乎整合了第 12 章所有的范例。另外，如果能结合语音识别，用说话代替打字，这样不论身处何时何地，人们都能够更方便地用手机与机器沟通，或是结合其他的软硬件，例如社群软件、智能音箱等，使得计算机可以更贴近用户的需求，提供人性化的服务，以往只能在电影里看见的各种科技场景正逐渐在我们的日常生活中变为现实。

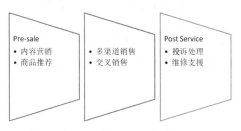

图 13.1　ChatBot 商业应用

话说回来，要开发一款功能完善的 ChatBot，除了技术之外，更要有良好的规划与设计作为基础，而其中有哪些重要的细节呢？现在就带大家来一探究竟。

## 13-1　ChatBot 类别

广义来说，ChatBot 不一定要具备 AI 的功能，只要能自动应答信息，基本上就称为 ChatBot。通常一说到聊天机器人，大家都会想到苹果公司的 Siri，它可以跟用户天南地北地聊天，话题不管是天气、金融，还是音乐、生活信息都难不倒它。但是，对于一般中小企业而言，这样的功能并不能带来商机，他们需要更直接的支持功能，因此我们把 ChatBot 分为以下类别。

(1) 不限话题的机器人：能够与人天南地北地闲聊，包括公开信息的查询与应答，比如温度、股市、播放音乐等，也包含日常寒暄，不需要精准的答案，只需有趣味性、实时回复。

(2) 任务型机器人：例如专家系统，具备特定领域的专业知识，服务范围可以是医疗、驾驶、航行、加密文件的解密等，着重在复杂的算法或规则式 (Rule Based) 的推理，需给予精准的答案，但不求实时的回复。

(3) 常见问答集 (Frequently Asked Questions, FAQ)：客服中心将长年累积的客户疑问集结成知识库，当客户询问时，可快速搜寻，找出相似的问题，并将对应的处理方式回复给客户，答复除了要求正确性与话术之外，也讲究内容是否浅显易懂与翔实周延，避免重复而空泛的回答，引发客户不耐烦与不满意。

(4) 信息检索：利用全文检索的功能，搜寻关键词的相关信息，比如 Google 搜寻，不需要完全精准的信息，也不需要单一的答案，而是提供所有可能的答案，由使用者自行做进一步判断。

(5) 数据库应用：借由 SQL 指令来查询、筛选或统计数据，例如，旅馆订房、餐厅订位、航班查询、订位、报价等，这是最传统的需求，但如果能结合 NLP，让输出/输入接口更友善，例如语音输入/输出，就可以引爆新一波的商机。

以上五种类别的 ChatBot 各有不同的诉求，功能设计方向也各有所差异，所以，在开工之前，务必要先搞清楚老板要的机器人是哪一种，免得到时候开发出来，老板才跟你说"这不是我要的"，那就欲哭无泪了。

## 13-2 ChatBot 设计

13-1 节谈到的 ChatBot 种类非常多元，本节仅针对共同的关键功能进行说明。ChatBot 的规划要点如下。

(1) 确定目标：根据规划的目标，选择适合的 ChatBot 类别，可以是多种类别的混合体。

(2) 收集应用案例 (Use Case)：收集应用的各种状况和场景，整理成案例，以航空机票的销售为例，就包括了每日空位查询、旅程推荐、订票、付款、退换票等，分析每个案例的现况与导入 ChatBot 后的场景与优点。

(3) 提供的内容：现在营销是内容为王 (Content is king) 的时代，有内容的信息才能吸引人潮并带来钱潮，这就是大家常听到的内容营销 (Content Marketing)。因此，要评估哪些信息是有效的、又该如何生产、并以何种方式呈现 (Video、PodCast、博客文章等)。

(4) 挑选开发平台，有下列四种方式可供选择。

①软件包：现在已有许多厂商提供某些行业的解决方案，像是金融、保险等，技术也从传统的 IVR 顺势转为 ChatBot，提供更便利的使用接口。

② ChatBot 平台：许多大型系统厂商都有提供 ChatBot 平台，他们利用独有的 NLP 技术以及大量的 NER 信息，整合各种社群软件，用户只要直接设定，就可以在云端使用 ChatBot 并享有相关的服务。厂商包括 Google DialogFlow、微软的 QnA Maker 等。

③开发工具：许多厂商提供开发工具，方便工程师快速完成一个 ChatBot，例如 Microsoft Bot Framework、Wit.ai 等，可参阅 *10 Best Chatbot Development Frameworks to Build Powerful Bots*[1]，另外 Google、Amazon 智能音箱也提供了 SDK。

④自行构建：可以利用框架加速开发，像 TextBlob、Gemsim、SpaCy、Transforms 等 NLP 函数库，或 Rasa、ChatterBot 等 ChatBot Open Source。

(5) 部署平台：可选择云端或本地端，云端可享有全球服务或以微服务的方式运作，以使用次数计费，可节省初期的高资本支出，因此，若 ChatBot 不是数据库交易类别的话，有越来越多企业采用云端方案。

(6) 用户偏好 (Preference) 与面貌 (Profile)：考虑要存储哪些与业务相关的用户信息。
ChatBot 的术语定义如下。
(1) 技能 (Skill)：例如银行的技能包含存提款、定存、换汇、基金购买、房贷等，每一个应用都称为一种技能。
(2) 意图 (Intent)：技能中每一种对谈的用意，例如，技能是旅馆订房，意图则是有查询某日是否有空的双人房、订房、换日期、退房、付款等。
(3) 实体 (Entity)：关键的人事时地物，利用前面所提的命名实体识别 (NER) 找出实体，每一个意图可指定必要的实体，例如，旅馆订房必须指定日期、房型、住房天数、身份证号码等。
(4) 例句 (Utterance)：因为不同的人表达同一意图会有各种的表达方式，所以需要收集大量的例句来训练 ChatBot，例如"我要订 3 月 21 日双人房""明天、双人房一间"等。
(5) 行动 (Action)：所需信息均已收集完整后，即可做出响应 (Response) 与相关的动作，例如订房，若已确定日期、房型、住房天数、身分证号码后，即可采取行动，为客人保留房间，并且响应客户"订房成功"。
(6) 开场白 (Opening Message)：例如欢迎词 (Welcome)、问候语 (Greeting) 等，通常要有一些例句供随机使用，避免一成不变，流于枯燥。

对话设计有些注意事项如下。
(1) 对话管理：有两种处理方式，有限状态机 (Finite-State Machine, FSM) 和槽位填充 (Slot Filling)。
①有限状态机：传统的自动语音应答系统 (IVR) 大多采取这种方式，事先设计问题顺序，确认每一个问题都能得到适当的回答，才会进入到下一状态，如果中途出错，就退回到前一状态重来，银行 ATM 操作 ( 如图 13.2 所示 )、计算机报修专线等也都是这种设计方式。

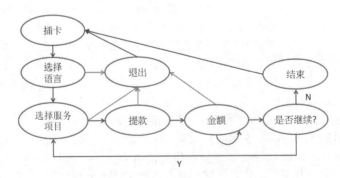

图 13.2　ATM 提款的有限状态机

②槽位填充 (Slot Filling)：有限状态机的缺点是必须按顺序回答问题，并且每次只能回答一个信息，而且要等到系统念完问题才能回答，对娴熟的使用者来说会很不耐烦，若能引进 NLP 技术，就可以让使用者用自然对谈的方式提供信息，例如"我要订 3 月 21 日双人房"，客户说一句话，系统就能够直接处理，若发现信息有欠缺，系统再询问欠缺的信息即可，与真人客服对谈一般，不必像往常一样，"中文请按 1，英文请按 2"，只是订个房还要过五关斩六将。

(2) 整合社群媒体，譬如 Line、Facebook Messenger、Twitter 等，用户无须额外安装软件，且不用教学，直接在对话群组加入官方账号，即可开始与 ChatBot 对话。

(3) 人机整合：ChatBot 设计千万不能原地打转，重复问相同的问题，必须设定跳出条件，一旦察觉对话不合理，就应停止或转由客服人员处理，避免引起使用者不快，造成反效果，使用有限状态机设计方式，常会发生这种错误，若状态已重复两次以上，就可能是漏洞(Bug)。几年前，微软聊天机器人 Tay 推出后不到 24 小时，就因为学会骂人、讲脏话，导致微软紧急将它下架。

除了技术层面之外，ChatBot 也称为"Conversational AI"，因此对话的过程，需注意使用者的个人资料保护，包含像是对话文件的存取权、对话中敏感信息的保密，并且让使用者清楚地知道 ChatBot 的能力与应用范围。

# 13-3  ChatBot 实操

本节先自行设置 ChatBot 的出发点，来看看几个范例。

**范例1. NLP加上相似度比较，制作简单ChatBot。**

**下列程序代码请参考【13_01_simple_chatbot.ipynb】。**

(1) 加载相关套件。

```
1  import spacy
2  import json
3  import random
4  import pandas as pd
```

(2) 加载训练数据：数据来自"Learn to build your first chatbot using NLTK & Keras"[2]。

```
1   # 训练数据
2   data_file = open('./chatbot_data/intents.json').read()
3   intents = json.loads(data_file)
4
5   intent_list = []
6   documents = []
7   responses = []
8
9   # 读取所有意图、例句、回应
10  for i, intent in enumerate(intents['intents']):
11      # 例句
12      for pattern in intent['patterns']:
13          # adding documents
14          documents.append((pattern, intent['tag'], i))
15
16          # adding classes to our class list
17          if intent['tag'] not in intent_list:
18              intent_list.append(intent['tag'])
19
20      # 回应(responses)
21      for response in intent['responses']:
22          responses.append((i, response))
23
24  responses_df = pd.DataFrame(responses, columns=['no', 'response'])
25
26  print(f'例句个数:{len(documents)}, 意图个数:{len(intent_list)}')
27  responses_df
```

执行结果：例句个数有 47 个，意图 (Intent) 个数有 9 个。

(3) 载入词向量。

```
1  nlp = spacy.load("en_core_web_md")
```

(4) 定义前置处理函数：去除停用词、词形还原。

```python
1  from spacy.lang.en.stop_words import STOP_WORDS
2
3  # 去除停用词函数
4  def remove_stopwords(text1):
5      filtered_sentence =[]
6      doc = nlp(text1)
7      for word in doc:
8          if word.is_stop == False:  # 停用词检查
9              filtered_sentence.append(word.lemma_) # lemma_：词形还原
10     return nlp(' '.join(filtered_sentence))
11
12 # 结束用语
13 def say_goodbye():
14     tag = 1 # goodbye 项次
15     response_filter = responses_df[responses_df['no'] == tag][['response']]
16     selected_response = response_filter.sample().iloc[0, 0]
17     return selected_response
18
19 # 结束用语
20 def say_not_understand():
21     tag = 3 # 不理解的项次
22     response_filter = responses_df[responses_df['no'] == tag][['response']]
23     selected_response = response_filter.sample().iloc[0, 0]
24     return selected_response
```

(5) 测试：相似度比较，为防止选出的问题相似度过低，可设定相似度下限，低于下限即调用 say_not_understand()，回复"我不懂你的意思，请再输入一次"，高于下限，才回答问题。

```python
1  # 测试
2  prob_thread =0.6 # 相似度下限
3  while True:
4      max_score = 0
5      intent_no = -1
6      similar_question = ''
7
8      question = input('请输入:\n')
9      if question == '':
10         break
11
12     doc1 = remove_stopwords(question)
13
14     # 比对：相似度比较
15     for utterance in documents:
16         # 两句话的相似度比较
17         doc2 = remove_stopwords(utterance[0])
18         if len(doc1) > 0 and len(doc2) > 0:
19             score = doc1.similarity(doc2)
20             # print(utterance[0], score)
21         # else:
22             # print('\n', utterance[0],'\n')
23
24         if score > max_score:
25             max_score = score
26             intent_no = utterance[2]
27             similar_question = utterance[1] +', '+utterance[0]
28
29     # 若找到相似问题，且高于相似度下限，才回答问题
30     if intent_no == -1 or max_score < prob_thread:
31         print(say_not_understand())
32     else:
33         print(f'你问的是：{similar_question}')
34         response_filter = responses_df[responses_df['no'] == intent_no][['response']]
35         # print(response_filter)
36         selected_response = response_filter.sample().iloc[0, 0]
37         # print(type(selected_response))
38         print(f'回答：{selected_response}')
39
40 # say goodbye!
41 print(f'回答：{say_goodbye()}')
```

将回答转成 Pandas DataFrame，便于筛选与抽样 (Sample)，针对相同问题，可作不同的回复。

执行结果：经过程序调校后，响应的结果很令人满意。

```
请输入：
hello
你问的是：greeting, Hello
回答：Hello, thanks for asking
请输入：
How you could help me
你问的是：options, What help you provide?
回答：I can guide you through Adverse drug reaction list, Blood pressure tracking, Hospitals and Pharmacies
请输入：
Adverse drug reaction
你问的是：adverse_drug, How to check Adverse drug reaction?
回答：Navigating to Adverse drug reaction module
请输入：
blood pressure result
你问的是：blood_pressure_search, Show blood pressure results for patient
回答：Patient ID?
请输入：
123
我不懂你的意思，请再输入一次.
请输入：
pharmacy
你问的是：pharmacy_search, Find me a pharmacy
回答：Please provide pharmacy name
请输入：
hospital
你问的是：hospital_search, Hospital lookup for patient
回答：Please provide hospital name or location
请输入：

回答：Bye! Come back again soon.
```

利用第 12 章所学的知识，只要短短数十行的程序代码，就可以完成一个具体而微小的 ChatBot，当然，它还可以加强的地方还很多，例如：

(1) 中文语料库测试。

(2) 可视化接口：可以利用 Streamlit、Flask、Django 等套件制作网页，提供使用者测试。

(3) 整合社群软件：例如 LINE，直接在手机上测试。

(4) 使用更完整的语料库，测试 ChatBot 的功能。目前使用 SpaCy 的分词速度有点慢，应该是词向量的转换和前置处理花了一些时间，可以改用 NLTK 试试看。

(5) 整合数据库：例如查询数据库，检查旅馆是否有空房、保留订房等。

(6) 利用 NER 提取实体：有了人、事、时、地、物的信息，可进一步整合数据库。

## 13-4 ChatBot 工具套件

网络上有许多的 ChatBot 工具套件，技术架构也相当多样化，笔者测试了一些套件。

(1) ChatterBot[3]：采用适配器模式 (Adapter Pattern)，它是一个可扩充式的架构，支持多语系。

(2) ChatBotAI[4]：以样板 (Template) 语法确定各种样板，接着再将变量嵌入样板中，除了原本内建的样板外，也可以自定义样板和变量来扩充 ChatBot 的功能。

(3) Rasa[5]：以 Markdown 格式确定意图 (Intent)、故事 (Story)、响应 (Response)、实体 (Entity) 与对话管理等功能，用户可以编辑各个组态文件 (*.yml)，重新训练后，就可以提供给 ChatBot 使用。

### 13-4-1 ChatterBot 实操

ChatterBot 采用适配器模式 (Adapter Pattern)，内建多种适配器 (Adapters)，主要分为两类：Logic adapters 和 Storage adapters，也能自制适配器，是一个扩充式的架构，也支持多语系。它本身并没有 NLP 的功能，只是单纯的文字比对功能。

**范例 2. ChatterBot测试。**

下列程序代码请参考【13_02_ChatterBot_test.ipynb】。

(1) 加载相关套件。

```
1  from chatterbot import ChatBot
2  from chatterbot.trainers import ListTrainer
```

(2) 训练：将后一句作为前一句的回答，例如，使用者输入"Hello"后，ChatBot则回答"Hi there!"，它会使用到 NLTK 的语料库。

```
1   chatbot = ChatBot("QA")
2
3   # 训练数据：将后一句作为前一句的回答
4   conversation = [
5       "Hello",
6       "Hi there!",
7       "How are you doing?",
8       "I'm doing great.",
9       "That is good to hear",
10      "Thank you.",
11      "You're welcome."
12  ]
13
14  trainer = ListTrainer(chatbot)
15
16  trainer.train(conversation)
```

(3) 简单测试。

```
1  response = chatbot.get_response("Good morning!")
2  print(f'回答：{response}')
```

执行结果：由于"Good morning!"不在训练数据中，所以 ChatBot 就从过往的对话中随机抽取一笔数据出来回答。

(4) 测试另一句在训练数据中的句子：ChatBot 通常会回答后一句，偶尔会回答过往的对话。

```
1  response = chatbot.get_response("Hi there")
2  print(f'回答：{response}')
```

(5) 加入内建的适配器 (Adapters)。

MathematicalEvaluation：数学公式运算，检视原始码后发现它是使用 mathparse 函数库[6]。

TimeLogicAdapter：有关时间的函数。

BestMatch：从设定的句子中找出最相似的句子。

```
1   bot = ChatBot(
2       'Built-in adapters',
3       storage_adapter='chatterbot.storage.SQLStorageAdapter',
4       logic_adapters=[
5           'chatterbot.logic.MathematicalEvaluation',
6           'chatterbot.logic.TimeLogicAdapter',
7           'chatterbot.logic.BestMatch'
8       ],
9       database_uri='sqlite:///database.sqlite3'
10  )
```

storage_adapter 参数指定对话记录存储在 SQL 数据库或 MongoDB。

(6) 测试时间的问题：问现在的时间。

```
1  response = bot.get_response("What time is it?")
2  print(f'回答：{response}')
```

问法可检视原始码 time_adapter.py。

```
self.positive = kwargs.get('positive', [
    'what time is it',
    'hey what time is it',
    'do you have the time',
    'do you know the time',
    'do you know what time it is',
    'what is the time'
])

self.negative = kwargs.get('negative', [
    'it is time to go to sleep',
    'what is your favorite color',
    'i had a great time',
    'thyme is my favorite herb',
    'do you have time to look at my essay',
    'how do you have the time to do all this',
    'what is it'
])
```

执行结果：回答"The current time is 04:37 PM."。

(7) 算术式测试。

```
1  # 7 + 7
2  response = bot.get_response("What is 7 plus 7?")
3  print(f'回答：{response}')
4
5  # 8 - 7
6  response = bot.get_response("What is 8 minus 7?")
7  print(f'回答：{response}')
8
9  # 50 * 100
10 response = bot.get_response("What is 50 * 100?")
11 print(f'回答：{response}')
12
13 # 50 * (85 / 100)
14 response = bot.get_response("What is 50 * (85 / 100)?")
15 print(f'回答：{response}')
```

执行结果。

```
回答：7 plus 7 = 14
回答：8 minus 7 = 1
回答：50 * 100 = 5000
回答：50 * ( 85 / 100 ) = 42.50
```

(8) 加入自定义的适配器 (Adapters)：自定义适配器为 my_adapter.py，类别名称为 MyLogicAdapter。

```
1  bot = ChatBot(
2      'custom_adapter',
3      storage_adapter='chatterbot.storage.SQLStorageAdapter',
4      logic_adapters=[
5          'my_adapter.MyLogicAdapter',
6          'chatterbot.logic.MathematicalEvaluation',
7          'chatterbot.logic.BestMatch',
8      ],
9      database_uri='sqlite:///database.sqlite3'
10 )
```

(9) 测试自定义适配器。

```
1  response = bot.get_response("我要订位")
2  print(f'回答：{response}')
```

执行结果：会回答"订位日期、时间及人数"或"哪一天、几点、人数"，这是程序中随机指定的。

自定义适配器必须实现三个函数：

　　__init__：初始化对象。

can_process：确定何种问题由此适配器处理，笔者设定的条件为 statement.text.find(' 订位 ') ≥ 0。

process：处理回答的函数。

以上范例是一个很简单的架构，虽然没有太多 NLP 的功能，提问的句子还必须与训练数据完全相同，但是，它提供一个可扩充式的架构，让读者可以利用各自的适配器开发技能，再结合相似度辨识，找出意图最相似的问题，最后做出对应的回答，程序就更适用了。

## 13-4-2　chatbotAI 实操

chatbotAI 同时提供 AI 的功能与对话管理的架构，可以通过 REST API 与社群软件进行整合，比方 Facebook Messenger，并且以样板语言为主，提供学习 (Learn)、记忆 (Memory) 设定、条件判断 (Conditional Switch)、主题式对话 (Topic-Based Conversation Handling) 等功能，如图 13.3 所示。

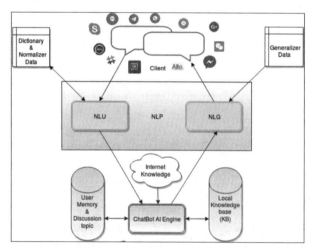

图 13.3　chatbotAI 的架构图

图片来源：chatbotAI 官网[4]

**范例 3. chatbotAI测试。**

下列程序代码请参考【13_03_ chatbotAI_test.ipynb】。

(1) 加载相关套件。

```
1  from chatbot import demo
```

(2) 功能展示。

```
1  demo()
```

执行结果：">"后面为使用者输入的问题，一般问题的回答还算得体，要结束的话需输入"quit"。

```
Hi, how are you?
> fine
  Nice to know that you are fine. What else?
> what is your name?
My name is ChatBot
> are you male?
I may be male? -- what do you think?
> how old are you?
Perhaps you can answer your own question.
> are you a kid?
Perhaps you believe I am a kid.
> should I get out tomorrow?
Why do you ask that?
> weather
Let's change focus a bit... Tell me about your family.
> I have one brother
Now that you have one brother, what will you do next?
> quit
Thank you for talking with me.
```

最新版已改为窗口接口，如图 13.4 所示。

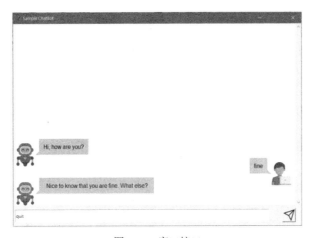

图 13.4　窗口接口

(3) 使用维基百科作为语料库，需安装 wikipedia 套件：

pip install wikipedia

使用这个套件可以搜寻维基百科的数据，输入关键词后，就可进行多功能的查寻，相关使用说明可参考 wikipedia 套件的官网 [7]。

(4) 注册可接收的关键词及负责响应的模块为"维基百科"。

```
1   # 载入相关套件
2   from chatbot import Chat, register_call
3   import wikipedia
4
5   # 注册可接收的关键字及负责回应的模组
6   @register_call("whoIs")
7   def who_is(session, query):
8       try:
9           # 回应
10          return wikipedia.summary(query)
11      # 例外处理
12      except Exception:
13          for new_query in wikipedia.search(query):
14              try:
15                  return wikipedia.summary(new_query)
16              except Exception:
17                  pass
18      return "I don't know about "+query
```

(5) 指定样板，开始对话。样本文件内容如下。

```
{% block %}
    {% client %}(Do you know about|what is|who is|tell me about) (?P<query>.*){% endclient %}
    {% response %}{% call whoIs: %query %}{% endresponse %}
{% endblock %}
```

client：使用者。

response：ChatBot 的回应。

(Do you know about|what is|who is|tell me about)：可接收的问句开头。

call whoIs: %query：指定注册的 whoIs 模块响应。

```
1  # 第一个问题
2  first_question="Hi, how are you?"
3
4  # 使用的样板
5  Chat("chatbot_data/Example.template").converse(first_question)
```

执行结果：询问一些比较专业的问题，都可以应答无碍。

(6) 使用中文关键词发问。

```
1  first_question="你好吗?"
2  Chat("chatbot_data/Example.template").converse(first_question)
```

执行结果：中文关键词能够正确回答。

(7) 记忆 (Memory) 模块定义：以下使用变量来记忆一个字符串或累计值，例如访客人数，并且使用 key/value 进行存储。

```
1   # 记忆(memory)模组定义
2   @register_call("increment_count")
3   def memory_get_set_example(session, query):
4       # 一律转成小写
5       name=query.strip().lower()
6       # 取得记忆的次数
7       old_count = session.memory.get(name, '0')
8       new_count = int(old_count) + 1
9       # 设定记忆次数
10      session.memory[name]=str(new_count)
11      return f"count {new_count}"
```

(8) 记忆 (Memory) 设定测试。

```
1   # 记忆(memory)测试
2   chat = Chat("chatbot_data/get_set_memory_example.template")
3   chat.converse("""
4   Memory get and set example
5
6   Usage:
7     increment <name>
8     show <name>
9     remember <name> is <value>
10    tell me about <name>
11
12  example:
13    increment mango
14    show mango
15    remember sun is red hot star in our solar system
16    tell me about sun
17  """)
```

chat.converse()：内含用法说明。

increment <name>：变数值加 1。

show <name>：显示变量值。

remember <name> is <value>：记忆变量与对应值。

tell me about <name>：显示变量对应值。

执行结果。

```
> increment mango
count  1
> increment mango
count  2
> show mango
2
> remember sun is red hot star in our solar system
I will remember sun is red hot star in our solar system
> tell me about sun
sun is red hot star in our solar system
> remember PLG 5/2 比赛结果 is 梦想家胜
I will remember plg 5/2 比赛结果 is 梦想家胜
> tell me about plg 5/2
I don't know about plg 5/2
> tell me about plg 5/2比赛结果
I don't know about plg 5/2比赛结果
> tell me about plg 5/2 比赛结果
plg 5/2 比赛结果 is 梦想家胜
> quit
Thank you. Have a good day!
```

chatbotAI 也提供一个扩充性的架构，可通过注册的模块和样板，以外挂的方式衔接各种技能。

### 13-4-3　Rasa 实操

Rasa 是一个 Open Source 的工具软件，也有付费版本，相当多的文章有提到它。它以 Markdown 格式确定意图 (Intent)、故事 (Story)、响应 (Response)、实体 (Entity) 以及对话管理等功能，用户可以编辑各个组态文件 (*.yml)，重新训练过后，就可以提供给 ChatBot 使用。

安装过程有点挫折，依照 Rasa 官网指示操作的话，会出现错误，正确的安装指令如下：

(1) Windows 操作系统：

pip install rasa --ignore-installed ruamel.yaml --user

** --user 参数：会让 Rasa 被安装在用户目录下。若不加此选项，则会出现权限不足的错误信息，表示 Python site-packages 目录不能安装。

(2) Linux、macOS 操作系统：

pip install rasa --ignore-installed ruamel.yaml

在 Windows 操作系统下，Rasa 安装成功后，程序会存储在 C:\users\<user_name>\appdata\roaming\python\python38\scripts\ 目录下。接着，测试步骤如下：

(1) 新增一个专案：

C:\users\<user_name>\appdata\roaming\python\python38\scripts\rasa.exe init --no-prompt

产生一个范例项目，子目录和文件列表如图 13.5 所示。

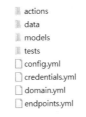

图 13.5　子目录和文件列表

会依据以上项目文件，同时进行训练，完成后建立模型文件，存储在 models 子目录内。

data 子目录内有几个重要的文件：

nlu.yml：NLU 训练数据，包含各类的意图 (Intent) 和例句 (Utterance)。

rules.yml：包含各项规则的意图和行动。

stories.yml：包含各项故事情节，描述多个意图和行动的顺序。

根目录的文件：

domain.yml：包含 Bot 各项的回应 (Response)。

config.yml：NLU 训练的管线 (Pipeline) 与策略 (Policy)。

(2) 测试：

C:\users\<user_name>\appdata\roaming\python\python38\scripts\rasa.exe shell

对话过程如下：并没有太大的弹性，必须完全按照 nlu.yml 提问问题，如图 13.6 所示。

```
Bot loaded. Type a message and press enter (use '/stop' to exit):
Your input -> hello
Hey! How are you?
Your input -> I am fine
Great, carry on!
Your input -> what is you name
I am a bot, powered by Rasa.
Your input -> I am disappointed
Here is something to cheer you up:
Image: https://i.imgur.com/nGF1K8f.jpg
Did that help you?
Your input -> yes
Great, carry on!
Your input -> great
Great, carry on!
Your input -> bye
Bye
Your input -> /stop
2021-05-04 22:15:07 INFO     root  - Killing Sanic server now.
```

图 13.6　对话过程

(3) 故事情节可视化：

C:\users\<user_name>\appdata\roaming\python\python38\scripts\rasa.exe visualize

执行结果：对应 stories.yml 的内容。

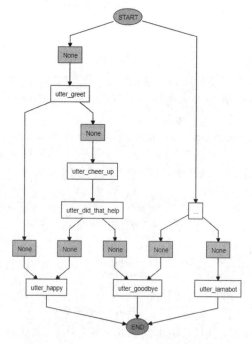

(4) 训练模型：可以修改上述的 .yml 文件后，重新训练模型。
C:\users\<user_name>\appdata\roaming\python\python38\scripts\rasa.exe train

**范例 3.** 建立自定义的行动(Custom action)。以下新增一个行动，询问姓名，并单纯回答 "Hello <name>"。

(1) 安装 Rasa SDK：
pip install rasa_core_sdk
(2) 在 domain.yml 文件内增加以下内容，请参阅范例文件：
inform 意图。
entities、actions。
entities:
 - name

actions:
 - action_save_name
responses 增加以下内容：
 utter_welcome:
 - text: "Welcome!"
 utter_ask_name:
 - text: "What's your name?"
(3) 在 data\nlu.yml 增加以下内容：
- intent: inform
  examples: |
    - my name is [Michael](name)
    - [Philip](name)
    - [Michelle](name)
    - [Mike](name)
    - I'm [Helen](name)
(4) 在 data\ stories.yml 的每一段 action: utter_ask_name 后面增加以下内容：
- intent: inform
  entities:
  - name: "name"
- action: action_save_name
(5) 在 endpoints.yml 解除批注：
action_endpoint:
 url: http://localhost:5055/webhook
(6) 增加一段 action 处理程序。

```python
from typing import Any, Text, Dict, List
from rasa_sdk import Action, Tracker
from rasa_sdk.executor import CollectingDispatcher

class ActionSaveName(Action):
    def name(self):
        return "action_save_name"
    def run(self, dispatcher: CollectingDispatcher,
           tracker: Tracker,
           domain: Dict[Text, Any]) -> List[Dict[Text, Any]]:

        name = \
        next(tracker.get_latest_entity_values("name"))
        dispatcher.utter_message(text=f"Hello, {name}!")
        return []
```

(7) 启动 action 程序：

C:\users\<user_name>\appdata\roaming\python\python38\scripts\rasa.exe run actions

(8) 重新训练模型：

C:\users\<user_name>\appdata\roaming\python\python38\scripts\rasa.exe train

(9) 测试：

C:\users\<user_name>\appdata\roaming\python\python38\scripts\rasa.exe shell

对话过程如图 13.7 所示，确实有回应"Hello <name>"。

```
Your input -> hello
What's your name?
Your input -> michael
Hello, michael!
Hey! How are you?
```

图 13.7　对话过程

Rasa 比较像传统的 AIML ChatBot，是以问答例句作为训练数据，算是相对生硬的方式，且需要大量的人力维护，但好处是可以精准控制回答的内容。

# 13-5　Dialogflow 实操

现在已经有许多厂商都推出成熟的 ChatBot 产品，只要经过适当的设定，就可以上线了，例如 Google Dialogflow、Microsoft QnA Maker、Azure Bot Service、IBM Watson Assistant 等。以下我们以 DialogFlow 为例介绍整个流程。

依据 Dialogflow 的官网说明 [8]，它是一个 NLU 平台，可将对话功能整合至网页、手机、语音响应接口，而输入/输出接口可以是文字或语音。就笔者试验结果，它主要是以槽位填充 (Slot Filling) 为出发点，并搭配完整的 NER 功能，例如时间，可输入 today、tomorrow、right now，系统会自动转换为日期，另外全世界的城市也能辨识，算是一个可轻易上手的产品，如图 13.8 所示。

它有两个版本：Dialogflow CX 和 Dialogflow ES，前者为进阶版本，后者为标准版本，均可免费试用，两者功能的比较表可参阅 Google Dialogflow [9]。

Dialogflow 的术语定义如下。

(1) Agent：即 ChatBot，每间公司可建立多个 ChatBot，各司其职。

(2) 意图 (Intent)：与之前定义相同，但更细致，包含以下内容。

训练的词组 (Training phrases)：定义使用者表达意图的词组，不必列举所有可能的词组，Dialogflow 有内建的机器学习智能，会自动加入类似的词组。

行动 (Action)：ChatBot 接收到意图后采取的行动。

参数 (Parameter)：定义槽位填充所需的信息，包括必填或选填的参数，Dialogflow

可以从使用者的表达中找出对应的实体 (Entity)。

回应 (Response)：行动完毕后，响应用户的文字或语音。

图 13.8　Dialogflow 可从意图中找出时间和地点

图片来源：Dialogflow 的官网说明[8]

(3) 实体 (Entity)：Dialogflow 已内建许多系统实体 (System Entity) 类别，包括日期、时间、颜色、Email 等，还包括多国语系的实体，详情可参阅官网[10]。

(4) 上下文 (Context)：如图 13.9 所示，Dialogflow 从第一句话中察觉意图是"查询账户信息" (CheckingInfo)，接着会问"何种信息"，用户回答"账户余额"后，Dialogflow 即将余额告诉使用者。Dialogflow 会先辨识意图，再根据缺乏的信息进一步询问，直到所有信息都满足为止，才会将答案回复给使用者，这就是槽位填充 (Slot Filling) 的机制。

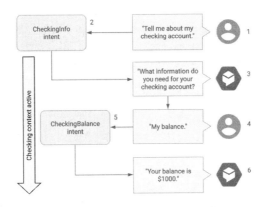

图 13.9　Dialogflow 上下文对话的流程

图片来源：Dialogflow 的官网说明[8]

(5) 追问意图 (Follow-up Intent)：可依据使用者的回答定义不同的回答方式，以追问意图，通过此功能可以建立有限状态机，对于复杂的流程有很大的帮助，Dialogflow 一样有内建追问意图的识别，如 Yes/No，可参阅 Google Dialogflow 官网说明[11]，例如，yes，回答 sure、exactly 也可以，如图 13.10 所示。

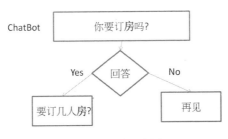

图 13.10　追问意图

(6) 履行 (Fulfillment)：ChatBot 除了响应文字之外，也能够与数据库或社群软件整合，开发者可以撰写一个服务，整合各种软硬件，如图 13.11 所示。

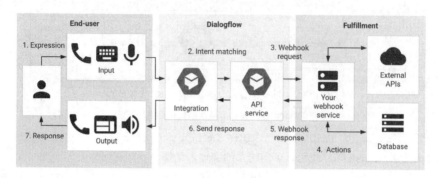

图 13.11　履行 (Fulfillment)

图片来源：Dialogflow 的官网说明 [8]

## 13-5-1　Dialogflow 基本功能

Dialogflow 无须安装，直接进入 Dialogflow 设定相关画面。

(1) 建立 Agent：浏览 https://dialogflow.cloud.google.com/#/newAgent，输入相关信息，支持多语系，单击"CREATE"按钮即可，如图 13.12 所示。

图 13.12　建立 Agent

(2) 内建意图：Dialogflow 会预先建立两个意图：

Default Fallback Intent：不理解使用者的意图时，会归属于此，通常会要求使用者再输入其他用语。

Default Welcome Intent：Agent 的欢迎词。

(3) 建立意图：单击"Create Intent"按钮，输入意图名称，并单击"Add Training Phrases"超链接，就可输入多组问句与响应，如图 13.13 所示。

图 13.13 建立意图

回应：单击"Add Response"超链接，如图 13.14 所示。

图 13.14 回应

存储：单击"Save"按钮。

(4) 测试：存盘，并确定训练完成的信息出现之后，即可在画面右侧测试，亦可直接用语音输入。

输入"what is your name"，如图 13.15 所示。

图 13.15 输入"what is your name"

输入"hello"，如图 13.16 所示。

图 13.16　输入"hello"

(5) 参数：输入的例句如果包含内建的实体 (Entity)，则会被解析出来，作为参数，可进一步设定参数属性。单击画面左侧 Intent 旁的"+"按钮，如图 13.17 所示。

图 13.17　设定参数属性

例如，输入"I know English"，Dialogflow 侦测到"English"是内建的语言 Entity(@sys.language)，系统自动新增一个参数 (Parameter)，如图 13.18 所示。

可针对参数设定属性：

Required：是否必要输入。

Parameter Name：参数名称。

Entity：选择 Entity 类别，可修改为其他类别。

Value：参数的名称，响应 (Response) 可以由此字段取得参数值。

Is List：参数值是否为 List，即一参数含多个值。

Prompts：若输入的问句或回答欠缺此参数，Agent 会显示此提示，询问用户。

图 13.18　新增一个参数

输入响应 (Response):"Wow! I didn't know you knew $language.",其中 $language 会自用户的问句取得变量值。

存储。

(6) 测试:输入"I speak english",响应的 $language = english,如图 13.19 所示。

图 13.19　输入"I speak english"

(7) 可建立自定义的实体 (Entity):单击画面左侧 Entities 旁的 + 按钮,如图 13.20 所示。

输入 Entity 和同义字 (Synonym)。

存储。

图 13.20　建立自定义实体

(8) 使用 english 的实体 (Entity):在原来的"set-language"Entity,输入例句 (Training phrase)。

I know javascript.。

I write the logic in python.

double click"javascript",选取"language-programming"。

输入响应 (Response):"$language-programming is an excellent programming language."。

存储。

(9) 测试:输入"you know js?"。

参数 language 若勾选"Required",会出现"What is the language?",如图 13.21、图 13.22 所示。

图 13.21　勾选"Required"

图 13.22　输出"What is the language？"

全部参数若都不勾选"Required"，则会出现"JavaScript is an excellent programming language."，如图 13.23、图 13.24 所示。

图 13.23　不勾选"Required"

图 13.24　输出"Java Script is an excellent programming language"

(10) 追问意图 (Follow-up intent)：若需考虑上文回答，可追问详细意图。

测试：修改回应为"Wow! I didn't know you knew \$language. How long have you known \$language?"。

加追问意图：单击画面左侧"intents"，将鼠标指针移至"set-language"，会出现"Add follow-up intent"，单击即可，会增加"set-language - custom"追问意图，如图 13.25 所示。

图 13.25　加追问意图

单击"set-language - custom"，输入例句 (Training phrase)：

3 years。

about 4 days。

for 5 years。

寿命 (Lifespan)：一般意图的预设寿命为 5 个对话，追问意图的预设寿命则为 2 个对话，超过 20 分钟，所有意图均不保留，即相关的对话状态会被重置。

测试：加入响应"I can't believe you've known #set-language-followup.language for $duration！"。

『#』：意图。

『$』：参数值。

(11) 测试追问意图：

输入"I know French"，如图 13.26 所示。

图 13.26　输入"I know French"

输入"for 5 years"，如图 13.27 所示。

图 13.27　输入"for 5 years"

## 13-5-2　履行

履行 (Fulfillment) 是在收集完整信息后采取的行动，撰写程序完成商业逻辑和交易，可与社群软件、硬件整合。Dialog 提供以下两种履行类型。

(1) Webhook：撰写一个网页服务 (Web Service)，Dialog 通过 POST 请求送给 Webhook，并接收响应。

(2) Inline Editor：通过 GCP 建立 Cloud Functions，使用 Node.js 执行环境，这是比较简单的方式，不过如果是正式的项目开发，还是要选择"Webhook"。

**范例4. Inline Editor须建立GCP付费账号才能使用，以下针对Webhook实操。**

(1) 建立一个新的意图：输入两个例句"order a room in Tainan at 2021/02/05""I want a double room in Taipei at 2021-01-01"，如图 13.28 所示。

图 13.28　建立一个新的意图

(2) 参数"geo-city""date-time"均设为必要字段，如图 13.29 所示。

图 13.29　设置必要字段

(3) 履行 (Fulfillment)：启用"Enable webhook call for this intent"。

(4) 撰写 Webhook 程序：可使用多种语言撰写，这里我们使用 Python 加上 Flask 框架，撰写 Web 程序，完整程序请参考 dialogflow\webhook\app.py，程序代码后续说明。

(5) 程序必须部署到因特网上，Dialogflow 才能存储到 app.py。可以使用 ngrok.exe 将内部网址对应到外部网址，这样就可以先在本机测试，等到测试成功后，再将程序部署到 Heroku 或其他网站测试，Heroku 是免费的网站部署平台，试用后可升级为付费账户。

(6) 启动 app.py：预设网址为 http://127.0.0.1:5000/。

python app.py

(7) 执行下列指令，取得对应的 https 网址。

ngrok http 5000

(8) 接着设定 Fulfillment：单击画面左侧的 Fulfillment 旁的 + 按钮，启用 Webhook，并设定上一步骤所取得的 https 网址，后面须加上 /webhook，单击下方"Save"按钮即可，如图 13.30 所示。

图 13.30　设定 Fulfillment

(9) 测试：在画面右侧输入"order a room in Tainan at 2021/02/05"，如图 13.31 所示。

图 13.31　测试

(10) 再查看 dialogflow\webhook\test.db SQLite 数据库，就可以看到每重复执行一次，

tainan/2021-02-05 的订房数 (room_count) 就会加 1。而输入不同的城市或日期则会新增一笔记录。SQLite 数据库可使用 SQLitespy.exe 或其他工具软件开启，如图 13.32 所示。

图 13.32　查看数据库

dialogflow\webhook\app.py，程序代码说明如下。

(1) 安装套件：

pip install flask

pip install sqlalchemy

(2) 加载相关套件。

```
from flask import Flask, request, jsonify, make_response
from sqlalchemy import create_engine
```

(3) 声明 Flask 对象。

```
app = Flask(__name__)
```

(4) 定义函数：必须为 @app.route('/webhook', methods=['POST'])。可取得请求、意图、实体等。

```
.route('/webhook', methods=['POST'])
hotel_booking():
# 取得请求
req = request.get_json(force=True)
# 取得意图 set-language
intent = req.get('queryResult').get('intent')['displayName']
# 取得实体
entityCity = req.get('queryResult').get('parameters')["geo-city"].lo
```

(5) 开启数据库联机。

```
33    # 开启数据库连线
34    engine = create_engine('sqlite:///test.db', convert_unicode=True)
35    con = engine.connect()
```

(6) 更新数据库：先根据城市、日期查询数据，若记录存在，则订房数加 1，反之，则新增一笔新的记录，最后回传 OK 信息。

```
37    if intent == 'booking':
38        # 根据城市、日期查询
39        sql_cmd = f"select room_count from hotels "
40        sql_cmd += f"where city = '{entityCity}' and order_date = '{entityDate}'"
41        result = con.execute(sql_cmd)
42        list1 = result.fetchall()
43
44        # 增修记录
45        if len(list1) > 0: # 订房数加 1
46            sql_cmd = f"update hotels set 'room_count' = {list1[-1][-1]+1} "
47            sql_cmd += f"where city = '{entityCity}' and order_date = '{entityDate}'"
48            result = con.execute(sql_cmd)
49        else: # 新增一笔记录
50            sql_cmd = "insert into hotels('city', 'order_date', 'room_count')"
51            sql_cmd += f" values ('{entityCity}', '{entityDate}', 1)"
52            result = con.execute(sql_cmd)
```

(7) 以上只是示范程序，在实际情况中，我们必须做例外处理，包括程序代码错误、意图、Entity 检查等。

Dialogflow 还可以整合语音交换机、社群媒体、Spark 等，详情可参阅 "Dialogflow Integrations 说明"[12]。另外，Dialogflow 也内建许多应用程序，可参阅 "DialogFlow Prebuilt Agents 说明"[13]。

## 13-5-3 整合

除了与后端数据库的连接外，DialogFlow 也可以整合 (Integration) 许多的前端设备与社群软件，官网罗列各种类别如下：

(1) 电话交换机 (Telephony)，如图 13.33 所示。

图 13.33　电话交换机

(2) 文字类的社群软件，如图 13.34 所示。

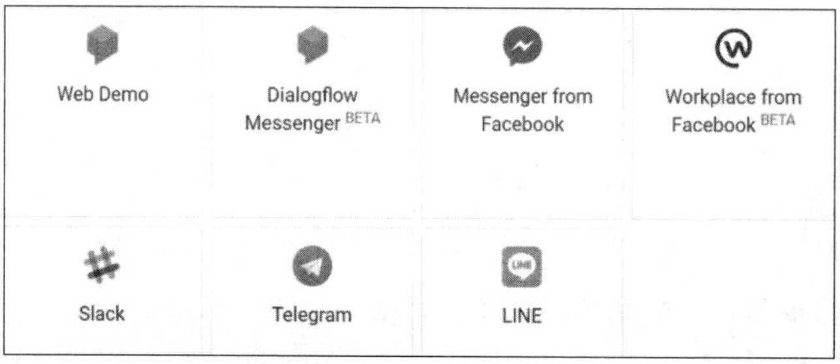

图 13.34　社群软件

(3) 开放原始码，如图 13.35 所示。

图 13.35　开放原始码

接下来说明如何整合 LINE App 的程序。

(1) 先至 LINE Developer Console (https://developers.line.biz/console/register/messaging-

api/provider) 开通。

(2) 建立 Provider：输入提供服务的个人或公司相关数据。

(3) 建立 Channel：选择要提供的服务类别，单击 "Messaging API"，输入图片文件 (Icon)、名称、说明等，如图 13.36 所示。

图 13.36　建立 Channel

(4) 接着至 Channel 的基本设定 (Basic settings) 标签，复制 Channel ID、Channel Secret，再至 Messaging API 标签，单击 "Issue" 按钮，产生 Channel access token，一并复制。

(5) 再回到 DialogFlow，单击左边的 "Integration" 菜单，选择 "LINE"，出现弹出式窗口，输入上一步骤的 Channel ID、Channel Secret、Channel access token，并复制 Webhook URL，单击下方的 "Start" 按钮。

(6) 切换至 LINE Developer Console，至 "Messaging API" 标签，输入 "Webhook URL"，并启用 "Use webhook"。

(7) 使用手机或计算机的 LINE，扫描 "Messaging API" 标签的 QR code，将连接此聊天机器人 channel 的 LINE 官方账号加为好友。

(8) 测试的 Intent 内容如图 13.37 所示。

(a)

(b)

图 13.37　测试

(9) 在 LINE 输入"订台北明天 4 人房",会得到响应消息"你的订房信息:台北市,2022-04-07, 4 人房",即 Text Response 设定的内容。

测试成功,一行程序都不用写,就可以整合 DialogFlow 与 LINE。LINE API 的详细说明可参阅"LINE Developers 文件"[14]。

# 13-6 总结

以上只局限于聊天机器人的功能介绍,实际上,我们还可以结合情绪分析、spaCy/Transformers、推荐等技术,例如当侦测到使用者情绪是负面时,可转由人工客服回答,或者读者连续三次问题都无法回答,落入 Fallback,也可转由人工客服介入。另外,除了意图的参数外,也可以使用 Transformers 分析语句,找到关键的命名实体 (NER)。

此外,对于大部分的企业来说,聊天机器人的实用度相当高,可以提供售前支持、销售甚至是售后服务等多方面的功能,但如何整合既有的业务流程及系统,使得聊天机器人能无缝接轨,使用者/员工也能快速上手,是设置系统的一大课题,最后,千万不要忘记系统要有自我学习的特性,随着服务时间的累积,系统要越来越聪明。

# 第 14 章
# 语音识别

近几年影像、语音等自然用户接口 (Natural User Interface, NUI) 有了突破性的发展，譬如 Apple Face ID 以脸部辨识登录，手机、智能音箱可以使用语音输入，这类操作方式大幅降低了输入的难度，尤其对于中老年人十分友好。根据统计，人们讲话的速度约为每分钟150~200字，而打字输入大概只有每分钟60字，如果能提高语音识别的能力，语音输入就会逐渐取代键盘打字了，此外，键盘在携带方便性与亲和力上也远不及语音。由此可见，要消弭人类与机器之间的隔阂，语音识别扮演相当重要的角色，接下来我们就来认识语音识别的发展。

回归现实，语音识别并不简单，必须要克服以下挑战。

(1) 说话者的个别差异：包括口音、音调的高低起伏，如男性和女性的音频差异就很大。

(2) 环境噪声：各种环境会有不同的背景音源，因此辨识前必须先去除噪声。

(3) 语调的差异：人在不同的情绪下讲话的语调会有所不同，譬如悲伤时讲话速度可能较慢，声音较小而低沉，反之，兴奋时，讲话速度快，声音较大。

光是"No"一个简单的词，不同的人说就有各种各样的声波，如图 14.1 所示。

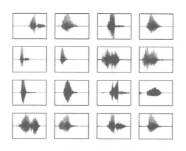

图 14.1　1000 种"No"的声波之部分撷取

图片来源：哥伦比亚大学语音识别课程讲义 [1]

因此，要能辨识不同人的声音，计算机必须先对收到的信号做前置处理，之后才能运用各种算法和数据库进行辨识，过程中所需的基础知识包括：

(1) 信号处理 (Signal processing)。

(2) 概率与统计 (Probability and statistics)。

(3) 语音学和语义学 (Phonetics; linguistics)。

(4) 自然语言处理 (Natural Language Processing, NLP)。

(5) 机器学习与深度学习。

## 14-1 语音基本认识

以说话为例，人类以胸腔压缩和变换嘴唇、舌头形状的方式，产生空气压缩与伸张的效果，形成声波，然后以每秒约 340 米的速度在空气中传播，当此声波传递到另一个人的耳朵时，耳膜就会感受到一伸一压的压力信号，接着内耳神经再将此信号传递到大脑，并由大脑解析与判读，分辨此信号的意义，详细说明可参阅 *Audio Signal Processing and Recognition*[2]。

声音的信号通常如图 14.2 一般是不规则的模拟信号，必须先经过数字化，才能交由计算机处理，做法是每隔一段时间衡量振幅，得到一个数字，这个过程称为"取样"(Sampling)，如图 14.3 所示，之后，再把所有数字记录下来变成数字音频，这个过程就是所谓的将模拟 (Analog) 信号转为数字 (Digital) 信号。

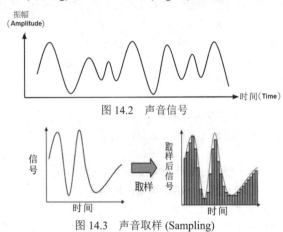

图 14.2　声音信号

图 14.3　声音取样 (Sampling)

图片来源：台湾大学普通物理试验室 [3]

信号可由波形的振幅 (Amplitude)、频率 (Frequency) 及相位 (Phase) 来表示。

(1) 振幅：波的高度，可以形容声音的大小。

(2) 频率：为 1 秒钟内波动的周期数，可以形容声音的高低。

(3) 相位 (Phase)：描述信号波形变化的度量，通常以度 ( 角度 ) 作为单位，也称为相角或相。当信号波形以周期的方式变化，波形循环一周即为 360°。

Gfycat 有一个动画 [4] 说明振幅、频率及相位所代表的意义 ( 如图 14.4 所示 )，简单易懂，值得一看。

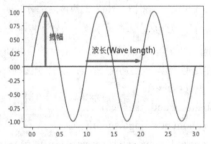

图 14.4　振幅 (Amplitude) 与波长 (Wave Length)

可参阅程序【**14_01_Amplitude_Frequency_Phase.ipynb**】，做一简单的测试。

频率是以赫兹 (Hz) 为单位，赫兹为信号每秒振动的周期数，通常人耳可以听到的频率在 20 Hz ~ 20 kHz 的范围内，但随着年龄的增长，人们会对高频信号越来越不敏感，如图 14.5 所示。

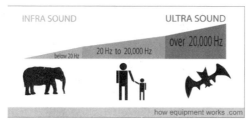

图 14.5　动物可听见的音频范围
图片来源：Audio Signal Processing[5]

根据奈奎斯特 (Nyquist) 定理，信号重建只有取样频率 (Sample Rate) 的一半，以传统电话为例，通常接收的音频范围约为 4kHz，因此取样频率通常是 8kHz，其他常见装置的取样频率如下。

(1) 网络电话：16kHz。

(2) CD：单声道为 22.05kHz，立体 ( 双 ) 声道为 44.1kHz。

(3) DVD：单声道为 48kHz，蓝光 DVD 为 96kHz。

(4) 其他装置的取样频率可参阅"Sampling (signal processing) 维基百科"[6]。

将信号转为数字时，若数字的精度不足也会造成信号的损失，因此，精度可分为 8、16、32 位不同的整数精度，这个过程称为"量化"(Quantization)，传统电话采用 8 位，网络电话采用 16 位。

信号经过数字化后，通常会把它存盘或通过网络传输给另一端，接着再把数字信号还原为模拟信号播放，即可原音重现，如图 14.6 所示，要以最小的数据量存储或传输，就牵涉到信号压缩的算法，即编码 (Encoding) 的机制。

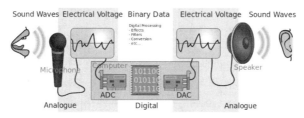

图 14.6　信号的数字化与还原
图片来源：File:CPT-Sound-ADC-DAC.svg[7]

常见的编码方式有：

(1) 脉冲编码调变 (Pulse-Code Modulation,PCM)：直接将每一个取样的振幅存储或传输至对方，这种编码方式效率不高。

(2) 非线性 PCM(Non-Linear PCM)：因人类对高频信号较不敏感，故可以把高频信号以较低精度编码，反之，低频信号采用较高精度，可降低编码量。

(3) 可调变 PCM(Adaptive PCM)：由于信号片段高低不一，因此不必统一编码，可以将信号切成很多片段，并把每一段都个别编码，进行正规化 (Regularization) 后，再作 PCM 编码。

最常见的语音文件应该是 wav 文件，它支持各种的精度与编码，最常见的是 16 位精度与 PCM 编码。

以上的过程可由示波器 (Oscilloscope) 观察，如图 14.7 所示，也能以程序实操。

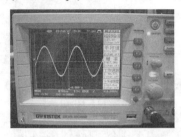

图 14.7 示波器 (Oscilloscope)

图片来源：台湾大学普通物理试验室[3]

**范例1. 音频文件解析。**

**下列程序代码请参考【14_02_audio_parsing.ipynb】。**

(1) 加载相关套件：Jupyter Notebook 本身就支持影像显示、语音播放。

```
1  import IPython
```

(2) 播放音频文件 (wav)：文件来源为如下[22]，autoplay 设定为 True 时，执行即会自动播放，无须另外按 PLAY 键。

```
1  # 文件来源：https://github.com/maxifjaved/sample-files
2  wav_file = './audio/WAV_1MG.wav'
3
4  # autoplay=True：自动播放，无须按 PLAY 键
5  IPython.display.Audio(wav_file, autoplay=False)
```

执行结果：可终止，显示文件长度为 33 秒。

(3) 取得音频文件的属性：可使用 Python 内建的模块 wave，取得音频文件的属性，相关说明可参阅 wave 说明文件[9]。

```
1  import wave
2
3  f=wave.open(wav_file)
4  print(f'取样频率={f.getframerate()}, 帧数={f.getnframes()}, ' +
5        f'声道={f.getnchannels()}, 精度={f.getsampwidth()}, ' +
6        f'文件秒数={f.getnframes() / f.getframerate():.2f}')
7  f.close()
```

执行结果：取样频率 =8000, 帧数 =268237, 声道 =2, 精度 =2, 文件秒数 =33.53。

getframerate()：取样频率。

getnframes()：音频文件总帧数。

getnchannels()：声道。

getsampwidth()：量化精度。

文件秒数 = 音频文件总帧数 / 取样频率。

(4) 使用 PyAudio 函数库串流播放：每读一个区块，就立即播放。PyAudio 在

Windows 操作系统下不能使用 pip install PyAudio 顺利安装，请直接至"Unofficial Windows Binaries for Python Extension Packages"（https://www.lfd.uci.edu/~gohlke/pythonlibs/#pyaudio）下载 PyAudio-0.2.11-cp39-cp39-win_amd64.whl，再执行 pip install PyAudio-0.2.11-cp39-cp39-win_amd64.whl。

注意，上述为 Python 3.9 对应的文件名，请依照本机安装的 Python 版本下载及安装。

```python
# 使用 PyAudio 串流播放
import pyaudio

def PlayAudio(filename, seconds=-1):
    # 定义串流区块大小(stream chunk)
    chunk = 1024
    
    # 开启音频文件
    f = wave.open(filename,"rb")
    
    # 初始化 PyAudio
    p = pyaudio.PyAudio()
    
    # 开启串流
    stream = p.open(format = p.get_format_from_width(f.getsampwidth()),
                    channels = f.getnchannels(), rate = f.getframerate(), output = True)
    
    # 计算每秒区块数
    sample_count_per_second = f.getframerate() / chunk
    
    # 计算总区块数
    if seconds > 0 :
        total_chunk = seconds * sample_count_per_second
    else:
        total_chunk = (f.getnframes() / (f.getframerate() * f.getnchannels())) \
                        * sample_count_per_second
    
    print(f'每秒区块数={sample_count_per_second}, 总区块数={total_chunk}')

    # 每次读一区块
    data = f.readframes(chunk)
    no=0
    while data:
        # 播放区块
        stream.write(data)
        data = f.readframes(chunk)
        no+=1
        if seconds > 0 and no > total_chunk :
            break
    
    # 关闭串流
    stream.stop_stream()
    stream.close()
    
    # 关闭 PyAudio
    p.terminate()
```

（5）播放音频文件。

```python
PlayAudio(wav_file, -1)
```

执行结果：每秒区块数 =7.8125，总区块数 =130.97509765625。

（6）绘制波形：多声道 wav 文件格式是交错存储的，先说明比较单纯的单声道 wav 文件读取。

```python
import numpy as np
import wave
import sys
import matplotlib.pyplot as plt

# 单声道绘制波形
def DrawWavFile_mono(filename):
    # 开启音频文件
    f = wave.open(filename, "r")
    
    # 字符串转换整数
    signal = f.readframes(-1)
```

```
13      signal = np.frombuffer(signal, np.int16)
14      fs = f.getframerate()
15
16      # 非单声道无法解析
17      if f.getnchannels() == 1:
18          Time = np.linspace(0, len(signal) / fs, num=len(signal))
19
20          # 绘图
21          plt.figure(figsize=(12,6))
22          plt.title("Signal Wave...")
23          plt.plot(Time, signal)
24          plt.show()
25      else:
26          print('非单声道无法解析')
```

(7) 测试。

```
1  wav_file = './audio/down.wav'
2  DrawWavFile_mono(wav_file)
```

执行结果。

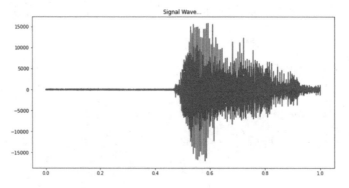

(8) 多声道绘制波形函数。

```
1   def DrawWavFile_stereo(filename):
2       # 开启音频文件
3       with wave.open(filename,'r') as wav_file:
4           # 字符串转换整数
5           signal = wav_file.readframes(-1)
6           signal = np.frombuffer(signal, np.int16)
7
8           # 为每一声道准备一个list
9           channels = [[] for channel in range(wav_file.getnchannels())]
10
11          # 将数据放入每个list
12          for index, datum in enumerate(signal):
13              channels[index % len(channels)].append(datum)
14
15          # 计算时间
16          fs = wav_file.getframerate()
17          Time=np.linspace(0, len(signal)/len(channels)/fs,
18                           num=int(len(signal)/len(channels)))
19
20          f, ax = plt.subplots(nrows=len(channels), ncols=1,figsize=(10,6))
21          for i, channel in enumerate(channels):
22              if len(channels)==1:
23                  ax.plot(Time,channel)
24              else:
25                  ax[i].plot(Time,channel)
```

(9) 测试。

```
1  wav_file = './audio/WAV_1MG.wav'
2  DrawWavFile_stereo(wav_file)
```

执行结果:双声道分别如下。

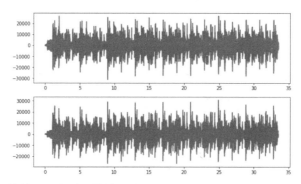

(10) 将前面的单、多声道函数整合在一起。

```
1  def DrawWavFile(wav_file):
2      f=wave.open(wav_file)
3      channels = f.getnchannels() # 声道
4      f.close()
5
6      if channels == 1:
7          DrawWavFile_mono(wav_file)
8      else:
9          DrawWavFile_stereo(wav_file)
```

(11) 测试。

```
1  wav_file = './audio/down.wav'
2  DrawWavFile(wav_file)
3  wav_file = './audio/WAV_1MG.wav'
4  DrawWavFile(wav_file)
```

执行结果。

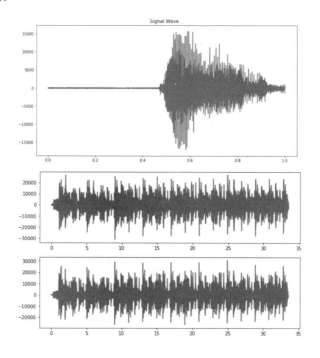

(12) 产生音频文件：以随机数生成音频文件，随机数介于 (-32767, 32767) 之间。

```python
import wave, struct, random

sampleRate = 44100.0 # 取样频率
duration = 1.0 # 秒数

wav_file = './audio/random.wav'
obj = wave.open(wav_file,'w')
obj.setnchannels(1) # 单声道
obj.setsampwidth(2)
obj.setframerate(sampleRate)
for i in range(99999):
    value = random.randint(-32767, 32767)
    data = struct.pack('<h', value) # <h: short, big-endian
    obj.writeframesraw(data)
obj.close()

IPython.display.Audio(wav_file)
```

执行结果：产生音频文件 random.wav，并播放。

(13) 取得音频文件的属性。

```python
f=wave.open(wav_file)
print(f'取样频率={f.getframerate()}, 帧数={f.getnframes()}, ' +
      f'声道={f.getnchannels()}, 精度={f.getsampwidth()}, ' +
      f'文件秒数={f.getnframes() / (f.getframerate() * f.getnchannels()):.2f}')
f.close()
```

执行结果：取样频率 =44100, 帧数 =99999, 声道 =1, 精度 =2, 文件秒数 =2.27，与设置一致。

(14) 将双声道音频文件转换为单声道。

```python
import numpy as np

wav_file = './audio/WAV_1MG.wav'
# 打开音频文件
with wave.open(wav_file,'r') as f:
    # 将字符串转换为整数
    signal = f.readframes(-1)
    signal = np.frombuffer(signal, np.int16)

    # 为每一声道准备一个 list
    channels = [[] for channel in range(f.getnchannels())]

    # 将数据放入每个 list
    for index, datum in enumerate(signal):
        channels[index % len(channels)].append(datum)

    sampleRate = f.getframerate() # 取样频率
    sampwidth = f.getsampwidth()

wav_file_out = './audio/WAV_1MG_mono.wav'
obj = wave.open(wav_file_out,'w')
obj.setnchannels(1) # 单声道
obj.setsampwidth(sampwidth)
obj.setframerate(sampleRate)
for data in channels[0]:
    obj.writeframesraw(data)
obj.close()
```

(15) 取得音频文件的属性。

```python
import wave

f=wave.open(wav_file)
print(f'取样频率={f.getframerate()}, 帧数={f.getnframes()}, ' +
      f'声道={f.getnchannels()}, 精度={f.getsampwidth()}, ' +
      f'文件秒数={f.getnframes() / f.getframerate():.2f}')
f.close()
```

执行结果：取样频率 =8000, 帧数 =268237, 声道 =2, 精度 =2, 文件秒数 =33.53。

除了读取文件之外，要如何才能直接从麦克风接收音频或是录音存盘呢？我们马上就来看看该怎么做。

**范例2. 录音与存盘。**

**下列程序代码请参考【14_03_Record.ipynb】。**

(1) SpeechRecognition 套件提供录音及语音识别，以下列指令安装套件：

pip install SpeechRecognition

(2) 另外，将文字转换语音 (Text To Speech, TTS) 的技术也已非常成熟，故一并安装 pyttsx3 套件，下面程序代码会使用到：

pip install pyttsx3

(3) 加载相关套件。

```
1 import speech_recognition as sr
2 import pyttsx3
```

(4) 列出计算机中的说话者 (Speaker)。

```
1 speak = pyttsx3.init()
2 voices = speak.getProperty('voices')
3 for voice in voices:
4     print("Voice:")
5     print(" - ID: %s" % voice.id)
6     print(" - Name: %s" % voice.name)
7     print(" - Languages: %s" % voice.languages)
8     print(" - Gender: %s" % voice.gender)
9     print(" - Age: %s" % voice.age)
```

执行结果：注意每位说话者是讲英文或中文。

```
Voice:
 - ID: HKEY_LOCAL_MACHINE\SOFTWARE\Microsoft\Speech\Voices\Tokens\TTS_MS_ZH-TW_HANHAN_11.0
 - Name: Microsoft Hanhan Desktop - Chinese (Taiwan)
 - Languages: []
 - Gender: None
 - Age: None
Voice:
 - ID: HKEY_LOCAL_MACHINE\SOFTWARE\Microsoft\Speech\Voices\Tokens\TTS_MS_EN-US_ZIRA_11.0
 - Name: Microsoft Zira Desktop - English (United States)
 - Languages: []
 - Gender: None
 - Age: None
```

(5) 指定说话者：每台计算机安装的说话者均不同，请以 ID 从中指定一位。

```
1 speak.setProperty('voice', voices[1].id)
```

(6) 录音：含文字转语音 (Text To Speech, TTS)，程序会等到持续静默一段时间 (预设是 0.8s) 后才结束。详细可参阅 SpeechRecognition 官方说明[10]。

```
1  r = sr.Recognizer()
2  with sr.Microphone() as source:
3      # 文字转语音
4      speak.say('请说话...')
5      # 等待说完
6      speak.runAndWait()
7
8      #降噪
9      r.adjust_for_ambient_noise(source)
10     # 麦克风收音
11     audio = r.listen(source)
```

(7) 录音存盘。

```
1  wav_file = "./audio/woman.wav"
2  with open(wav_file, "wb") as f:
3      f.write(audio.get_wav_data(convert_rate=16000))
```

(8) 语音识别：需以参数 language 指定要辨识的语系。

```
1  try:
2      text=r.recognize_google(audio, language='zh-tw')
3      print(text)
4  except e:
5      pass
```

笔者念了一段新闻，内容如下：

"受锋面影响，今天下午大雨，有些道路甚至发生积淹，曾文水库上游也传来好消息。"

辨识结果如下：

"受封面影响今天下午大雨有些道路甚至发曾记殷曾文水库上游也传来好消息。"

辨识结果大部分是对的，错误的文字均为同音异字。

(9) 检查输出文件：播放录音。

```
1  import IPython
2
3  # autoplay=True : 自动播放，无须按 PLAY 键
4  IPython.display.Audio(wav_file, autoplay=True)
```

(10) 取得音频文件的属性。

```
1  import wave
2
3  f=wave.open(wav_file)
4  print(f'取样频率={f.getframerate()}, 帧数={f.getnframes()}, '
5        f'声道={f.getnchannels()}, 精度={f.getsampwidth()}, '
6        f'文件秒数={f.getnframes() / (f.getframerate() * f.getnchannels()):.2f}')
7  f.close()
```

执行结果：取样频率 =16000，帧数 =173128，声道 =1，精度 =2，文件秒数 =10.82，与设置一致。

(11) 读取音频文件，转为 SpeechRecognition 音频格式，再进行语音识别。

```
1  import speech_recognition as sr
2
3  # 读取音频文件，转为音频
4  r = sr.Recognizer()
5  with sr.WavFile(wav_file) as source:
6      audio = r.record(source)
7
8  # 语音辨识
9  try:
10     text=r.recognize_google(audio, language='zh-tw')
11     print(text)
12 except e:
13     pass
```

执行结果如下：错误的文字均为同音异字。

"受封面影响今天下午大雨有些道路甚至发曾记殷曾文水库上游也传来好消息。"

(12) 显示所有可能的辨识结果及信赖度。

```
1  # 显示所有可能的辨识结果及信赖度
2  dict1=r.recognize_google(audio, show_all=True, language='zh-tw')
3  for i, item in enumerate(dict1['alternative']):
4      if i == 0:
5          print(f"信赖度={item['confidence']}\n{item['transcript']}")
6      else:
7          print(f"{item['transcript']}")
```

- 执行结果：

信赖度 =0.89820588。

所有可能的辨识结果：

"受封面影响今天下午大雨有些道路甚至发曾记殷曾文水库上游野传来好消息"
"受封面影响今天下午大雨有些道路甚至发生技烟曾文水库上游野传来好消息"

"受封面影响今天下午大雨有些道路甚至发生记烟曾文水库上游野传来好消息"
"受封面影响今天下午大雨有些道路甚至发生气烟曾文水库上游野传来好消息"
"受封面影响今天下午大雨有些道路甚至发生记燕曾文水库上游野传来好消息"

## 14-2 语音前置处理

另外还有一个非常棒的语音处理套件不得不提,那就是 Librosa,它可以将音频做进一步的解析和转换,我们会在后面实操相关功能,更多内容请参阅 Librosa 说明文件[11]。事实上,PyTorch 也支持类似的功能,我们两者都会介绍。

在开始测试之前,还有一些关于音频的概念需要先了解。由于音频通常是一段不规则的波形,很难分析,因此,学者提出傅里叶变换 (Fourier Transform),可以把不规则的波形变成多个规律的正弦波形 (Sinusoidal) 相加,如图 14.8 所示。

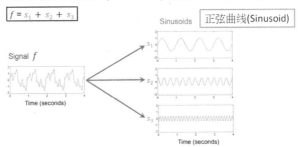

图 14.8　傅里叶变换 (Fourier transform)

图片来源:*Introduction Basic Audio Feature Extraction*[12]

每段正弦波形可以被表示为:

$$s(A, \omega, \varphi)(t) = A \cdot \sin(2\pi(\omega t - \varphi))$$

其中:

$A$:振幅。

$\omega$:频率。

$\varphi$:相位。

如图 14.9 所示,可以观察到振幅、频率、相位是如何影响正弦波形的。

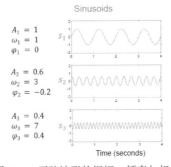

图 14.9　正弦波形的振幅、频率与相位

图片来源:*Introduction Basic Audio Feature Extraction*[12]

转换后的波形振幅和频率均相同,原来的 $X$ 轴为时域 (Time Domain) 就转换为频域 (Frequency Domain),如图 14.10 所示。

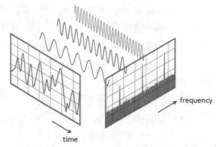

图 14.10　傅里叶变换后，时域 (Time Domain) 转换为频域 (Frequency Domain)

图片来源：*Audio Data Analysis Using Deep Learning with Python (Part 1)*[3]

不同的频率混合在一起称为频谱 (Spectrum)，而绘制的图表就称为频谱图 (Spectrogram)，如图 14.11 所示，通常 $X$ 轴为时间，$Y$ 轴为频率，可以从图表中观察到各种频率的能量。

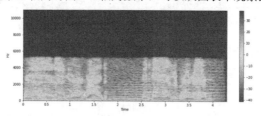

图 14.11　频谱图 (Spectrogram)

为了方便进行语音识别，与处理影像一样，我们会对音频进行特征提取 (Feature Extraction)，常见的有 FBank(Filter Banks)、MFCC(Mel-frequency Cepstral Coefficients) 两种，特征抽取前须先对声音做前置处理，如图 14.12 所示。

(1) 分帧：通常每帧是 25ms，帧与帧之间重叠 10ms，避免遗漏边界信号，如图 14.13 所示。

(2) 信号加强：针对高频信号做加强，使信号更清楚。

(3) 加窗 (Window)：目的是消除各个帧的两端信号可能不连续的现象，常用的窗函数有方窗、汉明窗 (Hamming Window) 等。有时候为了考虑上下文，会将相邻的帧合并成一个帧，这种处理方式称为"帧迭加" (Frame Stacking)。

(4) 去除噪声 (denoising or noise reduction)。

音频 → 取样 → 分帧 → 加窗(Window) → 特征提取 → MFCC

图 14.12　音频前置处理

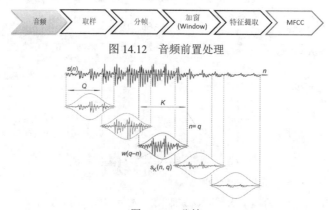

图 14.13　分帧

在计算频谱时，会将以上的前置处理，包含分帧、加窗、离散傅里叶变换 (Discrete Fourier Transform, DFT) 合并为一个步骤，称为短时傅里叶变换 (Short-Time Fourier Transform, STFT)，SciPy 支持此功能，函数名称为 stft。

**范例3.** 频谱图(Spectrogram)实时显示。由于程序是以动画呈现，无法在 Jupyter Notebook 上展示，故以 Python 文件执行。另外，【14_04_waves.py】可显示实时的波形。这两个程序均源自"Python audio spectrum analyzer"[14]。

下列程序代码请参考【14_05_spectrogram.py】。

(1) 加载相关套件。

```
2  import pyaudio
3  import struct
4  import matplotlib.pyplot as plt
5  import numpy as np
6  from scipy import signal
```

(2) 开启麦克风，设置收音相关参数。

```
12  # 参数设置
13  FORMAT = pyaudio.paInt16 # 精度
14  CHANNELS = 1 # 单声道
15  RATE = 48000 # 取样频率
16  INTERVAL = 0.32 # 缓冲区大小
17  CHUNK = int(RATE * INTERVAL) # 接收区块大小
18
19  # 开启麦克风
20  stream = mic.open(format=FORMAT, channels=CHANNELS, rate=RATE,
21              input=True, output=True, frames_per_buffer=CHUNK)
```

(3) 频谱图 (Spectrogram) 实时显示：调用 signal.spectrogram()，显示频谱图，设置显示满 100 张图表即停止，可依需要弹性调整。

```
23  i=0
24  while i < 100: # 显示100次即停止
25      data = stream.read(CHUNK, exception_on_overflow=False)
26      data = np.frombuffer(data, dtype='b')
27
28      # 绘制频谱图
29      f, t, Sxx = signal.spectrogram(data, fs=CHUNK)
30      dBS = 10 * np.log10(Sxx)
31      plt.clf()
32      # 设定X/Y轴标签
33      plt.ylabel('Frequency [Hz]')
34      plt.xlabel('Time [sec]')
35
36      plt.pcolormesh(t, f, dBS)
37      plt.pause(0.001)
38      i+=1
```

执行结果。

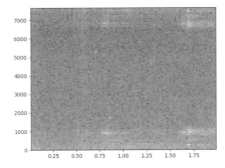

(4) 关闭所有装置。

```
40  # 关闭所有装置
41  stream.stop_stream()
42  stream.close()
43  mic.terminate()
```

**范例4. 音频前置处理**：利用**Librosa**函数库了解音频的前置处理程序。

**下列程序代码请参考【14_06_Preprocessing.ipynb】。**

(1) 加载相关套件。

```
1  import IPython
2  import pyaudio
3  import struct
4  import matplotlib.pyplot as plt
5  import numpy as np
6  from scipy import signal
7  import librosa
8  import librosa.display  # 一定要加
9  from IPython.display import Audio
```

(2) 载入文件：调用 librosa.load()，回传数据与取样频率。

可设定参数如下：

hq=True，表示加载时采用高质量模式 (high-quality mode)。

sr=44100，指定取样频率。

res_type='kaiser_fast'，表示快速载入文件。

```
1  # 文件来源：https://github.com/maxifjaved/sample-files
2  wav_file = './audio/WAV_1MG.wav'
3
4  # 载入文件
5  data, sr = librosa.load(wav_file)
6  print(f'取样频率={sr}, 总样本数={data.shape}')
```

执行结果：取样频率 =22050, 总样本数 =(739329,)。

(3) 绘制波形。

```
1  librosa.display.waveplot(data, sr)
```

执行结果。

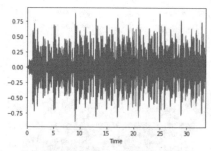

(4) 显示频谱图：先调用 melspectrogram() 取得梅尔系数 (Mel)，再调用 power_to_db() 转为分贝 (db)，最后调用 specshow() 显示频谱图。

```
1  # 载入频谱图
2  spec = librosa.feature.melspectrogram(y=data, sr=sr)
3
4  # 显示频谱图
5  db_spec = librosa.power_to_db(spec, ref=np.max,)
6  librosa.display.specshow(db_spec,y_axis='mel', x_axis='s', sr=sr)
7  plt.colorbar()
```

执行结果：颜色越明亮代表该频率能量越强。

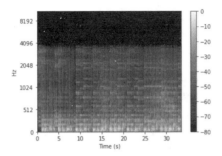

(5) 存盘：v0.8 版本之后已不支持 librosa.output.write_wav 函数了，须改用 soundfile 套件。

```
1  sr = 22050 # sample rate
2  T = 5.0    # seconds
3  t = np.linspace(0, T, int(T*sr), endpoint=False) # time variable
4  x = 0.5*np.sin(2*np.pi*220*t)# pure sine wave at 220 Hz
5
6  #playing generated audio
7  Audio(x, rate=sr) # Load a NumPy array
8
9  # v0.8后已不支持
10 # librosa.output.write_wav('generated.wav', x, sr) # writing wave file in .wav format
11 import soundfile as sf
12 sf.write('./audio/generated.wav', x, sr, 'PCM_24')
```

(6) 接着进行特征提取的实操，可作为深度学习模型的输入。

(7) 短时傅里叶变换(Short-time Fourier transform)：包括分帧、加窗、离散傅里叶变换，合并为一个步骤。

```
1  # Short-time Fourier transform
2  # return complex matrix D[f, t], which f is frequency, t is time (frame).
3  D = librosa.stft(data)
4  print(D.shape, D.dtype)
```

回传一个矩阵 **D**，其中包含频率、时间。

执行结果：(1025, 1445) complex64。

(8) MFCC：参数 n_mfcc 可指定每秒要回传几个 MFCC frame，通常是 13、40 个。

```
1  # mfcc
2  mfcc = librosa.feature.mfcc(y=data, sr=sr, n_mfcc=40)
3  mfcc.shape
```

执行结果：(40, 1445)。

(9) Log-Mel Spectrogram。

```
1  # Log-Mel Spectrogram
2  melspec = librosa.feature.melspectrogram(data, sr, n_fft=1024,
3                                            hop_length=512, n_mels=128)
4  logmelspec = librosa.power_to_db(melspec)
5  logmelspec.shape
```

执行结果：(128, 1445)。

(10) 接着说明 Librosa 内建音频加载的方法。

```
1  librosa.util.list_examples()
```

执行结果。

```
AVAILABLE EXAMPLES
--------------------------------------------------------
brahms       Brahms - Hungarian Dance #5
choice       Admiral Bob - Choice (drum+bass)
fishin       Karissa Hobbs - Let's Go Fishin'
nutcracker   Tchaikovsky - Dance of the Sugar Plum Fairy
trumpet      Mihai Sorohan - Trumpet loop
vibeace      Kevin MacLeod - Vibe Ace
```

(11) 加载 Librosa 默认的内建音频文件。

```
1  # 载入 librosa 内建音频
2  y, sr = librosa.load(librosa.util.example_audio_file())
3  print(f'取样频率={sr}, 总样本数={y.shape}')
```

执行结果：取样频率 =22050, 总样本数 =(1355168,)。

(12) 播放：利用 IPython 模块播放音频，设置 autoplay=True 会自动播放。

```
1  Audio(y, rate=sr, autoplay=True)
```

(13) 指定 ID 加载内建音频文件。

```
1  # 另一内建音频 nutcracker
2  filename = librosa.example('nutcracker')
3  y, sr = librosa.load(filename)
4  print(f'取样频率={sr}, 总样本数={y.shape}')
```

(14) 音频处理与转换：Librosa 支持多种音频处理与转换功能，我们逐一来试验。

(15) 重取样 (Resampling)：从既有的音频中重取样，通常是从高质量的取样频率，通过重取样，转换为较低取样频率的数据。

```
1  # 重取样
2  sr_new = 11000
3  y = librosa.resample(y, sr, sr_new)
4
5  print(len(y), sr_new)
6
7  Audio(y, rate=sr_new, autoplay=True)
```

(16) 将和音与敲击音分离 (Harmonic/Percussive Separation)：调用 librosa.effects.hpss() 可将和音与敲击音分离，从敲击音可以找到音乐的节奏 (Tempo)。

```
1  # 和音与敲击音分离(Harmonic/Percussive Separation)
2  y_h, y_p = librosa.effects.hpss(y)
3  spec_h = librosa.feature.melspectrogram(y_h, sr=sr)
4  spec_p = librosa.feature.melspectrogram(y_p, sr=sr)
5  db_spec_h = librosa.power_to_db(spec_h,ref=np.max)
6  db_spec_p = librosa.power_to_db(spec_p,ref=np.max)
7
8  plt.subplot(2,1,1)
9  librosa.display.specshow(db_spec_h, y_axis='mel', x_axis='s', sr=sr)
10 plt.colorbar()
11
12 plt.subplot(2,1,2)
13 librosa.display.specshow(db_spec_p, y_axis='mel', x_axis='s', sr=sr)
14 plt.colorbar()
15
16 plt.tight_layout()
```

执行结果。

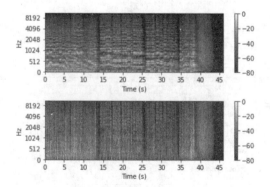

(17) 取得敲击音每分钟出现的样本数。

```
1  print(librosa.beat.tempo(y, sr=sr))
```

执行结果：143.5546875。

(18) 可分别播放和音与敲击音。

```
1  IPython.display.Audio(data=y_p, rate=sr)
```

(19) 绘制色度图 (Chromagram)：chroma 为半音 (Semitones)，可提取音准 (Pitch) 信息。

```
1   # 提取音准(pitch)资讯
2   chroma = librosa.feature.chroma_cqt(y=y_h, sr=sr)
3   plt.figure(figsize=(18,5))
4   librosa.display.specshow(chroma, sr=sr, x_axis='time', y_axis='chroma', vmin=0, vmax=1)
5   plt.title('Chromagram')
6   plt.colorbar()
7
8   plt.figure(figsize=(18,5))
9   plt.title('Spectrogram')
10  librosa.display.specshow(chroma, sr=sr, x_axis='s', y_axis='chroma', )
```

执行结果：Y 轴显示 12 个半音，Pitch 是有周期的循环。

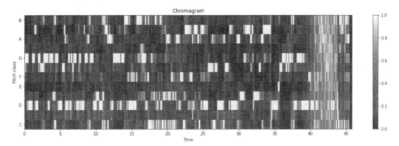

(20) 可任意分离频谱，例如将频谱分为 8 个成分 (Component)，以非负矩阵分解法 (NMF) 分离频谱，NMF 类似主成分分析 (PCA)。

```
1  # 将频谱分为 8 个成分(Component)
2  # Short-time Fourier transform
3  D = librosa.stft(y)
4
5  # Separate the magnitude and phase
6  S, phase = librosa.magphase(D)
7
8  # Decompose by nmf
9  components, activations = librosa.decompose.decompose(S, n_components=8, sort=True)
```

(21) 显示成分 (Component) 与 Activations。

```
1   # 显示成分(Component)与 Activations
2   plt.figure(figsize=(12,4))
3
4   plt.subplot(1,2,1)
5   librosa.display.specshow(librosa.amplitude_to_db(np.abs(components)
6                           , ref=np.max), y_axis='log')
7   plt.xlabel('Component')
8   plt.ylabel('Frequency')
9   plt.title('Components')
10
11  plt.subplot(1,2,2)
12  librosa.display.specshow(activations, x_axis='time')
13  plt.xlabel('Time')
14  plt.ylabel('Component')
15  plt.title('Activations')
16
17  plt.tight_layout()
```

执行结果：X 轴为 8 个成分。

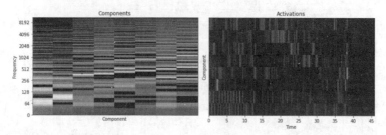

(22) 再以分离的 Components 与 Activations 重建音频。

```
1  # 以 Components 与 Activations 重建音频
2  D_k = components.dot(activations)
3
4  # invert the stft after putting the phase back in
5  y_k = librosa.istft(D_k * phase)
6
7  # And playback
8  print('Full reconstruction')
9
10 IPython.display.Audio(data=y_k, rate=sr)
```

执行结果：播放与原曲一致，这部分的功能可用于音乐合成或修改。

(23) 只以第一 Component 与 Activation 重建音频：播放与原曲大相径庭。

```
1  # 只以第一 Component 与 Activation 重建音频
2  k = 0
3  D_k = np.multiply.outer(components[:, k], activations[k])
4
5  # invert the stft after putting the phase back in
6  y_k = librosa.istft(D_k * phase)
7
8  # And playback
9  print('Component #{}'.format(k))
10
11 IPython.display.Audio(data=y_k, rate=sr)
```

(24) Pre-emphasis：用途为高频加强，前面说过，人类对高频信号较不敏感，所以能利用此技巧，增强音频里高频的部分。

```
1  # Pre-emphasis : 增强高频的部分
2  import matplotlib.pyplot as plt
3
4  y, sr = librosa.load(wav_file, offset=30, duration=10)
5
6  y_filt = librosa.effects.preemphasis(y)
7
8  # 比较原音与修正的音频
9  S_orig = librosa.amplitude_to_db(np.abs(librosa.stft(y)), ref=np.max)
10 S_preemph = librosa.amplitude_to_db(np.abs(librosa.stft(y_filt)), ref=np.max)
11
12 # 绘图
13 plt.subplot(2,1,1)
14 librosa.display.specshow(S_orig, y_axis='log', x_axis='time')
15 plt.title('Original signal')
16
17 plt.subplot(2,1,2)
18 librosa.display.specshow(S_preemph, y_axis='log', x_axis='time')
19 fig=plt.title('Pre-emphasized signal')
20
21 plt.tight_layout()
```

执行结果：可以很明显看到高频已被增强。

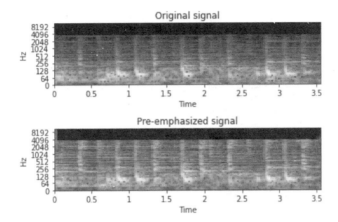

(25) 常态化 (Normalization)：在输入机器学习之前，我们通常会先进行特征缩放，除了能提高准确率外，也能加快优化求解的收敛速度，具体方式就是直接使用 SciKit-Learn 的 minmax_scale 函数。

```
1  # 常态化
2  from sklearn.preprocessing import minmax_scale
3
4  wav_file = './audio/WAV_1MG.wav'
5  data, sr = librosa.load(wav_file, offset=30, duration=10)
6
7  plt.subplot(2,1,1)
8  librosa.display.waveplot(data, sr=sr, alpha=0.4)
9
10 plt.subplot(2,1,2)
11 fig = plt.plot(minmax_scale(data), color='r')
```

执行结果。

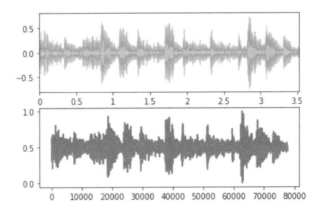

除了 Librosa 套件之外，也有 python_speech_features 套件，提供读取音频文件特征的功能，包括 MFCC/FBank 等，安装指令如下：

pip install python_speech_features

**范例5. 特征提取MFCC、Filter bank向量。**

下列程序代码请参考【14_07_python_speech_features.ipynb】。

(1) 加载相关套件。

```
1  import matplotlib.pyplot as plt
2  from scipy.io import wavfile
3  from python_speech_features import mfcc, logfbank
```

(2) 加载音乐文件。

```
1  sr, data = wavfile.read("./audio/WAV_1MG.wav")
```

(3) 读取 MFCC、Filter bank 特征。

```
1  mfcc_features = mfcc(data, sr)
2  filterbank_features = logfbank(data, sr)
3
4  # Print parameters
5  print('MFCC 维度:', mfcc_features.shape)
6  print('Filter bank 维度:', filterbank_features.shape)
```

执行结果：

MFCC 维度 : (6705, 13)。

Filter bank 维度 : (6705, 26)。

(4) MFCC、Filter bank 绘图。

```
1   plt.subplot(2,1,1)
2   mfcc_features = mfcc_features.T
3   plt.imshow(mfcc_features, cmap=plt.cm.jet,
4       extent=[0, mfcc_features.shape[1], 0, mfcc_features.shape[0]], aspect='auto')
5   plt.title('MFCC')
6
7   plt.subplot(2,1,2)
8   filterbank_features = filterbank_features.T
9   plt.imshow(filterbank_features, cmap=plt.cm.jet,
10      extent=[0, filterbank_features.shape[1], 0, filterbank_features.shape[0]], aspect='auto')
11  plt.title('Filter bank')
12  plt.tight_layout()
13  plt.show()
```

执行结果。

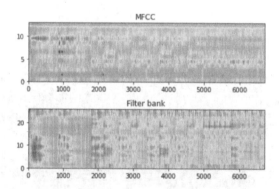

有关音频的转换还有另一个选择，可以下载 FFmpeg 工具程序[15]，它的功能非常多，包括裁剪、取样频率、编码等，详情可参阅 FFmpeg 官网说明[16]。举例来说，以下指令是将 input.wav 转为 output.wav，并改变取样频率、声道、编码：

ffmpeg.exe -i output.wav -ar 44100 -ac 1 -acodec pcm_s16le output.wav

## 14-3 PyTorch 语音前置处理

了解前面音频处理与转换的内容后，我们算是做好热身了，接下来，就正式实操几个与深度学习相关的应用。

PyTorch 有一个 Torch Audio 模块，不仅支持深度学习功能，也含语音的前置处理的函数，可以与前面介绍的 Librosa 套件相媲美。

**范例6. PyTorch语音的前置处理**，此范例程序修改自**PyTorch Audio教学文件**[17]。

**下列程序代码请参考【14_08_torchaudio.ipynb】。**

(1) 加载相关套件。

```
1  import torch
2  import torchaudio
3  import IPython
4  from IPython.display import Audio
5  import matplotlib.pyplot as plt
6  import os
7  import math
```

(2) 下载音频文件。

```
1  import requests
2
3  path = "./audio/steam-train-whistle-daniel_simon.wav"
4  url = "https://pytorch-tutorial-assets.s3.amazonaws.com/steam-train-whistle-daniel_simon.wav"
5  with open(path, 'wb') as file_:
6      file_.write(requests.get(url).content)
```

(3) 取得音频文件的属性 (Metadata)。

```
2  wav_file = './audio/steam-train-whistle-daniel_simon.wav'
3
4  metadata = torchaudio.info(wav_file)
5  print(metadata)
```

执行结果：包括取样率、帧数、声道数、精度、编码。

AudioMetaData(sample_rate=44100, num_frames=109368, num_channels=2, bits_per_sample=16, encoding=PCM_S)。

PyTorch 支持的编码有很多种，可参阅 PyTorch Audio 教学文件。

(4) 定义操作音频文件相关的函数。

```
1   # 取得一段语音的描述统计量
2   def print_stats(waveform, sample_rate=None):
3       if sample_rate:
4           print("Sample Rate:", sample_rate)
5       print("维度:", tuple(waveform.shape))
6       print("数据型态:", waveform.dtype)
7       print(f" - 最大值:         {waveform.max().item():6.3f}")
8       print(f" - 最小值:         {waveform.min().item():6.3f}")
9       print(f" - 平均数:         {waveform.mean().item():6.3f}")
10      print(f" - 标准差: {waveform.std().item():6.3f}")
11      print()
12      print(waveform)
13      print()
```

```
15  # 绘制语音的波形
16  def plot_waveform(waveform, sample_rate, title="Waveform", xlim=None, ylim=None):
17      waveform = waveform.numpy()
18
19      num_channels, num_frames = waveform.shape
20      time_axis = torch.arange(0, num_frames) / sample_rate
21
22      figure, axes = plt.subplots(num_channels, 1)
23      if num_channels == 1:
24          axes = [axes]
25      for c in range(num_channels):
26          axes[c].plot(time_axis, waveform[c], linewidth=1)
27          axes[c].grid(True)
28          if num_channels > 1:
29              axes[c].set_ylabel(f'Channel {c+1}')
30          if xlim:
31              axes[c].set_xlim(xlim)
32          if ylim:
33              axes[c].set_ylim(ylim)
34      figure.suptitle(title)
35      plt.show(block=False)
```

```
37  # 绘制语音的频谱
38  def plot_specgram(waveform, sample_rate, title="Spectrogram", xlim=None):
39      waveform = waveform.numpy()
40
41      num_channels, num_frames = waveform.shape
42      time_axis = torch.arange(0, num_frames) / sample_rate
43
44      figure, axes = plt.subplots(num_channels, 1)
45      if num_channels == 1:
46          axes = [axes]
47      for c in range(num_channels):
48          axes[c].specgram(waveform[c], Fs=sample_rate)
49          if num_channels > 1:
50              axes[c].set_ylabel(f'Channel {c+1}')
51          if xlim:
52              axes[c].set_xlim(xlim)
53      figure.suptitle(title)
54      plt.show(block=False)
```

```
56  # 播放语音
57  def play_audio(waveform, sample_rate):
58      waveform = waveform.numpy()
59
60      num_channels, num_frames = waveform.shape
61      if num_channels == 1:
62          display(Audio(waveform[0], rate=sample_rate))
63      elif num_channels == 2:
64          display(Audio((waveform[0], waveform[1]), rate=sample_rate))
65      else:
66          raise ValueError("不支持超过双声道的音频文件.")
67
68  # 取得文件资讯
69  def inspect_file(path):
70      print("-" * 10)
71      print("Source:", path)
72      print("-" * 10)
73      print(f" - File size: {os.path.getsize(path)} bytes")
74      print(f" - {torchaudio.info(path)}")
```

(5) 测试: 显示语音的描述统计量。

```
1  waveform, sample_rate = torchaudio.load(wav_file)
2  print_stats(waveform, sample_rate=sample_rate)
```

执行结果: waveform 回传 Tensor 数据格式。

```
Sample Rate: 44100
维度: (2, 109368)
数据型态: torch.float32
 - 最大值:         0.508
 - 最小值:        -0.449
 - 平均数:         0.000
 - 标准差:  0.118

tensor([[ 0.0027,  0.0063,  0.0092,  ...,  0.0032,  0.0047,  0.0052],
        [-0.0038, -0.0015,  0.0013,  ..., -0.0032, -0.0012, -0.0003]])
```

(6) 测试: 显示波形。

```
1  plot_waveform(waveform, sample_rate)
```

执行结果: 每一声道显示一个波形。

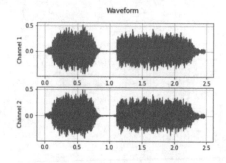

(7) 测试：绘制频谱。

```
1  plot_specgram(waveform, sample_rate)
```

执行结果：每一声道显示一个频谱。

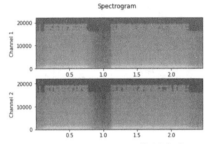

(8) 存盘：以 16-bit signed integer Linear PCM 编码存盘。

```
1  # 以 16-bit signed integer Linear PCM 编码存盘
2  path = "save_example_PCM_S16.wav"
3  torchaudio.save(
4      path, waveform, sample_rate,
5      encoding="PCM_S", bits_per_sample=16)
6  inspect_file(path)
```

(9) 重抽样：可重新设定抽样率，取得较低质量的音频。

```
1  import torchaudio.functional as F
2
3  # 重抽样率
4  resample_rate = 4000
5  resampled_waveform = F.resample(waveform, sample_rate, resample_rate)
6
7  # 绘制频谱
8  plot_specgram(resampled_waveform, resample_rate)
```

执行结果：因重抽样率只有 4000，故比原来的频谱线条更明显。

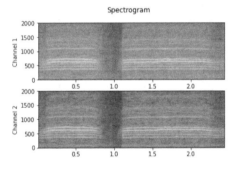

(10) 保存后播放测试：音质明显较粗糙。

```
1  path = "./audio/resample.wav"
2  torchaudio.save(path, resampled_waveform, resample_rate)
3
4  # autoplay=True：自动播放，无须按 PLAY 键
5  IPython.display.Audio(wav_file, autoplay=False)
```

(11) 音频数据增强 (Data Augmentation)：Torch Audio 利用 sox 函数库进行音频转换，可做到类似影像的数据增强效果，不过，sox 在 Windows 操作系统下测试有问题[18]，可参阅 sox_test.ipynb，笔者单独安装 sox 工具软件[19]及 pysox 套件[20]，运行是正常的。pysox 的官网有几个简单的范例可测试。

```
1  import sox
2  # create transformer
3  tfm = sox.Transformer()
4
5  # 裁剪原音频文件5 ～ 10.5 秒的片段。
6  # tfm.trim(5, 10.5)
7
8  # 压缩
9  tfm.compand()
10
11 # 应用 fade in/fade out 效果
12 tfm.fade(fade_in_len=1.0, fade_out_len=0.5)
13
14 # 产生输出文件
15 path = "audio/steam-train-whistle-daniel_simon.wav"
16 out_path = "audio/test.wav" #path.split('.')[0]+'.aiff'
17 if os.path.exists(out_path):
18     os.remove(out_path)
19 tfm.build_file(path, out_path)
20
21 # 输出至记忆体
22 array_out = tfm.build_array(input_filepath=path)
23
24 # 显示应用的效果
25 tfm.effects_log
```

执行结果：显示使用压缩及淡入 (Fade in)/ 淡出 (Fade out) 的特效。

['compand', 'fade']

语音与文字类似，也是要将原始数据转换为向量，才能交由深度学习的模型辨识，因此会利用特征提取 (Feature Extraction)，取得频谱、FBank、MFCC 等语音特征，Torch Audio 各种转换可参考 "TorchAudio Transforms"[21]，以下只举几种转换说明。

(1) 加载样本语音文件。

```
1  import torchaudio.functional as F
2  import torchaudio.transforms as T
3  import librosa
4
5  wav_file = './audio/speech.wav'
6  waveform, sample_rate = torchaudio.load(wav_file)
```

(2) 定义时频 (Spectrogram) 绘图函数。

```
1  def plot_spectrogram(spec, title=None, ylabel='freq_bin', aspect='auto', xmax=None):
2      fig, axs = plt.subplots(1, 1)
3      axs.set_title(title or 'Spectrogram (db)')
4      axs.set_ylabel(ylabel)
5      axs.set_xlabel('frame')
6      im = axs.imshow(librosa.power_to_db(spec), origin='lower', aspect=aspect)
7      if xmax:
8          axs.set_xlim((0, xmax))
9      fig.colorbar(im, ax=axs)
10     plt.show(block=False)
```

(3) 时频转换：即短时傅里叶变换 (Short-Time Fourier transform, STFT)，详细参数说明请参考 TorchAudio Transforms。

```
1  n_fft = 1024
2  win_length = None
3  hop_length = 512
4
5  # 时频转换定义
6  spectrogram = T.Spectrogram(
7      n_fft=n_fft,              # 快速傅里叶变换的长度(Size of FFT)
8      win_length=win_length,    # 视窗大小(Window size)
9      hop_length=hop_length,    # 视窗终非重叠的(Hop length)
10     center=True,              # 是否在音频前后补资料，使t时间点的框居中
11     pad_mode="reflect",       # 补数据的方式
12     power=2.0,                # 时频大小的指数(Exponent for the magnitude spectrogram)
13 )
14 # 进行时频转换
15 spec = spectrogram(waveform)
16
17 print_stats(spec)
18 plot_spectrogram(spec[0], title='torchaudio')
```

执行结果:因重抽样率只有 4000,故比原来的频谱线条更明显。

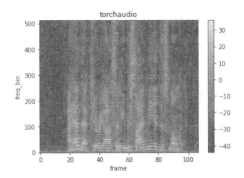

图 14.14 有助于理解上述参数的意义。

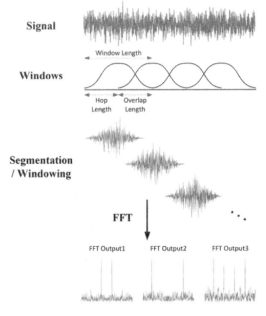

图 14.14 STFT 转换

图形来源:*Area-Efficient Short-Time Fourier Transform Processor for Time–Frequency Analysis of Non-Stationary Signals*[22]

(4) GriffinLim 转换:由频谱还原为原始音频。

```
1  # 原始音频
2  plot_waveform(waveform, sample_rate, title="Original")
3
4  griffin_lim = T.GriffinLim(
5      n_fft=n_fft,
6      win_length=win_length,
7      hop_length=hop_length,
8  )
9  waveform = griffin_lim(spec)
10
11 # 由频谱还原后的音频
12 plot_waveform(waveform, sample_rate, title="Reconstructed")
```

执行结果:左图为原始音频,右图为由频谱还原后的音频,几乎完全相同。

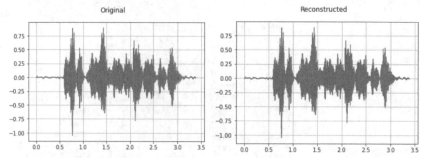

(5) 取得 FBank 特征向量。

```
1   # FBank 绘图
2   def plot_mel_fbank(fbank, title=None):
3       fig, axs = plt.subplots(1, 1)
4       axs.set_title(title or 'Filter bank')
5       axs.imshow(fbank, aspect='auto')
6       axs.set_ylabel('frequency bin')
7       axs.set_xlabel('mel bin')
8       plt.show(block=False)
9   
10  n_fft = 256
11  n_mels = 64
12  sample_rate = 6000
13  
14  mel_filters = F.melscale_fbanks(
15      int(n_fft // 2 + 1),  # 分成的组数(Number of frequencies to highlight)
16      n_mels=n_mels,        # FBank 个数
17      f_min=0.,             # 最小的频率
18      f_max=sample_rate/2., # 最大的频率
19      sample_rate=sample_rate, # 取样率
20      norm='slaney'         # 区域常态化(Area normalization)
21  )
22  plot_mel_fbank(mel_filters, "Mel Filter Bank - torchaudio")
```

执行结果。

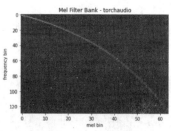

(6) 取得梅尔频谱 (MelSpectrogram)。

```
1   n_fft = 1024
2   win_length = None
3   hop_length = 512
4   n_mels = 128
5   
6   mel_spectrogram = T.MelSpectrogram(
7       sample_rate=sample_rate,  # 取样率
8       n_fft=n_fft,              # 快速傅里叶转换的长度(Size of FFT)
9       win_length=win_length,    # 视窗大小(Window size)
10      hop_length=hop_length,    # 视窗终非重叠的Hop length)
11      center=True,              # 是否在音频前后补数据,使t时间点的框居中
12      pad_mode="reflect",       # 补数据的方式
13      power=2.0,                # 时频大小的指数(Exponent for the magnitude spectrogram)
14      norm='slaney',            # 区域常态化(Area normalization)
15      onesided=True,            # 只传回一半得结果,避免重复
16      n_mels=n_mels,            # FBank 个数
17      mel_scale="htk",          # htk or slaney
18  )
19  
20  melspec = mel_spectrogram(waveform)
21  plot_spectrogram(melspec[0], title="MelSpectrogram - torchaudio", ylabel='mel freq')
```

执行结果。

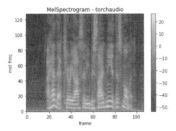

(7) 取得梅尔倒频谱 (MFCC)。

```
1  n_fft = 2048
2  win_length = None
3  hop_length = 512
4  n_mels = 256
5  n_mfcc = 256
6
7  mfcc_transform = T.MFCC(
8      sample_rate=sample_rate,
9      n_mfcc=n_mfcc,      # MFCC的个数
10     melkwargs={
11         'n_fft': n_fft,
12         'n_mels': n_mels,
13         'hop_length': hop_length,
14         'mel_scale': 'htk',
15     }
16 )
17
18 mfcc = mfcc_transform(waveform)
19
20 plot_spectrogram(mfcc[0])
```

执行结果。

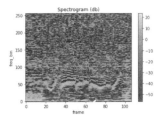

(8) 取得音高 (Pitch)：Pitch 即声音的高音、低音的表示法，音频的频率越高，Pitch 越高。

```
1  def plot_pitch(waveform, sample_rate, pitch):
2      figure, axis = plt.subplots(1, 1)
3      axis.set_title("Pitch Feature")
4      axis.grid(True)
5
6      end_time = waveform.shape[1] / sample_rate
7      time_axis = torch.linspace(0, end_time,  waveform.shape[1])
8      axis.plot(time_axis, waveform[0], linewidth=1, color='gray', alpha=0.3)
9
10     axis2 = axis.twinx()
11     time_axis = torch.linspace(0, end_time, pitch.shape[1])
12     ln2 = axis2.plot(
13         time_axis, pitch[0], linewidth=2, label='Pitch', color='green')
14
15     axis2.legend(loc=0)
16     plt.show(block=False)
17
18 # 侦测音高
19 pitch = F.detect_pitch_frequency(waveform, sample_rate)
20
21 # 绘制音高
22 plot_pitch(waveform, sample_rate, pitch)
```

执行结果。

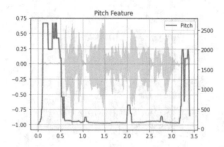

接着讨论如何针对以上衍生的特征，进行增强 (Feature Augmentation)。

(1) 先将音频进行时频转换。

```
1   n_fft = 400
2   win_length = None
3   hop_length = None
4
5   # 时频转换定义
6   spectrogram = T.Spectrogram(
7       n_fft=n_fft,              # 快速傅里叶转换的长度(Size of FFT)
8       win_length=win_length,    # 视窗大小(Window size)
9       hop_length=hop_length,    # 视窗终非重叠(Hop Length)
10      center=True,              # 是否在音频前后补数据，使t时间点的框居中
11      pad_mode="reflect",       # 补数据的方式
12      power=None,               # 时频大小的指数(Exponent for the magnitude spectrogram)
13  )
14  # 进行时频转换
15  spec = spectrogram(waveform)
```

(2) 音频拉长、缩短。

```
1   # 音频延伸转换
2   stretch = T.TimeStretch()
3
4   # 音频拉长1.2倍
5   rate = 1.2
6   spec_ = stretch(spec, rate)
7   plot_spectrogram(torch.abs(spec_[0]),
8                    title=f"Stretched x{rate}", aspect='equal', xmax=304)
9
10  plot_spectrogram(torch.abs(spec[0]),
11                   title="Original", aspect='equal', xmax=304)
12
13  # 音频缩短0.9倍
14  rate = 0.9
15  spec_ = stretch(spec, rate)
16  plot_spectrogram(torch.abs(spec_[0]),
17                   title=f"Stretched x{rate}", aspect='equal', xmax=304)
```

执行结果：左图音频拉长 1.2 倍，中间图形为原音频，右图音频缩短 0.9 倍。

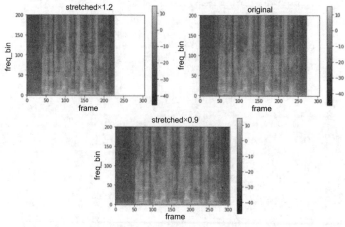

(3) 时间屏蔽 (Time Masking)：屏蔽一段时间 (X 轴) 的音频。

```
1  torch.random.manual_seed(4)
2
3  plot_spectrogram(spec[0], title="Original")
4
5  masking = T.TimeMasking(time_mask_param=40)
6  spec2 = masking(spec)
7
8  plot_spectrogram(spec2[0], title="Masked along time axis")
```

执行结果：屏蔽可设定时间长度 (time_mask_param)，TimeMasking 会随机屏蔽，故范例固定随机种子 (Seed)。

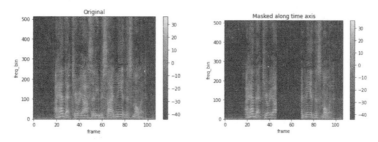

(4) 频率屏蔽 (Frequency Masking)：屏蔽一段频率 (Y 轴) 的音频。

```
1  torch.random.manual_seed(4)
2  plot_spectrogram(spec[0], title="Original")
3
4  masking = T.FrequencyMasking(freq_mask_param=80)
5  spec2 = masking(spec)
6
7  plot_spectrogram(spec2[0], title="Masked along frequency axis")
```

执行结果：屏蔽可设定频率长度 (freq_mask_param)，FrequencyMasking 会随机屏蔽，故范例固定随机种子 (Seed)。

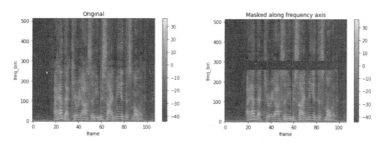

## 14-4　PyTorch 内建语音数据集

PyTorch 内建许多语音数据集，请参阅《PyTorch 内建语音数据集》[23]，每个数据集都有不同的用途，包括：

(1) CMU ARCTIC Dataset(CMUARCTIC)：是卡内基·梅隆大学建构的美式英语的单人演讲数据库，含 1150 个例句 (Utterances)，包括各种类别 (性别、语系、口音)：aew、ahw、aup、awb、axb、bdl、clb、eey、fem、gka、jmk、ksp、ljm、lnh、rms、rxr、slp 或 slt，预设为 aew，其中 bdl 为男性，slt 为女性。详细说明可参阅 *Pyroomacoustics CMU ARCTIC Corpus*[24]。

(2) CMU Pronouncing Dictionary(CMUDict)：发音字典，提供自动语音识别 (ASR) 使用，数据集格式为每个单词对应一个发音，若一字多音，使用多行表示。详细说明可参阅《CMU Pronouncing Dictionary 维基百科》[25]。

(3) Common Voice(COMMONVOICE)：多语系的数据集，由 Mozilla 公司发起的群众自发性的录音集合而成的。详细说明可参阅 Common Voice Datasets[26]，也有中文的语音，可参阅《Common Voice 繁体中文》[27]。

(4) GTZAN：音乐曲风的数据集，包含 10 类曲风，每类含 100 条 30 秒的音乐。详细说明可参阅 GTZAN Genre Collection[28]。

(5) YESNO：收录 60 段录音，以希伯来语发音，每段含 8 个 Yes/No 的发音。详细说明可参阅 Open Speech and Language Resources[29]。

(6) SPEECHCOMMANDS：Google 于 2018 年收集一个短指令的数据集，包括常用的词汇，譬如 stop、play、up、down、right、left 等，共有 35 个类别，每个语音文件约 1 秒。详细说明可参阅 Speech Commands: A Dataset for Limited-Vocabulary Speech Recognition[30]。

(7) 其他还有 LIBRISPEECH、LIBRITTS、LJSPEECH、SPEECHCOMMANDS、TEDLIUM、VCTK_092 及 DR_VCTK，请参阅《PyTorch 内建语音数据集》[23]。

**范例7. PyTorch内建语音数据集测试。**

下列程序代码请参考【14_09_torchaudio_dataset.ipynb】。

(1) 加载相关套件：会使用到【14_08_torchaudio_preprocessing.ipynb】的函数，将之存成 audio_util.py，以利重复使用。

```
1  import torch
2  import torchaudio
3  import IPython
4  from IPython.display import Audio
5  import matplotlib.pyplot as plt
6  import os
7  import math
8  import audio_util
```

(2) 下载 YES/NO 数据集，并建立 Dataset、DataLoader。

```
1  yesno_data = torchaudio.datasets.YESNO('./audio', download=True)
2  data_loader = torch.utils.data.DataLoader(yesno_data,
3                          batch_size=1, shuffle=True)
```

(3) 显示第一笔数据。

```
1  yesno_data[0]
```

执行结果：包含语音的特征向量、取样率及 8 个标记 (Label)，0:No, 1:Yes。

```
(tensor([[ 3.0518e-05,  6.1035e-05,  3.0518e-05,  ..., -1.8616e-03,
          -2.2583e-03, -1.3733e-03]]),
 8000,
 [0, 0, 0, 0, 1, 1, 1, 1])
```

(4) 显示频谱及播放：显示第 2、4、6 笔数据。

```
1  for i in [1, 3, 5]:
2      waveform, sample_rate, label = yesno_data[i]
3      audio_util.plot_specgram(waveform, sample_rate, title=f"Sample {i}: {label}")
4      audio_util.play_audio(waveform, sample_rate)
```

部分执行结果：8 个标记 (Label)，0:No, 1:Yes。

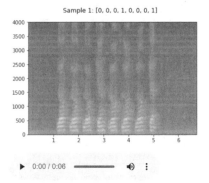

(5) 存盘：可使用多媒体播放器播放。

```
1  wav_file = "./audio/yesno1.wav"
2  torchaudio.save(
3      path, yesno_data[0][0], yesno_data[0][1])
4  inspect_file(path)
```

(6) 下载 GTZAN 数据集，并建立 Dataset、DataLoader：数据集有 1.2GB，下载时间有点久。

```
1  dataset1 = torchaudio.datasets.GTZAN('./audio', download=True)
2  data_loader = torch.utils.data.DataLoader(dataset1,
3                                  batch_size=1, shuffle=True)
```

(7) 显示第一笔数据。

```
1  dataset1[0]
```

执行结果：包含语音的特征向量、取样率及标记 (Label)，为蓝调曲风。

```
(tensor([[ 0.0073,  0.0166,  0.0076,  ..., -0.0556, -0.0611, -0.0642]]),
 22050,
 'blues')
```

(8) 显示频谱及播放：显示每一种曲风数据。

```
1  for i in range(0, 1000, 100):
2      waveform, sample_rate, label = dataset1[i]
3      audio_util.plot_specgram(waveform, sample_rate, title=f"Sample {i}: {label}")
4      audio_util.play_audio(waveform, sample_rate)
```

执行结果：非常棒的音乐。

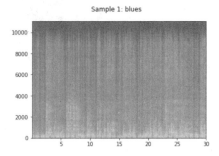

(9) 下载 CMU Pronouncing Dictionary 数据集，并建立 Dataset。

```
1  dataset2 = torchaudio.datasets.CMUDict('./audio', download=True)
```

(10) 显示 4 笔数据。

```
1  for i in range(0, 4):
2      print(dataset2[i])
```

执行结果：包含单词、发音的音素 (Phoneme)。

```
("'ALLO", ['AA2', 'L', 'OW1'])
("'APOSTROPHE", ['AH0', 'P', 'AA1', 'S', 'T', 'R', 'AH0', 'F', 'IY0'])
("'BOUT", ['B', 'AW1', 'T'])
("'CAUSE", ['K', 'AH0', 'Z'])
```

(11) 下载 Speech Commands 数据集，并建立 Dataset：数据集有 2.26GB，下载时间有点久。

```
1  dataset3 = torchaudio.datasets.SPEECHCOMMANDS('./audio', download=True)
```

(12) 显示第一笔数据。

```
1  dataset3[0]
```

执行结果：包含语音的特征向量、取样率、标记 (Label)、发音者代码及第 $N$ 个例句，发音应为 back，而非 backward。

```
(tensor([[-0.0658, -0.0709, -0.0753,  ..., -0.0700, -0.0731, -0.0704]]),
 16000,
 'backward',
 '0165e0e8',
 0)
```

(13) 显示频谱及播放：显示 10 笔指令。

```
1  for i in range(0, 20000, 2000):
2      waveform, sample_rate, label = dataset3[i]
3      audio_util.plot_specgram(waveform, sample_rate, title=f"Sample {i}: {label}")
4      audio_util.play_audio(waveform, sample_rate)
```

执行结果。

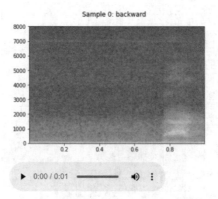

## 14-5 语音深度学习应用

接下来,就来实操两个应用。

(1) 音乐曲风的分类:依照语音特征对音乐分门别类。

(2) 短指令辨识:常用单词的辨识,例如游戏操控,Play、Stop、Up、Down、Left、Right 等。

**范例8. 音乐曲风的分类。**

**下列程序代码请参考【14_10_GTZAN_CNN.ipynb】。**

数据集:GTZAN 共有 10 个类别,每个类别各有 100 首歌,每首歌的长度均为 30 秒。

(1) 加载相关套件。

```
1  import torch
2  from torch import nn
3  import torchaudio
4  import torchaudio.transforms as T
5  from torch.nn import functional as F
6  from torch.utils.data import Dataset, DataLoader
7  import IPython
8  from IPython.display import Audio
9  import matplotlib.pyplot as plt
10 import os
11 import math
12 import audio_util
```

(2) 设定参数:笔者设定较大的批量,在本机或 Colab 上执行都会造成 GPU 内存不足。

```
1  PATH_DATASETS = "./audio" # 预设路径
2  BATCH_SIZE = 5  # 批量
3  device = torch.device("cuda" if torch.cuda.is_available() else "cpu")
4  "cuda" if torch.cuda.is_available() else "cpu"
```

(3) 下载 GTZAN 数据集,并建立 Dataset。

```
1  dataset_GTZAN = torchaudio.datasets.GTZAN(PATH_DATASETS, download=True)
```

(4) 共 10 个类别。

```
1  # label 类别
2  gtzan_genres = [
3      "blues",
4      "classical",
5      "country",
6      "disco",
7      "hiphop",
8      "jazz",
9      "metal",
10     "pop",
11     "reggae",
12     "rock",
13 ]
```

(5) 音频转换为 MFCC:也可以使用其他的特征,例如频谱、FBank 等。

```
1  n_fft = 2048
2  hop_length = 512
3  n_mels = 256
4  n_mfcc = 256
5
6  class GTZAN_DS(Dataset):
7      def __init__(self, dataset1):
8          self.dataset1 = dataset_GTZAN
9
10     def __len__(self):
```

```
11          return len(self.dataset1)
12
13      def __getitem__(self, n):
14          waveform , sample_rate, label = self.dataset1[n]
15          mfcc_transform = T.MFCC(
16              sample_rate=sample_rate,
17              n_mfcc=n_mfcc,    # MFCC 个数
18              melkwargs={
19                  'n_fft': n_fft,
20                  'n_mels': n_mels,
21                  'hop_length': hop_length,
22                  'mel_scale': 'htk',
23              }
24          )
25          mfcc = mfcc_transform(waveform)
26          # print(mfcc.shape)
27          mfcc = mfcc[:, :, :1280]
28          return mfcc, gtzan_genres.index(label)
29
30  dataset = GTZAN_DS(dataset_GTZAN)
```

(6) 数据分割并采取随机抽样：测试数据占 20%。

```
1  from torch.utils.data import random_split
2
3  test_size = int(len(dataset) * 0.2)
4  train_size = len(dataset) - test_size
5
6  train_ds, test_ds = random_split(dataset, [train_size, test_size])
7  len(train_ds), len(test_ds)
```

执行结果：训练数据 800 笔、测试数据 200 笔。

(7) CNN 模型：使用一般的"Conv+BatchNorm2d+MaxPool2d+Linear"神经层构成的 CNN 模型，网络上有采取较复杂的模型或使用预先训练模型，读者可自行测试。

```
1  # 建立模型
2  class ConvNet(nn.Module):
3      def __init__(self, num_classes=10):
4          super(ConvNet, self).__init__()
5          self.layer1 = nn.Sequential(
6              # Conv2d 参数: in-channel, out-channel, kernel size, Stride, Padding
7              nn.Conv2d(1, 16, kernel_size=5, stride=1, padding=2),
8              nn.BatchNorm2d(16),
9              nn.ReLU(),
10             nn.MaxPool2d(kernel_size=2, stride=2))
11         self.layer2 = nn.Sequential(
12             nn.Conv2d(16, 32, kernel_size=5, stride=1, padding=2),
13             nn.BatchNorm2d(32),
14             nn.ReLU(),
15             nn.MaxPool2d(kernel_size=2, stride=2))
16         self.fc1 = nn.Linear(655360, num_classes)
17         # self.fc2 = nn.Linear(1280, num_classes)
18
19     def forward(self, x):
20         out = self.layer1(x)
21         out = self.layer2(out)
22         out = out.reshape(out.size(0), -1)
23         out = self.fc1(out)
24         #out = self.fc2(out)
25         out = F.log_softmax(out, dim=1)
26         return out
27
28  model = ConvNet().to(device)
```

(8) 模型训练。

```
1  epochs = 10
2  lr=0.01
3
4  # 设定优化器(optimizer)
5  optimizer = torch.optim.Adam(model.parameters(), lr=lr)
6
7  model.train()
8  loss_list = []
9  for epoch in range(1, epochs + 1):
```

```
10    for batch_idx, (data, target) in enumerate(train_loader):
11        # if batch_idx == 0 and epoch == 1: print(type(data), type(target))
12        data, target = data.to(device), target.to(device)
13
14        optimizer.zero_grad()
15        output = model(data)
16        # if batch_idx == 0 : print(output.shape, target.shape)
17        loss = F.nll_loss(output, target)
18        loss.backward()
19        optimizer.step()
20
21        if (batch_idx+1) % 10 == 0:
22            loss_list.append(loss.item())
23            batch = (batch_idx+1) * len(data)
24            data_count = len(train_loader.dataset)
25            percentage = (100. * (batch_idx+1) / len(train_loader))
26            print(f'Epoch {epoch}: [{batch:5d} / {data_count}] ({percentage:.0f} %)' +
27                  f'  Loss: {loss.item():.6f}')
```

执行结果：使用的批量较小，损失波动比较剧烈。

(9) 对训练过程的损失绘图。

```
1  import matplotlib.pyplot as plt
2
3  plt.plot(loss_list, 'r')
```

执行结果：损失从一开始的 2389，到最后的 0.000751，收敛的效果非常好。

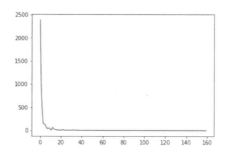

(10) 评分。

```
1  model.eval()
2  test_loss = 0
3  correct = 0
4  predictions = []
5  target_list = []
6  with torch.no_grad():
7      for data, target in test_loader:
8          data, target = data.to(device), target.to(device)
9          output = model(data)
10
11         # sum up batch loss
12         test_loss += F.nll_loss(output, target).item()
13
14         # 预测
15         output = model(data)
16
17         # 计算正确数
18         _, predicted = torch.max(output.data, 1)
19         predictions.extend(predicted.cpu().numpy())
20         target_list.extend(target.cpu())
21         correct += (predicted == target).sum().item()
22
23  # 平均损失
24  test_loss /= len(test_loader.dataset)
25  # 显示测试结果
26  batch = batch_idx * len(data)
27  data_count = len(test_loader.dataset)
28  percentage = 100. * correct / data_count
29  print(f'平均损失: {test_loss:.4f}, 准确率: {correct}/{data_count}' +
30        f' ({percentage:.2f}%)\n')
```

执行结果：训练 10 周期，准确率为 44.5%，并不高。

平均损失：0.3201, 准确率：89/200 (44.50%)。

(11) 显示混淆矩阵。

```
1  from sklearn.metrics import confusion_matrix, ConfusionMatrixDisplay
2  cm = confusion_matrix(target_list, predictions)
3  disp = ConfusionMatrixDisplay(confusion_matrix=cm, display_labels=gtzan_genres)
4  disp.plot()
5  plt.xticks(rotation=90);
```

执行结果：classical、metal 被错认的比例较低，country、disco、reggae 被错认的比例较高。

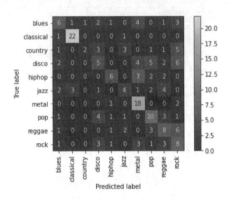

*Music Genre Recognition using Convolutional Neural Networks (CNN) — Part 1*[31] 一文提到两个改善的方向：

(1) 将音乐数据分段 (Segmentation)：每一段视为一笔数据，使用 pydub 套件调用 AudioSegment.from_wav() 加载文件，即可切割，PyTorch 范例可参考 prasad213 music-genre-classification[32]。

(2) 数据增强 (Data Augmentation)：利用数据增强产生更多的数据，可参考【14_08_torchaudio_preprocessing.ipynb】做法。

上文使用数据分段与数据增强确实能大幅提高准确率，训练 70 周期后，训练准确率高达 99.57%，而验证准确率也达到了 89.03%。范例采用 TensorFlow，而笔者改用 PyTorch 试验，但因 Sox 整合有问题 ( 见前文说明 )，未使用数据增强，发现准确率仅能达到 50% 左右，做法可参见程序【14_11_GTZAN_CNN_slicing.ipynb】。

接着再来看另一个范例，Google 收集一个短指令的数据集，包括常用的词汇，譬如 stop、play、up、down、right、left 等，共有 35 个类别，如果辨识率很高的话，我们就能应用到各种场域，像是玩游戏、控制机器人行进、简报播放等。

**范例9. 短指令辨识。**一样使用MFCC向量，输入CNN模型，即可进行分类。同时笔者也会结合录音实测，示范如何控制录音与训练样本一致(Alignment)，其中有些技巧将在后面说明。

下列程序代码请参考【14_12_Speech_Command.ipynb】。

数据集：Google's Speech Commands Dataset，也属 Torch Audio 内建数据集，为求缩短训练运行时间，我们只保留三个子目录 bed、cat、happy，其余均移置他处。

每个文件长度约为 1 秒，无杂音。它也附一个有杂音的目录，与无杂音的文件混合在一起，增加辨识的困难度。

(1) 加载相关套件。

```
1  import numpy as np
2  import matplotlib.pyplot as plt
3  import os
4  import torch
5  import torch.nn as nn
6  import torch.nn.functional as F
7  import torchaudio.transforms as T
8  import torch.optim as optim
9  import torchaudio
10 from torch.utils.data import Dataset, DataLoader
11 import sys
12 import audio_util
13 from IPython.display import Audio
14 from IPython.core.display import display
```

(2) 下载 Speech Commands 数据集，并建立 Dataset。

```
1  dataset = torchaudio.datasets.SPEECHCOMMANDS('./audio', download=True)
```

(3) 观察第一笔数据，共 5 个字段，分别为语音的特征向量、取样率、标记 (Label)、发音者代码 (Speaker Id) 及第 N 个例句 (Utterance)。

```
1  dataset[0]
```

执行结果。

```
(tensor([[ 9.1553e-05,  3.0518e-05,  1.8311e-04,  ..., -3.0518e-05,
          -9.1553e-05,  1.2207e-04]]),
 16000,
 'bed',
 '00176480',
 0)
```

(4) 也可直接选取一个文件测试，该文件发音为 happy。

```
1  train_audio_path = './audio/SpeechCommands/speech_commands_v0.02/'
2  wav_file = train_audio_path+'happy/0ab3b47d_nohash_0.wav'
3
4  # 播放语音
5  Audio(wav_file, autoplay=False)
```

执行结果：可直接单击 "Play" 按钮，播放语音。

- 可调用 waveform.shape 观察维度，窗体声道、16000 帧 (Frame)，再观察取样率 (sample_rate) 为 16000，故音频总长度 = 音频总帧数 / 每秒帧数 = 16000/16000=1( 秒 )。

(5) 绘制波形：加载的音频介于 [-1, 1] 之间，0 为静音，补足音频长度通常会补 0。

```
1  waveform, sample_rate = torchaudio.load(wav_file)
2  audio_util.plot_waveform(waveform, sample_rate)
```

执行结果。

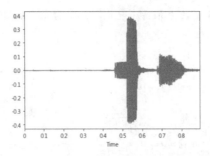

(6) 再选一个文件测试，该文件发音也为 happy：audio 函数不在最后一行，必须额外调用 IPython.core.display.display。

```
1  wav_file = train_audio_path+'happy/0b09edd3_nohash_0.wav'
2
3  # 播放语音
4  display(Audio(wav_file, autoplay=False))
5
6  # 绘制波形
7  waveform, sample_rate = torchaudio.load(wav_file)
8  audio_util.plot_waveform(waveform, sample_rate)
```

执行结果：与上图相比较，同样有两段振幅较大的声波，此代表 happy 的两个音节，但是，因为录音时每个人的起始发音点不同，因此，波形有很大差异，必须收集够多的训练数据才能找出共同的特征。

数据长度亦为 1 秒，与上一笔音频文件的数据长度相同。

可以播放音频文件观察其中的差异。

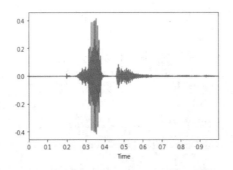

(7) 取得音频文件的属性：每个文件的长度不等，但都接近 1 秒。

```
1  info = torchaudio.info(wav_file)
2  print(f'取样率={info.sample_rate}, 帧数={info.num_frames}, ' +
3        f'声道={info.num_channels}, 精度={info.bits_per_sample}, ' +
4        f'文件秒数={info.num_frames / info.sample_rate:.2f}')
```

执行结果：取样频率 =16000，帧数 =16000，声道 =1，精度 =16，文件秒数 =1。

(8) 重抽样：可针对音频重抽样，以降低音频取样频率，缩小文件或使音频长度一致。

```
1  # 载入音频文件
2  waveform, sample_rate = torchaudio.load(wav_file)
3
4  # 重抽样率，每秒取 8000 个样本
5  resample_rate = 8000
6  resampled_waveform = torchaudio.functional.resample(
7                        waveform, sample_rate, resample_rate)
8  print(f'帧数={resampled_waveform.shape[1]}')
```

执行结果：帧数 =8000。

(9) 取得所有子目录名称，当作标记 (Y)。

```
1  labels=os.listdir(train_audio_path)
2  labels
```

执行结果：['bed', 'cat', 'happy']，共 3 个类别。

(10) 接着进行简单的数据探索 (EDA)：统计子目录的文件数。

```
1   no_of_recordings=[]
2   for label in labels:
3       waves = [f for f in os.listdir(train_audio_path + '/'+ label)
4                 if f.endswith('.wav')]
5       no_of_recordings.append(len(waves))
6
7   # 绘图
8   plt.rcParams['font.sans-serif'] = ['Microsoft JhengHei']
9   plt.rcParams['axes.unicode_minus'] = False
10
11  plt.figure(figsize=(10,6))
12  index = np.arange(len(labels))
13  plt.bar(index, no_of_recordings)
14  plt.xlabel('指令', fontsize=12)
15  plt.ylabel('文件数', fontsize=12)
16  plt.xticks(index, labels, fontsize=15, rotation=60)
17  plt.title('子目录的文件数')
18  print(f'文件数={no_of_recordings}')
19  plt.show()
```

- 执行结果：文件数目略有不同。
  文件数 = [2014, 2031, 2054]。

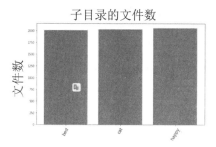

(11) 音频文件长度统计。

```
1  import seaborn as sns
2  length_list=[]
3  for x in dataset:
4      waveform, sample_rate, label, speaker_id, utterance_number = x
5      length_list.append(waveform.shape[1])
6  sns.histplot(length_list)
```

执行结果：以直方图的方式统计文件的长度，发现大部分的音频文件是接近 1 秒钟，但有少数文件的时间长度非常短，事实上这会影响训练的准确度，故统一长度有其必要性。

- X 轴为音频文件长度，Y 轴为笔数。

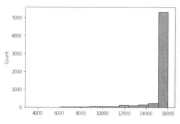

(12) 接着开始语音分类的工作：下载数据集，并建立 Dataset。

```
1  dataset = torchaudio.datasets.SPEECHCOMMANDS('./audio', download=True)
```

(13) 特征提取：音频截长补短，统一帧数为 16000，并转换为 MFCC。以下程序利用 F.pad 在长度不足时，右边补 0，也可以左右各补一半：
left = int(TOTAL_FRAME_COUNT-waveform.shape[1]/2)
right = TOTAL_FRAME_COUNT - int(TOTAL_FRAME_COUNT-waveform.shape[1]/2))
参数可改为 (left, right)。

```
1   TOTAL_FRAME_COUNT = 16000  # 统一帧数为 16000
2   n_mfcc = 40  # 提取 MFCC 个数
3
4   class SPEECH_DS(Dataset):
5       def __init__(self, dataset1):
6           self.dataset1 = dataset1
7
8       def __len__(self):
9           return len(self.dataset1)
10
11      def __getitem__(self, n):
12          waveform , sample_rate, label, _, _ = self.dataset1[n]
13          if waveform.shape[1] < TOTAL_FRAME_COUNT :  # 长度不足，右边补 0
14              waveform = F.pad(waveform,
15                  (0, TOTAL_FRAME_COUNT-waveform.shape[1]),'constant')
16          elif waveform.shape[1] > TOTAL_FRAME_COUNT :  # 长度过长则截断
17              waveform = waveform[:, :TOTAL_FRAME_COUNT]
18          if waveform.shape[1] != TOTAL_FRAME_COUNT:  # 确认帧数为 16000
19              print(waveform.shape[1])
20
21          mfcc_transform = T.MFCC(
22              sample_rate=sample_rate,
23              n_mfcc=n_mfcc,
24          )
25          mfcc = mfcc_transform(waveform)
26          # print(mfcc)
27          return mfcc, labels.index(label)
28
29  dataset_new = SPEECH_DS(dataset)
```

(14) 数据切割。

```
1   from torch.utils.data import random_split
2
3   test_size = int(len(dataset_new) * 0.2)
4   train_size = len(dataset_new) - test_size
5
6   train_ds, test_ds = random_split(dataset_new, [train_size, test_size])
7   len(train_ds), len(test_ds)
```

(15) 建立 DataLoader：测试时加大批量，以缩短运行时间。

```
1   train_loader = DataLoader(train_ds, BATCH_SIZE, shuffle=False)
2   test_loader = DataLoader(test_ds, BATCH_SIZE*2, shuffle=False)
```

(16) 建立 CNN 模型，与前一个范例相同，除了最后一层，类别数量改为 3，Linear_Input 的数值可由执行训练的错误信息获得。

```
1   Linear_Input = 6400
2   class ConvNet(nn.Module):
3       def __init__(self, num_classes=3):
4           super(ConvNet, self).__init__()
5           self.layer1 = nn.Sequential(
6               # Conv2d 参数：in-channel, out-channel, kernel size, Stride, Padding
7               nn.Conv2d(1, 16, kernel_size=5, stride=1, padding=2),
8               nn.BatchNorm2d(16),
9               nn.ReLU(),
10              nn.MaxPool2d(kernel_size=2, stride=2))
11          self.layer2 = nn.Sequential(
```

```
12          nn.Conv2d(16, 32, kernel_size=5, stride=1, padding=2),
13          nn.BatchNorm2d(32),
14          nn.ReLU(),
15          nn.MaxPool2d(kernel_size=2, stride=2))
16      self.fc = nn.Linear(Linear_Input, num_classes)
17
18  def forward(self, x):
19      out = self.layer1(x)
20      out = self.layer2(out)
21      out = out.reshape(out.size(0), -1)
22      out = self.fc(out)
23      out = F.log_softmax(out, dim=1)
24      return out
25
26  model = ConvNet(num_classes=3).to(device)
```

(17) 定义评分的函数。

```
1   def score_model():
2       model.eval()
3       test_loss = 0
4       correct = 0
5       prediction_list = []
6       target_list = []
7       with torch.no_grad():
8           for data, target in test_loader:
9               data, target = data.to(device), target.to(device)
10              # 预测
11              output = model(data)
12
13              # sum up batch loss
14              test_loss += F.nll_loss(output, target).item()
15
16              # 计算正确数
17              _, predicted = torch.max(output.data, 1)
18              correct += (predicted == target).sum().item()
19              prediction_list.extend(predicted.cpu().numpy())
20              target_list.extend(target.cpu().numpy())
21
22      # 平均损失
23      test_loss /= len(test_loader.dataset)
24      # 显示测试结果
25      batch = batch_idx * len(data)
26      data_count = len(test_loader.dataset)
27      percentage = 100. * correct / data_count
28      print(f'平均损失: {test_loss:.4f}, 准确率: {correct}/{data_count}' +
29          f' ({percentage:.2f}%)\n')
30      return prediction_list, target_list
```

(18) 模型训练：试采用动态排程，每 20 执行周期，学习率降低 10%，使梯度下降过程更细腻，若删除此排程，对本范例无显著影响。

```
1   epochs = 10
2
3   # 设定优化器(optimizer)
4   optimizer = optim.Adam(model.parameters(), lr=0.001, weight_decay=0.0001)
5   # 每 20 执行周期，学习率降低 10%
6   scheduler = optim.lr_scheduler.StepLR(optimizer, step_size=20, gamma=0.1)
7
8   model.train()
9   loss_list = []
10  for epoch in range(1, epochs + 1):
11      for batch_idx, (data, target) in enumerate(train_loader):
12          data, target = data.to(device), target.to(device)
13
14          optimizer.zero_grad()
15          output = model(data)
16          # if batch_idx == 0 : print(output, target)
17          loss = F.nll_loss(output, target)
18          loss.backward()
19          optimizer.step()
20
21          if (batch_idx+1) % 10 == 0:
22              loss_list.append(loss.item())
23              batch = (batch_idx+1) * len(data)
24              data_count = len(train_loader.dataset)
25              percentage = (100. * (batch_idx+1) / len(train_loader))
26              print(f'Epoch {epoch}: [{batch:5d} / {data_count}] ({percentage:.0f} %)' +
27                  f'  Loss: {loss.item():.6f}')
28      score_model()
29      scheduler.step()
```

执行结果:准确度由 70.22% 逐步爬升至 91.22%。

```
Epoch 1: [ 1000 / 4880] (20 %)   Loss: 1.294984
Epoch 1: [ 2000 / 4880] (41 %)   Loss: 1.038083
Epoch 1: [ 3000 / 4880] (61 %)   Loss: 1.007797
Epoch 1: [ 4000 / 4880] (82 %)   Loss: 0.917779
平均损失: 0.0047, 准确率: 856/1219 (70.22%)

Epoch 2: [ 1000 / 4880] (20 %)   Loss: 0.691356
Epoch 2: [ 2000 / 4880] (41 %)   Loss: 0.703693
Epoch 2: [ 3000 / 4880] (61 %)   Loss: 0.529919
Epoch 2: [ 4000 / 4880] (82 %)   Loss: 0.578867
平均损失: 0.0034, 准确率: 955/1219 (78.34%)

Epoch 3: [ 1000 / 4880] (20 %)   Loss: 0.450731
Epoch 3: [ 2000 / 4880] (41 %)   Loss: 0.417602
Epoch 3: [ 3000 / 4880] (61 %)   Loss: 0.347282
Epoch 3: [ 4000 / 4880] (82 %)   Loss: 0.366764
平均损失: 0.0025, 准确率: 1033/1219 (84.74%)
```

(19) 评分。

```
1  score_model();
```

执行结果:准确度高达 91.22%。

(20) 对训练过程的损失绘图。

```
1  import matplotlib.pyplot as plt
2
3  plt.plot(loss_list, 'r')
```

执行结果:损失逐步降低。

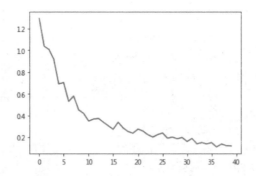

(21) 模型存盘与加载。

```
1  # 模型存盘
2  torch.save(model, 'Speech_Command.pth')
3  # 模型载入
4  model = torch.load('Speech_Command.pth')
```

(22) 定义一个预测函数,功能包括:
统一文件长度,右边补 0,过长则截掉。
转为 MFCC。
以 MFCC 预测词汇。

```
1   # 预测函数
2   def predict(wav_file):
3       waveform , sample_rate = torchaudio.load(wav_file)
4       
5       if waveform.shape[1] < TOTAL_FRAME_COUNT: # 长度不足，右边补 0
6           waveform = F.pad(waveform,(0,
7                           TOTAL_FRAME_COUNT-waveform.shape[1]),'constant')
8       elif waveform.shape[1] > TOTAL_FRAME_COUNT: # 长度过长则截断
9           waveform = waveform[:, :TOTAL_FRAME_COUNT]
10      if waveform.shape[1] != TOTAL_FRAME_COUNT:
11          print(waveform.shape[1])
12      
13      mfcc_transform = T.MFCC(
14          sample_rate=sample_rate,
15          n_mfcc=n_mfcc,     # MFCC 个数
16      )
17      mfcc = mfcc_transform(waveform)
18      mfcc = mfcc.reshape(1,*mfcc.shape)
19      # print(mfcc)
20      
21      #print(X_pred.shape, samples.shape)
22      # 预测
23      output = model(mfcc.to(device))
24      _, predicted = torch.max(output.data, 1)
25      return predicted.cpu().item()
```

(23) 任选一个文件进行预测，该文件发音为 bed。

```
1   predict(train_audio_path+'bed/0d2bcf9d_nohash_0.wav')
```

执行结果：0，正确判断为 bed。

(24) 再任选一个文件测试，该文件发音为 cat。

```
1   predict(train_audio_path+'cat/0ac15fe9_nohash_0.wav')
```

执行结果：1，正确判断为 cat。

(25) 接着再任选一个文件测试，该文件发音为 happy。

```
1   predict(train_audio_path+'happy/0ab3b47d_nohash_0.wav')
```

执行结果：2，正确判断为 happy。

(26) 自行录音测试：笔者开发一个录音程序【14_13_record.py】，用法如下：

python 14_13_record.py audio/happy.wav

最后的参数为存盘的文件名。程序录音长度设为 1 秒，也可以利用上述 predict 函数作法，录制较长的音频，再截取中间 1 秒钟的音频。

(27) 测试：分别录制三个音频文件测试，例如 bed.wav 文件。

```
1   predict('./audio/bed.wav')
```

执行结果：0，正确判断为 bed。

尽量模仿训练数据的发音，预测才会正确。

总结来说，如果用训练数据进行测试的话，准确率都还不错，但若是自行录音则准确率就差强人意了，可能原因有两点，第一是笔者发音欠佳，第二是录音的处理方式与训练方式不同，因此，建议还是要自己收集训练数据为宜，也建议读者发挥创意，多做试验，笔者光这个范例就花了好几天，才将结果弄得差强人意。

以上是参酌多篇文章后修改而成的程序，相关内容可参阅《Day 25：自动语音识别 (Automatic Speech Recognition)——观念与实践》[33]，此外，Kaggle 上也有一个关于这个数据集的竞赛，可参阅 *TensorFlow Speech Recognition Challenge*[34]。

上述试验只能辨识单词，假使要辨识更长的一句话或一段话，那就做来到了，因为讲话的方式千变万化，很难收集到完整的数据来训练，所以解决的办法则是把辨识目标切得更细，以音节或音素 (Phoneme) 为单位，并使用语言模型，考虑上下文才能精准预测，下一节我们就来探讨相关的技术。

## 14-6　自动语音识别

自动语音识别 (Automatic Speech Recognition, ASR) 的目标是将人类的语音转换为数字信号，之后计算机就可进一步理解说话者的意图，并做出对应的行动，譬如指令操控，应用于简报上 / 下页控制、车辆和居家装置开关的控制、产生字幕与演讲稿等。

英文的词汇有数万个，假如要进行分类，模型会很复杂，需要很长的训练时间，准确率也不会太高，这是因为相似音太多，所以自动语音识别多改以音素 (Phoneme) 为预测目标，依据维基百科[35]的说明如下：

音素 (Phoneme) 又称音位，是人类语言中能够区别意义的最小声音单位，一个词汇由一个至多个音节组成，每个音节又由一个至多个音段所组成，音素类似音段，但音素定义是要能区分语义。

举例来说，bat 由 3 个音素 /b/、/ae/、/t/ 所组成，连接这些音素，就是 bat 的拼音 (Pronunciation)，然后按照拼音就可以猜测到一个英文词汇 (Word)，当然，有可能发生同音异字的状况，这时就必须依靠上下文做进一步的推测了。如图 14.15 所示，Human 单词被切割成多个音素 HH、Y、UW、M、AH、N。

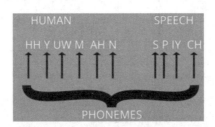

图 14.15　音素辨识示意图

图片来源：*Indian Accent Speech Recognition* [36]

各种语言的音素列表可参考"Amazon Polly 支持语言的音素"[37]，以美式英文 (en-US) 为例，音素列表主要包括元音和子音，共 40~50 个，而中文则另外包含声调 ( 一声、二声、三声、四声和轻声 )。

自动语音识别的流程可分成如图 14.16 所示四个步骤。

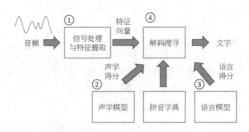

图 14.16　自动语音识别架构

(1) 信号处理与特征提取：先将音频进行傅里叶变换、去杂音等前置处理，接着转为特征向量，比如 MFCC、LPC(Linear Predictive Coding)。

(2) 声学模型 (Acoustic Model)：通过特征向量，转换成多个音素，再将音素组合成拼音，然后至拼音字典 (Pronunciation Dictionary) 里比对，找到对应的词汇与得分。

(3) 语言模型 (Language Model)：依据上一个词汇，猜测目前的词汇，事先以 n-gram 为输入，训练模型，之后套用此模型，计算一个语言的得分。

(4) 解码搜寻 (Decoding Search)：根据声学得分和语言得分来比对搜寻出最有可能的词汇。

经典的 GMM-HMM 算法是过去数十年来语音识别的主流，直到 2014 年 Google 学者使用双向 LSTM，以 CTC(Connectionist Temporal Classification) 为目标函数，将音频转换成文字，深度学习算法就此涉足这个领域，不过，目前大部分的工具箱依然以 GMM-HMM 算法为主，因此，我们还是要先来认识 GMM-HMM 的运作原理。

自动语音识别的流程可用贝式定理 (Bayes' Theorem) 来表示：

$$W = \arg\max P(W|O) \tag{14.1}$$

$$W = \arg\max \frac{P(O|W)P(W)}{P(O)} \tag{14.2}$$

$$W = \arg\max P(O|W)P(W) \tag{14.3}$$

其中：

$W$ 就是我们要预测的词汇 ($W_1$、$W_2$、$W_3$…)，$O$ 是音频的特征向量。

$P(W|O)$：已知特征向量，预测各个词汇的概率，故以 argmax 找到获得最大概率的 $W$，即为辨识的词汇。

14.2 公式的 $P(O)$ 不影响 $W$，可省略，故简化为 14.3 公式。

$P(O|W)$：通常就是声学模型，以高斯混合模型 (Gaussian Mixture Model, GMM) 算法建构。

$P(W)$：语言模型，以隐藏式马尔可夫模型 (Hidden Markov Model, HMM) 算法建构。

高斯混合模型 (Gaussian Mixture Model, GMM) 是一种非监督式的算法，假设样本是由多个常态分配混合而成的，则算法会利用最大似然法 (MLE) 推算出母体的统计量 ( 平均数、标准偏差 )，进而将数据分成多个集群 (Clusters)。应用到声学模型，就是以特征向量作为输入，算出每个词汇的可能概率，如图 14.17、图 14.18 所示。

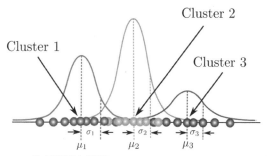

图 14.17　一维高斯混合模型 (Gaussian Mixture Model, GMM) 示意图
图片来源：*Gaussian Mixture Models Explained*[38]

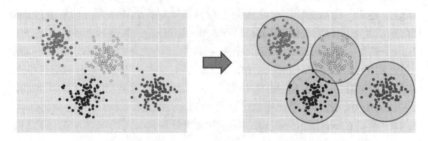

图 14.18　二维高斯混合模型 (Gaussian Mixture Model, GMM) 示意图

以声学模型推测出多个音素后，就可以比对拼音字典 (Pronunciation Dictionary)，找到相对的词汇，如图 14.19 所示。

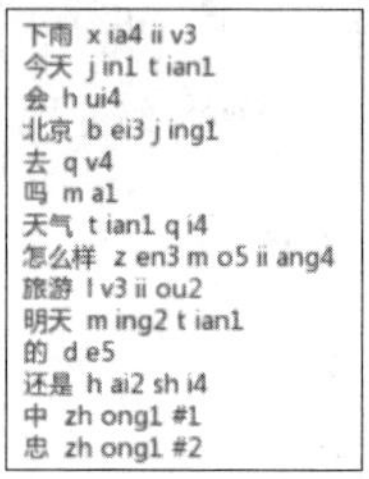

```
下雨   x ia4 ii v3
今天   j in1 t ian1
会     h ui4
北京   b ei3 j ing1
去     q v4
吗     m a1
天气   t ian1 q i4
怎么样 z en3 m o5 ii ang4
跷游   l v3 ii ou2
明天   m ing2 t ian1
的     d e5
还是   h ai2 sh i4
中     zh ong1 #1
忠     zh ong1 #2
```

图 14.19　拼音字典 (Pronunciation Dictionary) 示意图，左边是词汇，右边是对照的音素，
最后两个词汇同音，故标示 #1、#2

图片来源：语音识别系列 2——基于 WFST 译码器 [39]

隐藏式马尔可夫模型 (Hidden Markov Model，HMM) 系利用前面的状态 ($k$-1、$k$-2、…) 预测目前状态 ($k$)。应用到语言模型，就是以前面的词汇为输入，预测下一个词汇的可能概率，即前面提到 NLP 的 n-gram 语言模型。声学模型也可以使用 HMM，以前面的音素推测目前的音素，例如一个字的拼音为首是ㄅ，那么接着ㄆ就绝对不会出现。

$$P(\boldsymbol{w}) = \prod_{k=1}^{K} P(w_k | w_{k-1}, \ldots, w_1)$$

又比方，and、but、cat 三个词汇，采用 bi-gram 模型，我们就要根据上一个词汇预测下一个词汇的可能概率，并取其中概率最大者，如图 14.20 所示。

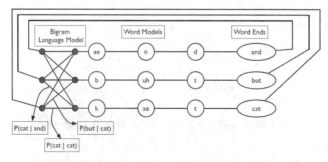

图 14.20　bi-gram 语言模型结合 HMM 的示意图

图片来源：《爱丁堡大学语音辨识课程》第 11 章 [40]

最后，综合 GMM 和 HMM 模型所得到的分数，再借由译码的方式搜寻最有可能的词汇。小型的词汇集可采用维特比译码 (Viterbi Decoding) 进行精确搜索 (Exact Search)，但大词汇连续语音辨识就会遭遇困难，所以一般会改采用光束搜寻 (Beam Search)、加权的有限状态转换机 (Weighted Finite State Transducers, WFST) 或其他算法 ( 如图 14.21 所示 )，如要获得较完整的概念可参阅《现阶段大词汇连续语音辨识研究之简介》[41] 一文的说明。

## 14-7 自动语音识别实操

Kaldi 是目前较为流行的语音识别工具箱，它囊括了上一节所介绍的声学模型、语言模型及译码搜寻的相关函数库实践，由 Daniel Povey 等研究人员所开发，源代码为 C++，其安装程序较繁复，必须安装许多公用程序和第三方工具，虽然可以在 Windows 操作系统上安装，但是许多测试步骤均使用 Shell 脚本 (*.sh)，因此，最好还是安装在 Linux 操作系统上，于是笔者在 PC 上另外安装 Ubuntu 操作系统，并从头设置 Python 环境，有兴趣的读者可参阅《Ubuntu 巡航记》系列文章，总共 5 篇，最后一篇介绍如何编译及设置 Kaldi 源代码，全程可能要花上一整天的时间，读者要有一点耐心。

安装完成之后就可以进行一些测试了，相关操作说明可参考 Kaldi 官网文件 [42]，内容相当多，需投入不少心力来研读，大家要有心理准备。

另外，各所大学都有开设整学期的课程，名称即为语音识别 (Speech Recognition)，网络上有许多开放性的教材，包括影片和投影片，有兴趣的读者可搜寻一下。

(1) 台湾大学李琳山教授 *Introduction to Digital Speech Processing* [43]。
(2) 哥伦比亚大学 *Speech Recognition* [44]。
(3) 爱丁堡大学 *Automatic Speech Recognition* [45]。

最后，语音识别必须要收集大量的训练数据，而 OpenSLR 就有提供非常多可免费下载的数据集和软件，详情可参考官网说明 [46]，其中也有中文的语音数据集 CN-Celeb，它包含了 1000 位著名华人的三十万条语音，而 VoxCeleb 则是知名人士的英语语音数据集，其他较知名的数据集还有：

(1) TIMIT[47]：美式英语数据集。
(2) LibriSpeech：电子书的英语朗读数据集，可在 OpenSLR 下载。
(3) 维基百科语音数据集 [48]。

## 14-8 总结

语音除了可以拿来语音识别之外，还有许多其他方面的应用，譬如：
(1) 声纹辨识：从讲话的声音分辨是否为特定人，属于生物识别技术 (Biometrics) 的一种，可应用在登录 (Sign In)、犯罪侦测、智能家居等领域。
(2) 声纹建模：模拟或创造特定人的声音唱歌或讲话，例如 Siri。
(3) 相似性比较：未来也许能够使用语音搜寻，类似现在的文字、图像搜寻。
(4) 音乐方面的应用：比方曲风辨识、模拟歌手的声音唱歌、编曲 / 混音等。

各位读者看到这里，应该能深刻了解到，文字、影像、语音识别是整个人工智能应用的三大基石，不论是自动驾驶、机器人、ChatBot，甚至医疗诊断通通都是建构在这些基础技术之上。如前文提到，现今的第三波 AI 浪潮之所以不会像前两波一样后继无力、不了了之，有一部分原因要归功于这些基本技术的开发，这就像盖房子的地基，技术的研发与应用实践并进，才不会空有理论，最后成为空中楼阁。相信在未来这三大基石还会有更进一步的发展，到时候又会有新的技能被解放，人们能通过 AI 完成的任务也就更多了，换言之，我们学习的脚步永远不会有停止的一天，笔者与大家共勉之。

# 第 5 篇 | 强化学习

强化学习 (Reinforcement Learning, RL) 相关的研究少说也有数十年的历史了，但与另外两类机器学习相比较冷门，直到 2016 年以强化学习为理论基础的 AlphaGo，先后击败世界围棋冠军李世乭、柯洁等人后，开始受到世人的注目，强化学习因此一炮而红，学者专家纷纷投入研发，接下来我们就来探究其原理与应用。

机器学习的分类如下。

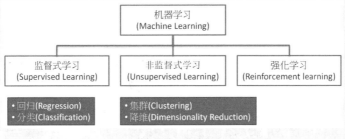

机器学习分类

# 第 15 章
# 强化学习原理及应用

强化学习是指机器与环境的互动过程中,人类不直接提供解决方案,而是通过计算机不断地尝试,并在错误中学习,称为试误法(Trial and Error),在尝试过程中,不断修正策略,最终计算机就可以学习到最佳的行动策略。例如,就像是训练小狗接飞盘一样,主人不会教狗如何接飞盘,而是不断地抛出飞盘让小狗去接,如果小狗成功接到飞盘,就给予食物奖励,反之就不给奖励,经过反复练习后,小狗就能练就一身好功夫了(如图 15.1 所示)。因此,强化学习并不是单一阶段(One Step)的算法,而是多阶段、反复求解,类似梯度下降法的优化过程。

图 15.1 训练小狗接飞盘,如果成功接住飞盘,就给予食物作为奖励

根据维基百科[1]的概述,强化学习涉及的学术领域相当多,包括博弈论(Game Theory)、自动控制、作业研究、信息论、仿真优化、群体智慧(Swarm Intelligence)、统计学以及遗传算法等,同时它的应用领域也是包山包海,例如:
(1) 下棋、电玩游戏策略 (game playing)。
(2) 制造/医疗/服务机器人的控制策略 (Robotic motor control)。
(3) 广告投放策略 (Ad-placement optimization)。
(4) 金融投资交易策略 (Stock market trading strategies)。
(5) 运输路线的规划 (Transportation Routing)。
(6) 库存管理策略 (Inventory Management)、生产排程 (Production scheduling)。

甚至于残酷的战争,只要应用场域能在模拟环境下 Trial and Error,并且需要人工智能提供行动的决策辅助,都是强化学习可以发挥的领域。

## 15-1 强化学习的基础

强化学习的理论基础为马尔可夫决策过程(Markov Decision Processes,MDP),主要是指所有的行动决策都会基于当时所处的状态(State)及行动后带来的奖励(Reward)

所影响，而状态与奖励是由环境所决定的，以图 15.2 所示示意图说明。

(1) 代理人行动后，环境会依据行动更新状态，并给予奖励。
(2) 代理人观察所处的状态及之前的行动，决定下次的行动。

图 15.2　马尔可夫决策过程的示意图

各个专有名词定义如下：

(1) 代理人或称智能体 (Agent)：也就是实际行动的主人翁，比如游戏中的玩家 (Player)、下棋者、机器人、金融投资者、接飞盘的狗等，主要的任务是与环境互动，并根据当时的状态 (State) 与预期会得到的奖励或惩罚，来决定下一步的行动。代理人可能不止一个，如果有多个则称为多代理人 (Multi-Agent) 的强化学习。

(2) 环境 (Environment)：根据代理人的行动 (Action)，给予立即或延迟的奖励／惩罚，同时会决定代理人所处的状态。

(3) 状态 (State)：指代理人所处的状态，譬如围棋的棋局、游戏中玩家／敌人／宝物的位置、能力和金额。有时候代理人只能观察到局部的状态，例如，扑克牌游戏 "21 点" (Black Jack)，庄家有一张牌盖牌，玩家是看不到的，这种情形称为部分观察 (Partial Observation)，因此状态也被称为观察 (Observation)。

(4) 行动 (Action)：代理人依据环境所提示的状态与奖励而做出的决策。

整个过程就是代理人与环境互动的过程，可以使用行动轨迹 (trajectory) 来表示：

$\{S_0, A_0, R_1, S_1, A_1, R_2, S_2, \cdots, S_t, A_t, R_{t+1}, S_{t+1}, A_{t+1}, R_{t+2}, S_{t+2}\}$

其中：

S：状态 (State)，$S_t$ 是 $t$ 时间点的状态。

A：行动 (Action)。

R：奖励 (Reward)。

行动轨迹：行动 ($A_0$) 后，代理人会得到奖励 ($R_1$)、状态 ($S_1$)，之后再采取下一步的行动 ($A_1$)，不停循环 ($A_t, R_{t+1}, S_{t+1}\cdots$) 直至游戏结束为止。

马尔可夫决策过程的演进，可依信息量分为三种模型 ( 如图 15.3 所示 )：

(1) 马尔可夫过程 (Markov Processes，MP)。
(2) 马尔可夫奖励过程 (Markov Reward Processes，MRP)。
(3) 马尔可夫决策过程 (Markov Decision Processes，MDP)。

图 15.3　马尔可夫决策过程的演进

先从马尔可夫过程 (Markov Process, MP) 开始讲起，它也称为马尔可夫链 (Markov Chain)，主要内容为描述状态之间的转换，例如，假设天气的变化状态只有两种，晴天和雨天，如图 15.4 所示就是一个典型的马尔可夫链的状态转换图。

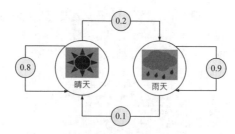

图 15.4　马尔可夫链 (Markov Chain) 的示意图

也可以使用表格说明，称为状态转换矩阵 (State Transition matrix)，见表 15.1。

表 15.1　状态转换矩阵

|  | 晴天 | 雨天 |
|---|---|---|
| 晴天 | 0.8 | 0.2 |
| 雨天 | 0.1 | 0.9 |

图 15.4 与表 15.1 要表达的信息如下：

(1) 今天是晴天，明天也是晴天的概率：0.8。

(2) 今天是晴天，明天是雨天的概率：0.2。

(3) 今天是雨天，明天是晴天的概率：0.1。

(4) 今天是雨天，明天也是雨天的概率：0.9。

也就是说，明日天气会受今日天气的影响，换言之，下一个状态出现的概率会受到目前状态的影响，符合这种特性的模型就称为马尔可夫性质 (Markov Property)，即目前状态 ($S_t$) 只受前 1 个状态 ($S_{t-1}$) 影响，与之前的状态 ($S_{t-2}$, $S_{t-3}$, …) 无关，也可以扩展为受 $n$ 个状态 ($S_{t-1}$, $S_{t-2}$, $S_{t-3}$, …, $S_{t-n}$) 影响，类似时间序列。

有了上面图表信息，我们就可以推测 $n$ 天后出现晴天或雨天的概率，例如：

(1) 今天是晴天，『后天』是晴天的概率 = 0.8 × 0.8 + 0.2 × 0.1 = 0.66，说明如下：

- 今天是晴天，『后天』是晴天有两种状况：
  - 晴天 ➜ 晴天 ➜ 晴天：
  今天是晴天，明天是晴天的概率：0.8。
  明天是晴天，后天是晴天的概率：0.8。
  当两者情况同时发生，而且事件独立：
  P(A ∩ B) = P(A) × P(B) = 0.8 × 0.8 = 0.64。
  - 晴天 ➜ 雨天 ➜ 晴天：
  今天是晴天，明天是雨天的概率：0.2。
  明天是雨天，后天是晴天的概率：0.1。
  P(A ∩ B) = P(A) × P(B) = 0.2 × 0.1 = 0.02。
- 两种状况的概率相加，0.64 + 0.02 = 0.66。

(2) 同理，今天是雨天，后天是晴天的概率 = 0.1 × 0.8 + 0.9 × 0.1 = 0.17。

依照上述的推理，我们可以预测 $n$ 天后是晴天或雨天的概率，$n$ = 1, 2, 3, …, ∞，这就是马尔可夫链的理论基础。

扩充一下，马尔可夫链加上奖励 (Reward)，就称为马尔可夫奖励过程 (Markov Reward Process, MRP)，即每个转换除了概率外，还带有奖励信息，如图 15.5 所示。

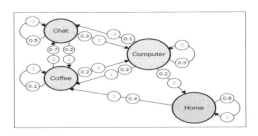

图 15.5  马尔可夫奖励过程 (Markov Reward Process, MRP)

图 15.5 是学生作息的状态转换，状态包括聊天 (Chat)、喝咖啡 (Coffee)、玩计算机 (Computer) 和在家 (Home)，依据转换概率和奖励，可算出每次转换后会获得的奖励期望值，加总起来就是状态的期望值，用以表达每个状态的价值，如下：

$$V(chat) = -1 * 0.5 + 2 * 0.3 + 1 * 0.2 = 0.3$$

$$V(coffee) = 2 * 0.7 + 1 * 0.1 + 3 * 0.2 = 2.1$$

$$V(home) = 1 * 0.6 + 1 * 0.4 = 1.0$$

$$V(computer) = 5 * 0.5 + (-3) * 0.1 + 1 * 0.2 + 2 * 0.2 = 2.8$$

状态期望值非常重要，它显露出代理人处在的状态相对有利的比较，可以作为行动决策的依据，例如我们玩游戏时，永远选择往最有利的状态前进，后续会有更详细的说明。

再扩充一下，马尔可夫奖励过程再加上行动转移矩阵 (Action Transition Matrix) 就是马尔可夫决策过程 (Markov Decision Processes, MDP)，那行动转移矩阵又是什么呢？举例来说，走迷宫时，玩家决定往上 / 下 / 左 / 右走的概率可能不相等，又如玩剪刀 / 石头 / 布时，每个人的猜拳偏好都不尽相同，假使第一次双方平手，第二次出手可能就会参考第一次的结果来出拳，因此出剪刀 / 石头 / 布的概率又会有所改变，这就是行动转移矩阵，它在每个状态的转移矩阵值可能都不一样，如图 15.6 所示。

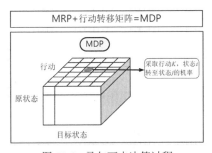

图 15.6  马尔可夫决策过程

马尔可夫决策过程 (MDP) 的假设是代理人的行动决策是希望获得的累计奖励最大化，并不是依据每一种状态下的立即奖励行动，比方说，下棋时我们会为了诱敌进入陷阱，而故意牺牲某些棋子，以求得最后的胜利，因此 MDP 是追求长期累计奖励的最大化，而非每一步骤的最大奖励 ( 短期利益 )，累计奖励称为报酬 (Return)。强化学习类似优化求解，目标函数是报酬，希望找到报酬最大化时应采取的行动策略 (Policy)，对比于神经网络求解，神经网络是基于损失函数最小化的前提下，求出各神经元的权重。

我们可将行动转移矩阵理解为策略 (Policy)，若策略是固定的常数，MDP 就等于 MRP，但是，我们面临的环境是多变的，策略通常不会一成不变，我们会因应状态不同，随之调整行动策略。总而言之，强化学习的目标就是在 MDP 的机制下，要找出最佳的行动策略，而目的是希望获得最大的报酬。

## 15-2 强化学习模型

接下来将 MDP 概念转化为数学模型，将马尔可夫决策过程的示意图转为数学符号如图 15.7(b) 所示。

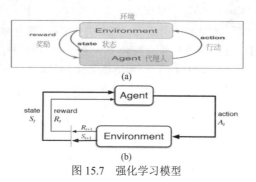

图 15.7　强化学习模型

MDP 行动轨迹 (trajectory) 如下：
$\{S_0, A_0, R_1, S_1, A_1, R_2, S_2, ..., S_t, A_t, R_{t+1}, S_{t+1}, A_{t+1}, R_{t+2}, S_{t+2}...\}$

(1) 在状态 $S_0$ 下，采取行动 $A_0$，获得奖励 $R_1$，并更新状态为 $S_1$，再采取行动 $A_1$，以此类推，形成轨迹 $S、A、R、S、A、R…$的循环。

(2) 状态转移概率：达到状态 $S_{t+1}$ 的概率如下：

$$p(S_{t+1}| S_t, A_t, S_{t-1}, A_{t-1}, S_{t-2}, A_{t-2}, S_{t-3}, A_{t-3}…)$$

依据马尔可夫性质的假设，$S_{t+1}$ 只与前一个状态 ($S_t$) 有关，上式简化为：

$p(S_{t+1}| S_t, A_t)$。

(3) 报酬：就以走迷宫当例子，到达终点时，所累积的奖励总和称为报酬，下式为从 $t$ 时间点走到终点 ($T$) 的累积奖励。

$$G_t = R_{t+1} + R_{t+2} + R_{t+3} + \cdots + R_T = \sum_{k=1}^{T} R_{t+k}$$

**范例1.** 以图15.8走迷宫为例，目标是以最短路径到达终点，故设定每走一步奖励为**-1**，即可算出每一个位置的报酬，计算方法是由终点倒推回起点，假设终点的报酬为**0**，倒推结果每个位置的报酬如图**15.8**中的数字。

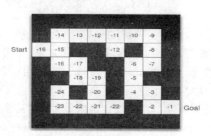

图 15.8　计算迷宫每一个位置的报酬

(4) 折扣报酬 (Discount Return)：模型目标是追求报酬最大化，若迷宫很大的话要

考虑的奖励 ($R_t$) 个数也会很多,为了简化模型,将每个时间的奖励乘以一个小于1的折扣因子 ($\gamma$),让越长远的奖励越不重要,避免考虑太多的状态,类似复利的概念,将未来奖励转换为现值,因此报酬公式修正为:

$$G_t = R_{t+1} + \gamma R_{t+2} + \gamma^2 R_{t+3} + \cdots + \gamma^{T-1} R_T = \sum_{k=1}^{T} \gamma^{k-1} R_{t+k}$$

同理:

$$G_{t+1} = R_{t+2} + \gamma R_{t+3} + \gamma^2 R_{t+4} + \cdots + \gamma^{T-2} R_T$$

两式相减:

$$G_t = R_{t+1} + \gamma G_{t+1}$$

即 $t$ 时间点的报酬可由 $t+1$ 时间点的报酬推算出来。

(5) 状态值函数 (State Value Function):以图15.9的迷宫为例,玩家所在的位置就是状态,图中的数字(报酬)即是状态值,代表每个状态的价值。但从起点走到终点的路径可能不止一种,故状态值函数是每条路径报酬的期望值(平均数)。

**范例2. 再看另一个迷宫游戏**,规则如下,起点为(1, 1),终点为(4, 3)或(4, 2),走到(4, 3)奖励为1,走到(4, 2)奖励为 -1,每走一步奖励均为 -0.04,如图15.9所示。

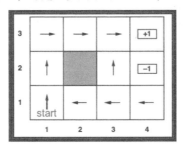

图 15.9  迷宫游戏

假设有三种走法:

① $(1,1) \to (1,2) \to (1,3) \to (1,2) \to (1,3) \to (2,3) \to (3,3) \to (4,3)$
　　-0.04　-0.04　-0.04　-0.04　-0.04　-0.04　-0.04　+1

② $(1,1) \to (1,2) \to (1,3) \to (2,3) \to (3,3) \to (3,2) \to (3,3) \to (4,3)$
　　-0.04　-0.04　-0.04　-0.04　-0.04　-0.04　-0.04　+1

③ $(1,1) \to (2,1) \to (3,1) \to (3,2) \to (4,2)$
　　-0.04　-0.04　-0.04　-0.04　-1

这三条走法的起点报酬计算如下:

① $1 - 0.04 \times 7 = 0.72$

② $1 - 0.04 \times 7 = 0.72$

③ $-1 - 0.04 \times 4 = -1.16$

因此 (1, 1) 状态期望值 = [ 0.72 + 0.72 + (-1.16) ] / 3 = 0.28 / 3 ≒ 0.09
状态值函数以 $V\pi(s)$ 为表示,其中 $\pi$ 为特定策略,故公式如下:

$$V\pi(s) = E(G|S=s)$$

MDP 依照公式计算出的值函数行动,每次行动后就重新计算所有状态值函数,周而复始,在每一个状态下都以最大化状态或行动值函数为准则,采取行动,以获取最大报酬,这就是最简单的策略,称为贪婪策略 (Greedy Policy)。

## 15-3　简单的强化学习架构

这一节我们把前面刚学到的理论统整一下，就能完成一个初阶的程序。回顾前面谈到的强化学习机制如图 15.10 所示。

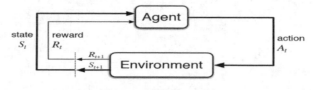

图 15.10　强化学习机制

由强化学习机制，制定面向对象设计 (OOP) 的架构如图 15.11 所示。

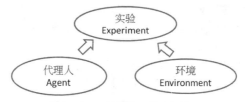

图 15.11　强化学习程序架构

程序架构大致分为以下三个类别或模块。

(1) 环境 (Environment)：比如迷宫、游戏或围棋，它会给予玩家奖励并负责状态转换，若是单人游戏，环境还要担任玩家的对手，像是计算机围棋、井字游戏。环境的职责 ( 方法 ) 如下：

- 初始化 (Init)：需定义状态空间 (State Space)、奖励 (Reward) 办法、行动空间 (Action Space)、状态转换 (State Transition definition)。
- 重置 (Reset)：每一回合 (Episode) 结束后，须重新开始，重置所有变量。
- 步骤 (Step)：代理人行动后，环境要驱动下一步的行动轨迹，更新状态、给予奖励，并判断游戏是否胜负已分，回合结束。
- Render ( 渲染 )：更新每次行动后的画面显示。

(2) 代理人 (Agent)：即玩家，职责如下：

- 行动 (Act)：代理人依据既定的策略与目前所处的状态，采取行动，例如上、下、左、右。
- 通常要确定特殊的策略，会继承基础类别 (Agent class)，在衍生的类别中，撰写行动函数 (Act)，并撰写策略逻辑。

(3) 试验 (Experiment)：建立两个类别对象，进行游戏。

接下来，我们就依照上述架构来开发程序。

**范例3. 建立简单的迷宫游戏**：共有5个位置，玩家一开始站在中间位置，每走一步扣0.02分，走到左端点得-1分，走到右端点得1分，走到左右端点该回合即结束，如图15.12所示。

图 15.12　建立简单的迷宫游戏

**下列程序代码请参考【RL_15_01_simple_game.py】。**

(1) 加载相关套件。

```
2  import random
```

(2) 建立环境类别。

__init__：初始化，每回合结束后比赛重置 (Reset)。
get_observation：回传状态空间 (State Space)，本游戏假设有五种，即 1、2、3、4、5。
get_actions：回传行动空间 (Action Space)，本游戏假设有两种，即 -1、1，只能往左或往右。
is_done：判断比赛回合是否结束。
step：触发下一步，根据传入的行动，更新状态，并计算奖励。

```
4  # 环境类别
5  class Environment:
6      def __init__(self): # 初始化
7          self.poistion = 3 # 玩家一开始站中间位置
8
9      def get_observation(self):
10         # 状态空间(State Space)，共有5个位置
11         return [i for i in range(1, 6)]
12
13     def get_actions(self):
14         return [-1, 1] # 行动空间(Action Space)
15
16     def is_done(self): # 判断比赛回合是否结束
17         # 是否走到左右端点
18         return self.poistion == 1 or self.poistion == 5
19
20     # 步骤
21     def step(self, action):
22         # 是否回合已结束
23         if self.is_done():
24             raise Exception("Game over")
25
26         self.poistion += action
27         if self.poistion == 1:
28             reward = -1
29         elif self.poistion == 5:
30             reward = 1
31         else:
32             reward = -0.02
33
34         return self.poistion, reward
```

(3) 建立代理人类别：主要是确定行动策略，本范例采取随机策略 ( 第 47 行 )。

```
37  # 代理人类别
38  class Agent:
39      # 初始化
40      def __init__(self):
41          pass
42
43      def action(self, env):
44          # 取得状态
45          current_obs = env.get_observation()
46          # 随机行动
47          return random.choice(env.get_actions())
```

(4) 定义好环境与代理人功能后，就可以进行试验了。

```
50  if __name__ == "__main__":
51      # 建立实验，含环境、代理人对象
52      env = Environment()
53      agent = Agent()
54
55      # 进行实验
56      total_reward=0   # 累计报酬
57      action_list = []
58      while not env.is_done():
```

```
59          # 采取行动
60          action = agent.action(env)
61          action_list += [action]
62
63          # 更新下一步
64          state, reward = env.step(action)
65
66          # 计算累计报酬
67          total_reward += reward
68
69          # 显示累计报酬
70      print(f"累计报酬: {total_reward:.4f}")
71      print(f"行动: {action_list}")
```

执行:【python RL_15_01_simple_game.py】。

执行结果：累计报酬为 -1.02，由于采随机策略，因此，每次结果均不相同。

**范例4.** 调用【RL_15_01_simple_game.py】，执行10回合。

**下列程序代码请参考【RL_15_02_simple_game_test.py】。**

(1) 载入【RL_15_01_simple_game.py】。

```
2 from RL_15_01_simple_game import Environment, Agent
```

(2) 试验：建立环境、代理人对象。

```
5 env = Environment()
6 agent = Agent()
```

(3) 进行试验。

```
8  # 进行试验
9  for _ in range(10):
10     env.__init__()    # 重置
11     total_reward=0    # 累计报酬
12     action_list = []
13     while not env.is_done():
14         # 采取行动
15         action = agent.action(env)
16         action_list += [action]
17
18         # 更新下一步
19         state, reward = env.step(action)
20
21         # 计算累计报酬
22         total_reward += reward
23
24     # 显示累计报酬
25     print(f"累计报酬: {total_reward:.4f}")
26     print(f"行动: {action_list}")
```

执行结果：可以看到每次的结果均不相同，这就是学习的过程。

```
累计报酬: -1.0600
行动: [-1, 1, -1, -1]
累计报酬: -1.0200
行动: [-1, -1]
累计报酬: -1.0600
行动: [-1, 1, -1, -1]
累计报酬: 0.8600
行动: [1, -1, -1, 1, -1, 1, 1]
累计报酬: 0.9800
行动: [1, 1]
累计报酬: 0.9800
行动: [1, 1]
累计报酬: 0.9000
行动: [1, -1, 1, -1, 1, 1]
累计报酬: -1.0200
行动: [-1, -1]
累计报酬: -1.1000
行动: [-1, -1, -1, 1, -1, -1]
累计报酬: 0.9800
行动: [1, 1]
```

**范例5.** 改以状态值函数最大者为行动依据，执行10回合，并将程序改成较有弹性，允许更多的节点。

下列程序代码请参考【RL_15_03_simple_game_with_state_value.ipynb】。

(1) 加载相关套件。

```
1  import numpy as np
2  import random
```

(2) 参数设定：使用 15 个节点数，可设定为任意奇数测试。

```
1  NODE_COUNT = 15       # 节点数
2  NORMAL_REWARD = -0.02 # 每走一步扣分 0.02
```

(3) 建立环境类别：增加以下函数。

- update_state_value：若回合结束，由终点倒推，利用 $G_t = R_{t+1} + \gamma G_{t+1}$ 公式 ($\gamma$=1) 更新状态值函数 ( 第 47~53 行 )，每一位置 ($G_{t+1}$) 加上当期奖励 ($R_{t+1}$)，即为上一位置的状态值函数 ($G_t$)。
- get_observation：取得状态值函数的期望值，即平均数 ( 第 56~63 行 )，代理人即可以状态值函数最大者为行动依据。

```
1   # 环境类别
2   class Environment():
3       # 初始化
4       def __init__(self):
5           # 存储状态值函数，索引值[0]:不用，从1开始
6           self.state_value = np.full((NODE_COUNT+1), 0.0)
7   
8           # 更新次数，索引值[0]:不用，从1开始
9           self.state_value_count = np.full((NODE_COUNT+1), 0)
10  
11      # 初始化
12      def reset(self):
13          self.poistion = int((1+NODE_COUNT) / 2)  # 玩家一开始站中间位置
14          self.trajectory=[self.poistion]  # 行动轨迹
15  
16      def get_states(self):
17          # 状态空间(State Space)
18          return [i for i in range(1, NODE_COUNT+1)]
19  
20      def get_actions(self):
21          return [-1, 1] # 行动空间(Action Space)
22  
23      def is_done(self): # 判断比赛回合是否结束
24          # 是否走到左右端点
25          return self.poistion == 1 or self.poistion == NODE_COUNT
26  
27      # 步骤
28      def step(self, action):
29          # 是否回合已结束
30          if self.is_done():
31              raise Exception("Game over")
32  
33          self.poistion += action
34          self.trajectory.append(self.poistion)
35          if self.poistion == 1:
36              reward = -1
37          elif self.poistion == NODE_COUNT:
38              reward = 1
39          else:
40              reward = NORMAL_REWARD
41  
42          return self.poistion, reward
43  
44      def update_state_value(self, final_value):
45          # 倒推，更新状态值函数
46          # 缺点：未考虑节点被走过两次或以上，分数会被重复扣分
47          for i in range(len(self.trajectory)-1, -1, -1):
```

```python
48            final_value += NORMAL_REWARD
49            self.state_value[self.trajectory[i]] += final_value
50            self.state_value_count[self.trajectory[i]] += 1
51
52    # 取得状态值函数期望值
53    def get_observation(self):
54        mean1 = np.full((NODE_COUNT+1), 0.0)
55        for i in range(1, NODE_COUNT+1):
56            if self.state_value_count[i] == 0:
57                mean1[i] = 0
58            else:
59                mean1[i] = self.state_value[i] / self.state_value_count[i]
60        return mean1
```

(4) 代理人类别：比较左右相邻的节点，以状态值函数最大者为行动方向，如果两个的状态值一样大，就随机选择一个。另外，避免随机选择时陷入重复一左一右的无限循环，加了简单的循环检查，如果侦测到循环，则采取相反方向行动，参见第 21~24 行，试验结果，简单的循环检查还是杜绝不了较大的循环。可以将第 23 行批注移除观察侦测到的循环。

```python
1  # 代理人类别
2  class Agent():
3      # 初始化
4      def __init__(self):
5          pass
6
7      def action(self, env):
8          # 取得状态值函数期望值
9          state_value = env.get_observation()
10
11         # 以左/右节点状态值函数大者为行动依据，如果两个状态值一样大，随机选择一个
12         if state_value[env.poistion-1] > state_value[env.poistion+1]:
13             next_action = -1
14         if state_value[env.poistion-1] < state_value[env.poistion+1]:
15             next_action = 1
16         else:
17             next_action = random.choice(env.get_actions())
18
19         # 如果侦测到循环，采反向行动
20  #        if len(env.trajectory)>=3 and \
21  #            env.poistion + next_action == env.trajectory[-2] and \
22  #            env.trajectory[-1] == env.trajectory[-3]:
23  #            # print('loop:', env.trajectory[-3:], env.poistion + next_action)
24  #            next_action = -next_action
25         return next_action
```

(5) 建立试验：程序逻辑与之前差不多，增加一些检查及信息显示。

若行动次数超过 100 次，应该已经陷入循环，提前终止该回合 ( 第 23~25 行 )。

提前终止的回合，不更新值函数，以免过度降低循环节点的值函数 ( 第 29 行 )。

可以将第 31 行批注移除观察每回合更新的值函数，借以了解下一回合的行动依据。

```python
1  # 建立实验，含环境、代理人物件
2  env = Environment()
3  agent = Agent()
4
5  # 进行实验
6  total_reward_list = []
7  for i in range(10):
8      env.reset()    # 重置
9      total_reward=0  # 累计报酬
10     action_count = 0
11     while not env.is_done():
12         # 采取行动
13         action = agent.action(env)
14         action_count+=1
15
16         # 更新下一步
17         state, reward = env.step(action)
18         #print(state, reward)
19         # 计算累计报酬
20         total_reward += reward
```

```
21          # 避免一直循环，跑不完
22        if action_count > 100:
23            env.poistion = int((1+NODE_COUNT) / 2)
24            break
25
26    print(f'trajectory {i}: {env.trajectory}')
27    # 未达终点不更新值函数，以免过度降低循环节点的值函数
28    if action_count <= 100:
29        env.update_state_value(total_reward)
30    # print(f"state value: {list(np.around(env.get_observation()[1:] ,2))}")
31    total_reward_list.append(round(total_reward, 2))
32
33 # 显示累计报酬
34 print(f"累计报酬: {total_reward_list}")
```

执行结果：可以看出每次的结果均不相同，基本上只要有一次往右走，后续几乎都会往右走，因为，右边节点值函数都高于左边节点，表示最佳策略就是一直往右走。

```
trajectory 0: [8, 9, 8, 7, 6, 7, 6, 5, 6, 7, 6, 7, 6, 7, 8, 9, 8, 7, 8, 9, 8, 9, 8, 9, 10, 9, 8, 9, 10, 11, 10, 11, 12, 1
1, 10, 11, 10, 11, 10, 11, 12, 13, 12, 13, 14, 15]
trajectory 1: [8, 9, 10, 11, 12, 13, 14, 15]
trajectory 2: [8, 9, 10, 11, 12, 13, 14, 15]
trajectory 3: [8, 9, 10, 11, 12, 13, 14, 15]
trajectory 4: [8, 9, 10, 11, 12, 13, 14, 15]
trajectory 5: [8, 9, 10, 11, 12, 13, 14, 15]
trajectory 6: [8, 9, 10, 11, 12, 13, 14, 15]
trajectory 7: [8, 9, 10, 11, 12, 13, 14, 15]
trajectory 8: [8, 9, 10, 11, 12, 13, 14, 15]
trajectory 9: [8, 9, 10, 11, 12, 13, 14, 15]
累计报酬: [-47, -5, -5, -5, -5, -5, -5, -5, -5, -5]
```

(6) 绘图。

```
1 import matplotlib.pyplot as plt
2
3 plt.figure(figsize=(10,6))
4 plt.plot(total_reward_list)
```

执行结果：各回合的报酬都很稳定，多训练几次，偶尔会出现一点波动。

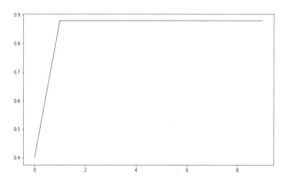

把节点数 (NODE_COUNT) 参数改为其他数字，执行结果大致相同。

作一简单结论如下：

(1) 以状态值函数最大者为行动依据，进行训练，果然可以找到最佳解，模型就是每个状态的值函数 (self.state_value)，可以将它存盘，之后实际上线时即可加载模型执行。

(2) 这个程序有以下缺点：

无法准确侦测所有循环。

更新状态值函数时，若某些节点被走过两次或以上，分数会被重复扣分，影响之后的行动决策。

以上缺点会造成模型不稳定，学习无法保证找到最佳行动策略，永远往右走，有时候会有无限循环或形成很长的轨迹。

**范例6.** 测试较复杂的迷宫游戏，规则如下，起点为(1, 1)，终点为(4, 3)或(4, 2)，走到(4, 3)奖励为1，走到(4, 2)奖励为 -1，每走一步奖励均为 -0.04，(2, 2)为柱子，不可驻留，如图15.13所示。

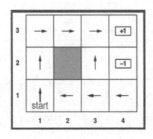

图 15.13　复杂的迷宫游戏

**下列程序代码请参考【 RL_15_04_maze.ipynb 】。**

(1) 加载相关套件。

```
1  import numpy as np
2  import random
```

(2) 参数设定：尽可能地将游戏参数化，方便之后调整设定测试。

```
1  ROW_COUNT, COLUMN_COUNT = 3, 4         # 3列 x 4行
2  NODE_COUNT = ROW_COUNT * COLUMN_COUNT  # 节点总数
3  NORMAL_REWARD = -0.04 # 每走一步扣分 0.04
4  WIN_REWARD = 1      # 终点(4, 3)的得分
5  LOSS_REWARD = -1    # 终点(4, 2)的失分
6
7  # 特殊节点
8  WIN_TERMINAL = NODE_COUNT-1           # 得分的终点
9  LOSS_TERMINAL = NODE_COUNT-1-COLUMN_COUNT # 失分的终点
10 WALL_NODES = [5]                      # 墙节点，不能驻留的节点
11 # 行动空间
12 (UP, DOWN, LEFT, RIGHT) = range(4) # 上/下/左/右
```

(3) 建立环境类别：大致上程序逻辑与上一范例相同，差异如下。

更新位置：第 31~46 行，迷宫有边界，不可超出界线，另外迷宫内有柱子，不可驻留或穿越。

考虑节点被走过两次或以上，分数会被重复扣分，采用首次访问 (first visit) 的报酬更新状态值，亦即某一节点在轨迹中出现多次，以第一次走过的状态值函数为准，请参阅 update_state_value 函数。也有人以平均值函数为准，在本例相差不大，读者可试试看，修改第 78 行即可。

```
1  # 环境类别
2  class Environment():
3      # 初始化
4      def __init__(self):
5          # 存储状态值函数
6          self.state_value = np.full(NODE_COUNT, 0.0)
7
8          # 更新次数
9          self.state_value_count = np.full(NODE_COUNT, 0)
10
11         # 初始化
12     def reset(self):
13         self.poistion = 0   # 玩家开始的位置
14         self.trajectory=[self.poistion] # 行动轨迹
15
16     def get_states(self):
17         # 状态空间(State Space)
```

```python
18         return [i for i in range(NODE_COUNT)]
19
20     def get_actions(self):
21         # 行动空间(Action Space)
22         return [UP, DOWN, LEFT, RIGHT] # 上/下/左/右
23
24     def is_done(self): # 判断比赛回合是否结束
25         # 是否走到终点
26         return self.poistion == WIN_TERMINAL or self.poistion == LOSS_TERMINAL
27
28     # 更新位置
29     def update_poistion(self, action):
30         if action == DOWN:
31             new_poistion = self.poistion - COLUMN_COUNT
32         if action == UP:
33             new_poistion = self.poistion + COLUMN_COUNT
34         if action == LEFT:
35             new_poistion = self.poistion - 1
36         if action == RIGHT:
37             new_poistion = self.poistion + 1
38
39         if new_poistion < 0 or new_poistion > NODE_COUNT \
40             or new_poistion in WALL_NODES:
41             return self.poistion
```

```python
43         return new_poistion
44
45     # 步骤
46     def step(self, action):
47         # 是否回合已结束
48         if self.is_done():
49             raise Exception("Game over")
50
51         self.poistion = self.update_poistion(action)
52         self.trajectory.append(self.poistion)
53         if self.poistion == WIN_TERMINAL:
54             reward = WIN_REWARD
55         elif self.poistion == LOSS_TERMINAL:
56             reward = LOSS_REWARD
57         else:
58             reward = NORMAL_REWARD
59
60         return self.poistion, reward
61
62     def update_state_value(self, final_value):
63         # 考虑节点被走过两次或以上，分数会被重复扣分
64         # 采首次访问(first visit)的报酬更新状态值
65         distinct_node_list = list(set(self.trajectory))
66         # print('distinct_node_list:', distinct_node_list)
67         distinct_state_value = np.full(len(distinct_node_list), 0.0)
68         # 倒推，更新状态值函数
69         reverse_trajectory = self.trajectory.copy()
70         reverse_trajectory.reverse()
71         for i in reverse_trajectory:
72             final_value += NORMAL_REWARD
73             # 如有访问多次的节点，状态值会被盖掉
74             index = distinct_node_list.index(i) # 取得索引值
75             distinct_state_value[index] = final_value # 暂存状态值函数
76
77         # 更新轨迹的状态值函数
78         # print('distinct_state_value:', distinct_state_value)
79         for index, val in enumerate(distinct_node_list):
80             # 更新状态值函数
81             self.state_value[val] += distinct_state_value[index]
82             self.state_value_count[val] += 1
```

```python
84     # 取得状态值函数期望值
85     def get_observation(self):
86         mean1 = np.full(NODE_COUNT, 0.0)
87         for i in range(NODE_COUNT):
88             if self.state_value_count[i] == 0:
89                 mean1[i] = 0
90             else:
91                 mean1[i] = self.state_value[i] / self.state_value_count[i]
92         return mean1
```

(4) 建立代理人类别：增加以下程序逻辑。

check_possible_action 函数：第 7~27 行，找出所在位置可采取的行动。

在可允许的行动中，依最大的值函数行动：第 33~43 行。

最大状态值函数的节点若有多个，则随机抽样。

若发生循环，采取随机抽样，选择其他的行动。

若要进一步了解程序代码的逻辑，可将相关的 print 取消批注，有益于观察选择行动的过程。

```python
1   # 代理人类别
2   class Agent():
3       # 初始化
4       def __init__(self):
5           pass
6   
7       # 取得可以行走的方向
8       def check_possible_action(self, env):
9           possible_actions = env.get_actions()
10          if env.poistion < COLUMN_COUNT:  # 最下一列不可向下
11              possible_actions.remove(DOWN)
12          if env.poistion % COLUMN_COUNT == 0:  # 第一行不可向左
13              possible_actions.remove(LEFT)
14          if env.poistion >= NODE_COUNT - COLUMN_COUNT :  # 最上一列不可向上
15              possible_actions.remove(UP)
16          if env.poistion % COLUMN_COUNT == COLUMN_COUNT -1 :  # 最右一行不可向右
17              possible_actions.remove(RIGHT)
18  
19          if env.poistion -1 in WALL_NODES :  # 向左若遇墙，不可向左
20              possible_actions.remove(LEFT)
21          if env.poistion +1 in WALL_NODES :  # 向右若遇墙，不可向右
22              possible_actions.remove(RIGHT)
23          if env.poistion + COLUMN_COUNT in WALL_NODES :  # 向上若遇墙，不可向上
24              possible_actions.remove(UP)
25          if env.poistion - COLUMN_COUNT in WALL_NODES :  # 向下若遇墙，不可向下
26              possible_actions.remove(DOWN)
27  
28          return possible_actions
29  
30      def action(self, env):
31          # 取得状态值函数期望值
32          state_value = env.get_observation()
33  
34          # 找到最大的状态值函数
35          max_value = -999
36          next_action_list = []
37          possible_actions = self.check_possible_action(env)
38          # print('possible_actions:', possible_actions)
39          for i in possible_actions:
40              if state_value[env.update_poistion(i)] > max_value:
41                  max_value = state_value[env.update_poistion(i)]
42                  next_action_list = [i]
43              elif state_value[env.update_poistion(i)] == max_value:
44                  next_action_list += [i]
45          # print('next_action:', next_action_list, ', max_value:', max_value)
46  
47          if len(next_action_list) == 0:
48              next_action = random.choice(possible_actions)
49          else:  # 有多个最大状态值函数的节点，随机抽样
50              next_action = random.choice(next_action_list)
51          new_poistion = env.update_poistion(next_action)
52  
53          # 若发生循环，随机抽样
54          while len(possible_actions) > 1 and len(env.trajectory)>=4 and \
55              new_poistion == env.trajectory[-2] and \
56              new_poistion == env.trajectory[-4] :
57              # print('loop:', env.trajectory[-4:], new_poistion)
58              possible_actions.remove(next_action)       # 去除造成循环的行动
59              next_action = random.choice(possible_actions)  # 选择其他的行动
60              new_poistion = env.update_poistion(next_action)
61              # print('change action:', new_poistion)
62  
63          # print('next_action:', next_action_list, ', max_value:', max_value)
64          return next_action
```

(5) 进行试验：与上一范例程序逻辑相同。

```python
1   # 建立试验，含环境、代理人对象
2   env = Environment()
3   agent = Agent()
4   
5   # 进行试验
6   total_reward_list = []
7   no = 0
8   done_no = 0
9   while no < 100 and done_no < 41:
10      no += 1
11      env.reset()  # 重置
12      total_reward=0  # 累计报酬
```

```
13      action_count = 0
14      while not env.is_done():
15          # 采取行动
16          action = agent.action(env)
17          action_count+=1
18
19          # 更新下一步
20          state, reward = env.step(action)
21          #print(state, reward)
22          # 计算累计报酬
23          total_reward += reward
24
25          # 避免一直循环，跑不完
26          if action_count > 100:
27              env.poistion = 0
28              break
29
30      print('trajectory', done_no, ':', env.trajectory)
31      # 未达终点不更新值函数，以免过度降低循环节点的值函数
32      if action_count <= 100:
33          env.update_state_value(total_reward)
34          total_reward_list.append(round(total_reward, 2))
35          done_no += 1
36
37      # state_value = np.around(env.get_observation().reshape(ROW_COUNT, COLUMN_COUNT),2)
38      # print(f"state value:\n{np.flip(state_value, axis=0)}") # 列反转，与图一致，便于观察
39      total_reward_list.append(round(total_reward, 2))
40
41  # 显示累计报酬
42  print(f"累计报酬: {total_reward_list}")
```

执行结果：采用首次访问 (first visit) 的报酬更新状态值，模型较为稳定，一旦走过胜利的 (4, 3)，就不会走到失败的 (4, 2)。

```
trajectory 0 : [0, 4, 0, 1, 2, 1, 0, 4, 0, 1, 2, 1, 2, 3, 7]
trajectory 1 : [0, 1, 2, 6, 10, 6, 10, 11]
trajectory 2 : [0, 1, 2, 6, 10, 11]
trajectory 3 : [0, 1, 2, 6, 10, 11]
trajectory 4 : [0, 1, 2, 6, 10, 11]
trajectory 5 : [0, 1, 2, 6, 10, 11]
trajectory 6 : [0, 1, 2, 6, 10, 11]
trajectory 7 : [0, 1, 2, 6, 10, 11]
trajectory 8 : [0, 1, 2, 6, 10, 11]
trajectory 9 : [0, 1, 2, 6, 10, 11]
trajectory 10 : [0, 1, 2, 6, 10, 11]
trajectory 11 : [0, 1, 2, 6, 10, 11]
trajectory 12 : [0, 1, 2, 6, 10, 11]
trajectory 13 : [0, 1, 2, 6, 10, 11]
trajectory 14 : [0, 1, 2, 6, 10, 11]
trajectory 15 : [0, 1, 2, 6, 10, 11]
trajectory 16 : [0, 1, 2, 6, 10, 11]
```

但还是存在以下一些缺点：

若未走过 (4, 3)，有可能一直如下走向失败的 (4, 2)，依照最大值函数 (state value) 导引，只有一条路通往失败的 (4, 2)。要显示最大值函数须第 37~38 行取消批注。

```
trajectory 0 : [0, 4, 8, 4, 0, 1, 0, 4, 8, 9, 8, 9, 10, 6, 10, 6, 2, 3, 7]
state value:
[[-2.36 -2.08 -1.96  0.  ]
 [-2.4   0.   -1.92 -1.72]
 [-2.44 -2.24 -1.8  -1.76]]
trajectory 1 : [0, 1, 2, 3, 7]
state value:
[[-2.36 -2.08 -1.96  0.  ]
 [-2.4   0.   -1.92 -1.44]
 [-1.88 -1.76 -1.52 -1.48]]
trajectory 2 : [0, 1, 2, 3, 7]
state value:
[[-2.36 -2.08 -1.96  0.  ]
 [-2.4   0.   -1.92 -1.35]
 [-1.69 -1.6  -1.43 -1.39]]
trajectory 3 : [0, 1, 2, 3, 7]
```

仍然会出现循环。

```
trajectory 22 : [0, 4, 8, 4, 8, 9, 10, 9, 8, 4, 8, 4, 0, 1, 0, 4, 8, 4, 8, 9, 10, 9, 8, 4, 8, 4, 0, 1, 0, 4, 8, 4, 8, 9, 10,
9, 8, 4, 8, 4, 0, 1, 0, 4, 8, 4, 8, 9, 10, 9, 8, 4, 8, 4, 0, 1, 0, 4, 8, 4, 8, 9, 10, 9, 8, 4, 8, 4, 0, 1, 0, 4, 8, 4, 8, 9,
10, 9, 8, 4, 8, 4, 0, 1, 0, 4, 8, 4, 8, 9, 10, 9, 8, 4, 8, 4, 0, 1, 0, 4, 8, 4]
trajectory 22 : [0, 4, 8, 4, 8, 9, 10, 9, 8, 4, 8, 4, 0, 1, 0, 4, 8, 4, 8, 9, 10, 9, 8, 4, 8, 4, 0, 1, 0, 4, 8, 4, 8, 9, 10,
9, 8, 4, 8, 4, 0, 1, 0, 4, 8, 4, 8, 9, 10, 9, 8, 4, 8, 4, 0, 1, 0, 4, 8, 4, 8, 9, 10, 9, 8, 4, 8, 4, 0, 1, 0, 4, 8, 4, 8, 9,
10, 9, 8, 4, 8, 4, 0, 1, 0, 4, 8, 4, 8, 9, 10, 9, 8, 4, 8, 4, 0, 1, 0, 4, 8, 4]
trajectory 22 : [0, 4, 8, 4, 8, 9, 10, 9, 8, 4, 8, 4, 0, 1, 0, 4, 8, 4, 8, 9, 10, 9, 8, 4, 8, 4, 0, 1, 0, 4, 8, 4, 8, 9, 10,
9, 8, 4, 8, 4, 0, 1, 0, 4, 8, 4, 8, 9, 10, 9, 8, 4, 8, 4, 0, 1, 0, 4, 8, 4, 8, 9, 10, 9, 8, 4, 8, 4, 0, 1, 0, 4, 8, 4, 8, 9,
10, 9, 8, 4, 8, 4, 0, 1, 0, 4, 8, 4, 8, 9, 10, 9, 8, 4, 8, 4, 0, 1, 0, 4, 8, 4]
trajectory 22 : [0, 4, 8, 4, 8, 9, 10, 9, 8, 4, 8, 4, 0, 1, 0, 4, 8, 4, 8, 9, 10, 9, 8, 4, 8, 4, 0, 1, 0, 4, 8, 4, 8, 9, 10,
9, 8, 4, 8, 4, 0, 1, 0, 4, 8, 4, 8, 9, 10, 9, 8, 4, 8, 4, 0, 1, 0, 4, 8, 4, 8, 9, 10, 9, 8, 4, 8, 4, 0, 1, 0, 4, 8, 4, 8, 9,
10, 9, 8, 4, 8, 4, 0, 1, 0, 4, 8, 4, 8, 9, 10, 9, 8, 4, 8, 4, 0, 1, 0, 4, 8, 4]
```

因为以上的范例完全依据最大值函数行动，这种策略称为贪婪策略 (Greedy Policy)，它有一个致命的缺点，若只有一条路径可选择，它就不会探索其他路径，因为其他未走过的节点值函数均未更新，永远不会受到贪婪策略的青睐，但是它们可能是更好的选择。该问题的解决办法是行动选择时，偶尔要保留一些机会用来探索未知路径，这种策略称为 ε-Greedy Policy，后续谈到算法时会详细说明。

强化学习要能依照自己所想的方式运作，必须将奖励、状态、环境规划妥当，才会出现预期的结果，上例如果将奖励调整为 -0.5，即使走过胜利的 (4, 3)，也可能会造成该轨迹的各个节点值函数均为负值，后续行动都不会再选择走向胜利的路径，因此这些超参数的规划是强化学习的关键，与算法一样重要。

## 15-4　Gym 套件

根据前面的练习，可以了解到强化学习是在各种环境中寻找最佳策略，因此，为了节省开发者的时间，网络上有许多套件设计了各种各样的环境，供大家试验，也能借由动画来展示训练过程，例如 Gym、Amazon SageMaker 等套件。

以下就来介绍 Gym 套件的用法，它是 OpenAI 开发的学习套件，提供数十种不同的游戏，Gym 官网[2] 首页上有展示一些游戏画面。不过请读者留意，有些游戏在 Windows 作业环境中并不能顺利安装，因为 Gym 是以 gcc 撰写的。

安装指令如下：

pip install gym

如需安装全部游戏，可执行以下指令，注意，这只能在 Linux 环境中执行，而且必须先安装相关软件工具，请参考 Gym GitHub 说明[3]，安装指令如下：

pip install "gym[all]"

网络上有一篇文章 *Install OpenAI Gym with Box2D and Mujoco in Windows 10*[4] 介绍如何在 Windows 作业环境中克服问题。

在 Windows 作业环境中可加装 Atari 游戏，Atari 为 1967 年开发的游戏机 ( 如图 15.14 所示 )，拥有几十种的游戏，例如《打砖块》(Breakout)《桌球》(Pong) 等，安装指令如下：

pip install "gym[atari, accept-rom-license]"

图 15.14　Atari 游戏机

Gym 提供的环境分为以下四类：

(1) 经典游戏 (Classic control) 和文字游戏 (Toy text)：属于小型的环境，适合初学者开发测试。

(2) 算法类 (Algorithmic)：像是多位数的加法、反转顺序等，这对计算机来说非常简单，但使用强化学习方式求解的话，是一大挑战。

(3) Atari：Atari 游戏机内的一些游戏。

(4) 2D 或 3D 机器人 (Robot)：机器人模拟环境，其中 Mujoco 是要付费的，可免费试用 30 天。

(5) 2021 年 DeepMind 自 OpenAI 买下 Mujoco 版权 [5]，并已改为免费，不过到 2022 年，Gym 并未更新相关安装程序。Mujoco 的操作可参阅官网文件 [6]。

接下来先认识一下 Gym 的架构，各位可能会觉得有点眼熟，这是由于前面的范例刻意模仿 Gym 的设计架构，因此两者的环境类别有些类似。

Gym 的环境类别包括以下方法：

(1) reset()：比赛一开始或回合结束时，调用此方法重置环境。

(2) step(action)：传入行动，触动下一步，回传下列信息。

observation：环境更新后的状态。

reward：行动后得到的奖励。

done：布尔值，True 表示比赛回合结束，False 表示比赛回合进行中。

info：为字典 (dict) 的数据类型，通常是除错信息。

(3) render()：渲染，即显示更新后的画面。

(4) close()：关闭环境。

**范例7. Gym入门实操。**

**下列程序代码请参考【RL_15_05_Gym.ipynb】。**

(1) 加载相关套件。

```
1  import gym
2  from gym import envs
```

(2) 显示已注册的游戏环境。

```
1  all_envs = envs.registry.all()
2  env_ids = [env_spec.id for env_spec in all_envs]
3  print(env_ids)
```

执行结果：每个字符串代表一种游戏。

```
['Copy-v0', 'RepeatCopy-v0', 'ReversedAddition-v0', 'ReversedAddition3-v0', 'DuplicatedInput-v0', 'Reverse-v0', 'CartPole-v0', 'CartPole-v1', 'MountainCar-v0', 'MountainCarContinuous-v0', 'Pendulum-v0', 'Acrobot-v1', 'LunarLander-v2', 'LunarLanderContinuous-v2', 'BipedalWalker-v3', 'BipedalWalkerHardcore-v3', 'CarRacing-v0', 'Blackjack-v0', 'KellyCoinflip-v0', 'KellyCoinflipGeneralized-v0', 'FrozenLake-v0', 'FrozenLake8x8-v0', 'CliffWalking-v0', 'NChain-v0', 'Roulette-v0', 'Taxi-v3', 'GuessingGame-v0', 'HotterColder-v0', 'Reacher-v2', 'Pusher-v2', 'Thrower-v2', 'Striker-v2', 'InvertedPendulum-v2', 'InvertedDoublePendulum-v2', 'HalfCheetah-v2', 'HalfCheetah-v3', 'Hopper-v2', 'Hopper-v3', 'Swimmer-v2', 'Swimmer-v3', 'Walker2d-v2', 'Walker2d-v3', 'Ant-v2', 'Ant-v3', 'Humanoid-v2', 'Humanoid-v3', 'HumanoidStandup-v2', 'FetchSlide-v1', 'FetchPickAndPlace-v1', 'FetchReach-v1', 'FetchPush-v1', 'HandReach-v0', 'HandManipulateBlockRotateZ-v0', 'HandManipulateBlockRotateZTouchSensors-v0', 'HandManipulateBlockRotateZTouchSensors-v1', 'HandManipulateBlockRotateParallel-v0', 'HandManipulateBlockRotateParallelTouchSensors-v0', 'HandManipulateBlockRotateParallelTouchSensors-v1', 'HandManipulateBlockRotateXYZ-v0', 'HandManipulateBlockRotateXYZTouchSensors-v0', 'HandManipulateBlockRotateXYZTouchSensors-v1', 'HandManipulateBlockFull-v0', 'HandManipulateBlock-v0', 'HandManipulateBlockTouchSensors-v0', 'HandManipulateBlockTouchSensors-v1', 'HandManipulateEggRotate-v0', 'HandManipulateEggRotateTouchSensors-v0', 'HandManipulateEggRotateTouchSensors-v1', 'HandManipulateEggFull-v0', 'HandManipulateEgg-v0', 'HandManipula
```

(3) 计算游戏个数。

```
1  len(env_ids)
```

执行结果：有 849 种游戏，个数依每台机器安装的情形而有所不同。

(4) 任意加载一个环境，例如《木棒台车》(CartPole) 游戏，并显示行动空间、状态空间 / 最大值 / 最小值。

```
1  # 载入《木棒台车》(CartPole) 游戏
2  env = gym.make("CartPole-v1")
3
4  # 环境的资讯
5  print(env.action_space)
6  print(env.observation_space)
7  print('observation_space 范围：')
8  print(env.observation_space.high)
9  print(env.observation_space.low)
```

执行结果：行动空间有两个离散值 (0: 往左，1: 往右 )、状态空间为 Box 数据类型，为连续型变量，维度大小为 4，分别代表台车位置 (Cart Position)、台车速度 (Cart Velocity)、木棒角度 (Pole Angle) 及木棒速度 (Pole Velocity At Tip)，另外显示 4 项信息的最大值和最小值。

```
Discrete(2)
Box([-4.8000002e+00 -3.4028235e+38 -4.1887903e-01 -3.4028235e+38], [4.8000002e+00 3.4028235e+38 4.1887903e-01 3.4028235e+38],
(4,), float32)
observation_space 范围：
[4.8000002e+00 3.4028235e+38 4.1887903e-01 3.4028235e+38]
[-4.8000002e+00 -3.4028235e+38 -4.1887903e-01 -3.4028235e+38]
```

(5) 加载《打砖块》(Breakout) 游戏，显示环境的信息。

```
1  env = gym.make("Breakout-v0")
2
3  # 环境的资讯
4  print(env.action_space)
5  print(env.observation_space)
6  print('observation_space 范围：')
7  print(env.observation_space.high)
8  print(env.observation_space.low)
```

执行结果：相较于《木棒台车》，《打砖块》就复杂许多，官网并未提供相关信息，可从 GitHub 下载源代码，再观看程序说明。或是查看安装目录，Atari 程序安装在 anaconda3\Lib\site-packages\atari_py 目录下，Breakout 原始程序为 ale_interface\src\games\supported\Breakout.cpp。

```
Discrete(4)
Box(0, 255, (210, 160, 3), uint8)
observation_space 范围：
[[[255 255 255]
  [255 255 255]
  [255 255 255]
  ...
  [255 255 255]
  [255 255 255]
  [255 255 255]]

 [[255 255 255]
  [255 255 255]
  [255 255 255]
  ...
  [255 255 255]
  [255 255 255]
  [255 255 255]]
```

(6) 试验《木棒台车》(CartPole) 游戏。

```
1   # 载入《木棒台车》(CartPole)游戏
2   env = gym.make("CartPole-v1")
3
4   # 比赛回合结束，重置
5   observation = env.reset()
6   # 将环境资讯写入日志文件
7   with open("CartPole_random.log", "w", encoding='utf8') as f:
8       # 执行 1000 次行动
9       for _ in range(1000):
10          # 更新画面
11          env.render()
12          # 随机行动
13          action = env.action_space.sample()
14          # 驱动下一步
15          observation, reward, done, info = env.step(action)
16          # 写入资讯
17          f.write(f"action={action}, observation={observation}," +
18              f"reward={reward}, done={done}, info={info}\n")
19          # 比赛回合结束，重置
20          if done:
21              observation = env.reset()
22  env.close()
```

执行结果：以随机的方式行动，可以看到台车时而前进，时而后退。

CartPole_random.log 日志文件的部分内容如下。

```
action=0, observation=[ 0.02797028 -0.203182    0.00185102  0.28630734],reward=1.0, done=F
action=1, observation=[ 0.02390664 -0.00808649  0.00757717 -0.00579121],reward=1.0, done=F
action=1, observation=[ 0.02374491  0.18692597  0.00746134 -0.29607385],reward=1.0, done=F
action=0, observation=[ 0.02748343 -0.00830155  0.00153987 -0.00104711],reward=1.0, done=F
action=0, observation=[ 0.0273174  -0.20344555  0.00150128  0.29212127],reward=1.0, done=F
action=1, observation=[ 2.32484899e-02 -8.34528672e-03  7.36135014e-03 -8.22245954e-05],re
action=1, observation=[ 0.02308158  0.18667032  0.00735971 -0.2904335 ],reward=1.0, done=F
action=0, observation=[ 0.02681169 -0.00855579  0.00155104  0.00456148],reward=1.0, done=F
action=0, observation=[ 0.02664387 -0.20369996  0.00164227  0.29773338],reward=1.0, done=F
action=1, observation=[ 0.02256988 -0.00860145  0.00759693  0.00556884],reward=1.0, done=F
action=0, observation=[ 0.02239785 -0.20383153  0.00770831  0.30063898],reward=1.0, done=F
```

《木棒台车》(CartPole) 的游戏规则说明如下：

(1) 可控制台车往左 (0) 或往右 (1)。

(2) 每走一步得一分。

(3) 台车一开始定位在中心点，平衡杆是直立 (upright) 的，在行驶中要保持平衡。

(4) 符合以下任一条件，即视为失败：

平衡杆偏差超过 12 度。

离中心点 2.4 单位。

(5) 胜利：依版本有不同的条件。

v0：行动超过 200 步。

v1：行动超过 500 步。

(6) 如果连续 100 回合的平均报酬超过 200 步，即视为解题成功。

Step 函数回传的内容：

(1)observation：环境更新后状态 ( 如图 15.15 所示 )。

台车位置 (Cart Position)。

台车速度 (Cart Velocity)。

- 木棒角度 (Pole Angle)。
- 木棒速度 (Pole Velocity At Tip)。

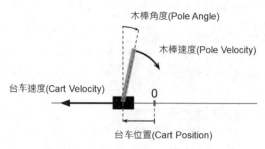

图 15.15　木棒台车的状态

(2) reward：行动后得到的奖励。

(3) done：布尔值，True 表示比赛回合结束，False 表示比赛回合进行中。

(4) info：为字典 (dict) 的数据类型，通常是除错信息，本游戏均不回传信息。

**范例8.《木棒台车》(CartPole)试验。**

**下列程序代码请参考【RL_15_06_CartPole.ipynb】。**

(1) 加载相关套件。

```
1  import gym
2  from gym import envs
```

(2) 设定比赛回合数。

```
2  no = 50          # 比赛回合数
```

(3) 试验：采取随机行动。

```
4   # 载入《木棒台车》(CartPole)游戏
5   env = gym.make("CartPole-v1")
6
7   # 重置
8   observation = env.reset()
9   all_rewards=[]  # 每回合总报酬
10  all_steps=[]    # 每回合总步数
11  total_rewards = 0
12  total_steps=0
13
14  while no > 0:   # 执行 50 比赛回合数
15      # 随机行动
16      action = env.action_space.sample()
17      total_steps+=1
18
19      # 触动下一步
20      observation, reward, done, info = env.step(action)
21      # 累计报酬
22      total_rewards += reward
23
24      # 比赛回合结束，重置
25      if done:
26          observation = env.reset()
27          all_rewards.append(total_rewards)
28          all_steps.append(total_steps)
29          total_rewards = 0
30          total_steps=0
31          no-=1
32
33  env.close()
```

执行结果：如下表所示，结果相当惨烈，没有一回合走超过 200 步，全军覆灭，显示对于强化学习而言这个游戏很有挑战性。

```
回合   报酬    结果
0     26.0   Loss
1     71.0   Loss
2     13.0   Loss
3     11.0   Loss
4     13.0   Loss
5     19.0   Loss
6     12.0   Loss
7     11.0   Loss
8     31.0   Loss
9     19.0   Loss
10    44.0   Loss
11    13.0   Loss
12    17.0   Loss
13    29.0   Loss
14    29.0   Loss
15    40.0   Loss
16    27.0   Loss
17    25.0   Loss
18    12.0   Loss
19    13.0   Loss
20    20.0   Loss
```

(4) 基于上述试验，可针对问题提出对策如下：

台车距离中心点大于 2.4 单位就算输了，所以设定每次行动采取一左一右的方式，尽量不偏离中心点。

由于台车的平衡杆角度偏差 12 度以上也算输，故而设定平衡杆角度若偏右 8 度以上，就往右前进，直到角度偏右小于 8 度为止。

反之，偏左也是类似处理。

(5) 首先建立 Agent 类别，撰写 act 函数实现以上逻辑。

```python
1  import math
2
3  # 参数设定
4  left, right = 0, 1     # 台车行进方向
5  max_angle = 8          # 偏右8度以上，就往右前进，偏左也是同样处理

1  class Agent:
2      # 初始化
3      def __init__(self):
4          self.direction = left
5          self.last_direction=right
6
7      # 自定策略
8      def act(self, observation):
9          # 台车位置、台车速度、平衡杆角度、平衡杆速度
10         cart_position, cart_velocity, pole_angle, pole_velocity = observation
11
12         '''
13         行动策略：
14         1. 设定每次行动采一左一右，尽量不离中心点。
15         2. 平衡杆角度偏8度以上，就往右前进，直到角度偏右小于8度。
16         3. 反之，偏左也是同样处理。
17         '''
18         if pole_angle < math.radians(max_angle) and \
19             pole_angle > math.radians(-max_angle):
20             self.direction = (self.last_direction + 1) % 2
21         elif pole_angle >= math.radians(max_angle):
22             self.direction = right
23         else:
24             self.direction = left
25
26         self.last_direction = self.direction
27
28         return self.direction
```

(6) 以 agent.act(observation) 取代 env.action_space.sample()。

```
1   no = 50          # 比赛回合数
2
3   # 载入《木棒台车》(CartPole) 游戏
4   env = gym.make("CartPole-v1")
5
6   # 重置
7   observation = env.reset()
8   all_rewards=[] # 每回合总报酬
9   all_steps=[]   # 每回合总步数
10  total_rewards = 0
11  total_steps=0
12
13  agent = Agent()
14  while no > 0:    # 执行 50 比赛回合数
15      # 行动
16      action = agent.act(observation) #env.action_space.sample()
17      total_steps+=1
18
19      # 触动下一步
20      observation, reward, done, info = env.step(action)
21      # 累计报酬
22      total_rewards += reward
23
24      # 比赛回合结束，重置
25      if done:
26          observation = env.reset()
27          all_rewards.append(total_rewards)
28          total_rewards = 0
29          all_steps.append(total_steps)
30          total_steps = 0
31          no-=1
```

(7) 显示执行结果：虽然比起随机行动的方式，改良后的报酬增加很多，但结果还是都失败了。

| 回合 | 报酬 | 结果 |
|---|---|---|
| 0 | 70.0 | Loss |
| 1 | 66.0 | Loss |
| 2 | 78.0 | Loss |
| 3 | 136.0 | Loss |
| 4 | 130.0 | Loss |
| 5 | 130.0 | Loss |
| 6 | 91.0 | Loss |
| 7 | 111.0 | Loss |
| 8 | 99.0 | Loss |
| 9 | 132.0 | Loss |
| 10 | 103.0 | Loss |
| 11 | 102.0 | Loss |
| 12 | 54.0 | Loss |
| 13 | 130.0 | Loss |
| 14 | 138.0 | Loss |
| 15 | 69.0 | Loss |
| 16 | 65.0 | Loss |
| 17 | 130.0 | Loss |
| 18 | 100.0 | Loss |
| 19 | 106.0 | Loss |
| 20 | 132.0 | Loss |

上述的解法还有以下一些缺点：

(1) 不具通用性：这个策略就算在《木棒台车》有效，也不能套用到其他游戏上。

(2) 无自我学习能力：无法随着训练次数的增加，使模型更加聪明、准确。

最近看到一篇文章 *From Scratch: AI Balancing Act in 50 Lines of Python*[7]，使用简单的策略行动可达到 500 分，很有意思，程序如下：

(1) 定义 Play 函数作为训练程序。

行动策略选择：第 20~21 行，行动策略 (policy) 与观察 (observation) 点积 (dot product)，产生一个数值，若数值 >0，则选择向右走，反之向左走。

```python
1   import numpy as np
2   
3   env = gym.make('CartPole-v1')
4   
5   def play(env, policy):
6       observation = env.reset()
7       
8       done = False
9       score = 0
10      observations = []
11      
12      # 训练5000步
13      for _ in range(5000):
14          observations += [observation.tolist()] # 记录历次状态
15          
16          if done: # 回合是否胜负已分
17              break
18          
19          # 行动策略选择
20          outcome = np.dot(policy, observation)
21          action = 1 if outcome > 0 else 0
22          
23          # 触发下一步
24          observation, reward, done, info = env.step(action)
25          score += reward
26          
27      return score, observations
```

(2) 训练 10 回合：产生 4 个随机变量作为策略，范围均介于 [0, 1]。

```python
1   # 训练 10 回合
2   max = (0, [], [])
3   for _ in range(10):
4       policy = np.random.rand(1,4) # 产生4个随机变量 [0, 1)
5       score, observations = play(env, policy) # 开始玩
6       
7       if score > max[0]: # 取最大分数
8           max = (score, observations, policy)
9   
10  print('Max Score', max[0])
```

执行结果：通常可达 500 分，执行多次偶尔会未达到 500 分。

(3) 原作者稍微改良一下，将随机变量范围改为介于 [-0.5, 0.5] 之间，会使 outcome > 0 的机会减少，故改为执行 100 回合，增加训练周期。

```python
1   # 最终版本
2   max = (0, [], [])
3   
4   for _ in range(100): # 训练 100 回合
5       policy = np.random.rand(1,4) - 0.5  # 改为 [-0.5, 0.5]
6       score, observations = play(env, policy)
7       
8       if score > max[0]:   # 取最大分数
9           max = (score, observations, policy)
10  
11  print('Max Score', max[0])
```

执行结果：每次均可达 500 分。

(4) 以最大分数的 policy 进行试验，验证最佳策略是否有效。

```python
1   # 取得最佳策略
2   policy = max[2]
3   policy
```

执行结果：最佳策略为 [ 0.05171018, -0.0624274 , 0.23256838, 0.22154222]。

(5) 以最佳策略取代随机 policy，进行 10 回合验证。

```python
1   for _ in range(10):
2       score, observations = play(env, policy)
3       print('Score: ', score)
```

执行结果：确实有效，虽然没有每次都 500 分以上。

```
Score:  500.0
Score:  210.0
Score:  500.0
Score:  205.0
Score:  143.0
Score:  500.0
Score:  500.0
Score:  215.0
Score:  500.0
Score:  500.0
```

以上的方法正符合尝试错误 (Trial and Error) 的精神，不断地尝试并修正行动策略，最终找到最佳策略，虽然《木棒台车》是一款规则很简单的游戏，不过，要能成功解题并不如想象中容易，上述试验若要采取状态值函数最大化策略，会碰到另一些难题：

(1) 状态非单一变量，而是四个变量，包括台车的位置和速度、木棒的角度和速度。

(2) 这四个变量都是连续性变量，然而计算状态值函数是针对每一个状态，因此状态空间必须是离散的，且状态必须是有限个数，才能倒推计算出状态值函数。

以上两个问题可以使用以下技巧处理：

(1) 四个变量混合列举出所有组合。

(2) 将连续性变量分组，变成有限个数。

处理的程序代码很简单，四个状态 ( 台车位置、台车速度、木棒角度、木棒速度 ) 均切成 N 等分 (bin_size)：

bins = [np.linspace(-4.8,4.8,bin_size), np.linspace(-4,4,bin_size),
    np.linspace(-0.418,0.418,bin_size), np.linspace(-4,4,bin_size)]

详细的解说及完整的程序可参阅 *Solving Open AI's CartPole Using Reinforcement Learning Part-1*[8]。

这里我们先不作说明，因为后续可以搭配更好的算法来解决上述问题，所以这题先搁在一边，待会再解决它。

## 15-5　Gym 扩充功能

虽然 Gym 有很多环境供开发者挑选，但万一还是没找到完全符合需求的环境该怎么办？这时可以利用扩充功能 Wrapper，客制化预设的环境，包括：

(1) 修改 step 函数回传的状态：预设只会回传最新的状态，我们可以利用 ObservationWrapper 达成此一功能。

(2) 修改奖励值：可以利用 RewardWrapper 修改预设的奖励值。

(3) 修改行动值：可以利用 ActionWrapper 输入行动值。

**范例9. ActionWrapper示范**：执行《木棒台车》游戏，原先台车固定往左走，但加上 ActionWrapper后，会有1/10的概率采取随机行动。

下列程序代码请参考【RL_15_07_Action_Wrapper_test.py】。

(1) 加载相关套件。

```
1  import gym
2  import random
```

(2) 建立一个 RandomActionWrapper 类别，继承 gym.ActionWrapper 基础类别，覆写 action() 方法。epsilon 变量为随机行动的概率，每次行动时 (step) 都会调用 RandomActionWrapper 的 action()。

```python
5   # 继承 gym.ActionWrapper 基础类别
6   class RandomActionWrapper(gym.ActionWrapper):
7       def __init__(self, env, epsilon=0.1):
8           super(RandomActionWrapper, self).__init__(env)
9           self.epsilon = epsilon # 随机行动的概率
10
11      def action(self, action):
12          # 随机乱数小于 epsilon，采取随机行动
13          if random.random() < self.epsilon:
14              print("Random!")
15              return self.env.action_space.sample()
16          return action
```

(3) 试验：step(0) 表示固定往左走，但却会被 RandomActionWrapper 的 action() 所拦截，结果如下，偶尔会出现随机行动。

```python
19  if __name__ == "__main__":
20      env = RandomActionWrapper(gym.make("CartPole-v0"))
21
22      for _ in range(50):
23          env.reset()
24          total_reward = 0.0
25          while True:
26              env.render()
27              # 固定往左走
28              print("往左走!")
29              obs, reward, done, _ = env.step(0)
30              total_reward += reward
31              if done:
32                  break
33
34          print(f"报酬: {total_reward:.2f}")
35          env.close()
```

执行结果：偶尔会出现随机行动 (Random)。

**范例10. 使用wrappers.Monitor录像**：可将训练过程存成多段影片文件(mp4)。

下列程序代码请参考【RL_15_08_Record_test.py】。

(1) 加载相关套件。

```python
4   import gym
```

(2) 录像：调用 Monitor()。

```
6   # 载入环境
7   env = gym.make("CartPole-v0")
8
9   # 录影
10  env = gym.wrappers.Monitor(env, "recording", force=True)
```

(3) 试验。

```
12  # 试验
13  for _ in range(50):
14      total_reward = 0.0
15      obs = env.reset()
16
17      while True:
18          env.render()
19          action = env.action_space.sample()
20          obs, reward, done, _ = env.step(action)
21          total_reward += reward
22          if done:
23              break
24
25      print(f"报酬: {total_reward:.2f}")
26
27  env.close()
```

执行结果：会将录像结果存在 recording 文件夹。

以上只是列举两个简单的范例，读者有兴趣可进一步参阅官网说明。

## 15-6 动态规划

为使上述 $G_t = R_{t+1} + \gamma G_{t+1}$ 公式更完善，贝尔曼 (Richard E. Bellman) 于 1957 年提出贝尔曼方程式 (Bellman Equation)，以数学公式表达状态值函数，使策略 ($\pi$)、行动转移概率 ($P$)、奖励 ($R$)、折扣因子 ($\gamma$) 及下一状态的值函数 ($V_\pi(s')$) 等参数，公式如下：

$$\upsilon_\pi(s) = \sum_a \pi(a|s) \sum_{s'} p_{ss'}^a \left[ R_{ss'}^a + \gamma \upsilon_\pi(s') \right]$$

其中：

$s'$：下一状态。

$\pi(a|s)$：采取特定策略时，在状态 $s$ 采取行动 $a$ 的概率。

$p_{ss'}^a$：为行动转移概率，即在状态 $s$ 采取行动 $a$，会达到状态 $s'$ 的概率。

后面括号 [] 的部分相当于 $G_t = R_{t+1} + \gamma G_{t+1}$。

Bellman 方程式让我们可以从下一状态的奖励及状态值函数推算出目前状态的值函数。看到这里，读者可能会愣一下，在目前状态下怎么会知道下一个状态的值函数？不要忘了，强化学习是以尝试错误 (Trial and Error) 的方式进行训练，因此可从之前的训练结果推算目前回合的值函数，譬如要算第 50 回合的值函数，可以从 1~49 回合累计的状态期望值推算，即 $V_\pi(s')$ 是之前回合的状态期望值，故 Bellman 方程式 $V_\pi(s)$ 是可以计算出来的。

另外，行动值函数 (Action Value Function) 是在特定状态下，采取某一行动的值函数，类似状态值函数，公式如下：

$$q_\pi(s,a) = \sum_{s'} p_{ss'}^a \left[ R_{ss'}^a + \gamma \sum_{a'} \pi(a'|s') q_\pi(a',s') \right]$$

$\sum_{a'} \pi(a'|s') q_\pi(a',s')$ 类似 $V_\pi(s)$ 公式。

$q_\pi(s, a)$ 可以从下一状态的奖励/行动值函数推算出来。

状态值函数与行动值函数的关系可以由倒推图 (Backup Diagram) 来表示，如图 15.16、图 15.17 所示。

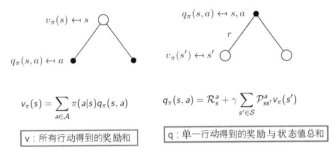

图 15.16　状态值函数与行动值函数的关系 (1)

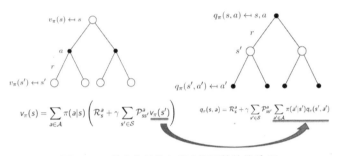

图 15.17　状态值函数与行动值函数的关系 (2)

如果上述的行动转移概率 ($\pi$)、状态转移概率 ($P$) 均为已知，亦即环境是明确的 (deterministic)，我们就可以利用 Bellman 方程式计算出状态值函数、行动值函数，以反复的方式求解，这种解法称为动态规划 (Dynamic Programming, DP)，类似程序【RL_15_03_simple_game_with_state_value.ipynb】，不过更为细致。

动态规划是将大问题切分成小问题，然后逐一解决每个小问题，由于每个小问题都很类似，因此整合起来就能解决大问题。譬如斐波那契数列 (Fibonacci)，其中 $F_n=F_{n-1}+F_{n-2}$，要计算整个数列的话，可以设计一个 $F_n=F_{n-1}+F_{n-2}$ 函数 ( 小问题 )，以递归的方式完成整个数列的计算 ( 大问题 )。

**范例11. 斐波那契数列计算。**

**下列程序代码请参考【RL_15_09_Fibonacci_Calculation.ipynb】。**

```
def fibonacci(n):
    if n == 0 or n ==1:
        return n
    else:
        return fibonacci(n-1)+fibonacci(n-2)

list1=[]
for i in range(2, 20):
    list1.append(fibonacci(i))
print(list1)
```

执行结果。

[1, 2, 3, 5, 8, 13, 21, 34, 55, 89, 144, 233, 377, 610, 987, 1597, 2584, 4181]

每个小问题的计算结果都会被存储下来,作为下个小问题的计算基础。

强化学习将问题切成两个步骤:

(1) 策略评估 (Policy Evaluation):当玩家走完一回合后,就可以更新所有状态的值函数,称为策略评估,即将所有状态重新评估一次,也称为预测 (prediction)。

(2) 策略改善 (Policy Improvement):依照策略评估的最新状态,采取最佳策略,以改善模型,也称为控制 (Control),通常都会依据最大的状态值函数行动,我们称为贪婪 (Greedy) 策略,但其实它是有缺陷的,后面会再详加说明。

最后,将两个步骤合并,循环使用,即是所谓的策略循环或策略迭代 (Policy Iteration),如图 15.18 所示。

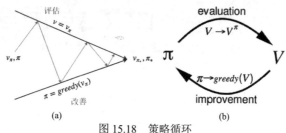

图 15.18　策略循环

图 (a) 表示先走一回合后,进行评估,接着依评估结果采取贪婪策略行动,再评估,一直循环下去,直到收敛为止。

图 (b) 强调经过行动策略后,状态值函数会由原来的 $V$ 变成 $V_\pi$,之后再以 $V_\pi$ 作为行动的依据。

**范例12.** Grid World 迷宫,左上角及右下角为终点,每走一步奖励为-1,如图15.19所示。

图 15.19　Grid World 迷宫

首先我们先来练习策略评估,计算每一格的状态值函数。

(1) 一开始设定所有位置的状态值函数均为 0,如图 15.20 所示。

| 0.0 | 0.0 | 0.0 | 0.0 |
|---|---|---|---|
| 0.0 | 0.0 | 0.0 | 0.0 |
| 0.0 | 0.0 | 0.0 | 0.0 |
| 0.0 | 0.0 | 0.0 | 0.0 |

图 15.20　设置所有位置状态值为 0

(2) 训练一周期,状态值函数更新如图 15.21 所示,例如 (0, 1) 状态值函数的计算,有 4 种可能:

往左走到 (0, 0):状态值函数 0+-1=-1。

往右走到 (0, 2):状态值函数 0+-1=-1。

往上走碰到边界仍停留在 (0, 1)：状态值函数 0+-1=-1。

往下走到 (0, 1)：状态值函数 0+-1=-1。

往左 / 右 / 上 / 下，概率均等，即 $\pi(a|s)$ = 0.25，故 (-1-1-1-1) * 0.25=-1。

其他位置状态值函数均同理可证，除了起点 / 终点不更新。

| 0.0 | -1.0 | -1.0 | -1.0 |
|---|---|---|---|
| -1.0 | -1.0 | -1.0 | -1.0 |
| -1.0 | -1.0 | -1.0 | -1.0 |
| -1.0 | -1.0 | -1.0 | 0.0 |

图 15.21　状态值函数更新 (1)

(3) 再训练一周期，状态值函数更新如图 15.22 所示，例如 (0, 1) 状态值函数的计算，有 4 种可能：

往左走到 (0, 0)：状态值函数 0+-1=-1。

往右走到 (0, 2)：状态值函数 -1+-1=-2。

往上走碰到边界仍停留在 (0, 1)：状态值函数 -1+-1=-2。

往下走到 (0, 1)：状态值函数 -1+-1=-2。

往左 / 右 / 上 / 下，概率均等，即 $\pi(a|s)$ = 0.25，故 (-1-2-2-2) * 0.25=-1.75，画面只取到小数点第 1 位≒ -1.7。

其他位置状态值函数均同理可证，请各位读者练习看看。

| 0.0 | -1.7 | -2.0 | -2.0 |
|---|---|---|---|
| -1.7 | -2.0 | -2.0 | -2.0 |
| -2.0 | -2.0 | -2.0 | -1.7 |
| -2.0 | -2.0 | -1.7 | 0.0 |

图 15.22　状态值函数更新 (2)

(4) 再训练一周期，状态值函数更新如图 15.23 所示，例如 (0, 1) 状态值函数的计算，有 4 种可能：

往左走到 (0, 0)：状态值函数 0+-1=-1。

往右走到 (0, 2)：状态值函数 -2+-1=-3。

往上走碰到边界仍停留在 (0, 1)：状态值函数 -1.75+-1=-2.75。

往下走到 (0, 1)：状态值函数 -2+-1=-3。

往左 / 右 / 上 / 下，概率均等，即 $\pi(a|s)$ = 0.25，故 (-1-3-2.75-3) * 0.25=-2.4375，画面只取到小数点第 1 位≒ -2.4。

其他位置状态值函数均同理可证，请各位读者练习看看。

| 0.0 | -2.4 | -2.9 | -3.0 |
|---|---|---|---|
| -2.4 | -2.9 | -3.0 | -2.9 |
| -2.9 | -3.0 | -2.9 | -2.4 |
| -3.0 | -2.9 | -2.4 | 0.0 |

图 15.23　状态值函数更新 (3)

策略评估就依上述方式不断更新，直至状态值函数不再显著变化为止，这时我们就认定已是最佳状态值函数，类似神经网络训练，正向传导/反向传导直至损失函数不再显著下降为止。接下来撰写程序实现状态值函数的计算。

此范例程序修改自 Denny Britz 网站[9]，是 *Reinforcement Learning: An Introduction*[10] 一书的习题，该书可说是强化学习的圣经。

**下列程序代码请参考【RL_15_10_Policy_Evaluation.ipynb】。**

(1) 加载相关框架。

```
1  import gym
2  import numpy as np
3  from lib.envs.gridworld import GridworldEnv
```

(2) 建立 Grid World 迷宫环境如图 15.24 所示，程序为 lib\envs\gridworld.py，定义游戏规则，注意，Grid World 在网络上有许多不同的版本，游戏规则略有差异。

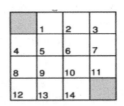

图 15.24　Grid World 迷宫环境

```
1  # 环境
2  env = GridworldEnv()
```

(3) 游戏的行动代码依顺时针设定：上 =0，右 =1，下 =2，左 =3。

(4) 定义 Grid World 迷宫环境的行动转移概率。

```
2  env.P
```

执行结果：Json 格式的内容为 { 状态 : { 行动 : [ ( 转移概率，下一个状态，奖励，是否到达终点 ),…]}}。下面截图为状态 0 及 1 的行动转移概率，观察第 2 段第一行：

1: {0: [(1.0, 1, -1.0, False)]}

表在状态 1 往上 (0) 的转移概率为 1，往上走碰到边界仍停留在 1，奖励为 1，未到达终点。

```
{0: {0: [(1.0, 0, 0.0, True)],
     1: [(1.0, 0, 0.0, True)],
     2: [(1.0, 0, 0.0, True)],
     3: [(1.0, 0, 0.0, True)]},
 1: {0: [(1.0, 1, -1.0, False)],
     1: [(1.0, 2, -1.0, False)],
     2: [(1.0, 5, -1.0, False)],
     3: [(1.0, 0, -1.0, True)]},
```

(5) 先定义一个简单的策略评估函数，计算状态值函数。

依下列状态值函数公式更新。

$$v_\pi(s) = \sum_a \pi(a|s) \sum_{s'} p_{ss'}^a \left[ R_{ss'}^a + \gamma v_\pi(s') \right]$$

```
1   def policy_eval(policy, env, epoch=1, discount_factor=1.0):
2       # 状态值函数初始化
3       V = np.zeros(env.nS)
4       V1 = np.copy(V)
5       no = 0
6       while no < epoch:
7           # 更新每个状态值的函数
8           for s in range(env.nS):
9               v = 0
10              # 计算每个行动后的状态值函数
11              for a, action_prob in enumerate(policy[s]):
12                  # 取得所有可能的下一状态值
13                  for  prob, next_state, reward, done in env.P[s][a]:
14                      # 状态值函数公式，依照所有可能的下一状态值函数加总
15                      v += action_prob * prob * (reward +
16                                  discount_factor * V[next_state])
17              V1[s] = v
18          V = np.copy(V1)
19          no+=1
20      return np.array(V)
```

(6) 训练 1 周期：采取随机策略，往上 / 下 / 左 / 右走的概率 ($\pi$) 均等。

```
1   # 随机策略，概率均等
2   random_policy = np.ones([env.nS, env.nA]) / env.nA
3   # 评估
4   v = policy_eval(random_policy, env, 1)
5   print("4x4 状态值函数:")
6   print(v.reshape(env.shape))
```

执行结果：与上述手工计算的结果相符。

```
[[ 0. -1. -1. -1.]
 [-1. -1. -1. -1.]
 [-1. -1. -1. -1.]
 [-1. -1. -1.  0.]]
```

(7) 训练 2 周期。

```
1   v = policy_eval(random_policy, env, 2)
2   print("4x4 状态值函数:")
3   print(v.reshape(env.shape))
```

执行结果：与上述手工计算的结果相符。

```
[[ 0.    -1.75 -2.   -2.  ]
 [-1.75 -2.   -2.   -2.  ]
 [-2.   -2.   -2.   -1.75]
 [-2.   -2.   -1.75  0.  ]]
```

(8) 训练 3 周期。

```
1   v = policy_eval(random_policy, env, 3)
2   print("4x4 状态值函数:")
3   print(v.reshape(env.shape))
```

执行结果：与上述手工计算的结果相符。

```
[[ 0.     -2.4375 -2.9375 -3.    ]
 [-2.4375 -2.875  -3.     -2.9375]
 [-2.9375 -3.     -2.875  -2.4375]
 [-3.     -2.9375 -2.4375  0.    ]]
```

(9) 接着进行完整的策略评估函数：不断更新，直至状态值函数不再显著变化为止，即前后两周期最大差值小于设定的门槛值才停止更新。

```python
1   # 策略评估函数
2   def policy_eval(policy, env, discount_factor=1.0, theta=0.00001):
3       # 状态值函数初始化
4       V = np.zeros(env.nS)
5       V1 = np.copy(V)
6       while True:
7           delta = 0
8           # 更新每个状态值的函数
9           for s in range(env.nS):
10              v = 0
11              # 计算每个行动后的状态值函数
12              for a, action_prob in enumerate(policy[s]):
13                  # 取得所有可能的下一状态值
14                  for prob, next_state, reward, done in env.P[s][a]:
15                      # 状态值函数公式，依照所有可能的下一状态值函数加总
16                      v += action_prob * prob * (reward +
17                              discount_factor * V[next_state])
18              # 比较更新前后的差值，取最大值
19              delta = max(delta, np.abs(v - V[s]))
20              V1[s] = v
21          V = np.copy(V1)
22          # 若最大差值小于门槛值，则停止评估
23          if delta < theta:
24              break
25      return np.array(V)
```

(10) 调用策略评估函数，显示状态值函数。

```python
1   # 随机策略，机率均等
2   random_policy = np.ones([env.nS, env.nA]) / env.nA
3   # 评估
4   v = policy_eval(random_policy, env)
5
6   print("4x4 状态值函数:")
7   print(v.reshape(env.shape))
```

执行结果。

```
[[  0.          -13.99989315 -19.99984167 -21.99982282]
 [-13.99989315 -17.99986052 -19.99984273 -19.99984167]
 [-19.99984167 -19.99984273 -17.99986052 -13.99989315]
 [-21.99982282 -19.99984167 -13.99989315   0.        ]]
```

(11) 验证答案是否正确：与书中的答案几乎相等。

```python
1   # 验证答案是否正确
2   expected_v = np.array([0, -14, -20, -22, -14, -18, -20, -20, -20, -20, -18, -14, -22, -20, -14, 0])
3   np.testing.assert_array_almost_equal(v, expected_v, decimal=2)
```

**范例13.** 以上述的策略评估结合策略改善，进行策略循环(Policy Iteration)。此范例程序修改自Denny Britz网站。

下列程序代码请参考【RL_15_11_Policy_Iteration.ipynb】。

(1) 加载相关套件：与前例相同。

```python
1   import numpy as np
2   from lib.envs.gridworld import GridworldEnv
```

(2) 建立 Grid World 迷宫环境：与前例相同。

```python
1   env = GridworldEnv()
```

(3) 策略评估函数：与前例相同。

```python
1   # 策略评估函数
2   def policy_eval(policy, env, discount_factor=1.0, theta=0.00001):
3       # 状态值函数初始化
4       V = np.zeros(env.nS)
5       V1 = np.copy(V)
6       while True:
7           delta = 0
8           # 更新每个状态值的函数
9           for s in range(env.nS):
10              v = 0
11              # 计算每个行动后的状态值函数
12              for a, action_prob in enumerate(policy[s]):
13                  # 取得所有可能的下一状态值
14                  for prob, next_state, reward, done in env.P[s][a]:
15                      # 状态值函数公式，依照所有可能的下一状态值函数加总
16                      v += action_prob * prob * (reward +
17                          discount_factor * V[next_state])
18              # 比较更新前后的差值，取最大值
19              delta = max(delta, np.abs(v - V[s]))
20              V1[s] = v
21          V = np.copy(V1)
22          # 若最大差值小于门槛值，则停止评估
23          if delta < theta:
24              break
25      return np.array(V)
```

(4) 定义策略改善函数：

one_step_lookahead：依下列公式计算每一种行动的值函数。

$$q_\pi(s,a) = \sum_{s'} p_{ss'}^a \left[ R_{ss'}^a + \gamma \sum_{a'} \pi(a'|s') q_\pi(a',s') \right]$$

一开始采取随机策略，进行策略评估，计算状态值函数。

调用 one_step_lookahead，计算下一步的行动值函数，找出最佳行动，并更新策略行动概率 ($\pi$)。

直到已无较佳的行动策略，则回传策略与状态值函数。

```python
1   def policy_improvement(env, policy_eval_fn=policy_eval, discount_factor=1.0):
2       # 计算行动值函数
3       def one_step_lookahead(state, V):
4           A = np.zeros(env.nA)
5           for a in range(env.nA):
6               for prob, next_state, reward, done in env.P[state][a]:
7                   A[a] += prob * (reward + discount_factor * V[next_state])
8           return A
9
10      # 一开始采随机策略，往上/下/左/右走的概率(π)均等
11      policy = np.ones([env.nS, env.nA]) / env.nA
12
13      while True:
14          # 策略评估
15          V = policy_eval_fn(policy, env, discount_factor)
16
17          # 若要改变策略，会设定 policy_stable = False
18          policy_stable = True
19
20          for s in range(env.nS):
21              # 依 P 选择最佳行动
22              chosen_a = np.argmax(policy[s])
23
24              # 计算下一步的行动值函数
25              action_values = one_step_lookahead(s, V)
26              # 选择最佳行动
27              best_a = np.argmax(action_values)
28
29              # 贪婪策略：若有新的最佳行动，修改行动策略
30              if chosen_a != best_a:
31                  policy_stable = False
32              policy[s] = np.eye(env.nA)[best_a]
33
34          # 如果已无较佳行动策略，则回传策略及状态值函数
35          if policy_stable:
36              return policy, V
```

(5) 执行策略循环。

```
1  policy, v = policy_improvement(env)
```

(6) 显示结果。

```
1  # 显示结果
2  print("策略概率分配:")
3  print(policy)
4  print("")
5
6  print("4x4 策略概率分配 (0=up, 1=right, 2=down, 3=left):")
7  print(np.reshape(np.argmax(policy, axis=1), env.shape))
8  print("")
9
10 print("4x4 状态值函数:")
11 print(v.reshape(env.shape))
```

执行结果。

```
策略机率分配:
[[1. 0. 0. 0.]
 [0. 0. 0. 1.]
 [0. 0. 0. 1.]
 [0. 0. 1. 0.]
 [1. 0. 0. 0.]
 [1. 0. 0. 0.]
 [1. 0. 0. 0.]
 [0. 0. 1. 0.]
 [1. 0. 0. 0.]
 [1. 0. 0. 0.]
 [0. 1. 0. 0.]
 [0. 0. 1. 0.]
 [1. 0. 0. 0.]
 [0. 1. 0. 0.]
 [0. 1. 0. 0.]
 [1. 0. 0. 0.]]

4x4 策略概率分配 (0=up, 1=right, 2=down, 3=left):
[[0 3 3 2]
 [0 0 0 2]
 [0 0 1 2]
 [0 1 1 0]]

4x4 状态值函数:
[[ 0. -1. -2. -3.]
 [-1. -2. -3. -2.]
 [-2. -3. -2. -1.]
 [-3. -2. -1.  0.]]
```

策略概率分配 ($\pi$)：代表采取上 / 右 / 下 / 左走的个别概率。

4×4 策略概率分配：是将策略概率分配转化而成的行动矩阵。

第三段是状态值函数 ($V$)。

$\pi$、$V$ 代表模型的参数，之后可依 $\pi$ 或 $V$ 行动，例如起点在 (1, 1) 的最佳路径如下：往上 (0)，再往左 (3)，即到达 (0, 0) 终点。读者可以测试其他起点，看看是否可顺利到达终点。

```
[0 3 3 2]
[0 0 0 2]
[0 0 1 2]
[0 1 1 0]
```

(7) 验证答案是否正确：与书中的答案相对照。

```
1 expected_v = np.array([ 0, -1, -2, -3, -1, -2, -3, -2, -2, -3, -2, -1, -3, -2, -1,  0])
2 np.testing.assert_array_almost_equal(v, expected_v, decimal=2)
```

笔者测试另一个游戏 WindyGridworldEnv(windy_gridworld.py)，在策略评估阶段一直无法收敛，接着再测试 Gym 套件内的 FrozenLake-v1 就没有问题，请参阅【RL_15_11_Policy_Iteration_FrozenLake.ipynb】，读者可多测试看看。

## 15-7 值循环

采用策略循环时，在每次策略改善前，必须先作一次策略评估，执行循环，更新所有状态值函数，直至收敛，非常耗时，所以，考虑到状态值函数与行动值函数的更新十分类似，干脆将其二者合并，以改善策略循环的缺点，这称为值循环(Value Iteration)。

**范例14.** 以上述的迷宫为例，使用值循环(Value Iteration)。此范例程序修改自Denny Britz网站。

下列程序代码请参考【RL_15_12_Value_Iteration.ipynb】。

(1) 加载相关套件，前两步与上例相同。

```
1 import numpy as np
2 from lib.envs.gridworld import GridworldEnv
```

(2) 建立 Grid World 迷宫环境。

```
1 env = GridworldEnv()
```

(3) 定义值循环函数：直接以行动值函数取代状态值函数，将策略评估函数与策略改善函数合而为一。

依下列行动值函数公式更新。

$$q_\pi(s,a) = \sum_{s'} p_{ss'}^a \left[ R_{ss'}^a + \gamma \sum_{a'} \pi(a'|s') q_\pi(a',s') \right]$$

```
1  # 值循环函数
2  def value_iteration(env, theta=0.0001, discount_factor=1.0):
3      # 计算行动值函数
4      def one_step_lookahead(state, V):
5          A = np.zeros(env.nA)
6          for a in range(env.nA):
7              for prob, next_state, reward, done in env.P[state][a]:
8                  A[a] += prob * (reward + discount_factor * V[next_state])
9          return A
10 
11     # 状态值函数初始化
12     V = np.zeros(env.nS)
13     while True:
14         delta = 0
15         # 更新每个状态值的函数
16         for s in range(env.nS):
17             # 计算下一步的行动值函数
18             A = one_step_lookahead(s, V)
19             best_action_value = np.max(A)
20             # 比较更新前后的差值，取最大值
21             delta = max(delta, np.abs(best_action_value - V[s]))
```

```
22          # 更新状态值函数
23          V[s] = best_action_value
24      # 若最大差值小于门槛值,则停止评估
25      if delta < theta:
26          break
27
28  # 一开始采随机策略,往上/下/左/右走的概率(π)均等
29  policy = np.zeros([env.nS, env.nA])
30  for s in range(env.nS):
31      # 计算下一步的行动值函数
32      A = one_step_lookahead(s, V)
33      # 选择最佳行动
34      best_action = np.argmax(A)
35      # 永远采取最佳行动
36      policy[s, best_action] = 1.0
37
38  return policy, V
```

(4) 执行值循环。

```
1  policy, v = value_iteration(env)
```

(5) 显示结果:与策略循环结果相同。

```
1   print("策略概率分配:")
2   print(policy)
3   print("")
4
5   print("4x4 策略概率分配 (0=up, 1=right, 2=down, 3=left):")
6   print(np.reshape(np.argmax(policy, axis=1), env.shape))
7   print("")
8
9   print("4x4 状态值函数:")
10  print(v.reshape(env.shape))
```

执行结果。

```
策略概率分配:
[[1. 0. 0. 0.]
 [0. 0. 0. 1.]
 [0. 0. 0. 1.]
 [0. 0. 1. 0.]
 [1. 0. 0. 0.]
 [1. 0. 0. 0.]
 [1. 0. 0. 0.]
 [0. 0. 1. 0.]
 [1. 0. 0. 0.]
 [1. 0. 0. 0.]
 [0. 1. 0. 0.]
 [0. 0. 1. 0.]
 [1. 0. 0. 0.]
 [0. 1. 0. 0.]
 [0. 1. 0. 0.]
 [1. 0. 0. 0.]]

4x4 策略概率分配 (0=up, 1=right, 2=down, 3=left):
[[0 3 3 2]
 [0 0 0 2]
 [0 0 1 2]
 [0 1 1 0]]

4x4 状态值函数:
[[ 0. -1. -2. -3.]
 [-1. -2. -3. -2.]
 [-2. -3. -2. -1.]
 [-3. -2. -1.  0.]]
```

(6) 验证答案是否正确：与书中的答案相对照。

```
1 expected_v = np.array([ 0, -1, -2, -3, -1, -2, -3, -2, -2, -3, -2, -1, -3, -2, -1,  0])
2 np.testing.assert_array_almost_equal(v, expected_v, decimal=2)
```

笔者测试另一个游戏 CliffWalkingEnv(cliff_walking.py)，一直无法收敛，接着再测试 Gym 套件内的 FrozenLake-v1 就没有问题，请参阅【RL_15_12_Value_Iteration_FrozenLake.ipynb】，读者可多测试看看。

综合以上测试，动态规划的优缺点如下：

(1) 适合定义明确的问题，即策略行动概率 ($\pi$)、状态转移概率 ($P$) 均为已知的状况。

(2) 适合中小型的模型，状态空间不超过百万个，比方说围棋，状态空间= $3^{19 \times 19}=1.74 \times 10^{172}$，状态值函数更新就会执行太久。另外，可能会有大部分的路径从未走过，导致样本代表性不足，造成维数灾难 (Curse of Dimensionality)。

# 15-8 蒙特卡罗

我们在玩游戏时，通常不会知道状态转移概率 ($P$)，其他应用领域也是如此，这称为无模型 (Model Free) 学习，在这样的情况下，动态规划的状态值函数就无法依公式计算。因此，有学者就提出蒙特卡罗 (Monte Carlo, MC) 算法，通过仿真的方式估计出状态转移概率。

**范例15. 先以蒙特卡罗算法求圆周率($\pi$)，见证算法的威力。**

**下列程序代码请参考【RL_15_13_MC_Pi.ipynb】。**

(1) 加载相关套件。

```
1 # 载入相关套件
2 import random
```

(2) 计算圆周率 ($\pi$)：

如图 15.25 所示，假设有一个正方形与圆形，圆形半径为 $r$，则圆形面积为 $\pi r^2$，正方形面积为 $(2r)^2=4r^2$。

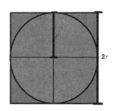

图 15.25　一个正方形和圆形

在正方形的范围内随机产生随机数一千万个点，计算落在圆形内的点数。
落在圆形内的点数 / 10000000 ≒ $\pi r^2 / 4r^2 = \pi/4$。
化简后 $\pi$= 4 × 落在圆形内的点数 / 10000000。

```
1   # 模拟一千万次
2   run_count=10000000
3   list1=[]
4
5   # 在 X:(-0.5, -0.5),Y:(0.5, 0.5) 范围内产生一千万个点
6   for _ in range(run_count):
7       list1.append([random.random()-0.5, random.random()-0.5])
8
9   in_circle_count=0
10  for i in range(run_count):
11      # 计算在圆内的点,即 (X^2 + Y^2 <= 0.5 ^ 2),其中 半径=0.5
12      if list1[i][0]**2 + list1[i][1]**2 <= 0.5 ** 2:
13          in_circle_count+=1
14
15  # 正方形面积:宽高各为2r,故面积=4*(r**2)
16  # 圆形面积:pi * (r ** 2)
17  # pi = 圆形点数 / 正方形点数
18  pi=(in_circle_count/run_count) * 4
19  pi
```

执行结果:得到 π = 3.1418028,与正确答案 3.14159 相去不远,如果仿真更多的点,就会更相近。

从以上的例子延伸,假如转移概率未知,我们也可以利用蒙特卡罗算法估计转移概率,利用随机策略去走迷宫,根据结果计算转移概率。

为了避免无聊,我们换另一款扑克牌游戏《21 点》(Blackjack)试验,读者如不熟悉游戏规则,可参酌维基百科关于《21 点》的说明[12]。

**范例16. 试验《21点》扑克牌(Blackjack)之策略评估。此范例程序修改自Denny Britz网站。**

**下列程序代码请参考【RL_15_14_Blackjack_Policy_Evaluation.ipynb】。**

(1) 加载相关套件。

```
1  import numpy as np
2  from lib.envs.blackjack import BlackjackEnv
3  from lib import plotting
4  import sys
5  from collections import defaultdict
6  import matplotlib
7
8  matplotlib.style.use('ggplot') # 设定绘图的风格
```

(2) 建立环境:程序为 lib\envs\blackjack.py。

```
1  env = BlackjackEnv()
```

(3) 试玩:采用的策略为如果玩家手上点数超过 ( ≥ )20 点,不补牌 (stick),反之都跟庄家要一张牌 (hit),策略并不合理,通常超过 16 点就不补牌了,若考虑更周延的话,会再视庄家的点数,决定是否补牌,读者可自行更改策略,观察试验结果的变化。采用这个不合理的策略,是要测试各种算法是否能有效提升胜率。

```
1  def print_observation(observation):
2      score, dealer_score, usable_ace = observation
3      print(f"玩家分数: {score} (是否持有A: {usable_ace}), 庄家分数: {dealer_score}")
4
5  def strategy(observation):
6      score, dealer_score, usable_ace = observation
7      # 超过20点,不补牌(stick),否则都跟庄家要一张牌(hit)
8      return 0 if score >= 20 else 1
9
10 # 试玩 20 次
11 for i_episode in range(20):
12     observation = env.reset()
13     # 开始依策略玩牌,最多 100 步骤,中途分出胜负即结束
14     for t in range(100):
```

```
15        print_observation(observation)
16        action = strategy(observation)
17        print(f'行动：{["不补牌", "补牌"][action]}')
18        observation, reward, done, _ = env.step(action)
19        if done:
20            print_observation(observation)
21            print(f"输赢分数：{reward}\n")
22            break
```

执行结果：读者若不熟悉玩法，可以观察下列过程。

```
玩家分数：13 (是否持有A: False)，庄家分数：3
行动：补牌
玩家分数：19 (是否持有A: False)，庄家分数：3
行动：补牌
玩家分数：24 (是否持有A: False)，庄家分数：3
输赢分数：-1

玩家分数：15 (是否持有A: False)，庄家分数：10
行动：补牌
玩家分数：25 (是否持有A: False)，庄家分数：10
输赢分数：-1

玩家分数：15 (是否持有A: False)，庄家分数：8
行动：补牌
玩家分数：20 (是否持有A: False)，庄家分数：8
行动：不补牌
玩家分数：20 (是否持有A: False)，庄家分数：8
输赢分数：1

玩家分数：19 (是否持有A: False)，庄家分数：10
行动：补牌
玩家分数：27 (是否持有A: False)，庄家分数：10
输赢分数：-1
```

(4) 定义策略评估函数：主要是通过既定策略计算状态值函数。

试验 1000 回合，记录玩牌的过程，然后计算每个状态的状态值函数。

状态为 Tuple 数据类型，内含玩家的总点数 0~31、庄家亮牌的点数 1~11(A)、玩家是否拿 A，维度大小为 (32, 11, 2)，其中 A 可为 1 点或 11 点，由持有者自行决定。

行动只有两种：0 为不补牌，1 为补牌。

注意，在一回合中每个状态有可能被走过两次以上，例如，一开始持有 A、5，玩家视 A 为 11 点，加总为 16 点，后来补牌后抽到 10 点，改视 A 为 1 点，加总也是 16 点，故 16 点这个状态被经历两次，如果倒推计算状态值函数，会被重复计算，造成该节点状态值函数特别大或特别小。解决的方式有两种：首次访问 (First Visit) 及每次访问 (Every Visit)。首次访问只计算第一次访问时的报酬，而每次访问则计算所有访问的平均报酬，本程序采用首次访问。

```
1   # 策略评估函数
2   def policy_eval(policy, env, num_episodes, discount_factor=1.0):
3       returns_sum = defaultdict(float)    # 记录每一个状态的报酬
4       returns_count = defaultdict(float)  # 记录每一个状态的访问个数
5       V = defaultdict(float) # 状态值函数
6
7       # 试验 N 回合
8       for i_episode in range(1, num_episodes + 1):
9           # 每 1000 回合显示除错信息
10          if i_episode % 1000 == 0:
11              print(f"\r {i_episode}/{num_episodes}回合。", end="")
12              sys.stdout.flush() # 清除画面
13
14          # 回合(episode)资料结构为阵列，每一项目含 state, action, reward
15          episode = []
16          state = env.reset()
17          # 开始依策略玩牌，最多 100 步骤，中途分出胜负则结束
18          for t in range(100):
19              action = policy(state)
20              next_state, reward, done, _ = env.step(action)
21              episode.append((state, action, reward))
22              if done:
23                  break
```

```
24          state = next_state
25
26      # 找出走过的所有状态
27      states_in_episode = set([tuple(x[0]) for x in episode])
28      # 计算每一状态的值函数
29      for state in states_in_episode:
30          # 找出每一步骤内的首次访问(First Visit)
31          first_occurence_idx = next(i for i,x in enumerate(episode)
32                                      if x[0] == state)
33          # 算累计报酬(G)
34          G = sum([x[2]*(discount_factor**i) for i,x in
35                   enumerate(episode[first_occurence_idx:])])
36          # 计算状态值函数
37          returns_sum[state] += G
38          returns_count[state] += 1.0
39          V[state] = returns_sum[state] / returns_count[state]
40
41  return V
```

(5) 确定策略：与前面策略相同。

```
1  # 采用相同策略
2  def sample_policy(observation):
3      score, dealer_score, usable_ace = observation
4      # 超过20点，不补牌(stick)，否则都跟庄家要一张牌(hit)
5      return 0 if score >= 20 else 1
```

(6) 分别试验 10000 与 500000 回合。

```
1  # 试验 10000 回合
2  V_10k = policy_eval(sample_policy, env, num_episodes=10000)
3  plotting.plot_value_function(V_10k, title="10,000 Steps")
4
5  # 试验 500,000 回合
6  V_500k = policy_eval(sample_policy, env, num_episodes=500000)
7  plotting.plot_value_function(V_500k, title="500,000 Steps")
```

执行结果：显示各状态的值函数，分成持有 A( 比较容易获胜 ) 及未持有 A。可以看到当玩家持有的分数很高时，胜率会明显提升。以下彩色图表可参考程序执行结果。

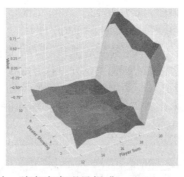

扫码看彩图

持有 A，但分数不高时，胜率也有明显提升。

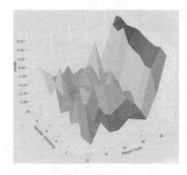

试验 500000 回合后的表现更明显。

接着进行值循环 ( 也可以使用策略循环 )，结合策略评估与策略改善。另外，还要介绍一个新的策略——ε-greedy policy，之前策略改善都是采用贪婪 (Greedy) 策略，它有一个弱点，就是一旦发现最大值函数的路径后，玩家就会一直重复相同的路径，这样便失去了寻找较佳路径的潜在机会，举例来说，家庭聚餐时，通常会选择最好的美食餐厅，若为了不踩雷，每次都去之前吃过最好吃的餐厅用餐，那新开的餐厅就永远没机会被发现了，这就是所谓的探索与利用 (Exploration and Exploitation)，如图 15.26 所示，而 ε-greedy 所采取的方式就是除了采最佳路径之外，还保留一定的比例去探索，刻意不走最佳路径。以下我们就尝试这种新策略与蒙特卡罗算法结合。

图 15.26　探索与利用

**范例17. 试验《21点》扑克牌游戏(Blackjack)之值循环。此范例程序修改自Denny Britz 网站。**

**下列程序代码请参考【RL_15_15_Blackjack_Value_Iteration.ipynb】。**

(1) 加载相关套件。

```
1  import numpy as np
2  from lib.envs.blackjack import BlackjackEnv
3  from lib import plotting
4  import sys
5  from collections import defaultdict
6  import matplotlib
7
8  matplotlib.style.use('ggplot') # 设定绘图的风格
```

(2) 建立环境：程序为 lib\envs\blackjack.py。

```
1  env = BlackjackEnv()
```

(3) 定义 ε-greedy 策略：若 ε =0.1，则 10 次行动有 1 次采取随机行动。

```
1  def make_epsilon_greedy_policy(Q, epsilon, nA):
2      def policy_fn(observation):
3          # 每个行动的概率初始化，均为 ε / n
4          A = np.ones(nA, dtype=float) * epsilon / nA
5          best_action = np.argmax(Q[observation])
6          # 最佳行动的概率再加上 1 - ε
7          A[best_action] += (1.0 - epsilon)
8          return A
9      return policy_fn
```

(4) 定义值循环函数：与上例的程序逻辑几乎相同，主要差异是将状态值函数改为行动值函数。

```python
1  def value_iteration(env, num_episodes, discount_factor=1.0, epsilon=0.1):
2      returns_sum = defaultdict(float)      # 记录每一个状态的报酬
3      returns_count = defaultdict(float)    # 记录每一个状态的访问个数
4      Q = defaultdict(lambda: np.zeros(env.action_space.n))  # 行动值函数
5
6      # 采用 ε-greedy策略
7      policy = make_epsilon_greedy_policy(Q, epsilon, env.action_space.n)
8
9      # 试验 N 回合
10     for i_episode in range(1, num_episodes + 1):
11         # 每 1000 回合显示除错信息
12         if i_episode % 1000 == 0:
13             print(f"\r {i_episode}/{num_episodes}回合.", end="")
14             sys.stdout.flush()  # 清除画面
15
16         # 回合(episode)资料结构为阵列，每一项目含 state, action, reward
17         episode = []
18         state = env.reset()
19         # 开始依策略玩牌，最多 100 步骤，中途分出胜负即结束
20         for t in range(100):
21             probs = policy(state)
22             action = np.random.choice(np.arange(len(probs)), p=probs)
23             next_state, reward, done, _ = env.step(action)
24             episode.append((state, action, reward))
25             if done:
26                 break
27             state = next_state
28
29         # 找出走过的所有状态
30         sa_in_episode = set([(tuple(x[0]), x[1]) for x in episode])
31         for state, action in sa_in_episode:
32             # (状态, 行动)组合初始化
33             sa_pair = (state, action)
34             # 找出每一步内的首次访问(First Visit)
35             first_occurence_idx = next(i for i,x in enumerate(episode)
36                                        if x[0] == state and x[1] == action)
37             # 算累计报酬(G)
38             G = sum([x[2]*(discount_factor**i) for i,x in
39                      enumerate(episode[first_occurence_idx:])])
40             # 计算行动值函数
41             returns_sum[sa_pair] += G
42             returns_count[sa_pair] += 1.0
43             Q[state][action] = returns_sum[sa_pair] / returns_count[sa_pair]
44
45     return Q, policy
```

(5) 执行值循环。

```python
1  Q, policy = value_iteration(env, num_episodes=500000, epsilon=0.1)
```

(6) 显示执行结果。

```python
1  V = defaultdict(float)
2  for state, actions in Q.items():
3      action_value = np.max(actions)
4      V[state] = action_value
5  plotting.plot_value_function(V, title="Optimal Value Function")
```

执行结果：上个范例只有当玩家的分数接近 20 分的时候，值函数特别高，然而，这个策略即使在低分时也有不差的表现，胜率比起上例明显提升。以下彩色图表可参考程序执行结果。

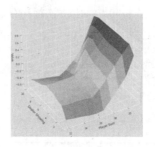

扫码看彩图

再来看另一种想法，目前为止的值循环在策略评估与策略改良上，均采用同一策略，即 ε-greedy，而这次两者则各自采用不同策略，即策略评估时采用随机策略，尽可能走过所有路径，在策略改良时，改用贪婪策略，尽量求胜。所以，当采用同一策略，我们称为On-policy，而采用不同策略则称为Off-policy。

**范例18.** 试验《21点》扑克牌游戏(Blackjack)之Off-policy值循环。此范例程序修改自Denny Britz网站。

下列程序代码请参考【**RL_15_16_Blackjack_Off_Policy.ipynb**】。

(1) 加载相关套件。

```python
import numpy as np
from lib.envs.blackjack import BlackjackEnv
from lib import plotting
import sys
from collections import defaultdict
import matplotlib

matplotlib.style.use('ggplot') # 设定绘图的风格
```

(2) 建立环境：程序为 lib\envs\blackjack.py。

```python
env = BlackjackEnv()
```

(3) 定义随机策略在评估时使用。

```python
def create_random_policy(nA):
    A = np.ones(nA, dtype=float) / nA
    def policy_fn(observation):
        return A
    return policy_fn
```

(4) 定义贪婪 (greedy) 策略在改良时使用。

```python
def create_greedy_policy(Q):
    def policy_fn(state):
        # 每个行动的概率初始化，均为 0
        A = np.zeros_like(Q[state], dtype=float)
        best_action = np.argmax(Q[state])
        # 最佳行动的概率 = 1
        A[best_action] = 1.0
        return A
    return policy_fn
```

(5) 定义值循环策略，使用重要性加权抽样。

重要性加权抽样 (Weighted Importance Sampling)：以值函数大小作为随机抽样比例的分母。

依重要性加权抽样，值函数公式如下：

$$Q(S_t, A_t) \leftarrow Q(S_t, A_t) + \frac{W}{C(S_t, A_t)} |G - Q(S_t, A_t)|$$

```python
# 定义值循环策略，使用重要性加权抽样
def mc_control_importance_sampling(env, num_episodes, behavior_policy, discount_factor=1.0):
    Q = defaultdict(lambda: np.zeros(env.action_space.n)) # 行动值函数
    # 重要性加权抽样(weighted importance sampling)的累计分母
    C = defaultdict(lambda: np.zeros(env.action_space.n))
    
    # 在策略改良时，采用贪婪策略
    target_policy = create_greedy_policy(Q)
    
    # 试验 N 回合
    for i_episode in range(1, num_episodes + 1):
        # 每 1000 回合显示除错信息
```

```
13      if i_episode % 1000 == 0:
14          print(f"\r {i_episode}/{num_episodes}回合.", end="")
15          sys.stdout.flush() # 清除画面
16
17      # 回合(episode)数据结构为阵列，每一项包含 state, action, reward
18      episode = []
19      state = env.reset()
20      # 开始依策略玩牌，最多 100 步骤，中途分出胜负即结束
21      for t in range(100):
22          # 评估时采用随机策略
23          probs = behavior_policy(state)
24          # 以值函数大小作为随机抽样比例的分母
25          action = np.random.choice(np.arange(len(probs)), p=probs)
26          next_state, reward, done, _ = env.step(action)
27          episode.append((state, action, reward))
28          if done:
29              break
30          state = next_state

32      G = 0.0 # 报酬初始化
33      W = 1.0 # 权重初始化
34      # 找出走过的所有状态
35      for t in range(len(episode))[::-1]:
36          state, action, reward = episode[t]
37          # 累计报酬
38          G = discount_factor * G + reward
39          # 累计权重
40          C[state][action] += W
41          # 更新值函数，公式参见书籍
42          Q[state][action] += (W / C[state][action]) * (G - Q[state][action])
43          # 已更新完毕，即跳出循环
44          if action != np.argmax(target_policy(state)):
45              break
46          # 更新权重
47          W = W * 1./behavior_policy(state)[action]
48
49      return Q, target_policy
```

(6) 执行值循环：评估时采用随机策略。

```
1  random_policy = create_random_policy(env.action_space.n)
2  Q, policy = mc_control_importance_sampling(env, num_episodes=500000,
3                                             behavior_policy=random_policy)
```

(7) 显示执行结果。

```
1  V = defaultdict(float)
2  for state, actions in Q.items():
3      action_value = np.max(actions)
4      V[state] = action_value
5  plotting.plot_value_function(V, title="Optimal Value Function")
```

执行结果：玩家分数在低分时胜率也明显提升。以下彩色图表可参考程序执行结果。

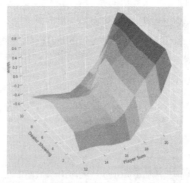

扫码看彩图

这一节我们学会了运用蒙特卡罗算法，还有探索与利用、On/Off Policy，使模型胜率提高了不少，这些概念不只可以应用在蒙特卡罗算法上，也能套用到后续其他的算法，读者可视项目不同的需求来选择。

## 15-9 时序差分

蒙特卡罗算法必须先完成一些回合后，才能计算值函数，接着依据值函数计算出状态转移概率，其算法有以下缺点：

(1) 每个回合必须走到终点，才能够倒推每个状态的值函数。

(2) 假使状态空间很大的话，还是一样要走到终点，才能开始下行动决策，速度实在太慢，例如围棋，根据统计，每下一盘棋平均约需 150 手，而且围棋共有 $3^{19 \times 19} \approx 1.74 \times 10^{172}$ 个状态，就算使用探索也很难测试到每个状态，计算值函数。

于是 Richard S. Sutton 提出时序差分 (Temporal Difference, TD) 算法，通过边走边更新值函数的方式，解决上述问题。值函数更新公式如下：

$$v(s_t) = v(s_t) + \alpha \left[ r_{t+1} + rv(s_{t+1}) - v(s_t) \right]$$

值函数每次加上下一状态值函数与目前状态值函数的差额，以目前的行动产生的结果代替 Bellman 公式的期望值，另外，再乘以 $\alpha$ 学习率 (Learning Rate)。这种走一步更新一次的作法称为 TD(0)，如果是走 $n$ 步更新一次的作法则称为 TD($n$)，进一步引进衰退因子 $\lambda$(Decay) 称为 TD($\lambda$)。

Sutton 在其著作 *Reinforcement Learning: An Introduction*[10] 中举了一个很好的例子，假设要开车回家，蒙特卡罗 (MC) 算法是回家才倒推各个中继点到家的时间，如图 15.27(a) 所示。时序差分 (TD) 算法则是每到一个中继点就修正预估到家的时间，好处就是实时更新，随时掌握目前动态，如图 15.27(b) 所示。

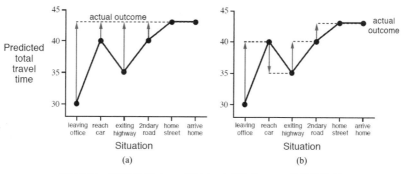

图 15.27　蒙特卡罗 (MC) 算法与时序差分 (TD) 算法的比较

例如，从办公室开车回家。

(1) 预计 30 分钟后到家。

(2) 下班 5 分钟后找到车发现下雨了，估计下雨堵车还要 35 分钟到家。

(3) 下班 20 分钟时下了高速公路发现没有堵车，估计还要 15 分钟到家。

(4) 下班 30 分钟时遇到堵车，估计还要 10 分钟到家。

(5) 下班 40 分钟后到了家附近的小路上，估计还要 3 分钟到家。

(6) 3 分钟后顺利到家。

(7) 通过预估分钟数及实际花费的时间，在每一中继点都可以修正实际到家总分钟数，如图 15.28 所示。

| State | Elapsed Time (minutes) | Predicted Time to Go | Predicted Total Time |
|---|---|---|---|
| leaving office, friday at 6 | 0 | 30 | 30 |
| reach car, raining | 5 | 35 | 40 |
| exiting highway | 20 | 15 | 35 |
| 2ndary road, behind truck | 30 | 10 | 40 |
| entering home street | 40 | 3 | 43 |
| arrive home | 43 | 0 | 43 |

图 15.28　开车回家，各个中继点实际花费的时间 (Elapsed Time)、预计到家的分钟数 (Predicted Time to Go) 及预估到家总分钟数 (Predicted Total Time)。

以倒推图比较动态规划、蒙特卡罗及时序差分算法的作法，如图 15.29 所示。

(1) 动态规划：逐步搜寻所有的下一个可能状态，计算值函数期望值。
(2) 蒙特卡罗：试走多个回合，再以回推的方式计算值函数期望值。
(3) 时序差分：每走一步更新一次值函数。

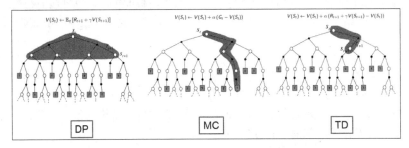

图 15.29　动态规划、蒙特卡罗及时序差分算法的倒推图

时序差分有两个种类的算法：
(1) SARSA 算法：On Policy 的时序差分。
(2) Q-learning 算法：Off Policy 的时序差分。

## 15-9-1　SARSA 算法

先介绍 SARSA 算法，它的名字是行动轨迹中 5 个元素的缩写 $s_t, a_t, r_{t+1}, s_{t+1}, a_{t+1}$，如图 15.30 所示，意味着每走一步更新一次。

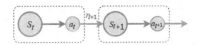

图 15.30　SARSA：$s_t, a_t, r_{t+1}, s_{t+1}, a_{t+1}$

接着我们就来进行算法的实操，再介绍一款新游戏 Windy Grid World，与原来的 Grid World 有些差异，变成 10×7 个格子，第 4、5、6、9 行的风力为 1 级，第 7、8 行的风力为 2 级，分别会把玩家往上吹 1 格和 2 格，而起点 (x) 与终点 (T) 的位置如图 15.31 所示。

图 15.31　起点与终点的位置

**范例19.** 试验Windy Grid World之SARSA策略。此范例程序修改自Denny Britz网站。

**下列程序代码请参考【RL_15_17_SARSA.ipynb】。**

(1) 加载相关套件。

```
1  import gym
2  import itertools
3  import matplotlib
4  import numpy as np
5  import pandas as pd
6  import sys
7  from collections import defaultdict
8  from lib.envs.windy_gridworld import WindyGridworldEnv
9  from lib import plotting
10
11 matplotlib.style.use('ggplot') # 设定绘图的风格
```

(2) 建立 Windy Grid World 环境。

```
1  env = WindyGridworldEnv()
```

(3) 试玩：一律往右走。

```
1   print(env.reset())   # 重置
2   env.render()         # 更新画面
3
4   print(env.step(1))   # 走下一步
5   env.render()         # 更新画面
6
7   print(env.step(1))   # 走下一步
8   env.render()         # 更新画面
9
10  print(env.step(1))   # 走下一步
11  env.render()         # 更新画面
12
13  print(env.step(1))   # 走下一步
14  env.render()         # 更新画面
15
16  print(env.step(1))   # 走下一步
17  env.render()         # 更新画面
18
19  print(env.step(1))   # 走下一步
20  env.render()         # 更新画面
```

执行结果。

30：第 30 个点，表示起始点为第 3 列第 0 行，索引值从 0 开始算。

o o o o o o o o o o

o o o o o o o o o o

o o o o o o o o o o

x o o o o o o T o o

o o o o o o o o o o

o o o o o o o o o o

o o o o o o o o o o

移至第 3 列第 1 行，奖励为 -1。

(31, -1.0, False, {'prob': 1.0})

o o o o o o o o o o

o o o o o o o o o o

o o o o o o o o o o

o x o o o o o o T o o
o o o o o o o o o o
o o o o o o o o o o
o o o o o o o o o o

...

(33, -1.0, False, {'prob': 1.0})

(24, -1.0, False, {'prob': 1.0}) ➔ 往上吹一格

(15, -1.0, False, {'prob': 1.0}) ➔ 往上吹一格

(4) 定义 ε-greedy 策略：与前面相同。

```
def make_epsilon_greedy_policy(Q, epsilon, nA):
    def policy_fn(observation):
        # 每个行动的概率初始化，均为 ε / n
        A = np.ones(nA, dtype=float) * epsilon / nA
        best_action = np.argmax(Q[observation])
        # 最佳行动的概率再加 1 - ε
        A[best_action] += (1.0 - epsilon)
        return A
    return policy_fn
```

(5) 定义 SARSA 策略：走一步算一步，然后采用 ε-greedy 策略，决定行动。

```
def sarsa(env, num_episodes, discount_factor=1.0, alpha=0.5, epsilon=0.1):
    # 行动值函数初始化
    Q = defaultdict(lambda: np.zeros(env.action_space.n))
    # 记录 所有回合的长度及奖励
    stats = plotting.EpisodeStats(
        episode_lengths=np.zeros(num_episodes),
        episode_rewards=np.zeros(num_episodes))

    # 使用 ε-greedy策略
    policy = make_epsilon_greedy_policy(Q, epsilon, env.action_space.n)

    # 试验 N 回合
    for i_episode in range(num_episodes):
        # 每 100 回合显示除错信息
        if (i_episode + 1) % 100 == 0:
            print(f"\r {(i_episode + 1)}/{num_episodes}回合.", end="")
            sys.stdout.flush() # 清除画面

        # 开始依策略试验
        state = env.reset()
        action_probs = policy(state)
        action = np.random.choice(np.arange(len(action_probs)), p=action_probs)

        # 每次走一步就更新状态值
        for t in itertools.count():
            # 走一步
            next_state, reward, done, _ = env.step(action)

            # 选择下一步行动
            next_action_probs = policy(next_state)
            next_action = np.random.choice(np.arange(len(next_action_probs))
                                        , p=next_action_probs)

            # 更新长度及奖励
            stats.episode_rewards[i_episode] += reward
            stats.episode_lengths[i_episode] = t

            # 更新状态值
            td_target = reward + discount_factor * Q[next_state][next_action]
            td_delta = td_target - Q[state][action]
            Q[state][action] += alpha * td_delta

            if done:
                break
```

```
45                action = next_action
46                state = next_state
47
48      return Q, stats
```

(6) 执行 SARSA 策略 200 回合。

```
1  Q, stats = sarsa(env, 200)
```

(7) 显示执行结果。

```
1  fig = plotting.plot_episode_stats(stats)
```

执行结果：共有三张图表。

每一回合走到终点的距离：刚开始的时候要走很多步才会到终点，不过执行到大约第 50 回合后就逐渐收敛了，每回合几乎都相同步数。

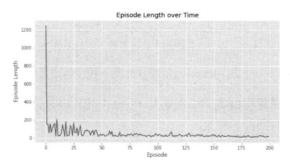

每一回合的报酬：每回合获得的报酬越来越高。

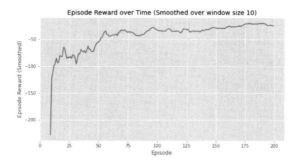

累计的步数与回合对比：呈现曲线上扬的趋势，即每回合到达终点的步数越来越少。

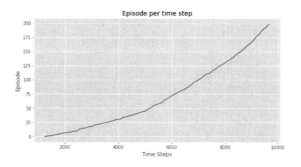

笔者测试另外一款游戏 Cliff Walking，程序为【RL_15_17_SARSA_CliffWalking.ipynb】，只需修改游戏对象(前两格)，其他程序代码均不需要更改，表示算法相关的程序代码具有通用性，可做广泛的应用。

## 15-9-2 Q-learning 算法

接着说明时序差分的第二种算法 Q-learning，它与 SARSA 的差别是采取 Off Policy，评估时使用 ε-greedy 策略，而改良时选择 greedy 策略。使用另一款游戏 Cliff Walking，同样也是迷宫，最下面一排除了起点 (x) 与终点 (T) 之外，其他都是陷阱 (C)，踩到陷阱即 Game Over，如图 15.32 所示。

```
o o o o o o o o o o o o
o o o o o o o o o o o o
o o o o o o o o o o o o
x C C C C C C C C C C T
```

图 15.32　最下面一排除了起点与终点，其他都是陷阱

**范例20.** 试验 Cliff Walking 之 Q-learning 策略。此范例程序修改自 Denny Britz 网站。

下列程序代码请参考【RL_15_18_Q_learning_CliffWalking.ipynb】。

(1) 加载相关套件。

```
1  import gym
2  import itertools
3  import matplotlib
4  import numpy as np
5  import pandas as pd
6  import sys
7  from collections import defaultdict
8  from lib.envs.windy_gridworld import WindyGridworldEnv
9  from lib import plotting
10
11 matplotlib.style.use('ggplot') # 设定绘图的风格
```

(2) 建立 Cliff Walking 环境。

```
1  env = CliffWalkingEnv()
```

(3) 试玩：随便走。

```
1  print(env.reset())      # 重置
2  env.render()            # 更新画面
3
4  print(env.step(0))      # 往上走
5  env.render()            # 更新画面
6
7  print(env.step(1))      # 往右走
8  env.render()
9
10 print(env.step(1))      # 往右走
11 env.render()
12
13 print(env.step(2))      # 往下走
14 env.render()
```

执行结果：

36：第 36 个点，表示起始点为第 3 列第 0 行，但程序却是在第 3 列第 6 行，应该是程序逻辑有问题，但不影响测试，就不除错了。

```
                              36
                              o o o o o o o o o o o o
                              o o o o o o o o o o o o
                              o o o o o o o o o o o o
                              x C C C C C C C C C C T
```

走到最后一步，移至第 3 列第 2 行，走到陷阱，奖励 -100。

```
                              (38, -100.0, True, {'prob': 1.0})
                              o o o o o o o o o o o o
                              o o o o o o o o o o o o
                              o o o o o o o o o o o o
                              o C x C C C C C C C C T
```

(4) 定义 ε-greedy 策略：与前面相同。

```python
def make_epsilon_greedy_policy(Q, epsilon, nA):
    def policy_fn(observation):
        # 每个行动的概率初始化，均为 ε / n
        A = np.ones(nA, dtype=float) * epsilon / nA
        best_action = np.argmax(Q[observation])
        # 最佳行动的概率再加 1 - ε
        A[best_action] += (1.0 - epsilon)
        return A
    return policy_fn
```

(5) 定义 Q-learning 策略：评估时采用 ε-greedy 策略，改良时选择 greedy 策略。

```python
def q_learning(env, num_episodes, discount_factor=1.0, alpha=0.5, epsilon=0.1):
    # 行动值函数初始化
    Q = defaultdict(lambda: np.zeros(env.action_space.n))
    # 记录 所有回合的长度及奖励
    stats = plotting.EpisodeStats(
        episode_lengths=np.zeros(num_episodes),
        episode_rewards=np.zeros(num_episodes))

    # 使用 ε-greedy策略
    policy = make_epsilon_greedy_policy(Q, epsilon, env.action_space.n)

    # 试验 N 回合
    for i_episode in range(num_episodes):
        # 每 100 回合显示除错信息
        if (i_episode + 1) % 100 == 0:
            print(f"\r {(i_episode + 1)}/{num_episodes}回合.", end="")
            sys.stdout.flush() # 清除画面

        # 开始依策略试验
        state = env.reset()
        # 每次走一步就更新状态值
        for t in itertools.count():
            # 使用 ε-greedy策略
            action_probs = policy(state)
            # 选择下一步行动
            action = np.random.choice(np.arange(len(action_probs)), p=action_probs)
            next_state, reward, done, _ = env.step(action)

            # 更新长度及奖励
            stats.episode_rewards[i_episode] += reward
            stats.episode_lengths[i_episode] = t

            # 选择最佳行动
            best_next_action = np.argmax(Q[next_state])
            # 更新状态值
            td_target = reward + discount_factor * Q[next_state][best_next_action]
            td_delta = td_target - Q[state][action]
            Q[state][action] += alpha * td_delta

            if done:
                break

            state = next_state

    return Q, stats
```

(6) 执行 Q-learning 策略 500 回合。

```
1  Q, stats = q_learning(env, 500)
```

(7) 显示执行结果。

```
1  fig = plotting.plot_episode_stats(stats)
```

执行结果：共有三张图表。

每一回合走到终点的距离：与 SARSA 相同，刚开始要走很多步后才会到终点，但执行到大概第 50 回合就逐渐收敛，之后的每个回合几乎都相同步数。由于这两个范例的游戏不同，不能比较 SARSA 与 Q-learning 的效能，若要比较效能，可改用同一款游戏进行比较。

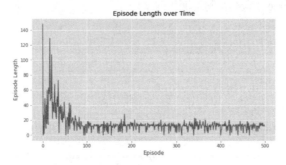

每一回合的报酬：每回合获得的报酬越来越高，不过尚未收敛。

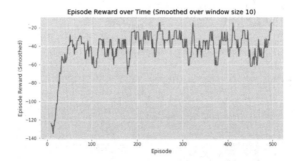

累计的步数与回合对比：呈现直线上扬的趋势，即每回合到达终点的步数差不多，这可能是因为有陷阱的关系，应该分成胜败两模拟，会更清楚。

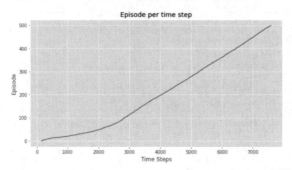

总体而言，SARSA 与 Q-learning 的比较如图 15.33 所示。

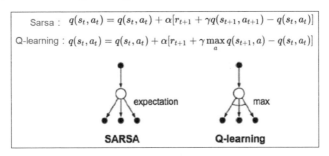

图 15.33　SARSA 与 Q-learning 策略的比较

笔者也测试了 Windy Grid World 游戏，程序为【RL_15_18_Q_learning.ipynb】，读者可以与上述程序比较。

看了这么多的算法，不管是策略循环 (Policy Iteration) 或值循环 (Value Iteration)，总体而言，它们的逻辑与神经网络优化求解，其实有那么一点相似，如图 15.34 所示。

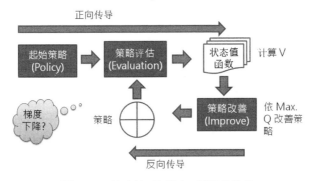

图 15.34　策略循环与梯度下降法的比较

## 15-10　井字游戏

接着我们就开始实战吧，拿最简单的井字游戏 (Tic-Tac-Toe)( 如图 15.35 所示 ) 来练习，包括如何把井字游戏转换为环境、如何定义状态及立即奖励、模型存盘等。

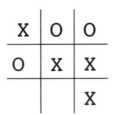

图 15.35　井字游戏

范例21. 试验井字游戏之Q-learning策略。此范例程序修改自"Reinforcement Learning — Implement TicTacToe"[13]，且程序撰写成类似Gym的架构。

**下列程序代码请参考【TicTacToe_1\ticTacToe.py】。**

(1) 加载相关套件。

```
2   import numpy as np
3   import pickle
4   import os
5
6   # 参数设定
7   BOARD_ROWS = 3 # 行数
8   BOARD_COLS = 3 # 列数
```

(2) 定义环境类别：与前面的范例类似，主要就是棋盘重置(reset)、更新状态、给予奖励及判断输赢，胜负未分的时候，计算机加 0.1 分，而玩家加 0.5 分，这样设定是希望计算机能尽速赢得胜利，读者可以试试看其他的给分方式，若胜负已定，则给 1 分。

(3) 环境初始化。

```
11  class Environment:
12      def __init__(self, p1, p2):
13          # 变数初始化
14          self.board = np.zeros((BOARD_ROWS, BOARD_COLS))
15          self.p1 = p1 # 第一个玩家
16          self.p2 = p2 # 第二个玩家
17          self.isEnd = False    # 是否结束
18          self.boardHash = None # 棋盘
19
20          self.playerSymbol = 1 # 第一个玩家使用X
21
22      # 记录棋盘状态
23      def getHash(self):
24          self.boardHash = str(self.board.reshape(BOARD_COLS * BOARD_ROWS))
25          return self.boardHash
```

(4) 判断输赢：取得胜利的情况包括连成一列、一行或对角线。

```
27      # 判断输赢
28      def is_done(self):
29          # 连成一列
30          for i in range(BOARD_ROWS):
31              if sum(self.board[i, :]) == 3:
32                  self.isEnd = True
33                  return 1
34              if sum(self.board[i, :]) == -3:
35                  self.isEnd = True
36                  return -1
37
38          # 连成一行
39          for i in range(BOARD_COLS):
40              if sum(self.board[:, i]) == 3:
41                  self.isEnd = True
42                  return 1
43              if sum(self.board[:, i]) == -3:
44                  self.isEnd = True
45                  return -1
46
47          # 连成对角线
48          diag_sum1 = sum([self.board[i, i] for i in range(BOARD_COLS)])
49          diag_sum2 = sum([self.board[i, BOARD_COLS - i - 1] for i in
50                           range(BOARD_COLS)])
51          diag_sum = max(abs(diag_sum1), abs(diag_sum2))
52          if diag_sum == 3:
53              self.isEnd = True
54              if diag_sum1 == 3 or diag_sum2 == 3:
55                  return 1
56              else:
57                  return -1
58
59          # 无空位置即算平手
60          if len(self.availablePositions()) == 0:
61              self.isEnd = True
62              return 0
63          self.isEnd = False
64          return None
```

(5) 定义显示空位置、更新棋盘、给予奖励等函数。

```python
66      # 显示空位置
67      def availablePositions(self):
68          positions = []
69          for i in range(BOARD_ROWS):
70              for j in range(BOARD_COLS):
71                  if self.board[i, j] == 0:
72                      positions.append((i, j))
73          return positions
74
75      # 更新棋盘
76      def updateState(self, position):
77          self.board[position] = self.playerSymbol
78          # switch to another player
79          self.playerSymbol = -1 if self.playerSymbol == 1 else 1
80
81      # 给予奖励
82      def giveReward(self):
83          result = self.is_done()
84          # backpropagate reward
85          if result == 1:  # 第一玩家赢，P1加一分
86              self.p1.feedReward(1)
87              self.p2.feedReward(0)
88          elif result == -1:  # 第二玩家赢，P2加一分
89              self.p1.feedReward(0)
90              self.p2.feedReward(1)
91          else:  # 胜负未分，第一玩家加 0.1分，第二玩家加 0.5分
92              self.p1.feedReward(0.1)
93              self.p2.feedReward(0.5)
```

(6) 棋盘重置。

```python
96      def reset(self):
97          self.board = np.zeros((BOARD_ROWS, BOARD_COLS))
98          self.boardHash = None
99          self.isEnd = False
100         self.playerSymbol = 1
```

(7) 训练：这是重点，本例训练 50000 回合后，可产生状态值函数表，将计算机和玩家的状态值函数表分别存盘 (policy_p1、policy_p2)，policy_p1 为先下子的策略模型，policy_p2 为后下子的策略模型。

```python
103     def play(self, rounds=100):
104         for i in range(rounds):
105             if i % 1000 == 0:
106                 print(f"Rounds {i}")
107
108             while not self.isEnd:
109                 # Player 1
110                 positions = self.availablePositions()
111                 p1_action = self.p1.chooseAction(positions,
112                                     self.board, self.playerSymbol)
113                 # take action and upate board state
114                 self.updateState(p1_action)
115                 board_hash = self.getHash()
116                 self.p1.addState(board_hash)
117
118                 # 检查是否胜负已分
119                 win = self.is_done()
120
121                 # 胜负已分
122                 if win is not None:
123                     self.giveReward()
124                     self.p1.reset()
125                     self.p2.reset()
126                     self.reset()
127                     break
128                 else:
129                     # Player 2
130                     positions = self.availablePositions()
131                     p2_action = self.p2.chooseAction(positions,
132                                         self.board, self.playerSymbol)
133                     self.updateState(p2_action)
134                     board_hash = self.getHash()
135                     self.p2.addState(board_hash)
```

(8) 比赛：与训练逻辑类似，差别是玩家要自行输入行动。

```
169            else:
170                # Player 2
171                positions = self.availablePositions()
172                p2_action = self.p2.chooseAction(positions)
173
174                self.updateState(p2_action)
175                self.showBoard()
176                win = self.is_done()
177                if win is not None:
178                    if win == -1 or win == 1:
179                        print(self.p2.name, " 胜!")
180                    else:
181                        print("平手!")
182                    self.reset()
183                    break
```

(9) 显示棋盘目前的状态。

```
186    def showBoard(self):
187        # p1: x  p2: o
188        for i in range(0, BOARD_ROWS):
189            print('-------------')
190            out = '| '
191            for j in range(0, BOARD_COLS):
192                if self.board[i, j] == 1:
193                    token = 'x'
194                if self.board[i, j] == -1:
195                    token = 'o'
196                if self.board[i, j] == 0:
197                    token = ' '
198                out += token + ' | '
199            print(out)
200        print('-------------')
```

(10) 计算机类别：包括计算机按最大值函数行动、比赛结束前存盘、比赛开始前载入文件。先进行初始化。

```
203    class Player:
204        def __init__(self, name, exp_rate=0.3):
205            self.name = name
206            self.states = []  # record all positions taken
207            self.lr = 0.2
208            self.exp_rate = exp_rate
209            self.decay_gamma = 0.9
210            self.states_value = {}  # state -> value
211
212        def getHash(self, board):
213            boardHash = str(board.reshape(BOARD_COLS * BOARD_ROWS))
214            return boardHash
```

(11) 计算机按最大值函数行动。

```
217    def chooseAction(self, positions, current_board, symbol):
218        if np.random.uniform(0, 1) <= self.exp_rate:
219            # take random action
220            idx = np.random.choice(len(positions))
221            action = positions[idx]
222        else:
223            value_max = -999
224            for p in positions:
225                next_board = current_board.copy()
226                next_board[p] = symbol
```

```
227            next_boardHash = self.getHash(next_board)
228            value = 0 if self.states_value.get(next_boardHash) is None \
229                    else self.states_value.get(next_boardHash)
230
231            # 按最大值函数行动
232            if value >= value_max:
233                value_max = value
234                action = p
235    # print("{} takes action {}".format(self.name, action))
236    return action
```

(12) 更新状态值函数。

```
238    # 更新状态值函数
239    def addState(self, state):
240        self.states.append(state)
241
242    # 重置状态值函数
243    def reset(self):
244        self.states = []
245
246    # 比赛结束,倒推状态值函数
247    def feedReward(self, reward):
248        for st in reversed(self.states):
249            if self.states_value.get(st) is None:
250                self.states_value[st] = 0
251            self.states_value[st] += self.lr * (self.decay_gamma * reward
252                                                - self.states_value[st])
253            reward = self.states_value[st]
```

(13) 存盘、载入文件。

```
255    # 存盘
256    def savePolicy(self):
257        fw = open(f'policy_{self.name}', 'wb')
258        pickle.dump(self.states_value, fw)
259        fw.close()
260
261    # 载入文件
262    def loadPolicy(self, file):
263        fr = open(file, 'rb')
264        self.states_value = pickle.load(fr)
265        fr.close()
```

(14) 玩家类别：自行输入行动。

```
267    # 玩家类别
268    class HumanPlayer:
269        def __init__(self, name):
270            self.name = name
271
272        # 行动
273        def chooseAction(self, positions):
274            while True:
275                position = int(input("输入位置(1~9):"))
276                row = position // 3
277                col = (position % 3) - 1
278                if col < 0:
279                    row -= 1
280                    col = 2
281                # print(row, col)
282                action = (row, col)
283                if action in positions:
284                    return action
285
286        # 状态值函数更新
287        def addState(self, state):
288            pass
289
290        # 比赛结束,倒推状态值函数
291        def feedReward(self, reward):
292            pass
293
294        def reset(self):
295            pass
```

(15) 画图说明输入规则：说明如何输入位置。

```python
297  # 画图说明输入规则
298  def first_draw():
299      rv = '\n'
300      no=0
301      for y in range(3):
302          for x in range(3):
303              idx = y * 3 + x
304              no+=1
305              rv += str(no)
306              if x < 2:
307                  rv += '|'
308          rv += '\n'
309          if y < 2:
310              rv += '-----\n'
311      return rv
```

执行结果：输入位置的号码如下。

```
1|2|3
-----
4|5|6
-----
7|8|9
```

(16) 主程序：

若已训练过，就不会再训练了，要重新训练的话可将 policy_p1、policy_p2 文件删除。

提供执行参数 2，让玩家可先下子，指令如下，否则一律由计算机先下子。

python ticTacToe.py 2

```python
313  if __name__ == "__main__":   # 主程式
314      import sys
315      if len(sys.argv) > 1:
316          start_player = int(sys.argv[1])
317      else:
318          start_player = 1
319  
320      # 产生对象
321      p1 = Player("p1")
322      p2 = Player("p2")
323      env = Environment(p1, p2)
324  
325      # 训练
326      if not os.path.exists(f'policy_p{start_player}'):
327          print("开始训练...")
328          env.play(50000)
329          p1.savePolicy()
330          p2.savePolicy()
331  
332      print(first_draw())   # 棋盘说明
333  
334      # 载入训练成果
335      p1 = Player("computer", exp_rate=0)
336      p1.loadPolicy(f'policy_p{start_player}')
337      p2 = HumanPlayer("human")
338      env = Environment(p1, p2)
339  
340      # 开始比赛
341      env.play2(start_player)
```

执行结果：笔者试了几回合，要赢计算机比较难。

```
-------------
| o | o | x |
-------------
| x | x | o |
-------------
| o | x | x |
-------------
平手!
```

## 15-11 连续型状态变量与 Deep Q-Learning 算法

通过 15-4 节的试验，我们了解到《木棒台车》的状态是连续型变量，状态个数无限多个，无法以数组存储所有状态的值函数，接下来介绍的 Deep Q-Learning 算法，它结合神经网络与强化学习，就可解决这个问题，所以我们继续以《木棒台车》试验。

之前介绍的大部分算法，包括 Q-Learning，都采用如表 15.2 记录 Q 值或 V 值，并在训练中不断更新表格，并以表格作为行动决策的依据，假设采取贪婪策略，例如在 S1 时就比较 Q11/Q12/Q13/Q14，以最大者作为下一步的行动。这种方式在离散型的状态且个数不多时非常好用，例如井字游戏或迷宫，但如果是木棒台车，不管是台车的位置或木棒倾斜的角度都是连续型变量，理论上状态个数有无限多个，就无法使用表格了，另外，围棋是 19×19 格的棋盘，状态个数等于 $3^{19×19}$=1.74×$10^{172}$，建构这么大的表格，要更新或搜寻都需要很长的运行时间，并不适合，因此，Deep Q-Learning 就以卷积神经网络取代表格。

表 15.2　记录 Q 值或 V 值

|    | A1(上) | A2(下) | A3(左) | A4(右) |
|----|--------|--------|--------|--------|
| S1 | Q11    | Q12    | Q13    | Q14    |
| S2 | Q21    | Q22    | Q23    | Q24    |
| S3 | Q31    | Q32    | Q33    | Q34    |
| S4 | Q41    | Q42    | Q43    | Q44    |
| S5 | Q51    | Q52    | Q53    | Q54    |

另外，Deep Q-Learning(简称 DQN) 也引进许多的想法，例如：
(1) Experience Replay Memory、Prioritised Replay。
(2) Huber 损失函数 (loss)。

以卷积神经网络取代表格，也就是建构一个模型，输入状态 S，预测所有行动的 Q 值，因此也称为 Q 网络，训练数据可依之前的算法，就是先采取 ε-greedy 或完全随机策略，累积一些训练数据，但是全盘接收会产生问题，因为状态 S 会有高度时序关联性，神经网络是假设数据是互相独立的，另外初期的行动较不准确，因此，使用一个队列 (Queue)，保持定量的数据，训练时再从队列中进行随机抽样，避免输入关联性，这种作法称为 Experience Replay Memory。

神经网络的目标定为时序差分，公式如下：
$$\delta = Q(s,a) - (r + \gamma \max_a Q(s',a))$$

为避免训练初期 Q 值不稳定，产生离群值，损失函数采用 Huber Loss，公式如下：
$$\mathcal{L} = \frac{1}{|B|} \sum_{(s,a,s',r) \in B} \mathcal{L}(\delta)$$

$$\text{where} \quad \mathcal{L}(\delta) = \begin{cases} \frac{1}{2}\delta^2 & \text{for } |\delta| \leq 1, \\ |\delta| - \frac{1}{2} & \text{otherwise.} \end{cases}$$

在时序差分 (δ) 很小时采用一般的误差平方和，但 δ 大于 1 时，误差平方和会放大，故采用绝对值。B 为批量，类似均方误差 (MSE) 的 n。

**范例22. 实操Deep Q-Learning算法。**

下列程序代码请参考【RL_15_19_DQN.ipynb】。

(1) 加载相关套件。

```
1   import gym
2   import torch
3   import torch.nn as nn
4   import torch.optim as optim
5   import torch.nn.functional as F
6   import torchvision.transforms as T
7   import math
8   import random
9   import numpy as np
10  import matplotlib
11  import matplotlib.pyplot as plt
12  from collections import namedtuple, deque
13  from itertools import count
14  from PIL import Image
15  from IPython import display
```

(2) 设定相关环境。

```
1   # 图表直接嵌入到 Notebook 之中
2   %matplotlib inline
3
4   # 采 non-block 模式,即 plt.show 不会暂停
5   plt.ion()
6
7   # 判断是否使用 gpu
8   device = torch.device("cuda" if torch.cuda.is_available() else "cpu")
```

(3) 载入《木棒台车》游戏。

```
1   env = gym.make('CartPole-v0').unwrapped
```

(4) 定义 Experience Replay Memory 机制:使用 Python 内建类别 deque 维护队列,capacity 为队列最大容量 ( 笔数 ),Transition 定义每一笔数据的域名。

```
1   # 定义训练数据栏位
2   Transition = namedtuple('Transition',
3                           ('state', 'action', 'next_state', 'reward'))
4
5   # 定义 Experience Replay Memory 机制
6   class ReplayMemory(object):
7
8       def __init__(self, capacity):
9           self.memory = deque([],maxlen=capacity)
10
11      def push(self, *args):
12          """Save a transition"""
13          self.memory.append(Transition(*args))
14
15      def sample(self, batch_size):
16          return random.sample(self.memory, batch_size)
17
18      def __len__(self):
19          return len(self.memory)
```

(5) 定义卷积神经网络模型:卷积层 + 批量正规化层 +ReLU 层。

```
1   class DQN(nn.Module):
2       def __init__(self, h, w, outputs):
3           super(DQN, self).__init__()
4           self.conv1 = nn.Conv2d(3, 16, kernel_size=5, stride=2)
5           self.bn1 = nn.BatchNorm2d(16)
6           self.conv2 = nn.Conv2d(16, 32, kernel_size=5, stride=2)
7           self.bn2 = nn.BatchNorm2d(32)
8           self.conv3 = nn.Conv2d(32, 32, kernel_size=5, stride=2)
```

```
 9            self.bn3 = nn.BatchNorm2d(32)
10
11            # 计算 Linear 神经层输入个数
12            def conv2d_size_out(size, kernel_size = 5, stride = 2):
13                return (size - (kernel_size - 1) - 1) // stride  + 1
14            convw = conv2d_size_out(conv2d_size_out(conv2d_size_out(w)))
15            convh = conv2d_size_out(conv2d_size_out(conv2d_size_out(h)))
16            linear_input_size = convw * convh * 32
17            self.head = nn.Linear(linear_input_size, outputs)
18
19        def forward(self, x):
20            x = x.to(device)
21            x = F.relu(self.bn1(self.conv1(x)))
22            x = F.relu(self.bn2(self.conv2(x)))
23            x = F.relu(self.bn3(self.conv3(x)))
24            return self.head(x.view(x.size(0), -1))
```

(6) 定义相关函数：包括影像转换为张量、取得台车在屏幕的位置、取得屏幕所有像素，并转换为张量，当作模型的输入，也可以直接使用"木棒台车"的四个状态值，不过，使用屏幕像素较具通用性，可适用于其他游戏。

```
 1  # 影像转换为张量
 2  resize = T.Compose([T.ToPILImage(),
 3                      T.Resize(40, interpolation=Image.CUBIC),
 4                      T.ToTensor()])
 5
 6  # 取得台车在荧幕的位置
 7  def get_cart_location(screen_width):
 8      world_width = env.x_threshold * 2
 9      scale = screen_width / world_width
10      return int(env.state[0] * scale + screen_width / 2.0)
11
12  # 取得荧幕所有像素，并转换为张量
13  def get_screen():
14      # 将荧幕像素格式转为三维张量：颜色、高度、宽度(CHW).
15      screen = env.render(mode='rgb_array').transpose((2, 0, 1))
16
17      # 台车影像只占荧幕下半部，故只撷取下半部像素
18      _, screen_height, screen_width = screen.shape
19      screen = screen[:, int(screen_height*0.4):int(screen_height * 0.8)]
20      view_width = int(screen_width * 0.6)
21      cart_location = get_cart_location(screen_width)
22      if cart_location < view_width // 2:
23          slice_range = slice(view_width)
24      elif cart_location > (screen_width - view_width // 2):
25          slice_range = slice(-view_width, None)
26      else:
27          slice_range = slice(cart_location - view_width // 2,
28                              cart_location + view_width // 2)
29      screen = screen[:, :, slice_range]
30
31      # 部分取值(slicing)后，阵列会变成不连续的储存，使用下列指令，改为连续
32      screen = np.ascontiguousarray(screen, dtype=np.float32) / 255
33      screen = torch.from_numpy(screen)
34      # 加一维度 (BCHW)
35      return resize(screen).unsqueeze(0)
36
37  # 重置游戏
38  env.reset()
```

(7) 超参数设定。

采用 ε-greedy 策略：初期采用随机的比例较高，之后随着模型准确率提高逐渐降低随机比例。

使用两个模型：一个作为训练 (Training Network)，选择行动，一个作为预测 (Target Network)，更新 Q 值，所以也称为 Double Q-learning，这是因为训练都是采取随机抽样，会不太稳定，因此，预测 Q 值的网络每隔一段时间才更新权值，以减少变异性 (Variance)，本范例采用每 100 回合自训练网络复制权值至 Target 网络。

```
1   BATCH_SIZE = 128         # 批量
2   GAMMA = 0.999            # 折扣率
3   EPS_START = 0.9          # ε greedy策略随机的初始比例
4   EPS_END = 0.05           # ε greedy策略随机的最小比例
5   EPS_DECAY = 200          # 随机的衰退比例
6   TARGET_UPDATE = 10       # 更新Q值的频率
7
8   # 取得荧幕宽高
9   init_screen = get_screen()
10  _, _, screen_height, screen_width = init_screen.shape
11
12  # 取得行动类别的个数
13  n_actions = env.action_space.n
14
15  # 定义2个网路
16  policy_net = DQN(screen_height, screen_width, n_actions).to(device)
17  target_net = DQN(screen_height, screen_width, n_actions).to(device)
18  target_net.load_state_dict(policy_net.state_dict())
19  target_net.eval()
20
21  # 定义优化器
22  optimizer = optim.RMSprop(policy_net.parameters())
23
24  # 定义伫列及容量
25  memory = ReplayMemory(10000)
26
27  # 初始化变数
28  steps_done = 0           # 完成的步骤个数
29  episode_durations = []   # 每回合的游戏时间
```

```
31  # 行动选择
32  def select_action(state):
33      global steps_done
34      sample = random.random()
35      eps_threshold = EPS_END + (EPS_START - EPS_END) * \
36          math.exp(-1. * steps_done / EPS_DECAY)
37      steps_done += 1
38      if sample > eps_threshold:
39          with torch.no_grad():
40              # t.max(1):取得每一列最大值
41              # [1]:取得索引值
42              return policy_net(state).max(1)[1].view(1, 1)
43      else:
44          return torch.tensor([[random.randrange(n_actions)]],
45                              device=device, dtype=torch.long)
46
47  # 绘制游戏时间的线图
48  def plot_durations():
49      plt.figure(2)
50      plt.clf()
51      durations_t = torch.tensor(episode_durations, dtype=torch.float)
52      plt.title('Training...')
53      plt.xlabel('Episode')
54      plt.ylabel('Duration')
55      plt.plot(durations_t.numpy())
56      # 每 100 回合取平均值绘图
57      if len(durations_t) >= 100:
58          means = durations_t.unfold(0, 100, 1).mean(1).view(-1)
59          means = torch.cat((torch.zeros(99), means))
60          plt.plot(means.numpy())
61
62      plt.pause(0.001)    # 暂停,让画面更新
63      display.clear_output(wait=True)   # 清画面
64      display.display(plt.gcf())        # 得到当前的figure并显示
```

(8) 定义神经网络训练函数。

```
1   def optimize_model():
2       if len(memory) < BATCH_SIZE:  # 累积够数据才训练
3           return
4       transitions = memory.sample(BATCH_SIZE)  # 随机抽样
5       batch = Transition(*zip(*transitions))   # 转换为输入格式
6
7       # 生成 next_states 栏位
8       non_final_mask = torch.tensor(tuple(map(lambda s: s is not None,
9                                     batch.next_state)), device=device, dtype=torch.bool)
10      non_final_next_states = torch.cat([s for s in batch.next_state
11                                         if s is not None])
12      state_batch = torch.cat(batch.state)
13      action_batch = torch.cat(batch.action)
14      reward_batch = torch.cat(batch.reward)
15
16      # 计算 Q(s_t, a),选择行动
```

```
17      state_action_values = policy_net(state_batch).gather(1, action_batch)
18
19      # 更新 Q 值
20      next_state_values = torch.zeros(BATCH_SIZE, device=device)
21      next_state_values[non_final_mask] = target_net(non_final_next_states
22                                                    ).max(1)[0].detach()
23      expected_state_action_values = (next_state_values * GAMMA) + reward_batch
24
25      # 计算 Huber loss
26      criterion = nn.SmoothL1Loss()
27      loss = criterion(state_action_values, expected_state_action_values.unsqueeze(1))
28
29      # 反向传导
30      optimizer.zero_grad()
31      loss.backward()
32      for param in policy_net.parameters():
33          param.grad.data.clamp_(-1, 1)
34      optimizer.step()
```

(9) 模型训练。

```
1   num_episodes = 300    # 训练 300 回合
2   for i_episode in range(num_episodes):
3       # Initialize the environment and state
4       env.reset()
5       last_screen = get_screen()
6       current_screen = get_screen()
7       state = current_screen - last_screen  # 目前画面像素与上一时间点之差
8       for t in count():  # 生成连续的变数值，即 0, 1, 2, ...
9           action = select_action(state)
10          _, reward, done, _ = env.step(action.item())
11          reward = torch.tensor([reward], device=device)
12
13          # 取得状态(荧幕像素)
14          last_screen = current_screen
15          current_screen = get_screen()
16          if not done:
17              next_state = current_screen - last_screen
18          else:
19              next_state = None
20
21          # 存入伫列
22          memory.push(state, action, next_state, reward)
23
24          # Move to the next state
25          state = next_state
26
27          # 训练
28          optimize_model()
29          if done:
30              episode_durations.append(t + 1)  # 记录每一回合的步数
31              #plot_durations()
32              break
33
34      # 复制权值至Target网路
35      if i_episode % TARGET_UPDATE == 0:
36          target_net.load_state_dict(policy_net.state_dict())
```

(10) 绘制训练过程。

```
1   plot_durations()    # 绘制训练过程
2   print('Complete')
3   env.render()        # 渲染画面
4   env.close()         # 结束游戏
5   plt.ioff()          # 恢复 Blocked 绘图模式
6   plt.show();         # 显示图形
```

执行结果：训练效果并不是很好，主要是因为使用屏幕像素作为输入的关系，模型需辨识微小的变化较困难。

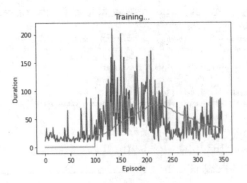

Deep Q-Learning 在 2015 年由 Mnih 等学者首度发表在 *Human-level control through deep reinforcement learning* 中[14]，后续有更多的学者提出改良模型，包括：

(1) Double Q-learning：双模型，上述范例已实操。

(2) Prioritized Experience Replay：使用目前画面像素与上一时间点画面像素之差，上述范例已实操。

(3) Dueling Network：卷积神经网络分别输出状态期望值 (V) 及行动值 (A)，再根据两者更新 Q，在某些游戏，玩家不需要知道每一时刻的 Q 值，以赛车为例，只有在撞车时，才需掌握当时的状态。

有兴趣的读者可继续深入研究各种模型。

## 15-12　Actor Critic 算法

Actor Critic 类似 GAN，主要分为两个神经网络，Actor( 行动者 ) 在评论者 (Critic) 的指导下，优化行动决策，而评论者则负责评估行动决策的好坏，并主导值函数模型的参数更新，详细的说明可参阅 Keras 官网说明[15]，以下的程序也来自该网页。

**范例23. Actor Critic算法。**

下列程序代码请参考【RL_15_20_Actor_Critic.py】。

执行指令：python RL_15_20_Actor_Critic.py。

执行结果：若报酬超过 195 分，即停止，表示模型已非常成熟，部分的执行结果如下，在 763 回合成功达成目标。

```
running reward: 173.41 at episode 680
running reward: 181.18 at episode 690
running reward: 188.73 at episode 700
running reward: 186.98 at episode 710
running reward: 179.31 at episode 720
running reward: 181.53 at episode 730
running reward: 187.06 at episode 740
running reward: 190.73 at episode 750
running reward: 194.45 at episode 760
Solved at episode 763!
```

官网也提供两段录制的动画，分别为训练初期与后期的比较，可以看出后期的木棒台车行驶得相当稳定。

训练初期：https://i.imgur.com/5gCs5kH.gif。

训练后期：https://i.imgur.com/5ziiZUD.gif。

【RL_15_20_Actor_Critic.py】程序系使用 TensorFlow/Keras。

这里再介绍一个套件 Stable Baselines3，它提供许多强化学习的进阶算法，直接指定即可，例如：

```
4  # 载入 A2C 演算法
5  model = A2C('MlpPolicy', env, verbose=0)
6  model.learn(total_timesteps=10000)
```

套件安装指令如下：

pip install stable-baselines3[extra]

pip install piglet

完整程序请参阅【RL_15_22_CartPole_Stable_Baselines3.ipynb】。套件详细说明请参考 Stable Baselines3 官网说明 [16]。

## 15-13　实际应用案例

前面我们都围绕在游戏的实操上，本节就利用 Q-Learning 算法实操仓库捡货系统，说明如何应用强化学习，教会机器人自动捡货。

**范例24.** 假设一间仓库的布置如图15.36所示，目标是教会机器人自动捡货，并以最短路径捡货。假设只有一台机器人，它的位置就是状态。本范例参考自*How to Automatize a Warehouse Robot*[17]。

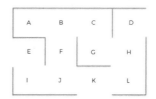

图 15.36　仓库布置图

下列程序代码请参考【RL_15_21_Warehouse.ipynb】。

(1) 载入套件。

```
1  import numpy as np
```

(2) 定义环境 (environment)：机器人的位置、行动空间、行动限制。定义机器人的位置如下。

```
1   # 位置编码
2   location_to_state = {'A': 0,
3                        'B': 1,
4                        'C': 2,
5                        'D': 3,
6                        'E': 4,
7                        'F': 5,
8                        'G': 6,
9                        'H': 7,
10                       'I': 8,
11                       'J': 9,
12                       'K': 10,
13                       'L': 11}
```

(3) 定义行动空间。

```
1  actions = [0,1,2,3,4,5,6,7,8,9,10,11]
```

(4) 定义行动限制，假设 G 点为终点，故奖励设为 1000。依照图 15.36 仓库布置图设定行动限制，列与行均为 A~L，矩阵界定两点间是否可通行，1: 可到达，0: 不可到达。

```
1  # 行动限制，1: 可到达，0: 不可到达
2  R = np.array([[0, 1, 0, 0, 0, 0, 0, 0, 0, 0, 0, 0],
3                [1, 0, 1, 0, 1, 0, 0, 0, 0, 0, 0, 0],
4                [0, 1, 0, 0, 0, 1, 0, 0, 0, 0, 0, 0],
5                [0, 0, 0, 0, 0, 0, 1, 0, 0, 0, 0, 0],
6                [0, 0, 0, 0, 0, 0, 0, 1, 0, 0, 0, 0],
7                [0, 1, 0, 0, 0, 0, 0, 0, 1, 0, 0, 0],
8                [0, 0, 1, 0, 0, 0, 1000, 1, 0, 0, 0, 0],
9                [0, 0, 0, 1, 0, 1, 0, 0, 0, 0, 0, 1],
10               [0, 0, 0, 0, 1, 0, 0, 0, 0, 1, 0, 0],
11               [0, 0, 0, 0, 0, 1, 0, 0, 1, 0, 1, 0],
12               [0, 0, 0, 0, 0, 0, 0, 0, 0, 1, 0, 1],
13               [0, 0, 0, 0, 0, 0, 1, 0, 0, 0, 1, 0]])
```

(5) 策略评估：依 TD(1) 算法更新行动值函数。

```
1  # 参数设定
2  gamma = 0.75
3  alpha = 0.9
4
5  # 行动值函数初始值为 0
6  Q = np.array(np.zeros([12,12]))
7
8  # 训练 1000 周期
9  for i in range(1000):
10     # 随机起始点
11     current_state = np.random.randint(0,12)
12     playable_actions = []
13     for j in range(12):
14         if R[current_state, j] > 0:
15             playable_actions.append(j)
16     # 任意行动
17     next_state = np.random.choice(playable_actions)
18     # 更新行动值函数
19     TD = R[current_state, next_state] + gamma*Q[next_state, \
20          np.argmax(Q[next_state,])] - Q[current_state, next_state]
21     Q[current_state, next_state] = Q[current_state, next_state] + alpha*TD
```

(6) 显示更新结果：越靠近 G 点，值函数越高。

```
1  import pandas as pd
2
3  q_values = pd.DataFrame(Q, columns=[location for location in location_to_state])
4  s = q_values.round().style.background_gradient(cmap='GnBu')
5  s
```

执行结果。

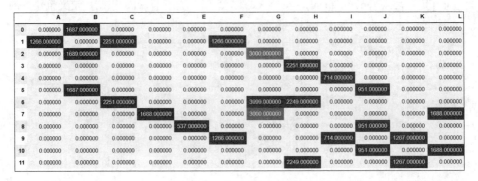

(7) 再加入策略改善：先重新定义行动限制，G 点不为终点，改在训练中指定。

```python
R = np.array([[0, 1, 0, 0, 0, 0, 0, 0, 0, 0, 0, 0],
              [1, 0, 1, 0, 0, 1, 0, 0, 0, 0, 0, 0],
              [0, 1, 0, 0, 0, 0, 1, 0, 0, 0, 0, 0],
              [0, 0, 0, 0, 0, 0, 0, 1, 0, 0, 0, 0],
              [0, 0, 0, 0, 0, 0, 0, 0, 1, 0, 0, 0],
              [0, 1, 0, 0, 0, 0, 0, 0, 0, 1, 0, 0],
              [0, 0, 1, 0, 0, 0, 0, 1, 1, 0, 0, 0],
              [0, 0, 0, 1, 0, 0, 1, 0, 0, 0, 0, 1],
              [0, 0, 0, 0, 1, 0, 0, 0, 0, 0, 1, 0],
              [0, 0, 0, 0, 0, 1, 0, 0, 0, 1, 0, 0],
              [0, 0, 0, 0, 0, 0, 0, 0, 1, 0, 0, 1],
              [0, 0, 0, 0, 0, 0, 0, 1, 0, 0, 1, 0]])
```

(8) 定义代码与位置对照表：训练函数中会用到。

```python
state_to_location = {state: location for location,
                     state in location_to_state.items()}
state_to_location
```

执行结果。

```
{0: 'A',
 1: 'B',
 2: 'C',
 3: 'D',
 4: 'E',
 5: 'F',
 6: 'G',
 7: 'H',
 8: 'I',
 9: 'J',
 10: 'K',
 11: 'L'}
```

(9) 定义路由训练函数。

```python
def route(starting_location, ending_location):
    # starting_location, ending_location：起点、终点
    # 位置转换为代码
    ending_state = location_to_state[ending_location]
    # 终点有最高优先度
    R_new = np.copy(R)
    R_new[ending_state, ending_state] = 1000

    # 策略评估：训练 1000 周期
    Q = np.array(np.zeros([12,12]))
    for i in range(1000):
        current_state = np.random.randint(0,12)
        playable_actions = []
        for j in range(12):
            if R_new[current_state, j] > 0:
                playable_actions.append(j)
        # 任意行动
        next_state = np.random.choice(playable_actions)
        # 更新行动值函数
        TD = R_new[current_state, next_state] + gamma * \
             Q[next_state, np.argmax(Q[next_state,])] - Q[current_state, next_state]
        Q[current_state, next_state] = Q[current_state, next_state] + alpha * TD

    # 策略改善：依TD找寻最佳路由
    route = [starting_location]
    next_location = starting_location
    while (next_location != ending_location):
        starting_state = location_to_state[starting_location]
        next_state = np.argmax(Q[starting_state,])
        next_location = state_to_location[next_state]
        route.append(next_location)
        starting_location = next_location
    return route
```

(10) 测试 E➔G 最佳路由。

```
1  route('E', 'G')
```

执行结果：['E', 'I', 'J', 'F', 'B', 'C', 'G']，如下图。

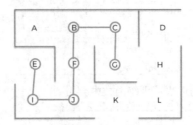

(11) 若需经过一个中继点，可以定义一个函数等于"起点 ➔ 中继点" + "中继点 ➔ 终点"。

```
1  # 3 个点的路由
2  def best_route(starting_location, intermediary_location, \
3                 ending_location):
4      # 3 个点的路由 = 2 个点的路由 + 2 个点的路由
5      return route(starting_location, intermediary_location) + \
6             route(intermediary_location, ending_location)[1:]
```

(12) 测试 E ➔ K ➔ G 最佳路由。

```
1  best_route('E', 'K', 'G')
```

执行结果：['E', 'I', 'J', 'K', 'L', 'H', 'G']。它可能不是全局的最佳路由 (Global Optimization)，但至少是区域的最佳解 (Local Optimization)。

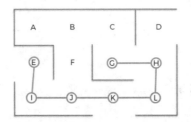

## 15-14 其他算法

不管是使用状态值函数还是行动值函数，前述的算法都是建立一个数组来记录所有的对应值，之后就从数组中选择最佳的行动，所以这类算法被称为表格型 (Tabular) 强化学习，做法简单直接，但只适合离散型的状态，像"木棒台车"这种的连续型变量或状态空间过大，就不适用，变通的方法有以下两种：

(1) 将连续型变量进行分组，转换成离散型变量。

(2) 使用概率分配或神经网络模型取代表格，以策略评估的训练数据，来估计模型的参数 ( 权重 )，选择行动时，就依据模型推断出最佳预测值，而 Deep Q-learning(DQN) 即是利用神经网络的 Q-learning 算法。

另外一个研究课题则是多人游戏的情境，不同于之前介绍的游戏，玩家都只有一位，现代的游戏设计重视人际之间的交流，可以有多位玩家同时参与一款游戏，这时就会产生协同合作或互相对抗的情境，玩家除了考虑奖励与状态外，也需观察其他玩家的状态，这种算法称为多玩家强化学习 (Multi-agent Reinforcement Learning)，比如许多扑克牌游戏都属于多玩家游戏。

近年来，强化学习研究的环境越来越复杂，各种算法相继推陈出新，可参阅维基百科关于强化学习的介绍[19]，见表 15.3。

**表 15.3　强化学习算法的比较**

| Algorithm | Description | Policy | Action Space | State Space | Operator |
| --- | --- | --- | --- | --- | --- |
| Monte Carlo | Every visit to Monte Carlo | Either | Discrete | Discrete | Sample-means |
| Q-learning | State–action–reward–state | Off-policy | Discrete | Discrete | Q-value |
| SARSA | State–action–reward–state–action | On-policy | Discrete | Discrete | Q-value |
| Q-learning - Lambda | State–action–reward–state with eligibility traces | Off-policy | Discrete | Discrete | Q-value |
| SARSA - Lambda | State–action–reward–state–action with eligibility traces | On-policy | Discrete | Discrete | Q-value |
| DQN | Deep Q Network | Off-policy | Discrete | Continuous | Q-value |
| DDPG | Deep Deterministic Policy Gradient | Off-policy | Continuous | Continuous | Q-value |
| A3C | Asynchronous Advantage Actor-Critic Algorithm | On-policy | Continuous | Continuous | Advantage |
| NAF | Q-Learning with Normalized Advantage Functions | Off-policy | Continuous | Continuous | Advantage |
| TRPO | Trust Region Policy Optimization | On-policy | Continuous | Continuous | Advantage |
| PPO | Proximal Policy Optimization | On-policy | Continuous | Continuous | Advantage |
| TD3 | Twin Delayed Deep Deterministic Policy Gradient | Off-policy | Continuous | Continuous | Q-value |
| SAC | Soft Actor-Critic | Off-policy | Continuous | Continuous | Advantage |

本书关于算法的介绍就此告一段落，想了解更多内容的读者，可详阅 *Reinforcement Learning: An Introduction*[10] 一书。

# 15-15　总结

上个范例的应用范围非常广泛，也可以使用在无人搬运车 (Automated Guided Vehicle, AGV) 或自动驾驶，它们都是利用摄影机侦测前方的障碍，但更核心的部分是利用强化学习，来采取最佳行动决策。以无人搬运车为例，我们只要模仿上个范例的做法，将办公室/工厂/医院平面图制成类似迷宫的路径，进行仿真训练后，将模型植入到机器人，就可以驱动机器人从 A 点送货至 B 点。

不仅如此，在股票投资、脑部手术等其他领域，也都看得到强化学习的身影。虽然强化学习理论较为艰深，且需要较扎实的程序基础，但是，它的好处是不用收集大量的训练数据，也不需要标记数据 (Labeling)。

# 第 6 篇 | 图神经网络

之前处理影像的神经网络，例如简单的神经网络、CNN、对象侦测、语义分割、脸部辨识等，都是以像素为输入特征，图神经网络 (Graph Neural Network, GNN) 则是以图形理论 (Graph Theory) 为基础，以向量作为输入，内含节点 (Node) 及边 (Edge)。

图神经网络可以应用到许多方面。

(1) 物理 (Physical System)，例如 3D 对象 (Particles) 的坐标与关联。

(2) 化学 (Chemistry)：例如分子结构 (Molecules)。

(3) 地理 (Geography)：地图、交通、道路等，可表达 A 点到 B 点的距离、时间、方向等。

(4) 医学：药物间的交互作用 (Drug-drug interaction)。

(5) 影像：向量图、影像分类等。

(6) 文字：文章引用、语义分析、知识图谱 (Knowledge Graph)，例如各种概率分配的关联等。

(7) 商品推荐：相似或互补商品的关联、组合包 (Sales kits) 等。

(8) 商业应用：产品制造列表 (BOM)、集团交叉持股等。

(9) 社群软件：社交圈 (Social Circle) 侦测、追踪 (Follow)/ 被追踪关系、假新闻的辨识与追踪。

图形理论应用的层面非常广泛，因此，近几年学者也试图结合神经网络，进行各种更进阶、复杂的任务。

# 第 16 章
# 图神经网络原理及应用

本章内容包括:
(1) 图形理论 (Graph Theory):介绍图形的基本理论。
(2) NetworkX 及 PyTorch Geometric(PyG) 套件功能介绍。
(3) 图神经网络 (Graph Neural Network, GNN):结合图形理论及神经网络。
(4) 范例及各种应用。

## 16-1 图形理论

一个图形包含多个节点 (Node) 及连接节点的边 (Edge),节点也称为顶点 (Vertex),边也称为连接 (Link),以数学公式表达:G=(V, E),如图 16.1 所示。

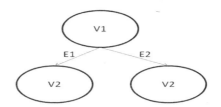

图 16.1 图形理论

依边的性质,图形又分为有向图 (Directed Graph) 及无向图 (Undirected Graph)。如果要表达人际关系,如图 16.2 所示,John、Mary、Helen、Tom 是朋友,只需以无向图表示。

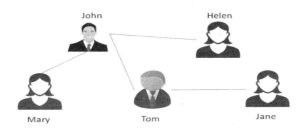

图 16.2 无向图

图形可以使用相邻矩阵 (Adjacency Matrix) 表示,例如图 16.2 可使用矩阵表示。

$$\begin{bmatrix} 0 & 1 & 1 & 1 & 0 \\ 1 & 0 & 0 & 0 & 0 \\ 1 & 0 & 0 & 0 & 0 \\ 1 & 0 & 0 & 0 & 0 \\ 0 & 0 & 1 & 0 & 0 \end{bmatrix}$$

列/行均为 John、Mary、Tom、Helen、Jane，1 代表朋友，0 则代表不是朋友。

又例如美国 66 号公路各地距离也可以矩阵表示。以洛杉矶 (Los Angeles) 为起点至各地距离分别为 198、303、736、871、1175、1475、1544、1913、2448 英里，以矩阵表示：

$$\begin{bmatrix} 0 & 198 & 303 & 736 & 871 & 1175 & 1475 & 1544 & 1913 & 2448 \\ 198 & 0 & 105 & 538 & 673 & 977 & 1277 & 1346 & 1715 & 2250 \\ 303 & 105 & 0 & 433 & 568 & 872 & 1172 & 1241 & 1610 & 2145 \\ 736 & 538 & 433 & 0 & 135 & 439 & 739 & 808 & 1177 & 1712 \\ 871 & 673 & 568 & 135 & 0 & 304 & 604 & 673 & 1042 & 1577 \\ 1175 & 977 & 872 & 439 & 304 & 0 & 300 & 369 & 738 & 1273 \\ 1475 & 1277 & 1172 & 739 & 604 & 300 & 0 & 69 & 438 & 973 \\ 1544 & 1346 & 1241 & 808 & 673 & 369 & 69 & 0 & 369 & 904 \\ 1913 & 1715 & 1610 & 1177 & 1042 & 738 & 438 & 369 & 0 & 535 \\ 2448 & 2250 & 2145 & 1712 & 1577 & 1273 & 973 & 904 & 535 & 0 \end{bmatrix}$$

矩阵中数字代表 A 点到 B 点的距离。

以上各点的距离可以利用 NumPy 套件的传播机制 (Broadcasting) 很容易计算出来，请看下面范例。

**范例1. 66号公路以洛杉矶(Los Angeles)为起点至各地距离分别为198、303、736、871、1175、1475、1544、1913、2448英里，试算各地之间的距离。**

程序：17_01_station_distance.ipynb。

(1) 载入套件。

```
1  import numpy as np
```

(2) 定义起点至各地距离。

```
1  mileposts = np.array([0, 198, 303, 736, 871, 1175, 1475, 1544, 1913, 2448])
2  mileposts
```

执行结果。

```
array([   0,  198,  303,  736,  871, 1175, 1475, 1544, 1913, 2448])
```

(3) 将数据转置，并转成二维矩阵。

```
1  mileposts[:, np.newaxis]
```

执行结果。

```
array([[   0],
       [ 198],
       [ 303],
       [ 736],
       [ 871],
       [1175],
       [1475],
       [1544],
       [1913],
       [2448]])
```

(4) 利用广播机制 (Broadcasting)，扩张为 $N \times N$ 矩阵。

```
1  distance_array = np.abs(mileposts - mileposts[:, np.newaxis])
2  distance_array
```

执行结果：下表中的每一个数字代表 A 点到 B 点的距离。

```
array([[   0,  198,  303,  736,  871, 1175, 1475, 1544, 1913, 2448],
       [ 198,    0,  105,  538,  673,  977, 1277, 1346, 1715, 2250],
       [ 303,  105,    0,  433,  568,  872, 1172, 1241, 1610, 2145],
       [ 736,  538,  433,    0,  135,  439,  739,  808, 1177, 1712],
       [ 871,  673,  568,  135,    0,  304,  604,  673, 1042, 1577],
       [1175,  977,  872,  439,  304,    0,  300,  369,  738, 1273],
       [1475, 1277, 1172,  739,  604,  300,    0,   69,  438,  973],
       [1544, 1346, 1241,  808,  673,  369,   69,    0,  369,  904],
       [1913, 1715, 1610, 1177, 1042,  738,  438,  369,    0,  535],
       [2448, 2250, 2145, 1712, 1577, 1273,  973,  904,  535,    0]])
```

了解数据结构后，再看看传统图形理论的算法如何应用数据。

(1) 搜寻：深度优先搜寻 (DFS) 和广度优先搜寻 (BFS)。

(2) 最短路径：Dijkstra Algorithm、最近邻 (Nearest Neighbor)。

(3) 最小生成树 (Minimum Spanning Tree)：Prim's Algorithm。

(4) 集群 (Clustering)。

话不多说，直接以范例取代文字说明，NetworkX 套件功能强大，不仅可以显示图形，也支持非常多图形理论的算法[3]，我们就先来研究这个套件的用法。

安装指令如下：

pip install networkx -U

**范例2. 初探NetworkX套件。**

程序：17_02_plot_graph.ipynb。

(1) 载入套件。

```
1  import numpy as np
2  import random
3  import networkx as nx
4  from IPython.display import Image
5  import matplotlib.pyplot as plt
```

(2) 建立图形。

```
1  G = nx.Graph()
```

(3) 加入节点的各种指令。

```
1  # 加一个节点
2  G.add_node(1)
3
4  # 一次加 2 个节点
5  G.add_nodes_from([2, 3])
6
7  # 加 2 个节点，并添加颜色属性
8  G.add_nodes_from([
9      (4, {"color": "red"}),
10     (5, {"color": "green"}),
11 ])
12
13 # 产生 0~9 共 10 个节点
14 H = nx.path_graph(10)
15 # 将 H 图形所有节点，并入 G 图形
16 G.add_nodes_from(H)
17
18 # 将 H 图形当作一个节点，并入 G 图形
19 G.add_node(H)
20
21 # 绘制图形
22 nx.draw(G, with_labels=True)
```

执行结果。

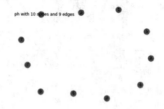

(4) 加入边的各种指令。

```
1  # 加边，连接节点 1 及 2
2  G.add_edge(1, 2)
3
4  # 另一种写法
5  e = (2, 3)
6  G.add_edge(*e)
7
8  # 一次加 2 条边
9  G.add_edges_from([(1, 2), (1, 3)])
10
11 # 绘制图形
12 nx.draw(G, with_labels=True)
```

执行结果：观察 1、2、3 节点互相连接。

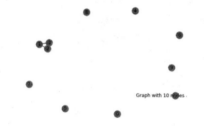

(5) 取得节点及边个数。

```
1  G.number_of_nodes(), G.number_of_edges()
```

执行结果：共 11 个节点、3 个边。

(6) 取得所有节点及边的信息。

```
1  G.nodes(), G.edges()
```

执行结果。

```
(NodeView((1, 2, 3, 4, 5, 0, 6, 7, 8, 9
EdgeView([(1, 2), (1, 3), (2, 3)]))
```

(7) 取得某一节点所有连接的节点。

```
1  # 加边，连接节点 3 及 4、4 及 5
2  G.add_edges_from([(3, 4), (4, 5)], color='red')
3
4  # 指定节点名称，取得连接的节点及属性
5  G[1], G[4]
```

执行结果：节点 1 连接的节点为 2、3。节点 1 连接的节点为 3、5，同时显示属性。

```
(AtlasView({2: {}, 3: {}}),
 AtlasView({3: {'color': 'red'}, 5: {'color': 'red'}}))
```

(8) 移除节点及边：移除边需要指定 2 个连接的节点。

```
1  # 移除节点 2
2  G.remove_node(2)
3
4  # 移除边 1-3
5  G.remove_edge(1, 3)
6
7  # 移除多个节点
8  G.remove_nodes_from([4, 5])
9
10 # 移除多个边
11 G.remove_edges_from([(1, 2), (2, 3)])
```

(9) 建立新图形：同时加节点、边及属性。

```
1  # 建立新图形，同时加节点、边及属性
2  G = nx.Graph([(1, 2, {"color": "yellow"})])
3
4  # 绘制图形
5  nx.draw(G, with_labels=True, cmap = plt.get_cmap('rainbow'))
```

(10) 清除所有节点及边。

```
1  G.clear()
```

(11) 边可以加入任意属性，例如权重，可代表距离、长度或其他衡量值，程序代码以权重属性作为线条的宽度。

```
1  # 建立图形
2  G = nx.Graph()
3
4  G.add_edge("1", "2")
5  G.add_edge("1", "6")
6  G.add_edges_from([("1", "3"),
7                    ("3", "4")])
8  G.add_edges_from([("1", "5", {"weight" : 3}),
9                    ("2", "4", {"weight" : 5})])
10
11 # 权重计算
12 weights = [1 if G[u][v] == {} else G[u][v]['weight'] for u,v in G.edges()]
13 # 以权重作为线条的宽度
14 nx.draw(G, with_labels=True, cmap = plt.get_cmap('rainbow'), width=weights)
```

执行结果：有 2 条线的宽度较粗。

(12) 可以加载 XML 文件，产生图形。

```
1  clothing_graph = nx.read_graphml("./graph/clothing_graph.graphml")
2  nx.draw_planar(clothing_graph,
3      arrowsize=12,
4      with_labels=True,
5      node_size=1000,
6      node_color="#ffff8f",
7      linewidths=2.0,
8      width=1.5,
9      font_size=14,
10 )
```

执行结果：clothing_graph.graphml 文件是标准的 XML 格式，<node> 为节点，

<edge> 为边,这个图是 Bumstead 教授有趣的举例,说明早上起床后的穿戴顺序[4]。

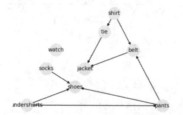

(13) 绘制图形:NetworkX 提供多种绘图布局的算法,详情可参阅 *NetworkX Graph Layout*[5],较重要的布局说明如下:

planar_layout:所有的边尽量不交叉,指令为 nx.draw_planar。

kamada_kawai_layout:以 Kamada-Kawai 路径长度为成本函数,力求路径长度最小化的情况下得到最佳图形,生成需要较长的运行时间,指令为 draw_kamada_kawai。

spring_layout:使用 Fruchterman-Reingold force-directed 算法,将边视作弹簧,尽量使节点靠近,指令为 draw_spring。

circular_layout:使图形成圆形状,指令为 draw_circular。

各种布局可参看以下程序代码执行结果。接下来,介绍各种图形分析、应用及相关的算法。

(1) 加载 NetworkX 套件内建数据 karate_club_graph,它是空手道俱乐部的社群连接,节点代表会员,边代表互相有往来。

```
1  # 载入内建数据
2  G_karate = nx.karate_club_graph()
3  # 指定布局,取得节点座标
4  pos = nx.spring_layout(G_karate)
5  # 绘制图形
6  nx.draw(G_karate, node_color="#ffff8f", with_labels=True, pos=pos)
```

执行结果。

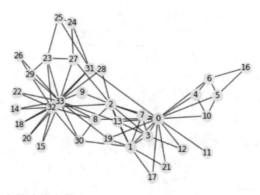

(2) 统计每个节点的连接个数。

```
2  G_karate.degree()
```

执行结果:0 有 16 个连接,1 有 9 个连接。

DegreeView({0: 16, 1: 9, 2: 10, 3: 6, 4: 3, 5: 4, 6: 4, 7: 4, 8: 5, 9: 2, 10: 3, 11: 1, 12: 2, 13: 5, 14: 2, 15: 2, 16: 2, 17: 2, 18: 2, 19: 3, 20: 2, 21: 2, 22: 2, 23: 5, 24: 3, 25: 3, 26: 2, 27: 4, 28: 3, 29: 4, 30: 4, 31: 6, 32: 12, 33: 17})

(3) 绘制连接个数直方图。

```
1  degree_freq = np.array(nx.degree_histogram(G_karate)).astype('float')
2  plt.figure(figsize=(10, 8))
3  plt.stem(degree_freq)    # 绘制垂直线
4  plt.ylabel("Frequence")
5  plt.xlabel("Degree")
6  plt.show()
```

执行结果：X 轴为连接个数，Y 轴为节点个数。

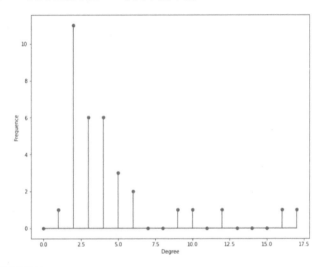

(4) 寻找最短路径。

```
1  # 回传每一条最短路径
2  nx.shortest_path(G_karate)
```

执行结果：回传任两点最短路径。

```
{0: {0: [0],
 1: [0, 1],
 2: [0, 2],
 3: [0, 3],
 4: [0, 4],
 5: [0, 5],
 6: [0, 6],
 7: [0, 7],
 8: [0, 8],
 10: [0, 10],
 11: [0, 11],
 12: [0, 12],
 13: [0, 13],
 17: [0, 17],
 19: [0, 19],
 21: [0, 21],
 31: [0, 31],
 30: [0, 1, 30],
```

(5) 指定起点与终点，可回传最短路径。

```
1  # 指定起点与终点，可回传最短路径
2  nx.shortest_path(G_karate)[0][23]
```

执行结果：0 为起点，23 为终点，最短路径为 [0, 2, 27, 23]。

(6) 一般路径会有距离，可以使用权重属性代表距离。

```
1   # 图的边，weight为距离
2   edges = [(1,2, {'weight':4}),
3            (1,3,{'weight':2}),
4            (2,3,{'weight':1}),
5            (2,4, {'weight':5}),
6            (3,4, {'weight':8}),
7            (3,5, {'weight':10}),
8            (4,5,{'weight':2}),
9            (4,6,{'weight':8}),
10           (5,6,{'weight':5})]
11  # 边的名称
12  edge_labels = {(1,2):4, (1,3):2, (2,3):1, (2,4):5, (3,4):8
13                 , (3,5):10, (4,5):2, (4,6):8, (5,6):5}
14
15  # 生成图
16  G = nx.Graph()
17  for i in range(1,7):
18      G.add_node(i)
19  G.add_edges_from(edges)
20
21  # 绘图
22  pos = nx.planar_layout(G)
23  nx.draw(G, node_color="#ffff8f", with_labels=True, pos=pos)
24
25  # 在边显示权重(weight)
26  labels = nx.get_edge_attributes(G,'weight')
27  nx.draw_networkx_edge_labels(G,pos,edge_labels=labels);
```

执行结果：边的标注为距离。

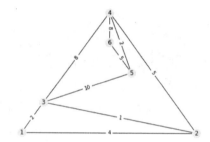

(7) 先计算起点为 1，到达其他节点的最短路径。

```
1   # 起点为 1，到达其他节点的最短路径
2   p1 = nx.shortest_path(G, source=1, weight='weight')
3   p1
```

执行结果：起点 1 到达其他节点的最短路径。例如最后一行，4→6 有两条路径，因为 4→5→6 = 2+5 = 7，而 4→6 = 8，故会选择前者。

```
{1: [1],
 2: [1, 3, 2],
 3: [1, 3],
 4: [1, 3, 2, 4],
 5: [1, 3, 2, 4, 5],
 6: [1, 3, 2, 4, 5, 6]}
```

(8) 可同时指定起点及终点。

```
1   # 起点为 1，终点为 6
2   p1to6 = nx.shortest_path(G, source=1, target=6, weight='weight')
3   p1to6
```

执行结果：起点 1 到达终点 6 的最短路径为 1, 3, 2, 4, 5, 6。

(9) 计算最短路径的总长度。

```
1  # 最短路径的总长度
2  length = nx.shortest_path_length(G, source=1, target=6, weight='weight')
3  length
```

执行结果：15。

**多数的算法也都支持使用权重属性代表距离，不再赘述。**

(10) 最小生成树 (Minimum Spanning Tree)：计算要将所有节点生成一棵树需要的连接。

```
1  from networkx.algorithms import tree
2
3  # 最小生成树
4  mst = tree.minimum_spanning_edges(G_karate, algorithm='prim', data=False)
5  edgelist = list(mst)
6  sorted(edgelist)     # 排序
```

执行结果：显示最小生成树的所有连接。

```
[(0, 1),
 (0, 2),
 (0, 3),
 (0, 4),
 (0, 5),
 (0, 6),
 (0, 7),
 (0, 8),
 (0, 10),
 (0, 11),
 (0, 12),
 (0, 13),
 (0, 17),
 (0, 19),
 (0, 21),
 (0, 31),
 (1, 30),
 (2, 9),
 (2, 27)]
```

(11) 极大团 (Maximal Clique)：团内每一个节点需与其他节点都相连，无法再找到任一节点可加入团内，并符合定义。

```
1  from networkx.algorithms import approximation as aprx
2
3  max_clique = aprx.max_clique(G_karate)
4  max_clique
```

执行结果：{0, 1, 2, 3, 7}。

(12) 绘制极大团 (Maximal Clique) 图形。

```
1  max_clique_subgraph = G_karate.subgraph(max_clique)
2  # 绘制图形
3  nx.draw_circular(max_clique_subgraph, node_color="#ffff8f", with_labels=True)
```

执行结果。

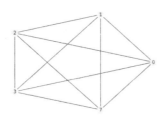

(13) 极大团生成：可利用 complete_graph 自动生成极大团，参数为节点个数。

```
1  G = nx.complete_graph(5) # 5 个节点
2  nx.draw(G, node_color="#ffff8f", with_labels=True)
```

执行结果。

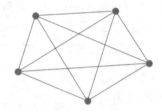

(14) 社群侦测 (Community Detection)：依照 Girvan 与 Newman 定义，社群内的节点是紧密连接，而社群间的连接是稀疏的，类似集群 (Clustering) 的概念。有非常多的算法支持分群，其中衡量社群的紧密程度称为 Modularity，它的范围介于 [-0.5,1) 之间，公式如下 [6][7]：

$$Q = \frac{1}{2m} \times \sum_{ij} \left[ A_{i,j} - \frac{k_i \times k_j}{2m} \right] \times \delta(C_i, C_j)$$

NetworkX 支持以下分群法：

Girvan/Newman。

Louvain。

Asynchronous Fluid。

Kernighan–Lin bipartition。

衡量分群的好坏，除了 Modularity，还有 partition_quality，内含 2 项指标：

coverage：衡量社群的覆盖程度，公式为社群内的边 / 总边数。

performance：公式为 (intra_edges + inter_edges) / ($n \times (n-1)$)，$n$ 为总边数，intra_edges 为社群内的边数，边的两端均需在社群内，inter_edges 为社群间的边。

```
1  from networkx.algorithms import community
2
3  # 内建数据，两个社群，各有 5 个节点,1个相连的节点
4  G = nx.barbell_graph(5, 1)
5  nx.draw_kamada_kawai(G, node_color="#ffff8f", with_labels=True)
```

- 执行结果：内建数据 barbell_graph，固定生成两个社群，参数为社群内各节点数 (5) 及非社群的节点数 (1)，可随意设定。

(15) Girvan/Newman 分群法：回传值是一个 Itereator，分次给值，第一次分为 2 群，

依次递增。

```
1  communities_generator = community.girvan_newman(G)
2  top_level_communities = next(communities_generator)
3  top_level_communities
```

执行结果：{0, 1, 2, 3, 4}、{5, 6, 7, 8, 9, 10} 两群。

(16) 再调用一次，分为 3 群。

```
1  next_level_communities = next(communities_generator)
2  next_level_communities
```

执行结果：{0, 1, 2, 3, 4}、{6, 7, 8, 9, 10}、{5}。

(17) 可使用循环产生不同个社群。

```
1  from networkx.algorithms import community
2  import itertools
3
4  k = 4 # 分成 2 ~ k+1 群
5  # Girvan Newman algorithm
6  comp = community.girvan_newman(G)
7  for communities in itertools.islice(comp, k):
8      print(tuple(sorted(c) for c in communities), ":\t\t",
9            community.modularity(G, communities), ":\t",
10           community.partition_quality(G, communities))
```

执行结果：分群，并显示分群的衡量指标 Modularity、Coverage、Performance，从 Performance 看分成 3 群最佳，其余指标判断分成 2 群最佳。

```
([0, 1, 2, 3, 4], [5, 6, 7, 8, 9, 10]) :         0.45351239669421484 :   (0.9545454545454546, 0.9090909090909091)
([0, 1, 2, 3, 4], [6, 7, 8, 9, 10], [5]) :       0.45144628099173555 :   (0.9090909090909091, 0.9636363636363636)
([0], [1, 2, 3, 4], [6, 7, 8, 9, 10], [5]) :     0.3398760330578512 :    (0.7272727272727273, 0.8909090909090909)
([0], [1], [2, 3, 4], [6, 7, 8, 9, 10], [5]) :   0.25723140495867763 :   (0.5909090909090909, 0.8363636363636363)
```

(18) Louvain 分群法：modularity 公式稍有不同 [8]。

```
1  G = nx.barbell_graph(5, 1)
2  community.louvain_communities(G)
```

执行结果：{0, 1, 2, 3, 4, 5}、{6, 7, 8, 9, 10}。

(19) Asynchronous Fluid 分群法：需要指定群的个数。

```
1  for k in range(2, 6):
2      comp = community.asyn_fluid.asyn_fluidc(G, k)
3      print(tuple(sorted(comp)))
```

执行结果：此分群法采用随机抽样，每次执行可能有不同的结果。

```
({0, 1, 2, 3, 4}, {5, 6, 7, 8, 9, 10})
({6, 7, 8, 9, 10}, {0, 1, 2, 3}, {4, 5})
({8, 9}, {10, 7}, {5, 6}, {0, 1, 2, 3, 4})
({9, 10}, {0, 3}, {1, 2, 4}, {8, 6, 7}, {5})
```

NetworkX 是一个很庞大的套件，提供了非常多的功能，足以写一本书介绍，在此仅做简单介绍，因为我们的重点还是放在图神经网络上。

# 16-2 PyTorch Geometric

上述传统的算法可以解决较简单的问题，例如寻找最短路径，如果要探索较复杂的问题，例如要比对指纹、图形分类等多个图形的比较，就要加入新的思维，这就是引进图神经网络的原因。

支持图神经网络的套件至少有三个：
(1) PyTorch Geometric(PyG)：设置在 PyTorch 之上。
(2) Deep Graph Library(DGL)：设置在 PyTorch、TensorFlow 及 MXNet 之上。
(3) Spektral：设置在 TensorFlow 2/Keras 之上。

因本书是以 PyTorch 为主，故我们只介绍 PyG，首先请参阅 PyTorch Geometric(PyG) 官网文件[9]产生安装指令，使用 conda 比较方便，它会侦测目前的环境，决定安装何种版本，指令如下：

conda install pyg -c pyg

**范例 3. PyG 基本功能。**

程序：17_03_PyG.ipynb。

(1) 加载相关套件。

```
1  import torch
2  from torch_geometric.data import Data
3  import networkx as nx
```

(2) 使用 PyG 建立图形：PyG 与 NetworkX 的图形格式不同，建立图形的指令也不同。

(3) 建立图形：先建立边及节点，再组合成图形。边的定义较特别，以二维数组定义，第一列为起点，第二列为终点，无向图须双向设定，即 (1, 0) 及 (0, 1)。

```
1  # 定义边，第一列为起点，第二列为终点，无向图须双向设定
2  edge_index = torch.tensor([[0, 1, 1, 2],
3                             [1, 0, 2, 1]], dtype=torch.long)
4  # 节点名称
5  x = torch.tensor([[-1], [0], [1]], dtype=torch.float)
6
7  # 建立新图形
8  data = Data(x=x, edge_index=edge_index)
9  data # 节点及边均为二维
```

执行结果：节点及边均为二维，有三个节点 (x)，四个边 (edge_index)，每个边含 ( 起点、终点 )。

```
Data(x=[3, 1], edge_index=[2, 4])
```

建立的图形如下。

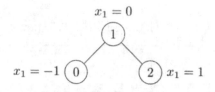

(4) 边的定义有另一种写法较直观，每一元素均为 ( 起点，终点 )，之后再转置 (.t)。

```
1  # 边有另一种写法较直观，每一元素均为(起点，终点)
2  edge_index = torch.tensor([[0, 1],
3                             [1, 0],
4                             [1, 2],
5                             [2, 1]], dtype=torch.long)
6
7  # 节点名称
8  x = torch.tensor([[-1], [0], [1]], dtype=torch.float)
9
10 # 要加 contiguous
11 data = Data(x=x, edge_index=edge_index.t().contiguous())
12 data # 节点及边均为二维
```

(5) 读取图形相关信息。

```
1  print(f'是否为有向图:{data.is_directed()}')
2  print(f'图形键值:{data.keys}')
3  print(f'节点名称:{data["x"]}')
4  print(f'节点个数:{data.num_nodes}')
5  print(f'边名称:{data["edge_index"]}')
6  print(f'边:{data.num_edges}')
7  print(f'节点属性个数:{data.num_node_features}')
8  print(f'未连接的节点个数:{data.has_isolated_nodes()}')
9  print(f'自我连接的节点个数:{data.has_self_loops()}')
10 print(f'节点属性个数:{data.num_node_features}')
```

执行结果。

```
是否为有向图:False
图形键值:['x', 'edge_index']
节点名称:tensor([[-1.],
        [ 0.],
        [ 1.]], device='cuda:0')
节点个数:3
边名称:tensor([[0, 1, 1, 2],
        [1, 0, 2, 1]], device='cuda:0')
边:4
节点属性个数:1
未连接的节点个数:False
自我连接的节点个数:False
节点属性个数:1
```

(6) 可复制图形数据至 GPU 内存。

```
1  device = torch.device('cuda')
2  data = data.to(device)
```

(7) 绘图：PyG 不提供绘图功能，需先转换为 NetworkX 格式，再利用 NetworkX 绘图，to_networkx/from_networkx 可进行 PyG/NetworkX 格式互转。

```
1  from torch_geometric.utils.convert import to_networkx
2  def draw_pyg(Data):
3      G = to_networkx(Data, to_undirected=True)
4      # 绘图
5      nx.draw(G,
6          with_labels=True,
7          node_size=1000,
8          node_color="#ffff8f",
9          width=0.8,
10         font_size=14,
11     )
12 draw_pyg(data)
```

执行结果。

(8) 笔者自己也写一个转换函数。

```
1  def draw_pyg2(data):
2      G = nx.Graph()
3  
4      # nodes
5      node_list = data["x"].cpu().numpy().reshape(data["x"].shape[0])
6      node_list = node_list.astype(int) # 节点名称改为整数
7      G.add_nodes_from(node_list)
8  
9      # edges
10     edges = data["edge_index"].cpu().numpy().T
11     edge_list = []
12     for item in edges:
13         edge_list.append((node_list[item[0]], node_list[item[1]]))
14     G.add_edges_from(edge_list)
15  
16     # 绘图
17     nx.draw(G,
18         with_labels=True,
19         node_size=1000,
20         node_color="#ffff8f",
21         width=0.8,
22         font_size=14,
23     )
24     # plt.savefig('grap.png')
25  
26 draw_pyg2(data)
```

(9) 载入内建数据集：PyG 也提供许多数据集，大部分来自 TUDatasets[10]，共有 120 个数据集，范围涵盖小分子 (Small molecules)、生物信息学 (Bioinformatics)、计算机视觉 (Computer vision) 及社群媒体 (Social networks) 等类别，也可直接至 TUDatasets 网站[11]下载。

```
1  from torch_geometric.datasets import TUDataset
2  
3  # 载入内建数据
4  dataset = TUDataset(root='./graph/ENZYMES', name='ENZYMES')
5  
6  # 数据集内含的图形个数
7  len(dataset)
```

执行结果：600 个图形。

(10) 取得类别 (Y) 个数, 特征 ( 属性 ) 个数。

```
1  dataset.num_classes, dataset.num_node_features
```

执行结果：6, 3。

(11) 读取数据集中的图形，读取第一个图形。

```
1  dataset[0]
```

执行结果：Data(edge_index=[2, 168], x=[37, 3], y=[1])。

(12) 随机抽样：可先洗牌，再抽样。

```
1  # 洗牌
2  dataset = dataset.shuffle()
3  # 读取第一个图形
4  dataset[0]
```

执行结果：Data(edge_index=[2, 218], x=[96, 3], y=[1])，与之前不同。

(13) 数据转换 (Data Transform)：类似 PyTorch 的数据转换，以下程序代码使用最近邻 (KNN) 算法，进行转换。

载入另一个数据集 ShapeNet。

运行时间较久，请耐心等候

```
1  import torch_geometric.transforms as T
2  from torch_geometric.datasets import ShapeNet
3
4  # KNNGraph：使用最近邻(KNN)演算法，每一点取6个最近的节点
5  dataset = ShapeNet(root='./graph/ShapeNet', categories=['Airplane'],
6                     pre_transform=T.KNNGraph(k=6))
7
8  dataset[0]
```

执行结果：Data(x=[2518, 3], y=[2518], pos=[2518, 3], category=[1], edge_index=[2, 15108])。

(14) 数据增强：也可以使用数据增强，产生更多样化的数据，例如 Random Translate 将节点位置随机移动，程序代码中的参数 0.01 表示变动范围在 (-0.01, 0.01)，更多的数据增强可参阅官网 Torch Geometric transforms 文件[12]。

```
1  # 数据增强：RandomTranslate
2  dataset = ShapeNet(root='./graph/ShapeNet', categories=['Airplane'],
3                     pre_transform=T.KNNGraph(k=6),
4                     transform=T.RandomTranslate(0.01))
5
6  dataset[0]
```

## 16-3 图神经网络

由于节点及边都是二维向量，我们会很自然地引进卷积神经网络，针对图形进行分类，这种针对图形的卷积神经网络即泛称图卷积神经网络 (Graph Convolutional Network, GCN)，算法的细节请参阅相关网站[13]。

**范例4.** 以图神经网络进行节点分类，实操论文分类，程序修改自 https://pytorch-geometric.readthedocs.io/en/latest/notes/colabs.html。

程序：17_04_Node_Classification.ipynb。

(1) 加载相关套件。

```
1  import torch
2  from torch_geometric.data import Data
3  import networkx as nx
```

(2) 加载内建数据集 Planetoid，它是一篇论文引用的关联图，节点代表论文，边代表引用关系，详细说明请看 Planetoid 类别说明[14]。

```
1  from torch_geometric.datasets import Planetoid
2
3  # 载入内建数据
4  dataset = Planetoid(root='./graph/Cora', name='Cora')
5
6  # 数据集内含的图形个数
7  len(dataset)
```

执行结果：此数据集只有一个图形。

(3) 数据切割：此数据集已切割训练、验证及测试数据，以屏蔽的方式筛选节点，形成多笔的数据。

```
1  data = dataset[0]
2  data.train_mask.sum().item(), data.val_mask.sum().item(), data.test_mask.sum().item()
```

执行结果：训练、验证及测试数据笔数分别为 140, 500, 1000。

(4) 标注 (Y) 共有 7 种类别，分别代表论文所属领域。

```
1  set(data.y.numpy())
```

执行结果：{0, 1, 2, 3, 4, 5, 6}。

(5) 判断是否使用 GPU

```
1  device = torch.device('cuda' if torch.cuda.is_available() else 'cpu')
```

(6) 定义模型：使用 2 层图卷积神经层 (GCNConv) + ReLU + Dropout + Log Softmax Activation Function。GCNConv 的细节可参阅 *Creating Message Passing Networks*[15]。

```
1  import torch.nn.functional as F
2  from torch_geometric.nn import GCNConv
3
4  class GCN(torch.nn.Module):
5      def __init__(self):
6          super().__init__()
7          self.conv1 = GCNConv(dataset.num_node_features, 16)
8          self.conv2 = GCNConv(16, dataset.num_classes)
9
10     def forward(self, data):
11         x, edge_index = data.x, data.edge_index
12
13         x = self.conv1(x, edge_index)
14         x = F.relu(x)
15         x = F.dropout(x, training=self.training)
16         x = self.conv2(x, edge_index)
17
18         return F.log_softmax(x, dim=1)
```

(7) 模型训练：训练 200 个周期。

```
1  model = GCN().to(device)
2  data = dataset[0].to(device)
3  optimizer = torch.optim.Adam(model.parameters(), lr=0.01, weight_decay=5e-4)
4
5  model.train()
6  for epoch in range(200):
7      optimizer.zero_grad()
8      out = model(data)
9      loss = F.nll_loss(out[data.train_mask], data.y[data.train_mask])
10     loss.backward()
11     optimizer.step()
12     print(f'Epoch: {epoch+1:03d}, Loss: {loss:.4f}')
```

执行结果：随着训练周期，损失越来越小。

```
Epoch: 001, Loss: 1.9499
Epoch: 002, Loss: 1.8468
Epoch: 003, Loss: 1.7267
Epoch: 004, Loss: 1.5937
Epoch: 005, Loss: 1.4304
Epoch: 006, Loss: 1.3332
Epoch: 007, Loss: 1.2139
Epoch: 008, Loss: 1.0807
Epoch: 009, Loss: 0.9255
Epoch: 010, Loss: 0.8363
Epoch: 011, Loss: 0.7731
Epoch: 012, Loss: 0.6450
Epoch: 013, Loss: 0.5537
Epoch: 014, Loss: 0.5057
Epoch: 015, Loss: 0.4355
Epoch: 016, Loss: 0.3986
Epoch: 017, Loss: 0.3341
Epoch: 018, Loss: 0.3536
```

(8) 模型评估。

```
1  model.eval()
2  pred = model(data).argmax(dim=1)
3  correct = (pred[data.test_mask] == data.y[data.test_mask]).sum()
4  acc = int(correct) / int(data.test_mask.sum())
5  print(f'Accuracy: {acc:.4f}')
```

执行结果：准确率为 0.7960。

(9) 显示混淆矩阵 (Confusion matrix)。

```
1  from sklearn.metrics import confusion_matrix
2
3  confusion_matrix(data.y[data.test_mask].cpu().numpy(),
4                  pred[data.test_mask].cpu().numpy())
```

执行结果：索引值 3 的错误分类比例较高。

```
[ 95,   4,   2,   8,   5,   6,  10]
[  3,  79,   3,   3,   1,   1,   1]
[  3,   5, 131,   4,   0,   1,   0]
[ 23,   7,   7, 240,  29,   8,   5]
[ 10,   1,   2,   8, 121,   6,   1]
[  8,   4,   4,   0,   0,  77,  10]
[  5,   0,   0,   1,   0,   5,  53]
```

(10) 绘图：使用 TSNE 算法降维至 2 个主成分、画散布图。

```
1  import matplotlib.pyplot as plt
2  from sklearn.manifold import TSNE
3
4  def visualize(h, color):
5      # 降维至2个主成分
6      z = TSNE(n_components=2).fit_transform(h.detach().cpu().numpy())
7
8      plt.figure(figsize=(10,10))
9      plt.xticks([])
10     plt.yticks([])
11
12     plt.scatter(z[:, 0], z[:, 1], s=70, c=color, cmap="Set2")
13     plt.show()
14
15 # 预测
16 model.eval()
17 out = model(data)
18 # 绘图
19 visualize(out.cpu(), color=data.cpu().y)
```

执行结果：每一类论文被有效隔开，表示分类效果不错。下图为彩色，请参阅程序执行结果。

扫码看彩图

**范例5.** 以图神经网络进行分子性质(Molecular Property)的预测，辨识特定分子结构是否可以抑制HIV病毒复制，就数据而言，范例4只有一个图形，程序针对图形内的节点进行分类，本例则有188张图形，每张图形是一笔数据，将整张图输入模型进行预测。

程序：17_05_Graph_Classification.ipynb，程序修改自 https://pytorch-geometric.readthedocs.io/en/latest/notes/colabs.html。

(1) 加载相关套件。

```python
import torch
from torch_geometric.data import Data
import networkx as nx
```

(2) 加载内建数据集 TUDataset MUTAG，它是关于分子性质的数据。

```python
from torch_geometric.datasets import TUDataset

# 载入内建数据
dataset = TUDataset(root='./graph/TUDataset', name='MUTAG')

print()
print(f'Dataset: {dataset}:')
print('====================')
print(f'Number of graphs: {len(dataset)}')
print(f'Number of features: {dataset.num_features}')
print(f'Number of classes: {dataset.num_classes}')

data = dataset[0]  # Get the first graph object.

print()
print(data)
print('=============================================================')

# Gather some statistics about the first graph.
print(f'Number of nodes: {data.num_nodes}')
print(f'Number of edges: {data.num_edges}')
print(f'Average node degree: {data.num_edges / data.num_nodes:.2f}')
print(f'Has isolated nodes: {data.has_isolated_nodes()}')
print(f'Has self-loops: {data.has_self_loops()}')
print(f'Is undirected: {data.is_undirected()}')
```

执行结果。

```
Dataset: MUTAG(188):
====================
Number of graphs: 188
Number of features: 7
Number of classes: 2

Data(edge_index=[2, 38], x=[17, 7], edge_attr=[38, 4], y=[1])
=============================================================
Number of nodes: 17
Number of edges: 38
Average node degree: 2.24
Has isolated nodes: False
Has self-loops: False
Is undirected: True
```

(3) 数据分割：洗牌后再分割。

```python
torch.manual_seed(12345)
dataset = dataset.shuffle()    # 洗牌

train_dataset = dataset[:150]  # 前 150 笔作为训练数据
test_dataset = dataset[150:]   # 后 38 笔作为测试数据

print(f'Number of training graphs: {len(train_dataset)}')
print(f'Number of test graphs: {len(test_dataset)}')
```

(4) 前置处理：由于每张图形的节点及边的个数均不相同，要输入模型前，必须要进行缩放 (Rescaling) 或补零 (Padding)，使每笔数据长度一致，GNN 采取另一种方式，将每批图的所有节点合并成一张大图，这样，每张图的尺寸就一样大了，如图 16.3 所示。

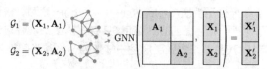

图 16.3　节点合并

图片来源：PyG 官网 Graph Classification with Graph Neural Networks 范例

例如，第一张图有编号 1~5 节点，第二张图有编号 3~7 节点，合并后，每张图都有编号 1~7 节点，只是合并进来的节点没有与其他节点相连接，PyG 的 DataLoader 可自动完成合并的功能，非常方便。

(5) 建立 DataLoader。

```
from torch_geometric.loader import DataLoader

train_loader = DataLoader(train_dataset, batch_size=64, shuffle=True)
test_loader = DataLoader(test_dataset, batch_size=64, shuffle=False)

# 显示每批数据的内容
for step, data in enumerate(train_loader):
    print(f'Step {step + 1}:')
    print('=======')
    print(f'一批内含图形的个数: {data.num_graphs}')
    print(data)
    print()
```

执行结果：每批图的节点会自动合并，每张图的节点及边的个数会一致。

```
Step 1:
=======
一批内含图形的个数: 64
DataBatch(edge_index=[2, 2636], x=[1188, 7], edge_attr=[2636, 4], y=[64], batch=[1188], ptr=[65])

Step 2:
=======
一批内含图形的个数: 64
DataBatch(edge_index=[2, 2506], x=[1139, 7], edge_attr=[2506, 4], y=[64], batch=[1139], ptr=[65])

Step 3:
=======
一批内含图形的个数: 22
DataBatch(edge_index=[2, 852], x=[387, 7], edge_attr=[852, 4], y=[22], batch=[387], ptr=[23])
```

(6) 判断是否使用 GPU。

```
device = torch.device('cuda' if torch.cuda.is_available() else 'cpu')
```

(7) 定义模型：使用 Readout 神经层，它有很多处理方式，这里使用嵌入向量平均值。

```
from torch.nn import Linear
import torch.nn.functional as F
from torch_geometric.nn import GCNConv
from torch_geometric.nn import global_mean_pool

class GCN(torch.nn.Module):
    def __init__(self, hidden_channels):
        super(GCN, self).__init__()
        torch.manual_seed(12345)
        self.conv1 = GCNConv(dataset.num_node_features, hidden_channels)
        self.conv2 = GCNConv(hidden_channels, hidden_channels)
        self.conv3 = GCNConv(hidden_channels, hidden_channels)
        self.lin = Linear(hidden_channels, dataset.num_classes)

    def forward(self, x, edge_index, batch):
        # 1. 转成嵌入向量
        x = self.conv1(x, edge_index)
        x = x.relu()
        x = self.conv2(x, edge_index)
        x = x.relu()
        x = self.conv3(x, edge_index)

        # 2. Readout Layer : 求向量平均值
        x = global_mean_pool(x, batch)  # [batch_size, hidden_channels]

        # 3. 分类
        x = F.dropout(x, p=0.5, training=self.training)
        x = self.lin(x)

        return x
```

(8) 模型训练及评估。

```
1   import numpy as np
2   
3   model = GCN(hidden_channels=64).to(device)
4   optimizer = torch.optim.Adam(model.parameters(), lr=0.01)
5   criterion = torch.nn.CrossEntropyLoss()
6   
7   def train():
8       model.train()
9       for data in train_loader:
10          data = data.to(device)
11          out = model(data.x, data.edge_index, data.batch)
12          loss = criterion(out, data.y)    # 计算损失
13          loss.backward()
14          optimizer.step()
15          optimizer.zero_grad()
16  
17  def test(loader):
18      model.eval()
19      correct = 0
20      pred_all = np.array([])
21      actual_all = np.array([])
22      for data in loader:
23          data = data.to(device)
24          out = model(data.x, data.edge_index, data.batch)
25          pred = out.argmax(dim=1)                    # 找最大概率
26          correct += int((pred == data.y).sum())      # 计算正确个数
27          correct_ratio = correct / len(loader.dataset)    # 计算正确比率
28          pred_all = np.concatenate((pred_all, pred.cpu().numpy()))
29          actual_all = np.concatenate((actual_all, data.y.cpu().numpy()))
30      return correct_ratio, pred_all, actual_all   # 正确比率，预测值，标注类别
31  
32  for epoch in range(1, 171):
33      train()
34      train_acc = test(train_loader)
35      test_acc = test(test_loader)
36      print(f'Epoch: {epoch:03d}, 训练准确率: {train_acc[0]:.4f}, ' +
37            f'测试准确率: {test_acc[0]:.4f}')
```

执行结果：训练准确率约 80%，测试准确率 79%。

```
Epoch: 155, 训练准确率: 0.8000, 测试准确率: 0.7895
Epoch: 156, 训练准确率: 0.8400, 测试准确率: 0.7632
Epoch: 157, 训练准确率: 0.8200, 测试准确率: 0.6316
Epoch: 158, 训练准确率: 0.8133, 测试准确率: 0.8158
Epoch: 159, 训练准确率: 0.7667, 测试准确率: 0.8158
Epoch: 160, 训练准确率: 0.8133, 测试准确率: 0.6579
Epoch: 161, 训练准确率: 0.8400, 测试准确率: 0.7368
Epoch: 162, 训练准确率: 0.7867, 测试准确率: 0.7895
Epoch: 163, 训练准确率: 0.8267, 测试准确率: 0.7632
Epoch: 164, 训练准确率: 0.8067, 测试准确率: 0.6842
Epoch: 165, 训练准确率: 0.7933, 测试准确率: 0.7895
Epoch: 166, 训练准确率: 0.7867, 测试准确率: 0.7895
Epoch: 167, 训练准确率: 0.8467, 测试准确率: 0.6842
Epoch: 168, 训练准确率: 0.8400, 测试准确率: 0.7105
Epoch: 169, 训练准确率: 0.8067, 测试准确率: 0.7895
Epoch: 170, 训练准确率: 0.8067, 测试准确率: 0.7895
```

(9) 显示混淆矩阵 (Confusion matrix)。

```
1   from sklearn.metrics import confusion_matrix
2   
3   confusion_matrix(test_acc[2], test_acc[1])
```

执行结果：错误率偏高。

```
[ 4,  6]
[ 2, 26]
```

(10) 绘图：使用 TSNE 算法降维至 2 个主成分、画散布图。

```
1   import matplotlib.pyplot as plt
2   from sklearn.manifold import TSNE
3
4   def visualize(h, color):
5       # 降维至2个主成分
6       z = TSNE(n_components=2).fit_transform(h.detach().cpu().numpy())
7
8       plt.figure(figsize=(10,10))
9       plt.xticks([])
10      plt.yticks([])
11
12      plt.scatter(z[:, 0], z[:, 1], s=70, c=color, cmap="Set2")
13      plt.show()
14
15  # 预测
16  model.eval()
17  test_loader_all = DataLoader(dataset[:], batch_size=len(dataset), shuffle=False)
18  for data in test_loader_all:
19      data = data.to(device)
20      out = model(data.x, data.edge_index, data.batch)
21      pred = out.argmax(dim=1)           # 找最大概率
22  # 绘图
23  visualize(out.cpu(), color=data.cpu().y)
```

执行结果：降维后辨识效果不佳，两个类别混在一起。

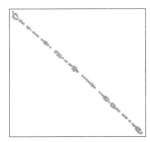

Christopher Morris 等学者后续采用 Neighborhood Normalization[16] 及 Skip-connection[17] 改良算法，PyG 也实操了该项功能，制作成 GraphConv 神经层，修改后的模型请参阅 17_06_Graph_Classification_improved.ipynb，准确率提高至 92%，降维后的效果也明显变好，如图 16.4 所示。

图 16.4　降维后效果明显变好

# 16-4　总结

图神经网络 (GNN) 提供另一种角度分析影像、文字、知识、社群等图形数据，网络上讨论的规模有渐增的趋势，应用的层面也很广泛，是值得深入研究的领域，同时相关的套件 NetworkX 及 PyG 功能也很强大，提供各种算法，入门也很轻松，不需要从零开始。